CLASSICAL MECHANICS

WITH MATLAB® APPLICATIONS

JAVIER E. HASBUN
UNIVERSITY OF WEST GEORGIA

JONES AND BARTLETT PUBLISHERS
Sudbury, Massachusetts
BOSTON   TORONTO   LONDON   SINGAPORE

World Headquarters

Jones and Bartlett Publishers	Jones and Bartlett Publishers	Jones and Bartlett Publishers
40 Tall Pine Drive	Canada	International
Sudbury, MA 01776	6339 Ormindale Way	Barb House, Barb Mews
978-443-5000	Mississauga, Ontario L5V 1J2	London W6 7PA
info@jbpub.com	Canada	United Kingdom
www.jbpub.com		

Jones and Bartlett's books and products are available through most bookstores and online booksellers. To contact Jones and Bartlett Publishers directly, call 800-832-0034, fax 978-443-8000, or visit our website www.jbpub.com.

Substantial discounts on bulk quantities of Jones and Bartlett's publications are available to corporations, professional associations, and other qualified organizations. For details and specific discount information, contact the special sales department at Jones and Bartlett via the above contact information or send an email to specialsales@jbpub.com.

Production Credits
Chief Executive Officer: Clayton Jones
Chief Operating Officer: Don W. Jones, Jr.
President, Higher Education and Professional Publishing: Robert W. Holland, Jr.
V.P., Sales and Marketing: William J. Kane
V.P., Design and Production: Anne Spencer
V.P., Manufacturing and Inventory Control: Therese Connell
Executive Editor, Science: Cathleen Sether
Managing Editor, Science: Dean W. DeChambeau
Acquisitions Editor, Science: Shoshanna Grossman
Associate Editor, Science: Molly Steinbach
Editorial Assistant, Science: Briana Gardell
Production Director: Amy Rose
Senior Production Editor: Jennifer Bagdigian
Senior Marketing Manager: Andrea DeFronzo
Cover Design: Timothy Dziewit
Cover Image: © Don Farrall/age fotostock
Composition: Northeast Compositors, Inc.
Printing and Binding: Malloy, Inc.
Cover Printing: Malloy, Inc.

Library of Congress Cataloging-in-Publication Data
Hasbun, Javier.
  Classical mechanics with MATLAB applications / Javier Hasbun. — 1st ed.
     p. cm.
  Includes bibliographical references and index.
  ISBN-13: 978-0-7637-4636-0
  ISBN-10: 0-7637-4636-3
  1. Mechanics. 2. MATLAB. I. Title.
  QC125.2.H37 2008
  531—dc22
                          2007031428

6048

Printed in the United States of America
12 11 10 09 08   10 9 8 7 6 5 4 3 2 1

Dedicated to Nancy, Ernest, and Amelia

Preface

Classical Mechanics with MATLAB® Applications presents the subject of classical mechanics at a level intended for undergraduate students. The text assumes students have taken a year of calculus-based physics. Although having completed course work in differential equations and mathematical physics will make comprehension of the material easier, such courses can be taken as co-requisites.

Mastery of classical mechanics is essential to the advanced study of physics and engineering. In addition to employing the traditional—and invaluable—analytical approach to problem solving, this text incorporates computational physics tools to build on those skills. In particular, it makes use of MATLAB applications to enhance student understanding.

Recent efforts to integrate computational physics into the undergraduate physics curriculum provided the impetus for writing this textbook. Why do students need exposure to computational physics? Presenting the tools of computational physics in the undergraduate physics curriculum allows students to apply computational techniques they have learned in the classroom as a means to explore realistic physics applications. In so doing, students become better prepared to perform research that will be useful throughout their scientific careers.

Classical Mechanics with MATLAB® Applications concentrates on mastering mechanics topics in analytical mechanics such as Newton's laws of motion, harmonic motion, motion beyond one dimension, vector calculus, systems of coordinates, central forces, gravitation, Rutherford scattering, systems of particles, rigid body motion, and Lagrangian dynamics.

Beginning with the first chapter, numerical techniques are introduced through classical harmonic oscillator time-dependent motion. A numerical algorithm is developed for solving differential equations and MATLAB scripts are introduced for students to run or to modify while experimenting with various parameters. Likewise, a study of motion of an object falling under the action of gravity, in the presence and absence of air resistance, is developed with a modification of an earlier programming script. Having gained insight from numerical solutions in Chapter 1, Chapter 2 follows with the analytical details associated with one-dimensional motion. This chapter studies forces that are constant and time-, position-, and velocity-dependent, as well as the approach to solving the related equations of motion. For an object falling under gravity, the book connects back to the numerical results of Chapter 1. In Chapter 3, the harmonic oscillator's analytic details are fully discussed and various programming scripts are developed to provide a full understanding of the analytic results. These results are compared to the corresponding numerical results of Chapter 1. Chapter 4 strengthens students' understanding of harmonic motion by considering various situations where small oscillations are characterized by natural frequencies with specific analytical forms. This chapter also tackles interacting spring systems with bimodal oscillation characteristics.

In Chapter 5, the concepts associated with the theorems of Gauss and Stokes are covered once a thorough review of vector analysis is carried out. Various programming scripts are also developed to help in the visualization of important mathematical concepts such as gradient, divergence, and curl. This is followed in Chapter 6 by the study of motion in two and three dimensions, along with details associated with potential energy functions. In addition to Cartesian coordinates, Chapter 7 builds on previous chapters' concepts to study motion for systems in cylindrical and spherical coordinates. These concepts are subsequently applied in the study of central forces in Chapter 8 with a connection to Kepler's laws and orbital transfers. Chapter 8 provides a smooth transition to the study of the force due to gravity in Chapter 9. Here many examples are presented and simulations on binary systems are carried out. Chapter 10 is an extension of concepts learned in Chapters 7 through 10, with a specific application to the important problem of charged-particle scattering or Rutherford scattering, followed by simulations and comparison with experimental findings.

Chapter 11 deals with multiparticle systems. Here, the importance of classical mechanics comes to bear on real applications. It is in this chapter where the concept of center of mass, variable mass rocket, collisions, and lab and center-of-mass frames are studied. This chapter also connects back to Rutherford scattering in the presence of a recoiling nucleus. Chapter 12 addresses the motion of rigid bodies. The classical mechanics concepts of Chapter 11 are applied to study the concept of moment of inertia and the associated inertia tensor, which leads directly into Euler's equations of motion and the study of Eulerian angles. Throughout Chapters 11 and 12, MATLAB scripts are developed to complement students' understanding and to develop further insight into the analytic subjects.

Finally, Chapter 13 builds on the concepts of Newtonian classical mechanics by applying the more sophisticated problem-solving machinery of Lagrangian dynamics. Its connection to Hamilton's principle is also considered here. Further, simulations are developed in this chapter to gain insight into these important physical concepts.

This text presents all of these fundamentals concepts while integrating numerical methods using MATLAB, an interactive software package that is widely used in academia and industry. MATLAB uses a modern language for problem solving and features sophisticated graphics, convenient editing and debugging tools, and high numerical efficiency. MATLAB is utilized in this text because it is easy for students to understand and use, and because it allows students to explore problems that are more realistic than those for which analytic formulas can be obtained. Learning classical mechanics while gaining proficiency in a modern calculational tool enables students to gain an upper hand in their future scientific endeavors.

The programming code used in the examples is included to make it easy for students to see how certain calculations are carried out. Computational physics is a tool that takes repeated use and practice to master. The examples are provided as a foundation for students to build upon. Programming code is valuable because it can be modified and extended to more complicated situations without having to start from scratch. While working through the analytics of a problem, students should take a look at the associated code and become familiar with the various techniques used. In general, the techniques learned, used, and practiced can be adapted to any other programming language. MATLAB is an easy-to-learn programming tool with many plotting and solving features that are difficult to obtain in higher-level programming languages such as Basic, FORTRAN, C, C++, and Java.

MATLAB is not a "black box" into which you put a question and then receive an unsupported answer in return. Rather, it is a language interpreter. MATLAB has its own language, but it is not a difficult language to learn. Appendix A contains a tutorial that introduces students to its language in an easy way. Students can also learn by doing—that is, when they read the text examples, they are free to modify the code, experiment with the parameters of the model, and explore various conditions. Students should be able to reproduce the text results easily and with understanding.

Although the language of MATLAB has a very easy learning curve, as with any programming language students do need to be careful how they organize the commands in order to carry out the desired calculations, according to the model's formulae.

The various concepts developed in the text are backed by MATLAB programs or scripts, which are included within the text for easy comprehension. These scripts can be easily modified in order to explore the initial conditions of the problems or experiment with the various parameters used. It is hoped that students will also learn by running the scripts in a "learn by example" approach. The Java applications available (listed below under Ancillary Materials) will enhance further students' understanding of the various topics presented in the text.

MATLAB is available from *http://www.mathworks.com* in both student and professional versions. The scripts provided to carry out the multitude of calculations presented in the textbook are self-contained and should easily run on a student version of

MATLAB release 11 and above. The scripts are fully commented so that students with minimal computational skills can easily understand them, modify them, and begin to make their own explorations of the various physics concepts presented.

In addition to the MATLAB tutorial in Appendix A, Appendices B, C, and D contain useful formulas employed in the text. For easy reference, constants or essential formulae often needed in computations are provided inside the front and back covers. Answers to selected problems are provided at the end of the text.

Ancillary Materials

A solutions manual is available to instructors who adopt the text. It is available from the publisher's website, *http://www.jbpub.com/*, which also contains the MATLAB code used throughout the text. Also downloadable from the publisher's website is a complete set of Java applications that complement the MATLAB code that can be run using a web browser that is Java-enabled. These applications will perform the text examples in a similar fashion to the MATLAB applications included in the text. Other necessary links and important information can be found on the publisher's website.

Acknowledgments

The author is thankful to the reviewers who provided input during the developmental stage of this textbook:

Stephen Adams, Widener University
John Carlsten, Montana State University
John Deisz, Iowa State University
Michael Jackson, Central Washington University
Porter Johnson, Illinois Institute of Technology
Peter Lemaire, Central Connecticut State College
Charles Leming, Henderson State University
John Powell, Reed College
Richard Prior, North Georgia College and State University
Claude Pruneau, Wayne State University
Jon Pumplin, Michigan State University
Michael Ram, State University of New York, Buffalo
Kausur Yasmin, California University of Ohio

The philosophy of this text is that computational physics is an essential extension of students' abilities just as much as calculus. To that end, this text is an effort toward strengthening students' understanding of the natural world in complementary ways.

Javier Hasbun
Carrollton, Georgia

Brief Contents

Contents

1

Review of Newton's Laws

■ 1.1 Introduction

In this chapter Newton's laws of motion are reviewed, and some applications are presented that introduce you to some basic computational tools. In addition to the simple harmonic oscillator, the forced harmonic oscillator with damping and the free-falling body under the action of gravity with air resistance are also considered. While you should have been exposed to some of the basic physics in this chapter, the material is presented with the goal that we will ultimately need to incorporate the formulae into a computer.

Before embarking on a journey regarding the application of Newton's laws, the basic foundation of this text, let's delve into a brief biography of Newton. Indeed, before Newton, there were Galileo (1564–1642) and Huygens (1629–1697), on whose great experimental works dealing with the inertia of a body, motions of projectiles, and the oscillations of pendulums Newton built his laws.

Isaac Newton (1642–1727) was a physicist and a mathematician who was born in Woolsthorpe, a sheep farm town in Lincolnshire, England. He studied at Cambridge University. In around 1665 the fall of an apple is said to have suggested the train of thought that led to the universal law of gravitation, which is duly named after him, and which we study in this text in Chapter 9. On his own, he also studied properties of light, concluding that white light is a mixture of colors that can be separated by refraction. A popular telescope, known today as a Newtonian reflecting telescope, was originally devised by him. It is a telescope based on a reflecting parabolic mirror rather than a refracting lens. He became professor of mathematics at Cambridge in 1669, where he resumed his work on gravitation, expounded finally in his famous *Philosophiae naturalis principia mathematica (Mathematical Principles of Natural Philosophy,* 1687). In 1696 he was appointed warden of the Mint, and was master of the Mint from 1699 until his death. He also sat in Parliament on two occasions, was elected President of the Royal Society in 1703, and was knighted in 1705. During his life he was involved in many controversies, notably with Leibniz over the question of priority in the discovery of calculus.

Newton laid the groundwork for many of his contributions in physics, mathematics, and astronomy at the age of 23. In his own words "I was in the prime of my age for invention, and minded mathematics and philosophy [science] more than any time since." He also said, "I do not know what I may appear to the world; but to myself I seem to have been only like a boy playing on the seashore, and diverting myself in now and then finding a smoother pebble or a prettier shell than ordinary, whilst the great ocean of truth lay all undiscovered before me." In his accomplishments, he seemed to have been guided by a simple principle: "Truth is ever to be found in the simplicity, and not in the multiplicity and confusion of things."

The French mathematician Joseph Louis Lagrange (1736–1813) had this to say: "Newton was the greatest genius who ever lived, and the most fortunate, for there cannot be more than once a system of the world to establish."

Newton's epitaph refers to a person who, "by vigor of mind almost divine, the motions and figures of the planets, the paths of comets, and the tides of the seas first demonstrated." Newton was buried at Westminster Abbey, where the inscription on his tomb reads: "Let Mortals rejoice that there has existed such and so great an ornament of the human race."

In this chapter, we aim only to write down Newton's laws for the simple harmonic oscillator, the forced harmonic oscillator with damping, and a falling body under the action of gravity with air resistance and to see how one can use the computer to solve the equations. While we could go ahead and obtain the analytic solutions to these popular problems, this task is postponed to Chapters 2 and 3. In this chapter we seek to obtain numerical solutions to these problems and explore them using the computational language of MATLAB. In Chapter 2 we seek to show that these numerical results do in fact agree with the analytic results obtained in that chapter.

■ 1.2 Basic Ideas of Newton's Laws of Motion

Newton's laws deal with classical mechanics, i.e., the area of physics that deals with large bodies, as opposed to the atomic world. The motion of a body refers to the change in position of that body as a function of time. Thus, there are three laws of motion.

First Law

This is commonly known as the law of inertia. It can be stated as follows: a body at rest remains at rest, or in motion, if in motion, in a straight line, unless it is applied an external net force. This can be stated mathematically as

$$\mathbf{F}_{net} = \sum_i \mathbf{F} = 0, \tag{1.2.1}$$

where, $\mathbf{F}_i$ is the ith external force. The total sum of all the present external forces results into the net force $\mathbf{F}_{net}$. This net force must equal zero for the first law to apply. When the first law holds for a given system, the system is said to form an *inertial frame of reference*. When a system accelerates, such system is said to form a *non-inertial frame of reference* and the first law no longer holds. In essence, the concept of acceleration enters through the second law of motion.

For example, consider the body shown in Figure 1.1a, which experiences no acceleration. If the forces $\mathbf{F}_1$ and $\mathbf{F}_2$ are known, then by Newton's first law we must have that $\mathbf{F}_1 + \mathbf{F}_2 + \mathbf{F}_3 = 0$. It follows that $\mathbf{F}_3 = -\mathbf{F}_1 - \mathbf{F}_2$.

Second Law

Figure 1.1(a) refers to a situation when the body is in equilibrium. It experiences no acceleration and therefore no net force. In the presence of an applied external net force, however, a body experiences an acceleration that is proportional to the net force and inversely proportional to the body's mass. Thus we write

$$\mathbf{a} = \mathbf{F}_{net}/m = \sum_i \mathbf{F}_i/m. \tag{1.2.2}$$

This is simply depicted in Figure 1.1(b). This also means that since $\mathbf{a} = d\mathbf{v}/dt$ then $m\mathbf{a} = d\mathbf{p}/dt$, where $\mathbf{p} = m\mathbf{v}$ is the linear momentum of the body moving with velocity $\mathbf{v}$. We will return to this shortly. For now, consider an example of the body shown in Figure 1.1(c), which experiences an acceleration in the positive x direction. While a full treatment of vectors will be made in Chapter 5, here we use the vector notation $\mathbf{F}_1 = (F_{1x}, F_{1y})$, $\mathbf{F}_2 = (F_{2x}, F_{2y})$ for the respective x and y components of the forces. Similarly, for the acceleration we write $\mathbf{a} = (a_x, a_y)$ and for the normal force we write $\mathbf{N} = (0, N_y)$. The weight of the object is $m\mathbf{g}$, due to the gravitational force, where $\mathbf{g} = (0, -g)$ whose direction is in the $-y$ direction. Because there is no acceleration in the y direction and if the forces $\mathbf{F}_1$ and $\mathbf{F}_2$ are known, then by Newton's first

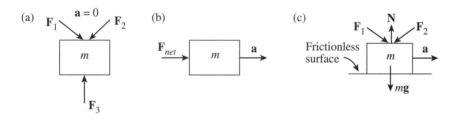

I FIGURE 1.1

law it follows that $F_{1y} + F_{2y} + N - mg = 0$ and so the magnitude of the normal force is $N = N_y = -F_{1y} - F_{2y} + mg$. However, we use Newton's second law to obtain the acceleration; that is, $F_{1x} + F_{2x} = ma_x$, and so the magnitude of the acceleration is $a = a_x = (F_{1x} + F_{2x})/m$. In this example, the components of the forces in the x and y directions would have to be known to obtain the acceleration and the normal force respectively. Also, the sign or direction of the acceleration will be according to the sign or direction of the net force as defined by (1.2.2).

This law is responsible for our ability to predict the position and velocity of a body as a function of time, given appropriate initial conditions. An accelerated system cannot represent an inertial frame of reference because it does not obey Newton's first law. The concept of a net force is very important. For example, imagine a body to be the only body in the universe and at rest. There are no external forces on it, so the first law applies, and the body will remain at rest forever. If this body had an initial velocity associated with it, it would still represent an inertial frame of reference moving at the same velocity forever because there are no forces acting on it. The body in Figure 1.1(a) has forces acting on it. But because the sum of those forces equals a zero net force, the body, moving or not, does not accelerate and remains an inertial frame of reference. Its velocity will remain the same always. However, if one considers a different situation, one in which a body rests on a table on Earth, the body, while at rest, can also have forces acting upon it. In other words, a body can be at rest even if there are forces present. To understand that kind of situation better, one applies Newton's third law of motion.

Third Law

Simply stated, for every action there is an equal and opposite reaction. To state this law mathematically, imagine two bodies interacting in free space and which we label i and j, with respective masses m_i and m_j. Referring to Figure 1.2(a), if we let $\mathbf{f}_{ij}$ be the force exerted on body i due to body j, and $\mathbf{f}_{ji}$ be the force exerted on body j due to body i, then the third law says that

$$\mathbf{f}_{ij} = -\mathbf{f}_{ji}. \tag{1.2.3a}$$

(a)   (b)   (c)

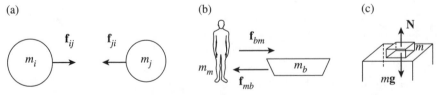

I FIGURE 1.2

A useful application of this law is that, in conjunction with the second law, one can obtain information about the relationship between the acceleration of each body. The acceleration of body i is

$$\mathbf{a}_i = \mathbf{f}_{ij}/m_i, \tag{1.2.3b}$$

and using (1.2.3a) the acceleration of body j becomes

$$\mathbf{a}_j = \mathbf{f}_{ji}/m_j = -\mathbf{f}_{ij}/m_i. \tag{1.2.3c}$$

Now, because the force each body exerts on the other is equal in magnitude but opposite in direction as in Figure 1.2(a), we can express the acceleration of body j in terms of the acceleration of body i from (1.2.3c) as

$$\mathbf{a}_j = -\mathbf{a}_i m_i/m_j. \tag{1.2.3d}$$

From this expression it is readily seen that if the masses are equal, the accelerations are equal in magnitude but opposite in direction.

Further, consider Figure 1.2(b). Suppose a 50-kg (m_m) man, who is standing on a frictionless surface, pushes on a 20-kg (m_b) boat, which is resting on a water surface, with a force of $f_{bm} = 30$ N. Assuming there is no fluid viscosity, the boat experiences an acceleration of $a_b = f_{bm}/m_b = 30/20 = 1.5$ m/s^2. However, the boat exerts an equal and opposite force on the man ($f_{mb} = -f_{bm}$), so that the man experiences an acceleration of $a_m = -a_b m_b/m_m = -1.5(20/50) = -0.6$ m/s^2. Over time, the lighter mass boat will be displaced by a larger distance than the more massive man.

It is noted that when we apply Newton's second law of motion on a particular system, it is also implied that the first and third laws are automatically to be taken into account. For example, if we apply the second law to an object that is not accelerating, this is equivalent to solving Newton's first law because the acceleration of the system is zero. We say that such a system is a *system in mechanical equilibrium* or simply in equilibrium. An example of this is shown in Figure 1.2(c). A book resting on a table is in equilibrium. The normal force acting on the book is a force acting on the surface of the book, at its interface between it and the table and perpendicular to it. It is a reaction of the table on the book. In the special case when there are no other forces present other than the ones shown in the figure, because the book is not accelerating, it is possible to find its value. Here there are two forces only; one is the normal force, which points up $\mathbf{N} = (0, N)$, and the other is the gravitational force, which points down $m\mathbf{g} = (0, -mg)$. Adding these two forces we have $\mathbf{N} + m\mathbf{g} = 0$, or $\mathbf{N} = -m\mathbf{g}$. Thus they are equal in magnitude and opposite in direction. Similarly, when a system is accelerating due to a net force, the third law must be considered carefully when the net force is accounted for. For example, in the case when friction (kinetic) is present, it is the reaction from a surface on an object and it is always pointed in a direction that opposes the motion of the object.

If we express the net force in terms of momentum, Newton's first law applied to systems in equilibrium leads to momentum conservation. More specifically, writing the net force as

$$\mathbf{F} = m\frac{d\mathbf{v}}{dt} = \frac{d\mathbf{p}}{dt}, \tag{1.2.4}$$

where $\mathbf{p} = m\mathbf{v}$ is the body's linear momentum and where we have taken the body's mass to be a constant, then applying Newton's first law, $\mathbf{F} = 0$, the right of (1.2.4) means that

$$\mathbf{p} = \text{constant}, \tag{1.2.5}$$

or that *momentum is conserved*.

The preceding three laws, along with the time derivative relations associated with the general position $\mathbf{r} = (x, y, z)$, velocity $\mathbf{v} = (v_x, v_y, v_z)$, and acceleration $\mathbf{a} = (a_x, a_y, a_z)$ of a body

$$\mathbf{v} = \frac{d}{dt}\mathbf{r}, \tag{1.2.6a}$$

$$\mathbf{a} = \frac{d}{dt}\mathbf{v} = \frac{d^2}{dt^2}\mathbf{r}, \tag{1.2.6b}$$

allow us to obtain information about the object's position, velocity, and acceleration as a function of time. For example, consider the motion of a particle in one dimension with an acceleration $\mathbf{a} = a_x\hat{i}$ and where we subsequently replace a_x with a. Its velocity as a function of time can be obtained directly by integrating (1.2.6b)

$$\int dv = \int a\,dt. \tag{1.2.7a}$$

If we assume the acceleration is constant, the integration of (1.2.7a) leads to the one-dimensional velocity,

$$v(t) = v_0 + at, \tag{1.2.7b}$$

where v_0 is the initial value of the velocity in the x direction. The displacement as a function of time is obtained by integrating the one-dimensional form of (1.2.6a),

$$\int dx = \int v(t)\, dt,$$ (1.2.7c)

or

$$x(t) = x_0 + v_0 t + \frac{1}{2} a t^2.$$ (1.2.7d)

If the acceleration is not constant, Equation (1.2.7a) can still be used to obtain $v(t)$ and the result can be used with Equation (1.2.7c) to obtain the corresponding $x(t)$. We will investigate this further in the next chapter. Sometimes, it is useful to be able to simulate the motion of a particle using a computational approach. The simplest way to do that is to use the Euler method. According to this method, suppose that we need the solution of a differential equation of the form

$$\frac{dy}{dx} = f(x, y) = y',$$ (1.2.8)

for $y(x)$ in the range $x_0 \leq x \leq x_f$. Then we first write this in the approximate form

$$y' \approx \frac{y(x + h) - y(x)}{h} = \frac{y_{i+1} - y_i}{h},$$ (1.2.9)

where $h = (x_f - x_0)/N$, with N the number of desired steps in the interval $[x_0, x_f]$. Here $y_i \equiv y(x_i)$ and we suppose that for $i = 0$, $y_0 = y(x_0)$ is given. Thus combining (1.2.8) and (1.2.9) we see that

$$\frac{y_{i+1} - y_i}{h} = f(x_i, y_i),$$ (1.2.10)

which gives the recursion formula for the value of y as a function of x as

$$y_{i+1} = y_i + hf(x_i, y_i), \text{ with } x_{i+1} = x_i + h.$$ (1.2.11)

If we apply the Euler method to our one-dimensional time-dependent problem for which we have

$$\frac{dv}{dt} = a \quad \text{and} \quad \frac{dx}{dt} = v,$$ (1.2.12)

then we get the Euler approach relation for the velocity by replacing $y_i \rightarrow v_i$, $x_i \rightarrow t_i$, $f \rightarrow a$, and $h \rightarrow \Delta t$ in (1.2.11) to obtain

$$v_{i+1} = v_i + a_i \Delta t, \tag{1.2.13}$$

with the initial velocity, $v_{i=0} = v_0$. Similarly, this resulting velocity along with the second equation of (1.2.12) can in turn be used to obtain the position, but this time we make the substitutions $y_i \rightarrow x_i$, $x_i \rightarrow t_i$, $f \rightarrow v$, and $h \rightarrow \Delta t$ in (1.2.11), to write

$$x_{i+1} = x_i + v_{i+1} \Delta t. \tag{1.2.14}$$

with initial position, $x_{i=0} = x_0$, and with time evolving simply as

$$t_{i+1} = t_i + \Delta t, \tag{1.2.15}$$

where $t_{i=0} = t_0 \equiv 0$. The process stops when the desired number of N steps is reached.

The actual Euler form of (1.2.14) uses v_i instead of v_{i+1}, but the presence of v_{i+1} produces more accurate results because a more recent value of the velocity is used. This slight modification of the Euler method is known as the Euler–Cromer method, which is equivalent to a so-called Verlet algorithm. The approximation (1.2.14) works well because, in essence, it involves a correction to x_{i+1} to second order in Δt (see Problem 1.12) as opposed to terms up to first order in Δt if we were to use v_i instead of v_{i+1}. Several algorithms exist to solve differential equations numerically, but the Euler–Cromer method is the simplest method that yields accurate solutions for oscillatory problems. Also, keep in mind that decreasing the step size and increasing the number of steps, in general, yields more accurate results, but not always. When the number of steps used is too large and the step size is too small, the calculational time can become too long, and it's probably wise to seek an alternate algorithm.

■ 1.3 Numerical Applications of Newton's Second Law Using the Modified Euler Method

In this section, Newton's laws of motion will be employed to carry out a numerical investigation of the motion of a mass under two different conditions. In the first case, a mass is attached to the end of a massless spring under the action of a damping as well as a driving force. In the second case, the mass is under the action of a constant gravitational force and a force due to air resistance. In this book, MATLAB will be

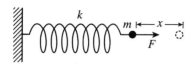

| FIGURE 1.3

used throughout for numerical applications. An introduction to MATLAB is available in Appendix A.

Spring Mass System

In its simplest form, one can consider a mass attached to the end of a massless spring while the opposite end of the spring is held fixed, as shown in Figure 1.3.

The figure shows that applying a force F on the mass has the effect of displacing it by amount x. By Hooke's law, the magnitude of the applied force is proportional to the displacement x. In the absence of damping and driving forces, according to Newton's third law, the spring responds with a force equal in magnitude but opposite in direction,

$$F_s = -kx(t). \tag{1.3.1}$$

According to Newton's second law, the bob's acceleration is $a(t) = -kx(t)/m = d^2x/dt^2$, which is a second-order linear differential equation for the bob's position as a function of time. The analytic solution will be investigated in Chapter 3. For now it is convenient to investigate it numerically using the Euler–Cromer method of the previous section.

The numerical solution can be carried out using Equations (1.2.13–1.2.15). Suppose we let $k = 1000$ N/m, $x_0 = 0.1$ m, $v_0 = 0.0$, and we wish to obtain $x(t)$ in the time interval $[0, t_{max}]$, with $t_{max} = 1$ s, and $N = 10$ steps, so that $\Delta t = t_{max}/N = 0.1$ s. Thus, specifically, our numerical equations for a mass of 5 kg become

$$a_i = -1000x_i/5, \quad v_{i+1} = v_i + 0.1a_i, \quad \text{and} \quad x_{i+1} = x_i + 0.1v_{i+1}. \tag{1.3.2a}$$

Table 1.1 has the results for the 10 steps. For the initial values we have,

$$t_0 \equiv 0, \quad v_0 \equiv 0, \quad x_0 \equiv 0.1, \quad a_0 = -1000(0.1)/5 = -200(0.1) = -20. \tag{1.3.2b}$$

Table 1.1   Results from the recursion formulas (1.2.15, 1.3.2a–c)

Initial Values: $t_0 \equiv 0$, $\Delta t = 0.1$, $v_0 \equiv 0$, $x_0 \equiv 0.1$, $a_0 = -20$					
i	$i+1$	$t_{i+1}=t_i+\Delta t$	$v_{i+1}=v_i+0.1a_i$	$x_{i+1}=x_i+0.1v_{i+1}$	$a_{i+1}=-200x_{i+1}$
0	1	0.1	−2.0	−0.1	20
1	2	0.2	0.0	−0.1	20
2	3	0.3	2.0	0.1	−20
3	4	0.4	0.0	0.1	−20
4	5	0.5	−2.0	−0.1	20
5	6	0.6	0.0	−0.1	20
6	7	0.7	2.0	0.1	−20
7	8	0.8	0.0	0.1	−20
8	9	0.9	−2.0	−0.1	20
9	10	1.0	0.0	−0.1	20

The next step, keeping $i = 0$ and using (1.3.2a and b), gives

$$t_1 = t_0 + 0.1 = 0.1, \quad v_1 = v_0 + 0.1(a_0) = 0 + 0.1(-20) = -2,$$

$$x_1 = x_0 + 0.1(v_1) = 0.1 + 0.1(-2) = -0.1, \tag{1.3.2c}$$

$$a_1 = -200(x_1) = -200(-0.1) = 20.$$

Table 1.1 shows the succeeding steps according to the formulas (1.2.15 and 1.3.2a). We notice that the position, velocity, and acceleration of the particle tend to be oscillatory, although not very smoothly so. We also notice that the sign of the acceleration is always opposite to that of the displacement. This is of course a consequence of any simple harmonic motion system, as expected. At this point, it is convenient to develop a MATLAB program that incorporates the above numerical algorithm and that can allow us to input the initial parameters of the model. By increasing the number of steps involved in the calculation, we can also increase the accuracy of the calculation. However, before developing such a program, we might as well include damping and driving forces to the model.

A standard way to model the damping experienced by a mass at the end of a spring is to notice that a resistive force tends to oppose the change in the motion; thus we write $F_R = -Cv$. If one were to attach a driving force on the mass at the end of the spring, it is common for such a driving force to be harmonic in nature. A possible form for such a force is $F_D = F_0 \sin(\omega t)$, where F_0 and ω are the amplitude and the frequency of the driving force. Including these two new forces to the original force of

Equation (1.3.1) associated with the spring results in the more general expression for
the differential equation of a mass-spring system

$$\frac{d^2x}{dt^2} = [-kx(t) - Cv(t) + F_0\sin(\omega t)]/m = a. \qquad (1.3.3)$$

While the analytic solution of this equation will be tackled in Chapter 3, for now we
continue with the numerical solution approach. We notice that depending on the values of the parameters such as k, C, F_0, and ω, a wide range of behaviors can be investigated. The MATLAB program, ho1.m, that is capable of investigating such motion
follows. Recall from Appendix A that the lines beginning with "%" are comment lines.
Some of these comments explain what the script does, and some give suggested values to use. Sample runs of this program are also shown.

SCRIPT

```
%ho1.m
%Calculation of position, velocity, and acceleration for a harmonic
%oscillator versus time. The equations of motion are used for small time intervals
clear;
%NPTS=100;TMAX=1.0;%example Maximum number of points and maximum time
TTL=input(' Enter the title name TTL:','s');%string input
NPTS=input(' Enter the number calculation steps desired NPTS: ');
TMAX=input(' Enter the run time TMAX: ');
NT=NPTS/10;%to print only every NT steps
%K=1000;M=5.0;C=0.0;E=0.0;W=0.0;x0=0.1;v0=0.0;% example Parameters
K=input(' Enter the Spring constant K: ');
M=input(' Enter the bob mass M: ');
C=input(' Enter the damping coefficient C: ');
E=input(' Enter the magnitude of the driving force E: ');
W=input(' Enter the driving force frequency W: ');
x0=input(' Enter the initial position x0: ');% Initial Conditions
v0=input(' Enter the initial velocity v0: ');% Initial Conditions
t0=0.0;% start at time t=0
dt=TMAX/NPTS;%time step size
fprintf(' Time step used dt=TMAX/NPTS=%7.4f\n',dt);%the time step being used
F=-K*x0-C*v0+E*sin(W*t0); % initial force
a0=F/M;% initial acceleration
fprintf('    t       x       v       a\n');%output column labels
v(1)=v0;
x(1)=x0;
a(1)=a0;
t(1)=t0;
fprintf('%7.4f %7.4f %7.4f %7.4f\n',t(1),x(1),v(1),a(1));%print initial values
```

```
for i=1:NPTS
    v(i+1)=v(i)+a(i)*dt;                    %new velocity
    x(i+1)=x(i)+v(i+1)*dt;                  %new position
    t(i+1)=t(i)+dt;                         %new time
    F=-K*x(i+1)-C*v(i+1)+E*sin(W*t(i+1));   %new force
    a(i+1)=F/M;                             %new acceleration
% print only every NT steps
    if(mod(i,NT)==0)
        fprintf('%7.4f %7.4f %7.4f %7.4f\n',t(i+1),x(i+1),v(i+1),a(i+1));
    end;
end;
subplot(3,1,1)
plot(t,x,'k-');
ylabel('x(t) (m)','FontSize',14);
h=legend('position vs time'); set(h,'FontSize',14);
title(TTL,'FontSize',14);
subplot(3,1,2)
plot(t,v,'b-');
ylabel('v(t) (m/s)','FontSize',14);
h=legend('velocity vs time'); set(h,'FontSize',14)
subplot(3,1,3)
plot(t,a,'r-');
ylabel('a(t) (m/s^2)','FontSize',14);
xlabel('time (sec)','FontSize',14);
h=legend('acceleration vs time'); set(h,'FontSize',14)
```

We have performed a first rough run of hol.m in order to reproduce the results of the simple harmonic oscillator of Table 1.1. While generally it is suggested to pick a large number of steps, the program's initial run that follows performs NPTS (=10) steps in the calculation. Should the value of NPTS be increased, the calculation is more accurate, but only 10 sets of instantaneous values are printed by the script as it is designed. However, the plots produced contain all the calculated results. The input/output of the first run follows and can be found in the file hol_1.txt, and is followed by the resulting MATLAB plot of Figure 1.4.

OUTPUT

```
hol_1.txt
>> hol
  Enter the title name TTL:Simple Harmonic Oscillator - rough
  Enter the number calculation steps desired NPTS: 10
  Enter the run time TMAX: 1
  Enter the Spring contant K: 1000
```

```
Enter the bob mass M: 5
Enter the damping coefficient C: 0
Enter the magnitude of the driving force E: 0
Enter the driving force frequency W: 0
Enter the initial position x0: 0.1
Enter the initial velocity v0: 0
Time step used dt=TMAX/NPTS= 0.1000
     t       x       v       a
0.0000  0.1000  0.0000 -20.0000
0.1000 -0.1000 -2.0000  20.0000
0.2000 -0.1000  0.0000  20.0000
0.3000  0.1000  2.0000 -20.0000
0.4000  0.1000  0.0000 -20.0000
0.5000 -0.1000 -2.0000  20.0000
0.6000 -0.1000  0.0000  20.0000
0.7000  0.1000  2.0000 -20.0000
0.8000  0.1000  0.0000 -20.0000
0.9000 -0.1000 -2.0000  20.0000
1.0000 -0.1000  0.0000  20.0000
```

The rough results shown in the ho1_1.txt file (reproduced here) agree with those calculated before in Table 1.1 and show how the program works.

Figure 1.4 shows that the acceleration is proportional to the negative of the displacement, as mentioned before. Also, from Table 1.1, we see that the velocity's maximum is

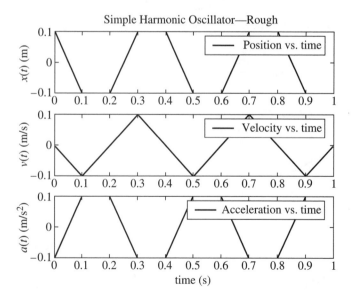

FIGURE 1.4 Rough calculation of position, velocity, and acceleration.

always shifted from the maximum of the displacement, similar to the behavior of the sine and cosine functions.

It is interesting to investigate the model further. The following are three sets of runs made with the program ho1.m that use more points for better accuracy. These are labeled as follows with the parameters used enclosed in parentheses:

Figure 1.5(a), "Simple Harmonic Oscillator" (NPTS=100, TMAX=10, K=1, M=1, C=0, E=0, W=0, x0=1, v0=0);

Figure 1.5(b), "Damped Harmonic Oscillator" (NPTS=200, TMAX=20, K=1, M=1, C=0.5, E=0, W=0, x0=1, v0=0);

Figure 1.5(c), "Forced Harmonic Oscillator with Damping" (NPTS=200, TMAX=20, K=1, M=1, C=0.5, E=0.1, W=0.8, x0=1, v0=0).

In the above runs, a higher value of NPTS is used when more accuracy is needed, which in turn decreases the step size, depending on the model. As a general rule, the more terms a model uses, the higher is the number of calculated points needed, along with a smaller step size, to minimize the error.

(a)

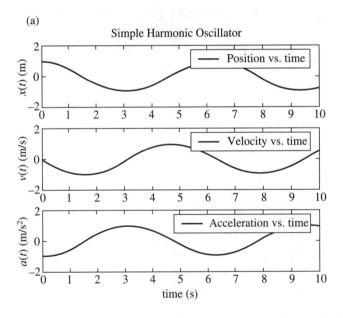

FIGURE 1.5 (a) "Simple Harmonic Oscillator" (NPTS=100, TMAX=10, K=1, M=1, C=0, E=0, W=0, x0=1, v0=0).

(b)

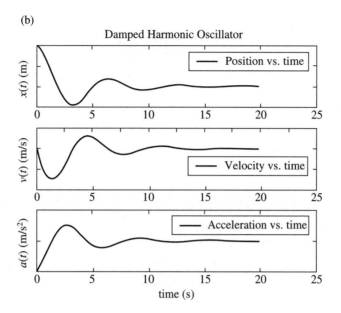

(c)

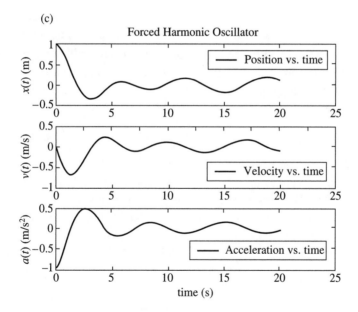

FIGURE 1.5 (b) "Damped Harmonic Oscillator" (NPTS=200, TMAX=20, K=1, M=1, C=0.5, E=0, W=0, x0=1, v0=0).

(c) "Forced Harmonic Oscillator with Damping" (NPTS=200, TMAX=20, K=1, M=1, C=0.5, E=0.1, W=0.8, x0=1, v0=0).

In Figure 1.5(a), we recognize the expected harmonic behavior obtained by increasing the number of points and using a smaller step value for dt. In Figure 1.5(b), the damped model results simply because the value of the damping coefficient is no longer zero. In Figure 1.5(c), the behavior is more complicated because the system tends to damp during a short time but after that, the system tends to respond to the driving force much more efficiently. Notice that the driving frequency is close to the value of 1, which for the present parameters corresponds to the natural frequency of the spring ($\sqrt{k/m}$). We will discuss this behavior later in Chapter 3 when we study the analytic solution.

Motion With and Without Air Resistance Under Constant Acceleration Due To Gravity

In the absence of air resistance and under a constant gravitational acceleration it is convenient to let $\mathbf{F} = -mg\,\hat{j} = m\mathbf{a}$, where $\hat{j}$ is a unit vector in the y-direction. Because only the y coordinate's motion is important we write

$$\frac{d^2y}{dt^2} = -g, \tag{1.3.4a}$$

where g is the constant value due to gravity. Because there is no air resistance, this equation refers to the motion of an object in *free fall*. The value of g is $9.80\,\text{m/s}^2$ on Earth and $1.63\,\text{m/s}^2$ on the moon. Integrating the preceding equation once over time gives the velocity, and integrating once more gives the displacement versus time

$$v = v_0 - gt, \quad \text{and} \quad y = y_0 + v_0 t - \frac{1}{2}gt^2, \tag{1.3.4b}$$

with the initial conditions v_0 for v, and y_0 for y. The inclusion of the effect due to air resistance on the motion is analytically more involved and is left for the next chapter. For now, a MATLAB program will be developed to include the resistive force due to air resistance. As Figure 1.6 shows, when the body is falling, the resistive force R points upward in a direction opposite to that of the velocity.

The net force associated with Figure 1.6 is written as

$$\mathbf{F}_{net} = [-mg + R(v)]\hat{j} = m\mathbf{a}, \tag{1.3.5}$$

where the force due to air resistance $R(v)$ is a function of velocity. In other words, this is a resistive force that is always pointed in a direction opposite to the direction of the motion, but its magnitude also depends on the magnitude of the velocity. We will

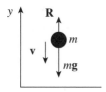

| FIGURE 1.6

consider two popular models for the magnitude of R, one for which $R \propto -v$, and the other for which $R \propto -v^2$. A realistic use of a specific model for the drag term may very well depend on the actual value of the speed, the object's geometry, its size, as well as its mass, in addition to the properties of the substance through which the object moves. Thus we write R in two ways

$$R = -C_1 v, \quad \text{and} \quad R = -C_2 v^2. \tag{1.3.6}$$

In particular situations it is possible to have expressions for C_1 and C_2. For example, when a spherical object of radius a travels through a fluid of viscosity η, it is standard to take the linear model with $C_1 = 6\pi a \eta$, which is commonly referred to as Stokes' law. For a large object, like a weight falling through the air with an open parachute, one could employ the quadratic model with $C_2 = C_D \rho A / 2$, where ρ is the air density, A is the object's cross-sectional area, and C_D is a drag coefficient in the range between 0.4 and 1.

Because the force due to gravity is constant, the velocity (or speed) of an object falling under gravity can achieve a high magnitude. There is a limit on how high such a speed can get, because referring to Figure 1.6, if the speed increases, then R will also increase. Therefore, there is a point where the net force will become zero, causing a zero acceleration and meaning that no further increase in speed is possible. In this case the speed reaches a terminal value v_t, commonly referred to as the *terminal velocity*. Thus in this limit from (1.3.5), we set

$$-mg + R(v_t) = 0. \tag{1.3.7}$$

Depending on the model for R used, two possible values of v_t can be found. Using (1.3.6–1.3.7), we can solve for the magnitude of the terminal velocity to get

$$v_{t_1} = mg/C_1, \quad \text{and} \quad v_{t_2} = \sqrt{mg/C_2}, \tag{1.3.8}$$

for the linear and the quadratic velocity models, respectively. Note that these are velocity magnitudes or speeds. Referring to Figure 1.6, the actual velocities are negative when objects are falling and positive when they are rising. For the quadratic model, the square of the speed is involved in the resistive force. In the analytic solution to the problem, in Chapter 2, the up and down motion is, therefore, treated separately. Here, the MATLAB program ho2.m solves this problem numerically. It accepts input from the user and produces results on a screen as well as the corresponding plots. The listing of the program follows. Thus, to incorporate the drag model of interest, we have included the "FLAG" input, which can take on the values of 0 or 1 depending on the drag model used, linear or quadratic, as mentioned previously.

SCRIPT

```
%ho2.m
%Calculation of position, velocity, and acceleration for a body in
%free fall with air resistance versus time.
%The equations of motion are used for small time intervals
clear;
%NPTS=200;TMAX=20.0;%example Maximum number of points and maximum time
TTL=input(' Enter the title name TTL:','s');%string input
NPTS=input(' Enter the number calculation steps desired NPTS: ');
TMAX=input(' Enter the run time TMAX: ');
NT=NPTS/10;%to print only every NT steps
%G=9.8;M=1.0;C=0.05;y0=0;v0=110;% example Parameters
G=input(' Enter value of gravity G: ');
M=input(' Enter the object mass M: ');
C=input(' Enter the drag coefficient C: ');
y0=input(' Enter the initial height y0: ');  % Initial Conditions
v0=input(' Enter the initial velocity v0: ');% Initial Conditions
FLAG=input(' Enter 0 (v drag) or 1 (v^2 drag) FLAG: ');
t0=0.0;% start at time t=0
dt=TMAX/NPTS;%time step size
if FLAG ==0
   F=-M*G-C*v0;          % initial force - case 1
   vt=abs(M*G/C);        % terminal velocity
   elseif FLAG==1
      F=-M*G-C*v0*abs(v0); % initial force - case 2
      vt=sqrt(M*G/C);     % terminal velocity
end;
%dt,FLAG, and vt used
fprintf(' FLAG=%1i, Time step dt=TMAX/NPTS=%5.2f, vt=%5.2f\n',FLAG,dt,vt);
a0=F/M;% initial acceleration
fprintf('    t       y       v       a\n');%output column labels
v(1)=v0;
```

```
y(1)=y0;
a(1)=a0;
t(1)=t0;
fprintf('%7.4f %7.4f %7.4f %7.4f\n',t(1),y(1),v(1),a(1));%print initial values
for i=1:NPTS
    v(i+1)=v(i)+a(i)*dt;                    %new velocity
    y(i+1)=y(i)+v(i+1)*dt;                  %new position
    t(i+1)=t(i)+dt;                         %new time
    if FLAG ==0
       F=-M*G-C*v(i+1);                     %new force - case 1
       elseif FLAG==1
         F=-M*G-C*v(i+1)*abs(v(i+1));       %new force - case 2
    end;
    a(i+1)=F/M;                             %new acceleration
% print only every NT steps
    if(mod(i,NT)==0)
        fprintf('%7.4f %7.4f %7.4f %7.4f\n',t(i+1),y(i+1),v(i+1),a(i+1));
    end;
end;
    plot(t,y,'k-',t,v,'b:',t,a,'r-.');
    ylabel('y (m), v (m/s), a (m/s^2)','FontSize',14);
    xlabel('time','FontSize',14);
    title(TTL,'FontSize',14);
    h=legend('position','velocity','acceleration',0); set(h,'FontSize',14)
```

An example run of this program, labeled "Falling with simple air resistance," including the input used, follows. In this run, we use the first drag model. This run produces the plot shown in Figure 1.7.

<hr>

OUTPUT

<hr>

```
ho2_1.txt
>> ho2
 Enter the title name TTL:Falling with simple air resistance
 Enter the number calculation steps desired NPTS: 200
 Enter the run time TMAX: 20
 Enter value of gravity G: 9.8
 Enter the object mass M: 1
 Enter the drag coefficient C: 0.05
 Enter the initial height y0: 0
 Enter the initial velocity v0: 110
 Enter 0 (v drag) or 1 (v^2 drag) FLAG: 0
 FLAG=0, Time step dt=TMAX/NPTS= 0.10, vt=196.00
```

```
    t         y          v         a
 0.0000   0.0000   110.0000  -15.3000
 2.0000  188.8649   80.8108  -13.8405
 4.0000  322.3215   54.4060  -12.5203
 6.0000  405.6549   30.5199  -11.3260
 8.0000  443.6466    8.9122  -10.2456
10.0000  440.6215  -10.6342   -9.2683
12.0000  400.4923  -28.3162   -8.3842
14.0000  326.7983  -44.3115   -7.5844
16.0000  222.7413  -58.7810   -6.8610
18.0000   91.2175  -71.8702   -6.2065
20.0000  -65.1530  -83.7109   -5.6145
```

In Figure 1.7, the acceleration tends to zero as the velocity begins to approach its terminal value, which for this particular set of parameters is 196 m/s.

Another run made using the same MATLAB program and labeled "Falling with v^2 air resistance," with the parameters: NPTS=200, TMAX=10, G=9.8, M=1, C=0.05, y0=0, v0=110, FLAG=1, is shown in Figure 1.8. The image was slightly magnified through MATLAB's figure viewer to show the curves better.

The calculated terminal velocity of Figure 1.8 is 14 m/s, a value much smaller than that of Figure 1.7. The difference is attributed to the stronger retardation effect

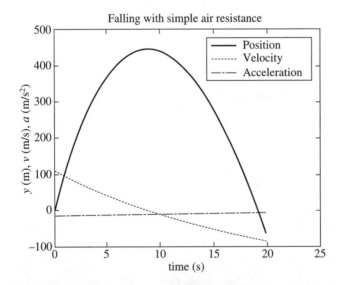

FIGURE 1.7 Falling Under Linear Air Resistance.

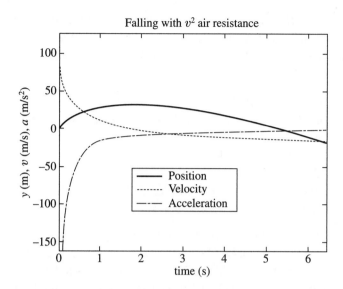

Falling with v^2 air resistance

I FIGURE 1.8 Falling under Quadratic Air Resistance.

associated with a drag force that's proportional to the square of the velocity. The acceleration decreases very rapidly to zero as well. Most of the parameters used in both Figures 1.7–1.8 are the same; the only difference in the runs is the value of the FLAG parameter, which determines the drag model used.

■ Chapter 1 Problems

1.1 Run the hol.m script with $m = 1\,\text{kg}$, $\Delta t = 0.001$, $k = 1\,\text{N/m}$, and the rest of the parameters compatible with the simple harmonic oscillator. At $t = 0$, take $x_0 = 1\,\text{m}$, $v_0 = 0.0$ m/s. From the output, find

a. the period.

b. the amplitude.

c. Are the position and velocity related? How?

1.2 Use the program hol.m along with the parameters $m = 1$ kg, $k = 1$ N/m, $x_0 = 1.5$ m, $v_0 = 0.0$ m/s, and a suitable value of Δt and do the following cases for the damped harmonic oscillator without a driving force: vary the value of C starting with 0.0 up to a value of 2.5 in steps of 0.5. Explain what happens to the period and amplitude as C varies.

1.3 Using a value of $C = 0.25$, for what value of frequency of the driving force does the highest amplitude occur in the damped harmonic oscillator? Use parameters similar to Problem 1.2 with a nonzero value of the driving force. What happens as C gets larger?

1.4 Use program ho2.m for free fall. Without air resistance obtain results for $x(t)$ and $v(t)$ and compare the results with the analytic expressions of Equations (1.3.4). Use the following initial values:

  a. $y_0 = 0$ m, $v_0 = 0.0$ m/s,

  b. $y_0 = 0$ m, $v_0 = 100.0$ m/s. Explain your results. Pick a reasonable value of mass, even though its value does not matter. Why?

1.5 A body falls with air resistance proportional to its velocity. A second body falls with air resistance proportional to the square of its velocity and a value of $C_2 = 0.01$. What value of C_1 should the first body have if it is to have the same terminal velocity as the second body? Run program ho2.m for both bodies and investigate the results for the two bodies with two sets of initial conditions:

  a. $y_0 = 0$ m, $v_0 = 0.0$ m/s,

  b. $y_0 = 0$ m, $v_0 = 100.0$ m/s. Assume $m = 1$ kg for both bodies and explain your results.

1.6 What are the units of the drag coefficients for the air resistance models that are linear and quadratic in velocity?

■ Additional Problems

1.7 Consider a body of mass m falling near the surface of the Earth and subject to a resistive force that is proportional to the cube of its velocity.

  a. Draw a vector diagram of the forces on it.

  b. Write Newton's second law for the body and obtain an expression for the terminal velocity. Explain all the units.

  c. If the acceleration can be treated as a constant for a short interval of time $\Delta t = 0.05$ s, calculate the acceleration, velocity, and position of the body for several intervals up to $t = 0.2$ s. Organize the calculations on a table for clarity. Use the numerical values $m = 1$ kg, $v_0 = 0$, $x_0 = 100$ m, $g = 10$ m/s^2, and a drag coefficient of $0.5\,\text{Ns}^3/\text{m}^3$.

1.8 Two blocks of masses m_1 and m_2 rest on a flat surface, as shown in Figure 1.9, while under an applied force F. The coefficient of static friction between the blocks and the surface is μ_s. Just before motion takes place,

   a. Draw a vector diagram showing the forces due to friction as well as the internal body forces, and obtain an expression for the value of F.

   b. Write the net force on block 1 and obtain an expression for the internal force exerted on block 1 due to block 2, F_{12}.

   c. Write the net force on block 2 and obtain an expression for the internal force on block 2 due to block 1, F_{21}. Finally, evaluate your expressions for F, F_{12}, and F_{21} using $m_1 = 1$ kg, $m_2 = 2$ kg, and $\mu_s = 0.4$.

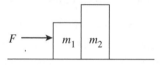

| FIGURE 1.9

1.9 Two blocks of masses m_1 and m_2 have an acceleration a, as shown in Figure 1.10, while under an applied force F. The coefficient of kinetic friction between the blocks and the surface is μ_k.

   a. Draw a vector diagram showing the forces due to friction as well as the internal body forces, and obtain an expression for the value of F that's responsible for the acceleration of both blocks.

   b. Write down Newton's second law for each block.

   c. Use the results of Parts (a) and (b) in order to obtain an expression for the value of the internal force between the blocks, say, F_{12}. Finally, evaluate your expressions for F and F_{12} using $m_1 = 2$ kg, $m_2 = 1$ kg, $\mu_s = 0.3$, and $a = 3.5$ m/s^2.

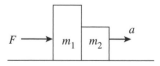

| FIGURE 1.10

1.10 A ball is dropped from a height h above the ground. In terms of h obtain expressions for

  a. The velocity of the ball just before it hits the ground;

  b. The force of impact, if when the ball hits the ground it takes time t_s for it to stop; and

  c. The time t_f the ball spends in the air before it reaches the ground.

  d. In order for the force of impact to be equal to the ball's weight, what should the value of t_s be, and what can you conclude from this result?

1.11 A spring of stiffness constant k has an unknown mass hanging from its free end. When a mass $m_1 = 2$ kg is added, the spring extends by 30 cm. If a mass $m_2 = 5$ kg is added instead, the spring extends by 85 cm. What are the values of the spring constant and the unknown mass?

1.12 The Verlet algorithm (see Giordano and Nakanishi, 2006) deals with a numerical method of solving a second-order differential equation. In the case of a constant acceleration problem in one dimension, $d^2x/dt^2 = a$, the Verlet method for $x(t)$ results in the discrete formula $x_{i+1} = 2x_i - x_{i-1} + a_i\Delta t^2$, which involves terms to second order in Δt. In Section 1.2, this equation is treated as a system of two first-order differential Equations (1.2.12) with the Euler–Cromer numerical solutions given by (1.2.13 and 1.2.14), wherein (1.2.14) v_{i+1} is used instead of v_i.

  a. Show that the Cromer–Euler method is identical to the Verlet algorithm.

  b. If in (1.2.14) v_i were to be used rather than v_{i+1}, how would the resulting expression for x_{i+1} compare to that of the preceding Verlet form? (*Hint:* Recall the approximate expression for the derivative Equation [1.2.9].)

2 Application of Newton's Second Law of Motion in One Dimension

■ 2.1 Introduction

In this chapter we consider the application of the general form of Newton's second law of motion, in one dimension, under less general circumstances, those for which the external net force takes on the specific forms:

$$F = Ca = const., \text{ is a constant,} \tag{2.1.1a}$$

$$F = F(t) \Rightarrow a = a(t), \text{ is a function of time,} \tag{2.1.1b}$$

$$F = F(x) \Rightarrow a = a(x), \text{ is a function of position,} \tag{2.1.1c}$$

$$F = F(v) \Rightarrow a = a(v), \text{ is a function of velocity.} \tag{2.1.1d}$$

The importance of these different force behaviors lies in the variable that visibly affects the force. An example of constant acceleration is the case of free fall, under the action of a constant gravitational force ($F = -mg$), which we reviewed in Chapter 1. A common application of a time-dependent force is the case of a driving harmonic force ($F = F_0 \sin(\omega t)$) similar to the one we considered in the forced harmonic oscillator in the previous chapter. Similarly, a force that depends on position can be found in naturally harmonic systems ($F = -kx$) like a pendulum or a spring-mass system. In both cases, the acceleration is proportional to the negative of the displacement. A simple and common example of a velocity-dependent force is an object that falls under the action of gravity with air resistance ($F = -mg - kv$). As we have seen before, here the force due to air resistance is taken to be proportional to the negative of the velocity raised to some power. Next, we examine each of these situations.

■ 2.2 Constant Force

This represents the simplest application of Newton's second law. We can in fact obtain a general solution for the motion of a particle as follows. We write the acceleration in terms of the net force, taking $F = C$, and

$$a = C/m, \tag{2.2.1a}$$

and using $a = \dfrac{dv}{dt}$, or integrating $\displaystyle\int_{v_0}^{v} dv = \int_{0}^{t} a\, dt$, to obtain

$$v = v_0 + at, \tag{2.2.1b}$$

for the velocity as a function of time. Here the object was taken to have an initial velocity v_0 at $t_0 = 0$. Using $v = dx/dt$ and integrating the preceding expression $\displaystyle\int_{x_0}^{x} dx = \int_{0}^{t} v\,dt = \int_{0}^{t} \{v_0 + at\}dt$, obtain

$$x = x_0 + v_0 t + \frac{a}{2}t^2, \tag{2.2.1c}$$

with the initial position indicated by x_0. This result is also familiar to us because if we replace the constant acceleration a with the value $-g$ we obtain the general kinematic equations on an object in free fall near Earth's surface. These are the same equations of motion (1.3.4) from Chapter 1, if we make the proper coordinate replacement of $x \rightarrow y$. While (2.2.1b) gives the velocity as a function of time, it is instructive to obtain the velocity as a function of position. This is accomplished if we write the acceleration as

$$a = \frac{dv}{dt} = \frac{dv}{dx}\frac{dx}{dt} = v'\dot{x} = v'v, \tag{2.2.2}$$

where we have now defined the spatial derivative with a prime, $v' \equiv \dfrac{dv}{dx}$, read v-prime, and the time derivative with a dot, $v = \dot{x} \equiv \dfrac{dx}{dt}$, read x-dot. We will use this notation later. Separating variables and integrating, $\int_{v_0}^{v} v\,dv = \int_{x_0}^{x} a\,dx$, gives the position-dependent velocity

$$v^2 = v_0^2 + 2a(x - x_0).\qquad(2.2.3)$$

We finally notice that if in (2.2.1b) we solve for a then substitute the result into (2.2.1c), we get the complementary useful expression for the position

$$x = x_0 + \frac{v + v_0}{2}t.\qquad(2.2.4)$$

EXAMPLE 2.1

In order to take off from a 94.2 m-long runway, a 945.5 kg WW1 SE5a British fighter needed to achieve a speed of 23.7 m/s. If the airplane started from rest and underwent a constant acceleration, what average force and what kind of engine was needed to achieve takeoff?

Solution

From (2.2.3), we can solve for the average force, $C = ma = mv^2/2x$, or $C = 945.5$ kg $\cdot (23.7$ m/s$)^2/2 \cdot 94.2$ m $= 2818.9$ N. The time rate of change of energy is the power $p(t)$. The average power p_{ave} can be obtained by looking at the energy used during takeoff

$$\int_{t_0}^{t} p(t)\,dt = p_{ave}\Delta t = \int_{x_0}^{x} F\,dx = C\Delta x,$$

$$\qquad(2.2.5a)$$

or

$$p_{ave} = C\frac{\Delta x}{\Delta t} = C\frac{x - x_0}{t}.\qquad(2.2.5b)$$

Solving for $x - x_0$ from (2.2.4) and substituting the result into the preceding expression, we see that the average power under a constant force is also given by

$$p_{ave} = C \, \frac{v + v_0}{2}. \tag{2.2.5c}$$

Thus the airplane requires an average power of $p_{ave} =$ 2818.9 N$(23.7 \text{ m/s} + 0 \text{ m/s})/2 = 33404.0$ W. A horsepower (hp) is equivalent to 746.3 watts, so this average power corresponds to a 44.8-hp engine. The actual engine used by this kind of aircraft was a Hispano Suiza 200-hp 8-cylinder engine. The difference between our calculated power and the actual engine's power is attributed to the neglect of drag or frictional forces and the assumption that the force is constant throughout. We can get a better value if we assume the engine is about 25% efficient, then using $e = p_{out}/p_{in}$, one obtains $p_{in} =$ 44.8-hp/0.25 = 179.2-hp, which is closer to the actual value. The actual engine efficiency is probably between 20% and 25%. Finally, notice that (2.2.5c) reduces to $p_{ave} = Cv$ when the velocity remains constant.

■ 2.3 Time-Dependent Force

In this case, the acceleration can be written as $a(t) = F(t)/m$, and since $a(t)dt = dv(t)$, we can integrate to obtain the velocity

$$\int_{v_0}^{v(t)} dv = \int_0^t a(t)dt \quad \Rightarrow \quad v(t) = v_0 + \int_0^t a(t)dt. \tag{2.3.1}$$

Similarly, since $v(t)dt = dx(t)$, we can integrate this expression to obtain the position as a function of time

$$\int_{x_0}^{x(t)} dx = \int_0^t v(t)dt \quad \Rightarrow \quad x(t) = x_0 + \int_0^t v(t)dt. \tag{2.3.2}$$

To proceed further, knowledge of the actual dependence of $a(t)$ is needed as shown in the following example.

EXAMPLE 2.2

A force $F(t) = F_0 \cos \omega t$ is exerted on a particle of mass m, find analytic expressions for $v(t)$ and $x(t)$. If the particle's initial speed is 0.05 m/s, and it starts from the origin, using values of $m = 1$ kg. $F_0 = 1$ N, and $\omega = 3$ rad/s, give plots of $a(t)$, $v(t)$, and $x(t)$ in the range $0 < t < 10$ s.

Solution

The acceleration is $a(t) = F_0 \cos \omega t / m$, and from (2.3.1) we have for the particle's velocity as a function of time,

$$v(t) = v_0 + \frac{F_0}{m} \int_0^t \cos \omega t \, dt = v_0 + \frac{F_0}{m\omega} \sin \omega t, \tag{2.3.3}$$

which is now used in (2.3.2) to obtain the position as a function of time,

$$x(t) = x_0 + \int_0^t \left\{ v_0 + \frac{F_0}{m\omega} \sin \omega t \right\} dt = x_0 + v_0 t - \frac{F_0}{m\omega^2} (\cos \omega t - 1). \tag{2.3.4}$$

The simple MATLAB script `foft.m` that follows can be used to produce the desired plots shown in Figure 2.1.

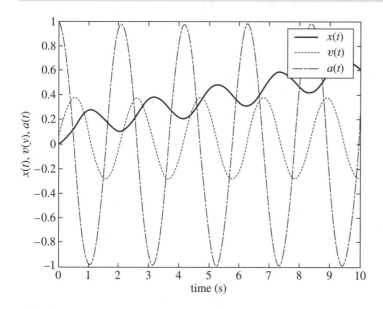

FIGURE 2.1 Example 2.2—Position, Velocity, and Acceleration Plot.

SCRIPT

```
%foft.m
clear;
m=1.0;                         %mass
f0=1.0;                        %force amplitude
w=3.0;                         %force angular frequency
x0=0.0;                        %initial position
v0=0.05;                       %initial velocity
t=[0:0.1:10];                  %time array
a=f0*cos(w.*t)/m;              %acceleration array
v=v0+f0*sin(w.*t)/m/w;         %velocity array
x=x0+v0.*t-f0*(cos(w.*t)-1)/m/w/w; %displacement array
plot(t,x,'k-',t,v,'b:',t,a,'r-.');
title('x(t),v(t),a(t) due to F=F_0*cos(wt)','FontSize',14)
ylabel('x(t),v(y),a(t)','FontSize',14);
xlabel('time(sec)','FontSize',14);
h=legend('x(t)','v(t)','a(t)'); set(h,'FontSize',14)
```

■ 2.4 Position-Dependent Force

In this case, the net force takes the form $F = F(x) = m\,a(x)$. We can write the acceleration as

$$a = \dot{v} = \frac{dv}{dx}\frac{dx}{dt} = v'\dot{x} = vv'. \tag{2.4.1}$$

This enables us to write the net force as

$$F(x) = mv(x)\frac{dv(x)}{dx} = \frac{1}{2}m\frac{d}{dx}\left[v(x)^2\right] = \frac{dT}{dx}, \tag{2.4.2}$$

where the mass is assumed constant and where we have used the definition of kinetic energy $T \equiv \frac{1}{2}mv^2$. The preceding expression for the force is significant, because if we separate variables and integrate

$$\int_{x_0}^{x} F(x)\,dx = \int_{T_0}^{T} dT = T - T_0, \tag{2.4.3}$$

which, if we define the work done by a net force on an object displaced from an initial position x_0 to a final position x as

$$W(x) \equiv \int_{x_0}^{x} F(x)\,dx, \tag{2.4.4}$$

then (2.4.3) is a representation of the work–energy theorem,

$$W(x) = \Delta T; \tag{2.4.5}$$

that is, the work done by a net force on an object is equal to the change in kinetic energy of that object. Thus, the approach we will take to analyze the case of a position-dependent force is to view the problem from an energy consideration point of view. It is convenient to define the quantity

$$u(x) \equiv \int_{x_r}^{x} F(x)\,dx, \tag{2.4.6a}$$

as the work done by a force on an object moving from a reference position x_r to position x. This also allows us to write (2.4.4) as

$$W(x) = u(x) - u(x_0). \tag{2.4.6b}$$

Furthermore, we define the potential energy at position x as

$$V(x) \equiv -u(x). \tag{2.4.7}$$

Using this with (2.4.4–2.4.6), we see the change in potential energy,

$$\Delta V(x) = V(x) - V(x_0) = -W(x) = -\int_{x_0}^{x} F(x)\,dx; \tag{2.4.8a}$$

that is, the change in potential energy of an object between initial and final positions equals the negative of the work done by the applied force between the same two points. From (2.4.8) we see that if we define the potential energy such that $V(x_0) = \text{constant} \equiv 0$, then

$$V(x) = -\int_{x_0}^{x} F(x)\,dx, \tag{2.4.8b}$$

which means that $dV(x)/dx = -[F(x) - F(x_0)]$, or

$$F(x) = -\frac{dV(x)}{dx}, \qquad (2.4.9)$$

where one has also and consistently taken $F(x_0) = -dV(x_0)/dx = 0$. This indicates that a position-dependent force has a potential energy function associated with it. Looking back to (2.4.3), we see that (2.4.8) implies $\Delta V = -\Delta T$; that is, the sum of potential and kinetic energies

$$V + T = V_0 + T_0 \equiv E, \qquad (2.4.10)$$

represents the total *mechanical energy*, E, of a system and that this total energy is constant. This is nothing more than an expression of the conservation of mechanical energy principle.

Given $F(x)$ and using (2.4.9) to obtain the function $V(x)$, the preceding energy concept suggests that one can use (2.4.10) to write $T = \frac{1}{2}m\dot{x}^2 = E - V(x)$, so that

$$\dot{x} = \frac{dx}{dt} = \sqrt{\frac{2[E - V(x)]}{m}}, \qquad (2.4.11)$$

where, for convenience, we've taken the positive root. By separating variables and integrating, we find a relationship between position and time as

$$\int_0^t dt = \int_{x_0}^x \frac{dx}{\sqrt{\dfrac{2[E - V(x)]}{m}}} = t. \qquad (2.4.12)$$

EXAMPLE 2.3

Consider that a body of mass m moving with velocity v_0 is applied a zero net force. Using Equation (2.4.12), obtain an expression of its position as a function of time.

Solution

Since $F = 0$, Equation (2.4.9) implies that $V(x)$ is a constant, which can be taken to be zero. Equation (2.4.12) simplifies to

$$t = \frac{1}{\sqrt{2E/m}} \int_{x_0}^{x} dx = \frac{x - x_0}{\sqrt{2E/m}}. \tag{2.4.13}$$

Because the initial energy of the body is purely kinetic $E = mv_0^2/2$, which is constant, Equation (2.4.13) gives, $t = (x - x_0)/v_0$, or

$$x = x_0 + v_0 t, \tag{2.4.14}$$

as we expect from Newton's first law for an object moving at constant speed in the absence of an external net force.

EXAMPLE 2.4

Consider a spring-mass simple harmonic oscillator. (a) Assuming the mass is initially at the origin and moving at a speed v_0, use Equation (2.4.12) to obtain an expression for the position of the mass as a function of time. Also, obtain expressions for $v(t)$, $a(t)$, and the total energy of the system. (b) Using the results of Part (a), for one oscillation period, plot the potential energy, the kinetic energy, the total energy, and the spring force as a function of position, and explain their relationship. For plotting purposes, use values of mass, spring constant, and initial velocity of $m = 1$ kg, $k = 0.01$ N/m, $x_0 = 0.0$ m, and $v_0 = 0.5$ m/s respectively.

Solution

(a) For the simple harmonic oscillator spring-mass system, the net force is provided by Hooke's law, $F(x) = -kx$. Using (2.4.9), the potential energy function is

$V(x) = -\displaystyle\int_{x_0}^{x} F(x)dx = \frac{1}{2}k(x^2 - x_0^2)$. Equation (2.4.12) simplifies to

$$t = \int_{x_0}^{x} \frac{dx}{\sqrt{\dfrac{2}{m}\left[E - \dfrac{1}{2}kx^2\right]}} = \int_{x_0}^{x} \frac{dx}{\sqrt{\dfrac{k}{m}\left[\dfrac{2E}{k} - x^2\right]}}, \qquad (2.4.15)$$

since $x_0 = 0$. Making the substitutions $u = x$, $b = \sqrt{2E/k}$, and using the integral of

Appendix B, $\displaystyle\int \frac{du}{\sqrt{b^2 - u^2}} = \sin^{-1}(u/b)$, (2.4.15) becomes

$$t = \sqrt{\frac{m}{k}} \int_{x_0}^{x} \frac{dx}{\sqrt{2E/k - x^2}} = \sqrt{\frac{m}{k}} \sin^{-1}\left(\sqrt{\frac{k}{2E}}x\right), \qquad (2.4.16)$$

which we can invert to obtain the position as a function of time as

$$x(t) = \sqrt{\frac{2E}{k}} \sin(\omega t) = A\sin(\omega t), \qquad (2.4.17)$$

where we have defined $\omega \equiv \sqrt{k/m}$. Also notice from here that $A \equiv \sqrt{2E/k}$ is the maximum vibration amplitude. This result has a harmonic behavior, in agreement with our discussion in Section 1.3 in the absence of damping and a driving force. Furthermore, we can obtain the velocity as a function of time by taking the time derivative to get

$$v(t) = \omega\sqrt{\frac{2E}{k}} \cos(\omega t). \qquad (2.4.18)$$

Because initially, at time $t = 0$, the mass is at the origin but moving at v_0, we can solve for the total energy by setting $\omega\sqrt{2E/k} = \sqrt{2E/m} = v_0$, which gives $E = mv_0^2/2$, the initial kinetic energy of the mass. The quantity ω in the argument of the sine and cosine functions is known as the *natural frequency* of the spring-mass system. The period of oscillation is defined to be $\tau = 2\pi/\omega$. Finally, the acceleration is obtained from the force $a(t) = F(x(t))/m = -kx(t)/m$, where $x(t)$ is given by (2.4.17.)

(b) The MATLAB script `fofx.m` with values of $m = 1$ kg, $k = 0.01$ N/m, $x_0 = 0.0$ m, and $v_0 = 0.5$ m/s follows. The actual plots are shown in Figure 2.2, where the total energy, along with the kinetic energy, the potential energy, and the force are shown versus position.

SCRIPT

```
%fofx.m
clear;
m=1.0;                          %mass
k=0.01;                         %spring constant
w=sqrt(k/m);                    %natural frequency
x0=0.0;                         %initial position
v0=0.5;                         %initial velocity
t=[0:0.05:2*pi/w];              %time array from zero to one oscillation period
E0=0.5*m*v0^2;                  %total initial energy
x=sqrt(2*E0/k)*sin(w.*t);       %position versus time array
v=v0*cos(w.*t);                 %velocity versus time array
%a=-k*x/m;                      %acceleration versus time array if needed
PE=0.5*k*x.^2;                  %potential energy array
KE=0.5*m*v.^2;                  %kinetic energy array
E=PE+KE;                        %total energy array
F=-k*x;                         %force array
plot(x,PE,'k-',x,KE,'b:',x,E,'r-.',x,F,'m-');
title('Spring-Mass Simple Harmonic Energy-Force Relation','FontSize',14)
ylabel('PE, KE, E, F','FontSize',14);
xlabel('x(m)','FontSize',14);
h=legend('PE','KE','E','F',3); set(h,'FontSize',14)
```

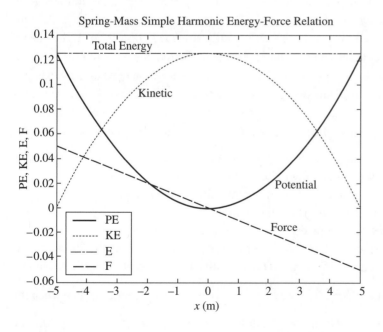

FIGURE 2.2 Example 2.4—Plot of Potential, Kinetic, and Total Energies.

(a)   (b)

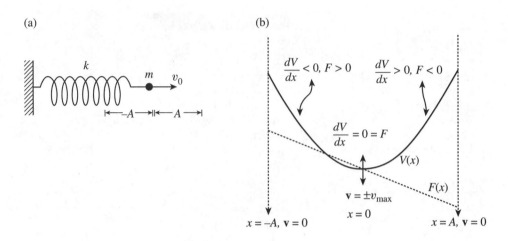

FIGURE 2.3   (a) Mass on a Spring, (b) Potential Energy and Force Diagram.

In Figure 2.3(a) the present initial conditions of the spring-mass system are shown. From the initial conditions given in the problem, we find that $E = mv_0^2/2 = 0.5(1.0)(0.5)^2 = 0.125$ J, which must be conserved. Thus, the amplitude, from (2.4.17) is $A = \sqrt{2E/k} = \sqrt{2(0.125)/0.01} = 5$ m, it follows from (2.4.17), the displacement's range is $-5 \leq x \leq 5$. As can be seen in Figure 2.2, initially, the energy is all kinetic, but when it's fully stretched at position A, the energy is all potential. After a time equal to half the system's period, the mass returns to its initial position at maximum velocity equal and opposite in direction to its initial value, and the energy is once again all kinetic (KE), but the total energy remains constant. The mass continues until the amplitude is $-A$, at which point the spring is fully compressed and the energy is all potential (PE). The natural frequency of the system has a value of $\omega = \sqrt{k/m} = 0.1$ rad/s. After a full period, $\tau = 2\pi/\omega = 62.832$ s, the mass returns to its initial position ready to begin a new cycle.

Also depicted in Figure 2.2 and in Figure 2.3(b) is the relationship played by the force and the potential energy. This relationship obeys Equation (2.4.9). At the origin, the PE is zero and similarly its derivative, the force is, therefore, zero. The value of x when the force is zero is referred to as the *equilibrium position*. As the spring begins to stretch, the PE begins to increase. The derivative of the potential increases, and by (2.4.9) the force must decrease since $\dfrac{dV(x)}{dx} > 0$, to reach a minimum value when the displacement equals A. At this point the force reaches its most negative value, the spring is fully stretched, which means that the mass experiences a maximum force toward the negative direction. After the maximum displacement is reached, the mass turns around, and begins to return to its equilibrium value, thereby decreasing its PE. The decrease in PE along with a decrease in position corresponds to positive derivative $\dfrac{dV(x)}{dx} = \dfrac{-|dV(x)|}{-|dx|} > 0$, which is equivalent to a negative force by (2.4.9), so the mass moves toward $-x$ and begins to gain speed until the force once again reaches zero at the origin, $\dfrac{dV(x)}{dx} = 0$. The mass continues toward the negative x direction with a maximum velocity of $-v_0$ as it passes through the origin. The PE begins to increase once again as the mass travels toward $-x$. The change in potential is positive, but the change in position is negative; by (2.4.9), since $\dfrac{dV(x)}{dx} = \dfrac{|dV(x)|}{-|dx|} < 0$, the force must become positive. This continues to be the trend until the displacement equals $-A$. At this point, the PE is a maximum, the spring is fully compressed, and the force reaches its maximum positive value. This is because the mass experiences the maximum spring force toward the positive direction, begins to turn around, and starts on its return trip toward the origin. In principle, the cycle is repeated indefinitely, unless a damping force is present. All this is shown in Figure 2.3(b). The points of maximum potential energy, i.e., $x = \pm A$, are known as the classical turning points. Finally, notice that Figure 2.2 does confirm that the total energy $E = PE + KE$ is conserved and remains constant throughout the whole cycle.

■ 2.5 Velocity-Dependent Force

As mentioned before, drag forces are velocity-dependent. In such cases we write the force as

$$F = F(v) = m\frac{dv}{dt} \Rightarrow dt = m\frac{dv}{F(v)}, \tag{2.5.1}$$

where the variables have been separated. Integrating the left side on $[0, t]$ and the right side on $[v_0, v]$, we obtain a relationship between velocity and time as

$$t = m\int_{v_0}^{v} \frac{dv}{F(v)}, \tag{2.5.2}$$

which gives the time it takes an object to reach the velocity v, under the action of $F(v)$, if it starts with an initial velocity v_0. If this result is inverted to express $v(t)$, then the position $x(t)$ can be obtained from the now-familiar expression

$$x = x_0 + \int_0^t v(t)\,dt. \tag{2.5.3}$$

An alternate route can be taken if we instead write the force as

$$F(v) = m\frac{dv}{dt} = m\frac{dv}{dx}\frac{dx}{dt} = mv\frac{dv}{dx} \Rightarrow \int_{x_0}^x dx = \int_{v_0}^v \frac{mv\,dv}{F(v)} \Rightarrow x = x_0 + m\int_{v_0}^v \frac{v\,dv}{F(v)}, \tag{2.5.4}$$

where the integration has been performed after separating variables. This gives an expression for $x(v)$ that when inverted yields an expression for $v(x)$. To obtain the $x(t)$ we can write $dt = \dfrac{dx}{v(x)}$, and integrating gives

$$t = \int_{x_0}^x \frac{dx}{v(x)}, \tag{2.5.5}$$

which gives an expression for $t(x)$. This resulting expression needs to be inverted to obtain the position as a function of time.

EXAMPLE 2.5

Consider the example of a body of mass m moving near Earth's surface and subject to an air resistive force $R = -Cv$ that was studied numerically in Chapter 1. (a) Obtain analytic expressions for the velocity and position as a function of time. (b) Show that in the limit of zero drag coefficient, the motion has the expected behavior. (c) Give plots of $y(t)$, $v(t)$, $a(t)$ versus t while making sure not to plot beyond the point where $y(t_{max}) = 0$, where t_{max} is the time the body reaches the ground. Use the following parameters for plotting purposes: $g = 9.8$ m/s^2, $m = 1$ kg, $C = 0.05$ kg/s, $v_0 = 20$ m/s, and $y_0 = 10.0$ m.

Solution

(a) From Chapter 1, Equation (1.3.5), the velocity-dependent force is written as

$$F(v) = -mg - Cv, \tag{2.5.6}$$

so that from (2.5.2) we have

$$t = m\int_{v_0}^v \frac{dv}{-mg - Cv} = -\frac{1}{g}\int_{v_0}^v \frac{dv}{1 + \dfrac{C}{mg}v}. \tag{2.5.7}$$

Making the substitution $u = 1 + \dfrac{C}{mg}v$, $du = \dfrac{C}{mg}dv$, we find $t = -\dfrac{m}{C}\displaystyle\int\dfrac{du}{u} = -\dfrac{m}{C}\ln(u)$, or

$$t = -\frac{m}{C}\left[\ln\left(1 + \frac{C}{mg}v\right)\right]_{v_0}^{v} = -\frac{m}{C}\ln\left[\frac{mg + Cv}{mg + Cv_0}\right], \qquad (2.5.8)$$

which can be inverted to obtain

$$v = \left(\frac{mg}{C} + v_0\right)e - \frac{C}{m}t - \frac{mg}{C}. \qquad (2.5.9)$$

Using (2.5.3) with the coordinate y replacing x, and performing the integration, we obtain the time-dependent position

$$y(t) = y_0 - \left[\frac{m}{C}\left(\frac{mg}{C} + v_0\right)e - \frac{C}{m}t + \frac{mg}{C}t\right]_0^t,$$

or

$$y(t) = y_0 - \frac{mg}{C}t - \frac{m}{C}\left(\frac{mg}{C} + v_0\right)\left(e - \frac{C}{m}t - 1\right). \qquad (2.5.10)$$

From (2.5.9) and (2.5.10) we notice that at $t = 0$, the initial conditions $y = y_0$ and $v = v_0$ result, as should be. Also, as $t \to \infty$, the velocity reaches the terminal value $v \to v_t = -mg/C$, as expected from Chapter 1.

(b) In the limit of small drag coefficient, the exponential in (2.5.9) is expanded $(e^{-x} \approx 1 - x + x^2/2 \ldots)$ to second order

$$v \approx \left(\frac{mg}{C} + v_0\right)\left(1 - \frac{C}{m}t + \frac{1}{2}\left[\frac{C}{m}t\right]^2\right) - \frac{mg}{C}$$

$$= v_0 - gt - \frac{v_0 C}{m}t + \frac{1}{2}\left(\frac{mg}{C} + v_0\right)\left[\frac{C}{m}t\right]^2, \qquad (2.5.11)$$

and in the limit as $C \to 0$ the result is that of the velocity in free fall in the absence of air resistance. Similarly for the position we can write from (2.5.10)

$$y(t) \approx y_0 - \frac{mg}{C}t - \frac{m}{C}\left(\frac{mg}{C} + v_0\right)\left(-\frac{C}{m}t + \frac{1}{2}\left[\frac{C}{m}t\right]^2\right)$$

$$= y_0 + v_0 t - \frac{g}{2}t^2 - \frac{v_0 C}{2m}t^2 \qquad (2.5.12)$$

and as $C \to 0$, there results the familiar expression for the position of an object performing free fall.

(c) For purposes of carrying out the desired plot, the MATLAB script fofv.m, shown here, is used. The script calculates the position, velocity, and acceleration of the body using the analytic formulas (2.5.6, 2.5.9, 2.5.10). In order to find the time the body reaches the ground, t_{max}, we solve the equation $0 = y_0 - \frac{mg}{C}t_{max} - \frac{m}{C}\left(\frac{mg}{C} + v_0\right)\left(e - \frac{C}{m}t - 1\right)$ numerically within the script using the MATLAB provided function fzero. The solution process requires an initial guess for t_{max}. This is obtained by choosing the positive root in the equation for the position without air resistance, i.e., $0 = y_0 + v_0 t - \frac{g}{2}t_{guess}^2$. This yields $t_{guess} = \frac{v_0}{g} + \sqrt{\left(\frac{v_0}{g}\right)^2 + \frac{2y_0}{g}}$.

SCRIPT

```
%fofv.m
clear;
g=input(' Enter value of gravity g: ');
m=input(' Enter the object mass m: ');
C=input(' Enter the drag coefficient C: ');
if C< 1.e-3 C=1.e-3; end              %prevent division by zero
y0=input(' Enter the initial height y0: ');  %Initial Conditions
v0=input(' Enter the initial velocity v0: ');%Initial Conditions
%if needed use next line to make NPTS an input
%NPTS=input(' Enter the number calculation steps desired NPTS: ');
NPTS=100;
% use next line to make TMAX a desired input
%TMAX=input(' Enter the run time TMAX: ');
%tz is the time for the frictionless free fall case to have y=0
tz=v0/g+sqrt((v0/g)^2+2*y0/g);
%define a function of several parameters using the inline method
```

```
f=inline('y0-m*(g*t+(m*g/C+v0)*(exp(-C*t/m)-1))/C','t','g','m','C','y0','v0');
%TMAX is the time for the case with drag to reach zero, use tz as guess
TMAX=fzero(f,tz,[],g,m,C,y0,v0);
t=[0:TMAX/NPTS:TMAX];%time array from zero to the time to reach ground
y=y0-m*(g*t+(m*g/C+v0)*(exp(-C*t/m)-1))/C;
v=(m*g/C+v0)*exp(-C*t/m)-m*g/C;
a=-g-C*v/m;
plot(t,y,'k-',t,v,'b:',t,a,'r-.');
title('Resistive Force Proportional to -v Example','FontSize',14)
ylabel('y, v, a','FontSize',14);
xlabel('t(sec)','FontSize',14);
h=legend('y','v','a',0); set(h,'FontSize',14)
```

A plot of the results is shown in Figure 2.4 with inputs of $g = 9.8$ m/s^2, $m = 1$ kg, $C = 0.05$ kg/s, $v_0 = 20$ m/s, and $y_0 = 10.0$ m. Notice that the acceleration is not constant, because the velocity is changing. Just before the body hits the ground, the velocity is about -23 m/s. This value is far from the terminal velocity

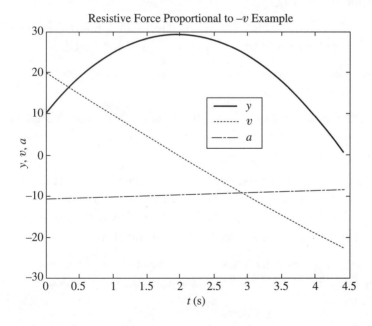

FIGURE 2.4 Example 2.5—Plot of Position, Velocity, and Acceleration for an Object Falling under Linear Air Resistance and Using Analytic Formulas.

value of $v_t = -mg/C = -1*9.8/0.05 = -196$ m/s, at which value the acceleration would approach zero. Here, the body hit the ground before it had a chance to reach the terminal velocity.

■ Chapter 2 Problems

2.1 What is the average power developed by a 200-kg motorcycle as it applies a constant acceleration from an initial velocity of 0 m/s to 25 m/s in 5 s?

2.2 A truck travels for a distance of 1 km at a constant velocity of 32 m/s. If the truck's engine uses half of its maximum power of 250 hp, what is the average force the engine applies in the process?

2.3 A ball is thrown vertically upward at a speed of 18 m/s.

    a. What is the maximum altitude of its flight?

    b. What is the ball's time of flight?

2.4 In Example 2.4, what would be a simple way to modify the results obtained for $x(t)$ and $v(t)$ if initially the mass is at rest with the spring stretched by x_0? What is the initial energy?

2.5 Use Equation (2.4.12) to obtain $x(t)$ for an object falling under the action due to gravity near the Earth's surface. Assume it starts from rest.

2.6 a. Write the differential equation for $x(t)$ of a spring-mass system performing simple harmonic motion.

    b. For this differential equation give a solution, describe its time-dependent behavior, and show that it satisfies the differential equation. Assume that at $t = 0$, $x = x_0$.

2.7 Modify the MATLAB script fofx.m in order to obtain a plot of $x(t)$, $v(t)$, and $a(t)$ for the undamped harmonic oscillator and use $x_0 = 0$, $v_0 = 0.5$ m/s for initial conditions. Compare the results with those of the MATLAB script hol.m of Chapter 1 under the same initial conditions.

2.8 A body of mass m initially moving with velocity v_0 experiences a decelerating force $F = -Cv^2$. Find expressions for the body's velocity as a function of time when

    a. going down.

    b. going up.

c. If the object is dropped from rest, obtain the expression for its position as a function of time. If the object is thrown upward with v_0,

d. what is its position as a function of time,

e. how long does it take the object to reach the maximum height, and

f. what is the maximum height reached?

2.9 Modify the MATLAB script ho2.m in order to incorporate the formulas developed in Example 2.5. Compare the numerical and the analytic approaches for the parameters used to obtain Figure 1.7 of Chapter 1. Discuss your results.

■ Additional Problems

2.10 In the solution for $x(t)$ in Example 2.4 it was assumed that $x_0 = 0$. Redo the problem without this assumption; that is, at $t = 0$, $x = x_0$, and $v = v_0$. Check your results to see that they give the correct limit at $t = 0$.

2.11 When a mass travels under gravity and a drag force due to air resistance, it is possible to model this drag force as $\mathbf{R} = -c\mathbf{v}$; that is, the drag is proportional to the negative of the velocity and c is the drag coefficient. Suppose that under these conditions a mass is thrown directly upward into the air.

a. Obtain an expression for the time (t_{top}) it takes to reach the maximum height.

b. Obtain an approximation of the time to reach the top by expansion methods to first order in the drag coefficient.

c. Show that in the limit of $c \to 0$, the expression for t_{top} simplifies to what is expected in the absence of any drag.

2.12 A ball of mass m falls under gravity while experiencing a force f due to friction. The ball falls from an original height h. It is able to bounce off the floor and reach a height that's $3/4$ of its original height.

a. Obtain an expression for the force f.

b. What is the velocity of the ball when it passes by $1/2$ of the original height after its first rebound?

2.13 A mass m is acted on by the force shown in Figure 2.5. What is the speed of a mass at the end of t_2? (Assume the mass starts from rest.)

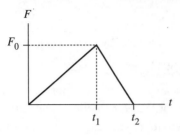

I FIGURE 2.5

2.14 A stunt motorcycle rider traveling with speed v_0 suddenly deploys a parachute. The decelerating force experienced is proportional to the square of his velocity.

   a. Give an expression for the velocity of the motorcycle as a function of time (let $t = 0$ be the time at which the parachute is deployed).

   b. Show that at $t = 0$, and your expression gives the correct limit.

   c. How long would it take before his velocity is one half of his initial velocity?

2.15 Find the maximum height an object thrown upward with an initial speed v_0 will reach if it is subject to gravity and to air resistance proportional to its velocity.

2.16 A spherical marble of radius a in a fluid experiences a drag force given by Stokes' law, $\mathbf{R} = -6\pi\eta a\mathbf{v}$, where η is the fluid viscosity and $\mathbf{v}$ its velocity. Suppose that this marble is placed on the surface of the ocean and let go.

   a. Write down Newton's second law for the marble.

   b. Derive an expression for the terminal speed of the marble in terms of its density ρ_0, its radius, the ocean water density ρ_w, gravity g, and the viscosity.

2.17 A force $F(t) = F_0 \cos\omega t + ma_0$ is exerted on a particle of mass m.

   a. Find analytic expressions for $v(t)$ and $x(t)$.

   b. If the particle's initial speed is 0.05 m/s, and it starts from the origin, using values of $m = 1$ kg, $F_0 = 1$ N, $a_0 = 0.02$ m/s^2, and $\omega = 3$ rad/s give plots of $a(t)$, $v(t)$, and $x(t)$ in the range $0 < t < 10$ s. Explain the observed behavior of the motion.

3 Harmonic Motion in One Dimension

■ 3.1 Introduction

Harmonic motion is oscillatory motion over time. Its behavior is characterized by an amplitude and a frequency (or a period). Many physical systems in nature have oscillatory properties. Some examples are musical notes, a suspension bridge, a child on a swing, a pendulum clock, a mass at the end of a spring, and even a planet going around the sun. In fact, harmonic motion has a very important characteristic, i.e., the acceleration associated with the system is proportional to the negative of its displacement. For example, in Figure 3.1, imagine that the point on the ring is going around with a speed v and that the ring's radius is r with position (x, y). The angular speed of the object is $\omega = v/r$ and the angular displacement is $\theta = \omega t$. The angular speed, measured in $rad/$s, is related to frequency, measured in $rev/$s by $\omega = 2\pi f$. The inverse of the frequency is the time it takes the object to make a revolution, i.e., the object's period $\tau = 1/f$.

We write the position of the object as

$$\mathbf{r} = (x, y) = (r\cos\theta, r\sin\theta) = (r\cos\omega t, r\sin\omega t). \tag{3.1.1}$$

Taking the time derivative of this we get the velocity

$$\mathbf{v} = (\dot{x}, \dot{y}) = (-r\omega\sin\omega t, r\omega\cos\omega t), \tag{3.1.2}$$

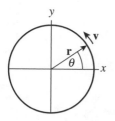

I FIGURE 3.1 An object performing circular motion.

and taking the derivative of this we get the acceleration

$$\mathbf{a} = (\ddot{x}, \ddot{y}) = (-r\omega^2\cos\omega t, -r\omega^2\sin\omega t). \tag{3.1.3}$$

Comparing (3.1.1) and (3.1.3) we see that

$$\mathbf{a} = -\omega^2\mathbf{r}. \tag{3.1.4}$$

This result has several meanings. First, because the position vector points away from the center of the circle, the acceleration points toward the center. Second, what we have obtained is the *centripetal* acceleration of the object whose magnitude is $a = \omega^2 r = v^2/r$. Third, the fact that the acceleration is proportional to the negative of the displacement, not surprisingly, indicates that circular motion properties can be used to describe harmonic motion. Suppose we look at a projection of $\mathbf{r}$ onto the x axis. This projection is $x = r\cos\omega t = r\cos(2\pi ft) = r\cos(2\pi t/\tau)$. This describes the horizontal position of the object as a function of time having a period τ. The maximum displacement of the object in the x direction is called the *amplitude*, which in our case happens to be the circle's radius r. A plot of $x(t)$ using the MATLAB code shown here, shows the familiar harmonic behavior in Figure 3.2, where the period was taken as $\tau = 2$ s, the amplitude $r = 2m$, and the time range is $[0, 2\tau]$.

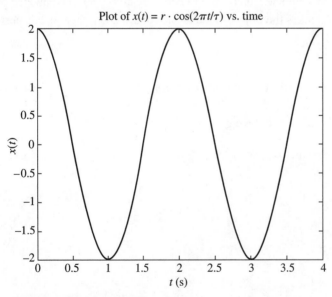

I FIGURE 3.2 Example of a harmonic behavior.

SCRIPT

```
%xoft.m
% plots x=r*cos(2pi*t/tau)
clear;
tau=2;                  %period
NPTS=100;               %number of points
t=[0:2/NPTS:2*tau];     %time array
r=2.0;                  %amplitude
x=r*cos(2*pi*t/tau);    %position array
plot(t,x);
title('Plot of x(t)=r\cdotcos(2\pit/\tau) vs time','FontSize',14)
ylabel('x(t)','FontSize',14);
xlabel('t(sec)','FontSize',14);
```

The figure shows two periods with a frequency of oscillation equal to 0.5 Hz and a vibrational amplitude of 2 units. Had we plotted the y coordinate as a function of time instead of x, the only difference would have been the value taken by the position at $t = 0$, but the characteristics of the harmonic motion would be the same. Sine and cosine functions naturally play a significant role in the description of oscillatory behavior because they themselves are harmonic.

■ 3.2 Hooke's Law and the Simple Harmonic Oscillator

Hooke's law is an expression of the elastic properties of a spring, as discussed in Chapter 2. However, in contrast to considering the problem from the viewpoint of a position-dependent force, we will look at the differential equation. In Figure 3.3, a mass is shown to be displaced by an amount x from the equilibrium position.

Recalling Hooke's law, the restoring force the spring applies is

$$F = m\ddot{x} = -kx \qquad (3.2.1)$$

or

$$\ddot{x} + \omega^2 x = 0, \qquad (3.2.2)$$

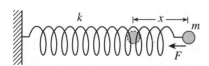

❚ FIGURE 3.3 A mass on a spring performs simple harmonic motion.

with the definition for the angular frequency, $\omega \equiv \sqrt{k/m}$. Equation (3.2.2) is a second-order differential equation, which can be written as

$$(D + i\omega)(D - i\omega)x = 0, \tag{3.2.3}$$

where $D \equiv d/dt$ and $i \equiv \sqrt{-1}$. We can think of (3.2.3) as two separate first-order differential equations

$$(D + i\omega)x = 0, \quad (D - i\omega)x = 0, \tag{3.2.4}$$

or

$$\frac{dx}{dt} = -i\omega x, \quad \frac{dx}{dt} = +i\omega x, \tag{3.2.5}$$

whose variables are separable and whose solutions are obtained as

$$x_1 = Ae^{-i\omega t}, \quad x_2 = Be^{i\omega t}. \tag{3.2.6}$$

This suggests that (3.2.2) can be satisfied by a linear combination of these solutions. We can, therefore, write in general

$$x = Ae^{-i\omega t} + Be^{i\omega t}$$

$$= (A + B)\cos(\omega t) + i(B - A)\sin(\omega t)$$

$$= C\sin(\omega t + \delta), \tag{3.2.7a}$$

where, on the right, to obtain the second expression, we have used the Euler identity

$$e^{ix} = \cos(x) + i\sin(x), \tag{3.2.7b}$$

and we have used the following relations for the constants

$$(A + B) \equiv C\sin(\delta), \quad i(B - A) = C\cos(\delta), \tag{3.2.7c}$$

in the third expression of (3.2.7a), with the assumption that $B = A^*$ (see Problem 3.8) in order for x to be real. The trigonometric identities of Appendix C have also been used. To obtain a particular solution to (3.2.2), however, the boundary conditions of the problem have to be satisfied. First, using the last of (3.2.7), we notice that the velocity is given by

$$v = \frac{dx}{dt} = C\omega\cos(\omega t + \delta),$$ (3.2.8)

and that the acceleration is $a = dv/dt = -C\omega^2\sin(\omega t + \delta) = -\omega^2 x$, as expected for harmonic motion. If at $t = 0$, the spring is displaced by amount x_0, with speed v_0, then (3.2.8) and the last of (3.2.7) can be used to write

$$x_0 = C\sin\delta, \quad v_0 = C\omega\cos\delta,$$ (3.2.9)

so that the expressions for the constants C and δ can be solved for and written in terms of the initial conditions as

$$C = \sqrt{\left(x_0^2 + \left(\frac{v_0}{\omega}\right)^2\right)}, \quad \delta = \arctan\left(\frac{\omega x_0}{v_0}\right).$$ (3.2.10)

Thus, the last of (3.2.7a) with (3.2.10) represents the particular solution to (3.2.2).

EXAMPLE 3.1

Write the particular time-dependent solution of the harmonic oscillator given that at time $t = 0$, the spring is initially stretched by x_0 with zero initial velocity. Also, write the form of the time-dependent velocity and identify each term in the expression.

Solution

From (3.2.10) we set $v_0 = 0$, so that $C = x_0$, and $\delta = \lim_{v_0 \to 0} \arctan(\omega x_0/v_0) \to \pi/2$, and from (3.2.7) we find the desired particular solution

$$x(t) = x_0\sin\left(\omega t + \frac{\pi}{2}\right) = x_0\cos\omega t.$$ (3.2.11a)

The time-dependent velocity is obtained by taking the derivative of this result, or

$$v(t) = -\omega x_0\sin\omega t = -v_{max}\sin\omega t,$$ (3.2.11b)

where $v_{max} = \omega x_0$ is the maximum velocity achieved by the mass as it passes by the equilibrium position. The velocity is negative because as soon as the mass is released from its initial positive position at $t = 0$, the mass will begin to move toward the equilibrium position in the negative direction and vice versa if x_0 were negative.

■ 3.3 Small Oscillations and the Potential Energy Function

In the previous chapter we discussed the connection between a position-dependent force and its relation with a potential energy function. There is also a connection between a general potential energy function and a harmonic oscillator. A general potential energy function of x can be expressed in the form of a Taylor series expansion about a point x_0, i.e.,

$$V(x) = V(x_0) + V'(x_0)(x - x_0) + \frac{1}{2!}V''(x_0)(x - x_0)^2 + \cdots$$

$$+ \frac{1}{(n-1)!}V^{n-1}(x_0)(x - x_0)^{n-1} + R_n, \tag{3.3.1}$$

where the point x_0 represents a minimum of the potential, as shown in Figure 3.4, and R_n is the remainder.

For small oscillations, that is, for x in the neighborhood of x_0 as shown by the dashed circle in Figure 3.4, the difference $(x - x_0)$ is small and we can approximate the potential of (3.3.1) as

$$V(x) \approx V(x_0) + V'(x_0)(x - x_0) + \frac{1}{2!}V''(x_0)(x - x_0)^2. \tag{3.3.2}$$

At x_0, the potential has a minimum, as shown in Figure 3.4, and so $V'(x_0) = 0$; furthermore, since $V(x_0)$ is a constant, we can redefine the zero of the potential so that we take $V(x_0) \equiv 0$; then we can write (3.3.2) as

$$V(x) \sim \frac{1}{2}V''(x_0)(x - x_0)^2 \equiv \frac{1}{2}k_{eff}(x - x_0)^2, \tag{3.3.3}$$

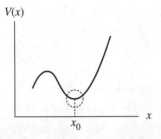

FIGURE 3.4 A general potential function can be expanded about its minimum point.

that is, for small oscillations, a general potential energy function has an effective spring constant associated with it and can be thought of as a harmonic oscillator. The oscillation frequency is given by

$$\omega_0 = \sqrt{\frac{k_{eff}}{m}} = \sqrt{\frac{V''(x_0)}{m}}.$$ (3.3.4)

EXAMPLE 3.2

Consider a two-atom molecule being held in a plane by a potential of the form $V(x) = \left(\frac{A}{x^3} - \frac{B}{x^2}\right)$. Given that $A = u_0 a_0^3$ and $B = u_0 a_0^2$, where u_0 is a unit of molecular energy, and a_0 is a unit of molecular distance, (a) obtain the molecule's bond length if one of the atoms is considered to be at the origin. (b) Find the molecule's frequency of vibration. (c) Give a plot of the potential and the related molecular force. Identify the potential's most significant features in the plot.

....

Solution

(a) Setting the derivative of the potential to zero

$$V'(x_b) = \left(-\frac{3A}{x^4} + \frac{2B}{x^3}\right)\Bigg|_{x=x_b} = 0, \text{ and solving for the bond length } x_b, \text{ we get}$$

$$x_b = \frac{3A}{2B} = \frac{3}{2}a_0.$$ (3.3.5)

(b) The effective molecular spring constant is $k_{eff} = V''(x_b) = \frac{12A}{x_b^5} - \frac{6B}{x_b^4}$.

Using this, and the value of x_b into (3.3.4), the frequency of vibration is found to be $\omega = \sqrt{\frac{32B^5}{81mA^4}} = \sqrt{\frac{32u}{81ma_0^2}}$.

(c) We note that our potential is $V(x) = \left(\frac{u_0 a_0^3}{x^3} - \frac{u_0 a_0^2}{x^2}\right)$. This means that for plotting purposes, we can write it in dimensionless form if we define $x = \bar{x}a_0$ and $V(x) = \bar{V}(\bar{x})u_0$ to obtain

$$\overline{V}(\overline{x}) = \left(\frac{1}{\overline{x}^3} - \frac{1}{\overline{x}^2}\right). \tag{3.3.6}$$

The minimum of the potential occurs at the bond length. Evaluating $V(x)$ at $x = x_b$ of (3.3.5) corresponds to evaluating $\overline{V}(\overline{x})$ at $\overline{x} = \overline{x}_b \equiv 3/2$, so that the minimum of the potential is

$$\overline{V}_{\min} \equiv \overline{V}(\overline{x}_b) = \left(\frac{1}{\overline{x}_b^3} - \frac{1}{\overline{x}_b^2}\right) = -\frac{4}{27}, \tag{3.3.7}$$

which means that the molecular ionization energy is $-(4/27)u_0$ in our units.

The related molecular force is from (2.4.9) $F(x) = -V'(x) =$ $\left(\frac{3u_0 a_0^3}{\overline{x}^4 a_0^4} - \frac{2u_0 a_0^2}{\overline{x}^3 a_0^3}\right) = \overline{F}(\overline{x})\frac{u_0}{a_0}$ in the present units, with

$$\overline{F}(\overline{x}) = \frac{3}{\overline{x}^4} - \frac{2}{\overline{x}^3}, \tag{3.3.8}$$

which has a value of zero at $\overline{x}_b$, as expected. The MATLAB script v_and_f.m performs the desired plots shown in Figure 3.5.

SCRIPT

```
%v_and_f.m
% plots V=(1/x^3-1/x^2) and the related force F=3/x^4-2/x^3
clear;
xb=3/2;               %the bond length
vmin=-4/27;           %potential min determines size of plotting window
xmin=0.5;             %minimum value of x
xmax=5*xb;            %use the bond length to determine the maximum x
NPTS=100;             %number of points
x=[xmin:2/NPTS:2*xmax]; %distance array
V=1./x.^3-1./x.^2;    %the potential array
F=3./x.^4-2./x.^3;    %the force array
plot(x,V,'b:',x,F,'r-.');
line([0,xmax],[0,0],'Color','k','LineStyle','-') %draw a zero line
axis([xmin, xmax,1.5*vmin,-1.5*vmin])
title('Plot of V(x)=(1/x^3-1/x^2) and F=3/x^4-2/x^3 vs position','FontSize',14)
ylabel('V(x), F(x) (u_0)','FontSize',14);
xlabel('x(a_0)','FontSize',14);
h=legend('V','F',0); set(h,'FontSize',14)
```

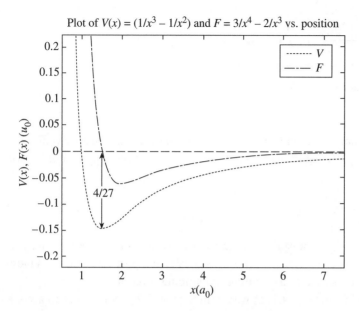

Plot of $V(x) = (1/x^3 - 1/x^2)$ and $F = 3/x^4 - 2/x^3$ vs. position

I FIGURE 3.5   Potential approximation of a two-atom molecule.

In Figure 3.5, we see that the force has a zero value when the potential reaches its minimum value. Also, the ionization energy value (4/27 in units of u_0) is shown to be the value between the minimum of the potential and zero. This is the amount of energy required for the molecule to separate into two individual atoms. This information was added to the figure using MATLAB's figure editor.

■ 3.4   The Harmonic Oscillator with Damping

The motion of the simple harmonic oscillator discussed in Section 3.2 is idealized to the extent that in reality a spring has mass as well as internal friction. A real spring is not a purely elastic body so that some of the system's energy is dissipated to frictional losses within it. Additionally, air molecules surround the spring-bob system, which is a source of friction and causes mechanical energy not to be conserved. One of the ways to incorporate friction is to separate the frictional losses from the ideal motion in the form of two terms in Newton's second law; that is,

$$F = m\ddot{x} = -kx + R, \tag{3.4.1}$$

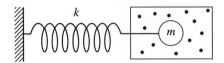

FIGURE 3.6 Harmonic oscillator with damping.

which is similar to what we did in Chapter 1. In its simplest form the resistive force is proportional to the negative of the velocity, $R = -c\dot{x}$, so that our equation of motion becomes

$$m\ddot{x} + c\dot{x} + kx = 0. \tag{3.4.2}$$

Effectively, we think of the spring as massless and frictionless. We also imagine that the mass is immersed in a molecular fluid, as shown in Figure 3.6. The mass moving through the fluid experiences a retarding or damping force R.

We could proceed to solve (3.4.2) as we did in Section 3.2. Another way to accomplish a similar task, however, is to assume a general solution to the problem. The assumed solution and its derivatives can then be substituted into (3.4.2). In our case, we let

$$x = Ae^{rt}, \quad \dot{x} = Are^{rt}, \quad \ddot{x} = Ar^2e^{rt}, \tag{3.4.3}$$

which when inserted into (3.4.2) and after dividing by Ae^{rt}, we get

$$r^2 + \frac{c}{m}r + \frac{k}{m} = 0, \tag{3.4.4a}$$

whose roots are

$$r_{1,2} = -\frac{c}{2m} \pm \sqrt{\left(\frac{c}{2m}\right)^2 - \frac{k}{m}}. \tag{3.4.4b}$$

Since r has two roots, x from (3.4.3) can be written as a linear combination as

$$x = Ae^{r_1 t} + Be^{r_2 t}. \tag{3.4.5}$$

Notice that when $c \to 0$, the roots (3.4.4b) are imaginary and (3.4.5) gives back the standard simple harmonic oscillator. The general solution falls into three categories, which we consider on the next page.

Case 1: Overdamping

This occurs whenever the condition $(c/2\,m)^2 > k/m$ under the square root in (3.4.4b) occurs. In this case, both roots $r_{1,2} = -\gamma_{1,2}$ are negative, with $\gamma_{1,2} \equiv c/2\,m \mp \sqrt{(c/2\,m)^2 - k/m}$ being positive. The overdamped (od) solution is

$$x_{od} = A_{od}e^{-\gamma_1 t} + B_{od}e^{-\gamma_2 t}, \tag{3.4.6}$$

and is characterized by a strong exponential decay behavior.

Case 2: Critically Damped

When the critical condition occurs at which $(c/2\,m)^2 = k/m$, once again we get two real roots, but this time they are equal $r_{1,2} = -\gamma$, where $\gamma \equiv c/2\,m$. In this case, we know one solution has the form

$$x_1 = Ae^{-\gamma t}, \tag{3.4.7a}$$

but, according to standard results of ordinary differential equations, the second solution must be linear in time (see Problem 3.9) and still obey the original differential equation (3.4.2). To this end we assume

$$x_2 = Bte^{-\gamma t}, \tag{3.4.7b}$$

and insert it into (3.4.2) to get

$$m(-2\gamma Be^{-\gamma t} + \gamma^2 Bte^{-\gamma t}) + c(Be^{-\gamma t} - t\gamma Be^{-\gamma t}) + kBte^{-\gamma t} = 0,$$

or

$$m(-2\gamma + \gamma^2 t) + c(1 - t\gamma) + kt = 0.$$

Collecting powers of t get

$$(c - 2\gamma m) + t(m\gamma^2 - c\gamma + k) = 0,$$

but since $\gamma = c/2\,m$ and also $(c/2\,m)^2 = k/m$, then each coefficient vanishes, indicating that (3.4.7b) is also a solution. Thus, a general solution for the critically damped (cd) case is a linear combination of (3.4.7),

$$x_{cd} = A_{cd}e^{-\gamma t} + B_{cd}te^{-\gamma t}. \tag{3.4.8}$$

It is instructive to compare the overdamped case with the critically damped under the same conditions. To this end, let's assume that at $t = 0$, $x = x_0$, and $v = v_0$. We need to obtain the constants to be used in the previous solutions. For the overdamped case, we set the solution and its first derivative to satisfy the initial conditions at $t = 0$; thus we have $x_0 = A_{od} + B_{od}$ and $v_0 = -\gamma_1 A_{od} - \gamma_2 B_{od}$. Solving for the coefficients we find

$$A_{od} = x_0 - \frac{v_0 + \gamma_1 x_0}{\gamma_1 - \gamma_2}, \quad B_{od} = \frac{v_0 + \gamma_1 x_0}{\gamma_1 - \gamma_2}. \tag{3.4.9}$$

Similarly, in the critically damped case, at $t = 0$, we set $x_0 = A_{cd}$, and $v_0 = -\gamma A_{cd} + B_{cd}$, yielding

$$A_{cd} = x_0, \quad B_{cd} = v_0 + \gamma x_0. \tag{3.4.10}$$

We are now ready to run the following MATLAB script over_crit_damp.m, to provide a graphical comparison between these two cases. The parameters used are $m = 1$ kg, $k = 1$ N/m, $c = 0.5$ N · s/m, $x_0 = 1.0$ m, and $v_0 = 5.0$ m/s.

SCRIPT

```
%over_crit_damp.m
%plots the overdamped and critically damped HO solutions
clear;
m=0.05;                             %mass
k=1;                                %spring constant
c=0.5;                              %drag coefficient
x0=1.0;                             %initial position
v0=5.0;                             %initial speed
gam=c/2/m;                          %critical gamma
desc=gam^2-k/m;                     %must be positive
if desc <= 0;                       %ensure appropriate problem conditions
    disp('gam needs to be greater');
    return;
end
gam1=gam+sqrt(desc);                %overdamped gamma1
gam2=gam-sqrt(desc);                %overdamped gamma2
Bo=(v0+gam1*x0)/(gam1-gam2);        %constant B for overdamped
Ao=x0-Bo;                           %constant A for overdamped
Ac=x0;                              %constant A for critically damped
Bc=v0+gam*x0;                       %constant B for critically damped
tmax=2;                             %maximum time
NPTS=100;                           %number of points
t=[0:tmax/NPTS:tmax];               %time array
xo=Ao*exp(-gam1*t)+Bo*exp(-gam2*t); %overdamped
```

```
xc=Ac*exp(-gam*t)+Bc*t.*exp(-gam*t);%critically damped
plot(t,xo,'b:',t,xc,'r-.');
title('Overdamped and Critically Damped Comparison','FontSize',14)
ylabel('Xoverdamped, Xcritically-damped','FontSize',14);
xlabel('t','FontSize',14);
h=legend('Overdamped','Critically Damped',1); set(h,'FontSize',14);
```

The plot of the overdamped and critically damped cases is shown in Figure 3.7. In particular, the critically damped case decays more quickly because the decay constant γ is larger than the smaller of the decay constants of the overdamped case, i.e., γ_1. The preceding damping properties are useful in real-world applications. Large damping constants are needed in, for example, building doors that are needed to open or close gently, or even automobiles with unneeded bouncing characteristics. This is in contrast to the case when the drag coefficient is much smaller as in the next case.

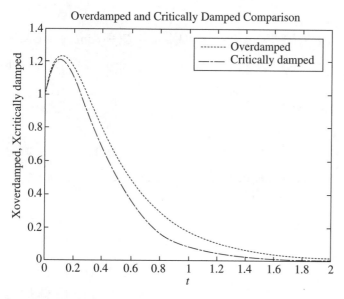

FIGURE 3.7 Comparison of the critical and overdamped solutions of a harmonic oscillator.

Case 3: Underdamped

When the drag coefficient of the harmonic oscillator system previously considered is small so that $(c/2\,m)^2 < k/m$, then the sign of the result under the square root term in (3.4.4b) becomes negative and, therefore, the term becomes imaginary. Under these circumstances, the roots are now more conveniently written in the complex form

$$r_{1,2} = -\gamma \pm i\omega, \tag{3.4.11a}$$

where $\omega = \sqrt{\omega_0^2 - \gamma^2}$ is the new frequency associated with an underdamped harmonic oscillator, $\omega_0 = \sqrt{k/m}$ is the natural frequency of the undamped simple harmonic oscillator, and $\gamma \equiv c/2\,m$ as before. For this case, Equation (3.4.5) becomes, in a similar way to (3.2.7),

$$x = e^{-\gamma t}(Ae^{i\omega t} + Be^{-i\omega t}) = Ce^{-\gamma t}\sin(\omega t + \theta), \tag{3.4.11b}$$

where C and θ depend on the initial conditions. For example, if at $t = 0$, $x = x_0$, and $v = v_0$, we see that

$$x_0 = C\sin\theta, \tag{3.4.12a}$$

and

$$v_0 = Ce^{-\gamma t}(-\gamma\sin(\omega t + \theta) + \omega\cos(\omega t + \theta))|_{t=0} = C(-\gamma\sin\theta + \omega\cos\theta). \tag{3.4.12b}$$

From these two, we see that

$$\frac{v_0 + \gamma x_0}{\omega} = C\cos\theta, \tag{3.4.13}$$

which when squared and added to the square of (3.4.12a), we get

$$C = \sqrt{x_0^2 + \left(\frac{v_0 + \gamma x_0}{\omega}\right)^2}; \tag{3.4.14a}$$

furthermore, dividing (3.4.12a) by (3.4.13) gives

$$\theta = \tan^{-1}\left(\frac{\omega x_0}{v_0 + \gamma x_0}\right). \tag{3.4.14b}$$

Thus, (3.14.14) fully specifies the underdamped solution (3.4.11b). We incorporate the preceding equations into a MATLAB script in order to obtain a plot of (3.4.11b) under a set of initial conditions similar to what we did before, except that, in contrast to the overdamped and critically damped cases, $\gamma^2 < \omega_0^2$ in the present case. The MATLAB script under_damp.m implements the underdamped solution equations with parameters used are $m = 1$ kg, $k = 1$ N/m, $c = 0.08$ N $\cdot$ s/m, $x_0 = 1.0$ m, and $v_0 = 5.0$ m/s.

SCRIPT

```
%under_damp.m
%plots the underdamped HO solution
clear;
m=0.05;                              %mass
k=1;                                 %spring constant
c=0.08;                              %drag coefficient
x0=1.0;                              %initial position
v0=5.0;                              %initial speed
gam=c/2/m;                           %critical gamma
wo=sqrt(k/m);                        %SHO natural frequency
desc=wo^2-gam^2;                     %must be positive this time
if desc <= 0;                        %ensure appropriate problem conditions
    disp('gam needs to be smaller');
    return;
end
w=sqrt(desc);                        %underdamped frequency
B=sqrt(x0^2+(v0+gam*x0)^2/w^2);      %constant B for underdamped
th=atan(w*x0/(v0+gam*x0));           %angle theta
tmax=5;                              %maximum time
NPTS=100;                            %number of points
t=[0:tmax/NPTS:tmax];                %time array
x=B*exp(-gam*t).*sin(w*t+th);        %underdamped solution
xe=B*exp(-gam*t);                    %the decay envelope
plot(t,x,'b:',t,xe,'r-.');
title('Underdamped Harmonic Oscillator','FontSize',14)
ylabel('Underdamped, Decay Envelope','FontSize',14);
xlabel('t','FontSize',14);
h=legend('Underdamped','Decay Envelope',1); set(h,'FontSize',14);
```

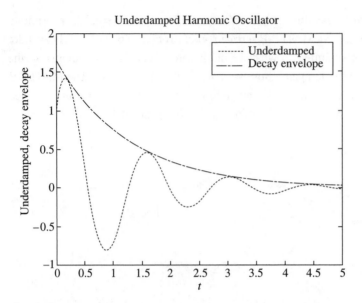

FIGURE 3.8 Underdamped harmonic oscillator and the decay envelope.

Figure 3.8 contains the results of the calculation. The decay envelope is the exponential prefactor in the solution (3.4.11b). While in this underdamped case, the oscillator performs a few oscillation periods with frequency ω, the drag coefficient γ determines the rate at which the oscillator damps. In the limit as $\gamma \to 0$, ω approaches the ideal harmonic oscillator frequency ω_0.

■ 3.5   The Forced Harmonic Oscillator with Damping

An external force can be added to the discussion of the dynamics of a spring-mass system. For example, if a motor is added to Figure 3.6, as shown in Figure 3.9, we would now have a different kind of motion. The motion is that of a driven harmonic oscillator with damping.

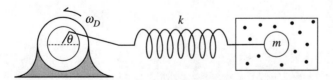

FIGURE 3.9 Harmonic oscillator with damping and a driving force.

The motor rotates with the driving frequency $\omega = \omega_D$ and the force it produces can be approximated as $F_{ext} = F_0 \cos(\omega_D t + \theta)$ where θ is an angle that depends on the initial conditions. For example, if at $t = 0$, the force happens to have a maximum value, then θ can be taken to be zero. Newton's second law for the problem is

$$F = m\ddot{x} = -kx + R + F_{ext}, \qquad (3.5.1)$$

where for convenience, we will take $R = -c\dot{x}$ as before. The simplest way to approach a solution to this equation is to use complex exponentials. The solution to the differential equation will be of a complex form. However, because $\cos(\omega_D t + \theta) = real[e^{i(\omega_D t + \theta)}]$, it is the real part of the solution that is of interest here. The equation of motion for the driven harmonic oscillator, including damping, can therefore be written as

$$m\ddot{x} + c\dot{x} + kx = F_0 e^{i(\omega_D t + \theta)}, \qquad (3.5.2)$$

which has both a homogeneous and an inhomogeneous solution. The homogeneous solution is also known as the transient solution and occurs when the right-hand side of (3.5.2) is zero. This has already been done in the previous section. The inhomogeneous solution is also the particular solution and is the one responsible for the steady state behavior of the system. We will begin by assuming a solution that has a similar behavior to the driving force but with a phase shift

$$x = A e^{i(\omega_D t + \delta)}. \qquad (3.5.3)$$

By substituting this into (3.5.2) and dividing by x, we get

$$-m\omega_D^2 + ic\omega_D + k = \frac{F_0}{A} e^{i(\theta - \delta)}. \qquad (3.5.4)$$

We next separate the real and imaginary parts to get

$$-m\omega_D^2 + k = \frac{F_0}{A} \cos(\theta - \delta), \qquad c\omega_D = \frac{F_0}{A} \sin(\theta - \delta). \qquad (3.5.5)$$

We can solve for the unknown constants A and δ. Dividing the second of these by the first, and solving for the phase shift, we get

$$\delta = \theta - \tan^{-1}\left(\frac{2\gamma\omega_D}{\omega_0^2 - \omega_D^2}\right), \qquad (3.5.6)$$

where $\omega_0 = \sqrt{k/m}$ and $\gamma = c/2\,m$ as in the previous section. Squaring of (3.5.5) and adding the results then solving for A gives

$$A = \frac{F_0/m}{\sqrt{(2\gamma\omega_D)^2 + (\omega_0^2 - \omega_D^2)^2}}. \tag{3.5.7}$$

The final particular solution to the original equation (3.5.1) is, as mentioned before, obtained by taking the real part

$$x_p = \mathrm{Re}(Ae^{i(\omega_D t + \delta)}) = A\cos(\omega_D t + \delta). \tag{3.5.8}$$

As the driving frequency ω_D is varied, the spring responds according to x_p. At a certain value of this frequency, the amplitude A increases dramatically. This condition is known as *resonance*. Let's find the resonant condition under which the maximum amplitude occurs as we vary the driving force frequency. We seek a maximum of A; that is,

$$\frac{dA}{d\omega_D}\bigg|_{\omega_D = \omega_r} = \frac{F_0}{m}\left\{ -\frac{1}{2}[(2\gamma\omega_r)^2 + (\omega_0^2 - \omega_r^2)^2]^{-3/2}[8\gamma^2\omega_r + 2(\omega_0^2 - \omega_r^2)(-2\omega_r)] \right\} = 0,$$

or

$$4\gamma^2 - 2\omega_0^2 + 2\omega_r^2 = 0 \quad \Rightarrow \quad \omega_r = \sqrt{\omega_0^2 - 2\gamma^2}. \tag{3.5.9}$$

Thus, when $\omega_D = \omega_r$, the amplitude of the vibrating system will be a maximum. The value of the maximum amplitude can be obtained by substituting ω_r for ω_D into (3.5.7), to get

$$A_{\max} = \frac{F_0/m}{2\gamma\sqrt{\omega_0^2 - \gamma^2}}. \tag{3.5.10}$$

We notice that in the limit of a small drag coefficient, $\gamma \to 0$, $\omega_r \to \omega_0$, and $A_{\max} \to \infty$, as should be expected. In the absence of any dissipative forces, when the driving frequency matches the natural frequency of the simple spring-mass harmonic oscillator, the vibration amplitude increases indefinitely. The presence of damping tends to shift the resonant frequency to a smaller value than that of the natural frequency. The MATLAB script `drive_amp.m` shows the behavior of the amplitude (3.5.7) versus ω_D for various values of the drag coefficient c. For this calculation, the parameters chosen are $m = 1/2$, $k = 1/2$, $F_0 = 1/2$, so that $\omega_0 = 1$, and $\gamma = c/2m = c$. We notice in the plot of Figure 3.10 that the maxima (3.5.10), which are based on the current parameters, have the numerical values of $A_{\max} = 1/2\gamma\sqrt{1 - \gamma^2}$ and are indicated by the dots. Thus, for example, when $\gamma = 0.2$, $A_{\max} = 1/2\gamma\sqrt{1 - \gamma^2}\Big|_{\gamma=0.2} \sim 2.55$. From (3.5.9) this maximum amplitude occurs at $\omega_r = \sqrt{1 - 2\gamma^2}\Big|_{\gamma=0.2} \sim 0.96$. For ω_r to be real, the maximum value of γ allowed by (3.5.9) is when $\gamma = \omega_0/\sqrt{2} \sim 0.707\ldots$ in our case, so that

$$A_{\max} = 1/2\gamma\sqrt{1 - \gamma^2}\Big|_{\gamma=0.707\ldots} = 1.0,$$

which occurs when $\omega_r = \sqrt{1 - 2\gamma^2}\Big|_{\gamma=0.707\ldots} = 0.0$. The rest of the curves shown in Figure 3.10 are for intermediate values of the drag coefficient whose values are shown. The higher the drag coefficient, the flatter the amplitudes become, which shows that large damping can moderate resonance.

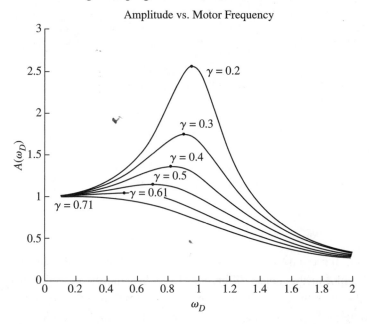

FIGURE 3.10 Amplitude of a forced harmonic oscillator versus driving frequency.

SCRIPT

```
%drive_amp.m
%plots the amplitude of the solution for a driven HO
clear;
m=0.5;                                  %mass
k=0.5;                                  %spring constant
F0=0.5;                                 %driving force amplitude
wo=sqrt(k/m);                           %SHO natural frequency
wmin=0.1;                               %minimum frequency
wmax=2;                                 %maximum frequency
NPTS=200;                               %number of points
w=[wmin:wmax/NPTS:wmax];                %w array
hold on                                 %get ready to superimpose plots
cmin=0.2;                               %minimum value of c
cmax=2*m*wo/sqrt(2);                    %maximum c so that om_res is real
cstep=(cmax-cmin)/5;                    %c step size
for c=cmin:cstep:cmax,                  %loop over the drag coefficient
    gam=c/2/m;                          %find gamma
    desc=(2*gam*w).^2+(wo^2-w.^2).^2;
    A=F0/m./sqrt(desc);                 %The driven ho amplitude
    plot(w,A);                          %plot amplitude
    om_res=sqrt(wo^2-2*gam^2);          %resonant frequency
    Amax=F0/2/m/gam/sqrt(wo^2-gam^2);   %Maximum amplitude at om_res
                                        %next, draw point at position of Amax
    line([om_res;om_res],[Amax;Amax],'Color','red','Marker','.');
                                        %num2str(c,p) c to string with p digits
                                        %cat(2,'a','b') concatenates a and b
    str=cat(2,'\gamma=',num2str(gam,2));
    text(om_res+0.1,Amax-0.06,str,'FontSize',10,'Color','red');
end
title('Amplitude versus Motor Frequency','FontSize',14)
ylabel('A(\omega_D)','FontSize',14);
xlabel('\omega_D','FontSize',14);
```

It is important to understand the role played by the phase δ in the solution (3.5.8). It is more convenient, however, to look at the phase difference, from (3.5.6)

$$\phi \equiv \theta - \delta = \tan^{-1}\left(\frac{2\gamma\omega_D}{\omega_0^2 - \omega_D^2}\right), \tag{3.5.11}$$

between the driving force and the oscillator's motion. The MATLAB script drive_phase.m does a plot of this phase difference. To understand the behavior of ϕ, we refer to Table 3.1, where three special cases are listed. Thus, while the driving frequency lies in the range $0 \leq \omega_D \leq \infty$ the phase difference between the external force

Table 3.1   Special Limits of ϕ vs. ω_D

ω_D	$2\gamma\omega_D/(\omega_0^2 - \omega_D^2)$	ϕ
0	0	0
ω_0	∞	$\pi/2$
∞	$\sim \dfrac{2\gamma}{\omega_D} \to -0$	π

and the harmonic oscillator spans the range $0 \le \phi \le \pi$. The MATLAB script drive_phase.m does the calculation of $\phi(\omega_D)$ for several values of γ. The parameters chosen are as in Figure 3.10 so that $\gamma = c/2\,m = c$.

SCRIPT

```
%drive_phase.m
%plots the phase difference between the driving force and
%the solution for a driven HO
clear;
m=0.5;                          %mass
k=0.5;                          %spring constant
wo=sqrt(k/m);                   %SHO natural frequency
wmin=0.0;                       %minimum frequency
n=2.5;                          %used to increase wmax and NPTS
wmax=n*wo;                      %maximum frequency in terms of wo
NPTS=n*33+1;                    %w points
wstep=(wmax-wmin)/NPTS;         %w step size
hold on                         %get ready to superimpose plots
cmin=0.01;                      %minimum value of c
cmax=1;                         %maximum value of c
cstep=(cmax-cmin)/5;            %c step size
for c=cmin:cstep:cmax,          %loop over the drag coefficient
  gam=c/2/m;                    %gamma
  for i=1:NPTS
    w(i)=wmin+(i-1)*wstep;
    den=wo^2-w(i)^2;
      if(w(i)<=wo)
        ph(i)=atan(2*gam*w(i)/den);    %The phase difference
      else
        ph(i)=pi+atan(2*gam*w(i)/den); %shift by pi needed if w>wo
      end
  end
plot(w,ph);            %plot phi
```

```
                    %num2str(c,p) converts c to a string with p digits
                    %cat(2,'a','b') concatenates a and b
      str=cat(2,'\gamma=',num2str(gam,2));
      text(w(20),ph(20)*(1+0.05),str,'FontSize',8,'Color','red');
      text(w(78),ph(78),str,'FontSize',8,'Color','red');
    end
    title('Phase Difference Between Driving Force and Solution','FontSize',14)
    ylabel('\phi(\omega_D)','FontSize',14);
    xlabel('\omega_D','FontSize',14);
```

Figure 3.11 contains the family of curves using the script. Notice that as γ increases, the closer to a straight line the shape of $\phi(\omega_D)$ becomes and reaches the maximum phase difference less sharply.

Thus, Figure 3.11 shows that as the driving frequency increases, the oscillator tends to lag behind the driving force until at a very large ω_D the oscillator is π radians out of phase. The amplitude of vibration is more dramatic when ω_D is in the neigh-

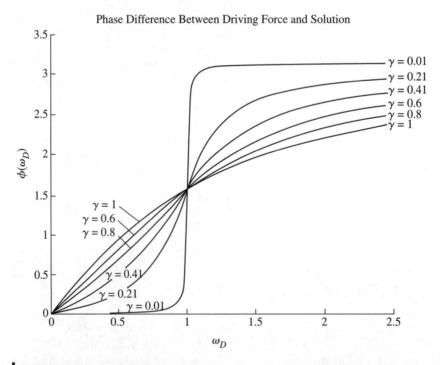

FIGURE 3.11 Phase difference versus driving frequency between the driving force and the driven mass of a harmonic oscillator with damping.

borhood of ω_0 with some attenuation from the drag coefficient, as shown in Figure 3.10. We plot the driving force $F_{ext} = F_0 \cos(\omega_D t + \theta)$ and the solution (3.5.8) versus time for various values of ω and c for visualization purposes. The amplitude (3.5.7) and the phase (3.5.11) are to be used again for this purpose, which makes the problem interesting. The MATLAB script drive_sol.m performs the desired plots. The script prompts the user to input the frequency and the drag coefficient. For convenience, the rest of the parameters used are the same as in Figure 3.11. The plots shown in Figure 3.12 on page 69 include the particular solution (3.5.8) and the full solution $x = x_h + x_p$, where x_h is the transient or homogeneous solution in the absence of an external force. For practical reasons we have taken the underdamped oscillator solution (3.4.11b) as the representative homogeneous solution.

SCRIPT

```
%drive_sol.m
%plots the solution for a driven HO and the applied external force
clear;
w=input(' Enter the driving frequency: ');
c=input(' Enter the drag coefficient value: ');
x0=1.0;                         %initial position for homogeneous part
v0=5.0;                         %initial speed for homogeneous part
m=0.5;                          %mass
k=0.5;                          %spring constant
F0=0.5;                         %driving force amplitude
theta=0;                        %driving force initial phase angle
wo=sqrt(k/m);                   %SHO natural frequency
dt=0.05;                        %time step
tau=2*pi/w;                     %force's period of rotation
tmax=5*tau;                     %maximum time in terms of tau
NPTS=tmax/dt;                   %number of points
gam=c/2/m;                      %find gamma
desc=(2*gam*w).^2+(wo^2-w^2).^2;
A=F0/m/sqrt(desc);              %The driven ho amplitude
den=wo^2-w^2;
if den==0, den=1.e-3; end
if(w <= wo)
   ph=atan(2*gam*w/den);        %phase difference between force and soln
else
   ph=pi+atan(2*gam*w/den);     %shift by pi needed if w > wo
end
%fprintf('gamma =%7.4f\n',gam); %uncomment to print value to screen
delta=theta-ph;                 %the forced solution's phase
t=[0:tmax/NPTS:tmax];
```

```
xf=A*cos(w*t+delta);              %the forced or particular solution
F=F0*cos(w*t+theta);
%========================================================================
% forced solution
subplot(2,1,1)          %setup 2 x 1 matrix plot - 1st window
plot(t,F,'k-',t,xf,'b:')
                        %num2str(c,p) converts c to a string with p digits
                        %cat(2,'a','b') concatenates a and b
str=cat(2,'\gamma=',num2str(gam,3),', \omega_D=',num2str(w,3),...
    ', \phi=',num2str(ph,3));
top=max(A,F0)*(1+0.3); %to create a window large enough
topt=max(A,F0)*(1+0.1);%to post parameters on the plot
axis([0 tmax -top  top]);
text(0,topt,str,'FontSize',12,'Color','red');
title('Driving Force and Driven HO Solution','FontSize',8)
ylabel('x_f(t), F(t)');
xlabel('t','FontSize',8);
legend('F(t)','x(t)');
%========================================================================
% homogeneous + forced solution
subplot(2,1,2)                      %setup 2 x 1 matrix plot - 2nd window
desc=wo^2-gam^2;                    %must be positive
if desc <= 0;                       %ensure homogeneous problem conditions
    disp('gam needs to be smaller');
    return;
end
wu=sqrt(desc);                     %underdamped homogeneous soln frequency
th=atan(wu*x0/(v0+gam*x0));        %angle theta
B=sqrt(x0^2+(v0+gam*x0)^2/wu^2);   %constant B for underdamped
xh=B*exp(-gam*t).*sin(wu*t+th);    %homogeneous solution
plot(t,xh+xf,'r');                 %plot homogeneous+particular solution
title('Homogenous+Particular solutions','FontSize',8)
ylabel('x(t)=x_h(t)+x_f(t)');
xlabel('t','FontSize',8);
axis([0 max(t) min(xh+xf) max(xh+xf)])
```

The plots from three separate runs as shown in Figures 3.12(a, b, c) are for driving frequency values of $\omega_D = 0.1, 0.8, 3.0\,\text{rad/s}$, respectively. The maximum time shown for each solution set is $5\tau (\tau = 2\pi/\omega)$. A drag coefficient $c = 0.1\,\text{N} \cdot \text{s/m}$ has been used. Figure 3.11 shows that for the small frequency of $\omega_D = 0.1\,\text{rad/s}$, the oscillator follows the driving force with a small phase difference of $\phi \sim 0.02$ rad, but as the driving force's frequency increases to a value of $\omega_D = 0.8\,\text{rad/s}$, the response of the oscillator begins to get out of phase with a value of $\phi \sim 0.42$ rad, and when the frequency increases further to a value of $\omega_D = 3\,\text{rad/s}$, the phase difference is

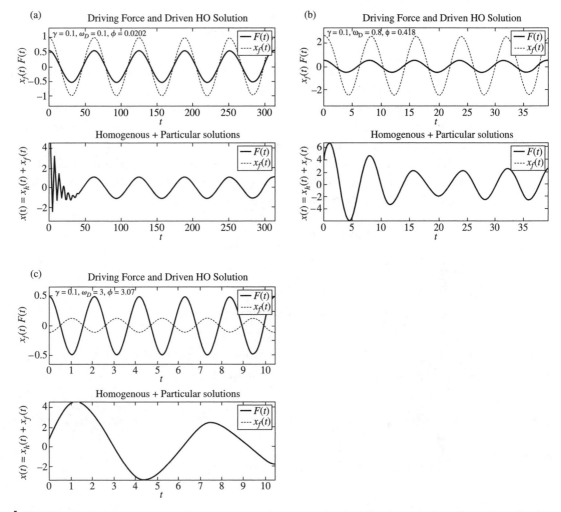

FIGURE 3.12 The upper plot is the inhomogeneous solution compared to the driving force of a forced harmonic oscillator. The lower plot is the full solution (homogeneous and inhomogeneous). (a) low frequency, (b) frequency near resonance, (c) high frequency.

$\phi \sim 3$ rad, which is near the maximum value of π. This is consistent with Table 3.1, as discussed before. The solution's amplitude shown in Figure 3.12 behaves according to Figure 3.10; it increases when its value is near the natural frequency of the system, ω_0, and decays when the frequency is away from it.

Figure 3.10 contains the forced solution as well as the full solution. The full solution is the sum of the homogeneous (Equation 3.4.11) and the inhomogeneous (equation 3.4.22), i.e., $x = x_{\text{homogeneous}} + x_{\text{forced}}$ has been plotted for each set of parameters below the plot of each forced solution. Both sets of plots are carried out by the MAT-LAB script drive_soln.m. We notice that after the homogeneous or transient solution decays, the forced motion takes over rapidly by about $t = 20$ s. The amplitude of the large frequency $(\omega = 3\,\text{rad/s})$ forced solution is smaller than the homogeneous solution, and because they are out of phase, the transient behavior experiences much interference from the forced solution. The least interference occurs at the smaller frequencies as expected. For the current parameters, the interference will not matter after about $t = 20$ s go by, however.

The phase difference between the motion and the force is quite important. It tells us how power is being absorbed by the system (supplied by the force). For the driven harmonic oscillator, the power is written as

$$\text{p} = F_{ext}v = F_0\cos(\omega_D t + \theta)\dot{x}, \tag{3.5.12a}$$

where

$$\dot{x} = \frac{d}{dt}[A\cos(\omega_D t + \delta)] = -v_m\sin(\omega_D t + \delta), \tag{3.5.12b}$$

where $v_m \equiv A\omega_D$. Using (3.5.11) to express δ in terms of θ and ϕ, in addition to expanding the sin function, find for the instantaneous power

$$\text{p}(t) = -AF_0\omega_D[\cos(\omega_D t + \theta)\sin(\omega_D t + \theta)\cos(-\phi) + \cos^2(\omega_D t + \theta)\sin(-\phi)]. \tag{3.5.13}$$

The average power is obtained by averaging over one cycle

$$\text{p}_{ave} = \frac{1}{T}\int_0^T \text{p}(t)dt. \tag{3.5.14}$$

Since $\dfrac{1}{T}\displaystyle\int_0^T \cos^2(\omega_D t + \theta)dt = \dfrac{1}{2}$, and $\dfrac{1}{T}\displaystyle\int_0^T \cos(\omega_D t + \theta)\sin(\omega_D t + \theta)dt = 0$, then from (3.5.13, 14) obtain

$$\text{p}_{ave} = \frac{AF_0\omega_D}{2}\sin\varphi, \tag{3.5.15}$$

with A and ϕ given by (3.5.7 and 3.5.11), respectively. The MATLAB script drive_power.m plots the average power (3.5.15) versus the driving frequency. The plots for various values of the drag coefficient are shown in Figure 3.13.

SCRIPT

```
%drive_power.m
%plots the power supplied by the driving force versus frequency
clear;
m=0.5;                              %mass
k=0.5;                              %spring constant
F0=0.5;                             %driving force amplitude
wo=sqrt(k/m);                       %SHO natural frequency
wmin=0.01;                          %minimum frequency
wmax=3;                             %maximum frequency
NPTS=200;                           %number of points
dw=(wmax-wmin)/(NPTS-1);            %step for w
hold on                             %get ready to superimpose plots
cmin=0.2;                           %minimum value of c
cmax=1;                             %maximum c
cstep=(cmax-cmin)/3;                %c step size
for c=cmin:cstep:cmax,              %loop over the drag coefficient
  gam=c/2/m;                        %find gamma
  for i=1:NPTS
    w(i)=wmin+(i-1)*dw;
    desc=(2*gam*w(i))^2+(wo^2-w(i)^2)^2;
    A=F0/m/sqrt(desc);              %The driven ho amplitude
    den=wo^2-w(i)^2;
    if den==0, den=1.e-3; end
    if(w(i) <= wo)
    ph=atan(2*gam*w(i)/den);    %the phase difference
    else
    ph=pi+atan(2*gam*w(i)/den);%shift by pi needed if w > wo
    end
    power(i)=0.5*F0*A*w(i)*sin(ph);
  end
  plot(w,power)
  [p,j]=max(power);                 %point where the power is maximum
  str=cat(2,'\gamma=',num2str(gam,3));
  text(w(j),power(j)+0.02,str,'FontSize',10,'Color','red');
end
title('Average Power Supply vs Drive Frequency','FontSize',14)
ylabel('Power','FontSize',14);
xlabel('\omega_D','FontSize',14);
```

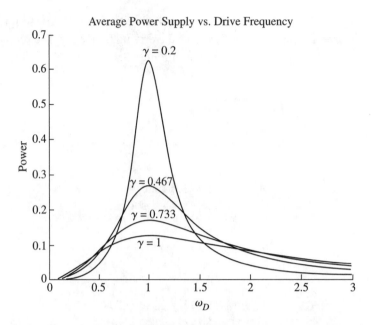

FIGURE 3.13 The driven harmonic oscillator power versus driving frequency.

The power peaks at $\omega_D = \omega_0$, as expected, because maximum power transmission occurs at $\phi = \pi/2$ from (3.5.15). The phase's behavior versus ω_D has been discussed in Table 3.1. The various curves in Figure 3.13 show that the average power tends to flatten as γ increases, making the power transmission less effective. It becomes small for large ω_D because (3.5.15) depends on the phase difference through $\sin\phi$. For very large frequency, the phase difference becomes close to π, as shown in Figure 3.11, making the power transmission less effective.

■ Chapter 3 Problems

3.1 Modify the MATLAB code for xoft.m so that the $y(t)$ function is plotted instead of $x(t)$, and explain your observations.

3.2 Write the particular time-dependent solution of the undamped harmonic oscillator, given that at time $t = 0$, the spring is initially at its equilibrium with an initial velocity v_0; identify each term in the expression. Also, write the form of the time-dependent velocity and show that it obeys the initial conditions.

3.3 The Lennard–Jones two-atom molecule potential model of a noble gas at absolute zero Kelvin is $V(x) = \left(\dfrac{A}{x^{12}} - \dfrac{B}{x^6}\right)$. Given that $A = 4\varepsilon\sigma^{12}$, and $B = 4\varepsilon\sigma^6$, where ε is a unit of energy, and σ is a unit of distance

  a. obtain the molecule's bond length if one of the atoms is considered to be at the origin.

  b. Find the molecule's frequency of vibration.

  c. Give a plot of the potential and the related molecular force. Identify the potential's most significant features in the plot.

3.4 A carpenter desires to install a screen door on a house in such a way that as the owner enters and releases the door, he will have enough time to clear out of the way without getting bumped by the door. If the screen door has a mass of 5 kg, and it takes 5 s for the door to close, what is the drag coefficient needed to accomplish the task?

3.5 A 1,000-kg automobile makes five full up-and-down bounces as the driver breaks to make a full stop. Before the springs deteriorated, the car springs' drag coefficient had a value of 0.6 Ns/m and it produced only a single bounce. What is the present value of the drag coefficient?

3.6 Assuming values of $\omega \sim 0.5$ rad/s and $c = 0.1$ Ns/m, compare the analytic results for driven harmonic oscillator solution, as calculated by the MATLAB script driv_sol.m, with the numerical results of Chapter 1 as obtained by hol.m. (*Hint:* Recall that $\sin(\theta + \pi/2) = \cos(\theta)$.) When compared with each other, do the results look as expected? Why?

3.7 The average power (3.5.15) of the driven harmonic oscillator is a function of frequency. Obtain the expressions for its peak average power and the frequency at which it occurs.

■ Additional Problems

3.8 Show that in order for Equations (3.2.7) to be true and for $x(t)$ to be real, we must have $B = A^*$. (*Hint:* Write $A = a_1 + ia_2$ and $B = b_1 + ib_2$ where the a_i's and the b_i's are real, and obtain their relationship.)

3.9 Show that the form of the critically damped harmonic oscillator solution Equations (3.4.7 and 3.4.8) can be obtained using a method similar to that employed in Section 3.2 for the undamped harmonic oscillator. (*Hints:* A differential equation of the form $(D - r)(D - r)x = 0$ can be written as $(D - r)u = 0$, where $u = (D - r)x$ and $D = d/dt$; also, the first-order linear differential equation $dy/dv + P(v)y = Q(v)$ has the solution

$$y(v) = e^{-Pv} \int Q(v) e^{Pv} + Ce^{-Pv}, \text{ where } C \text{ is a constant. See Boas (2006).}$$

3.10 Based on the initial conditions that at $t = 0$, $x = x_0$, and $v = v_0$

   a. show the steps leading to Equations (3.4.9 and 3.4.10) of the overdamped and the critically damped harmonic oscillator.

   b. Assuming the same initial conditions, which solution experiences a faster decay rate? Prove this.

3.11 By direct substitution into (3.4.2), show that $x = Be^{-\gamma t} \sin(\omega t + \theta)$ is a solution of the harmonic oscillator with damping. Be sure to identify the relationships that γ and ω must satisfy in order for the solution to be true.

3.12 Consider a driven harmonic oscillator for which the maximum amplitude of the driving force is 0.5 N with a driving frequency of 5 rad/ s; that is, $F = F_0 \cos(\omega_D t + \theta)$ and $\theta = 0$. The system's spring constant is 10 N/m, the mass at its end is 0.2 kg, and the coefficient of friction is 0.05 kg/ s. After a long time, the spring will oscillate according to $x(t) = A \cos(\omega_D t + \delta)$. Obtain the numerical values of (a) A, (b) δ, and (c) the average power transmitted by the driving motor to the system. Show all units.

3.13 The two-atom molecule potential of Example 3.2 is $V(x) = (A/x^3 - B/x^2)$ where $A = u_0 a_0^3$ and $B = u_0 a_0^2$, u_0 is a unit of molecular energy, and a_0 is a unit of molecular distance. Show that the force associated with this potential vanishes exactly when $x = x_b$ where x_b is the bond length.

3.14 Show that

   a. $\dfrac{1}{T} \displaystyle\int_0^T \cos^2(\omega_D t + \theta) \, dt = \dfrac{1}{2}$, and

   b. $\dfrac{1}{T} \displaystyle\int_0^T \cos(\omega_D t + \theta) \sin(\omega_D t + \theta) \, dt = 0$.

4 Examples of Harmonic Motion

◼ 4.1 Introduction

Harmonic motion has many applications in the physical world. Applications extend from oscillations involving planetary motion, to light waves, to charged particle motion, to oscillations involving atoms in a solid crystal, to swinging bridges, and even to earthquakes and swaying buildings. This chapter looks at some of the applications that find themselves useful in real-world conditions.

◼ 4.2 The Simple Pendulum

The simple pendulum is a swinging mass attached to the end of a string that is held at the opposite end by a stable support, as shown in Figure 4.1(a). In the analysis, the string is considered massless and it is assumed that it does not stretch as the mass swings back and forth. The pendulum string is under a tension, T, and the mass is acted on by the force due to gravity, as shown in Figure 4.1(b).

The instantaneous equations of motion for the x and y directions, respectively, from Newton's second law are

$$\sum F_y = T - mg\cos\theta = m\frac{v(\theta)^2}{\ell} \quad \text{and} \quad \sum F_x = -mg\sin\theta = m\frac{d^2x}{dt^2}. \qquad (4.2.1)$$

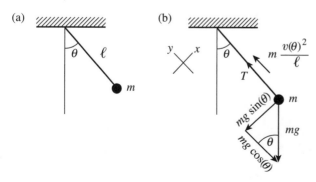

❙ FIGURE 4.1 (a) The simple pendulum and (b) the forces on it.

The idea is that if a force is applied in the positive x direction, so as to displace the bob, the pendulum's response in the opposite direction is due to a gravity component. However, the path described by the motion of the pendulum bob is a circular path, so that it is convenient to make the replacement $x = \ell\theta$, where ℓ is the length of the pendulum's string. Furthermore, the first equation of (4.2.1) only gives information about the tension on the string, which due to the component of gravity in the y direction and the centripetal term is a function of θ. The second gives information about the position as a function of time, which is what we concentrate on. Thus, our main equation of motion for the pendulum is

$$\frac{d^2\theta}{dt^2} + \frac{g}{\ell}\sin\theta = 0. \tag{4.2.2}$$

This equation happens to be a complicated differential equation for $\theta(t)$ without a close-form analytic solution. It can be easily solved numerically, however, as it will be shown later in this chapter. It is possible to consider oscillation for small angle deviations, generally less than about 15 degrees or 0.26 radians. The sine function can be expanded in a Taylor series, about $\theta = 0$:

$$\sin(\theta) = \theta - \frac{\theta^3}{3!} + \frac{\theta^5}{5!} - \cdots, \tag{4.2.3}$$

and for small angles, we take the leading first-order term, so that $\sin\theta \sim \theta$, to obtain the *simple pendulum* equation,

$$\frac{d^2\theta}{dt^2} + \frac{g}{\ell}\theta = 0. \tag{4.2.4}$$

A quick look at Chapter 3 tells us that this is a simple harmonic oscillator equation with solution,

$$\theta(t) = \theta_0\cos\omega_0 t, \tag{4.2.5}$$

where $\omega_0 \equiv \sqrt{g/\ell}$, and where at $t = 0$, $\theta(0) = \theta_0$ (i.e., some initial angle), and where we have assumed that $\dot{\theta}(0) = 0$.

EXAMPLE 4.1

A certain simple pendulum has a period of 1 s on Earth's surface. What is its period on the surface of Mars?

Solution

Since the period of a pendulum on Earth's surface is $\tau_E = 2\pi/\omega_0 = 2\pi\sqrt{\ell/g_E}$,

solving for its length gives $\ell = \dfrac{g_E\tau_E^2}{4\pi^2}$. Using this value for the length, one obtains

for the period on Mars $\tau_M = 2\pi\sqrt{\dfrac{g_E\tau_E^2}{4\pi^2}/g_M} = \sqrt{\dfrac{g_E}{g_M}}\tau_E$. We next recall that the

surface gravitational acceleration of a planet is given by $g_p = GM_p/R_p^2$, where G is the universal gravitational constant, M_p is the planet's mass, and R_p is the planet's radius. Using the values inside the text's front cover, on Earth $g_E = 9.83$ m/s^2,

and on Mars $g_M = \dfrac{6.67 \times 10^{-11}\dfrac{\text{Nm}^2}{\text{kg}^2}6.42 \times 10^{23}\,\text{kg}}{(3.37 \times 10^6\,\text{m})^2} = 3.77$ m/s^2. Therefore, the

period of the pendulum on Mars is $\tau_M = \sqrt{\dfrac{9.83}{3.77}}(1\,\text{s}) = 1.61$ s.

EXAMPLE 4.2

A simple pendulum makes a few swings before it stops. Write the equation of motion for this pendulum and obtain its angular behavior as a function of time.

Solution

The equation of motion for the damped pendulum is similar to (4.2.4) with a term involving the frictional effects. If we assume the resistive term is proportional to the angular speed, then we have

$$\frac{d^2\theta}{dt^2} + \frac{c}{m}\frac{d\theta}{dt} + \frac{g}{1}\theta = 0, \tag{4.2.6a}$$

where c is the coefficient due to friction. This equation can be written as

$$\frac{d^2\theta}{dt^2} + 2\gamma\frac{d\theta}{dt} + \omega_0^2\theta = 0, \tag{4.2.6b}$$

where $\gamma = c/2m$, $\omega_0 = \sqrt{g/1}$. This equation is similar to the damped harmonic oscillator problem of Chapter 3. Thus, the underdamped solution we seek is from Section 3.4,

$$\theta(t) = \theta_0 e^{-\gamma t}\cos\omega t, \tag{4.2.7}$$

so that at $t = 0$, $\theta(0) = \theta_0$ and where

$$\omega = \sqrt{\omega_0^2 - \gamma^2}. \tag{4.2.8}$$

This solution assumes that the pendulum's initial position is at θ_0 with an initial speed of $\dot{\theta}(t = 0) = (-\gamma\theta_0 e^{-\gamma t}\cos\omega t - \omega\theta_0 e^{-\gamma t}\sin\omega t)|_{t=0} = -\gamma\theta_0$.

■ 4.3 The Physical Pendulum

The physical pendulum is very similar to the simple pendulum; the main difference is that rather than dealing with the mass of the bob, one deals with the moment of inertia of the swinging object whose mass is located at its center of mass. Consider a rigid body of general shape as shown in Figure 4.2.

The weight of the object whose center of mass is located at ℓ produces a clockwise torque at that instant of time. As the object swings to the opposite side, the torque reverses direction and so on. Thus, the object performs harmonic motion similar to the pendulum. Its equation of motion is now the torque equation

$$I\alpha + mg\ell\sin\theta = 0 \tag{4.3.1}$$

where I is the moment of inertia and α is the angular acceleration, i.e., $\ddot{\theta}$. However, as with the pendulum, we can find a simple solution if we make the approximation $\sin\theta \sim \theta$, which is suitable for small angles, so that (4.3.1) becomes

$$\frac{d^2\theta}{dt^2} + \omega^2\theta = 0, \tag{4.3.2}$$

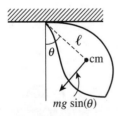

I **FIGURE 4.2** A general-shape physical pendulum.

where now, $\omega \equiv \sqrt{mg\ell/I}$. The solution is as in (4.2.5).

■ 4.4   The Bobbing Buoy

A buoy floating in a body of water performs simple harmonic motion. Consider, for example, a cylinder of radius r, height h, and mass m that is able to float in a liquid of density ρ. As shown in Figure 4.3, y measures the part of the cylinder that sticks out above water.

The cylinder floats by Archimedes' principle, which states that the buoyant force equals the weight of fluid displaced. To be at equilibrium, we must have $mg = \rho g V_s$, where $V_s = (h - y_0)A$ is the cylinder's submerged volume, and where the product ρV_s is the mass of the displaced fluid, and y_0 is the equilibrium position. If a cylinder that is initially at its equilibrium position is suddenly shifted by amount $\Delta y = y - y_0$, the buoyant force changes due to a change in volume of water displaced. Looking at the volume of cylinder exposed above the fluid, it experiences a change in volume as well. This is $\Delta V = V_{final} - V_{initial}$, where $V_{final} = A(h - y)$, and we take $V_{initial} = V_s$, so that $\Delta V = A[(h - y) - (h - y_0)] = -A\Delta y$, where the area $A = \pi r^2$. The buoyant force changes so as to increase if ΔV decreases and decrease if ΔV increases past what it takes to keep the cylinder at its equilibrium position. This is the necessary condition for simple harmonic motion; thus we have

$$m\frac{d^2y}{dt^2} = -\rho g A \Delta y. \tag{4.4.1a}$$

If in this equation we let $y' = y - y_0$, where y_0 is a constant, it can be rewritten as

$$\frac{d^2y'}{dt^2} = -\rho g A y'/m. \tag{4.4.1b}$$

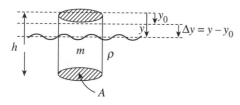

I FIGURE 4.3 A cylinder model of a buoy.

One therefore sees that the system behaves as a spring with an effective spring constant of $k_{eff} = \rho g A$ and an oscillatory frequency of $\omega = \sqrt{k_{eff}/m} = \sqrt{\rho g A/m}$.

■ 4.5 The Elongating Wire

When a wire of length L, cross-sectional area $A = \pi r^2$, and Young's modulus Y is applied a force F, the wire experiences a stress that causes it to strain. According to the theory of materials, in the elastic region *stress $\propto$ strain*; that is, $F/A = Yx/L$, where x measures the amount of strain. The wire reacts to the force applied by producing a restoring force that is proportional to its displacement. We can thus write the simple harmonic motion relation with a wire due to a mass hanging from its end, as shown in Figure 4.4.

| FIGURE 4.4 A wire straining under a stress.

The mass m will stretch the wire to an equilibrium position $x_e = \dfrac{mgL}{AY}$. Oscillations occur about x_e according to

$$m\frac{d^2x}{dt^2} = -\frac{YA}{L}(x - x_e),\qquad(4.5.1)$$

from which we see the effective spring constant $k_{eff} = YA/L$ and a frequency of vibration $\omega = \sqrt{k_{eff}/m} = \sqrt{YA/mL}$.

■ 4.6 The Torsional Pendulum

A torsional pendulum consists of a wire twisted in such as way that a permanent deformation does not occur. For example, if a disk of mass m, fixed at the free end of the wire, is made to rotate through a small angle θ, the wire produces a torque that is proportional to this angle; that is, a restoring torque $\Gamma = -\kappa\theta$ is responsible for the resulting angular oscillations, where κ is the torsional stiffness constant. Thus, we have the equation of motion

$$I\frac{d^2\theta}{dt^2} = -\kappa\theta, \qquad\qquad (4.6.1)$$

where $I = mR^2/2$ is the moment of inertia of the disk about its center of mass, which coincides with the center of rotation as shown in Figure 4.5.

Therefore, the frequency of oscillation in the plane of the disk is $\omega = \sqrt{\kappa/I}$; the frequency depends on the moment of inertia of the suspended body.

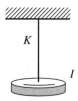

FIGURE 4.5   A torsional pendulum.

■ 4.7   The Parallel Beam Under a Shearing Stress

When a beam is acted on by a force perpendicular to its long axis as shown in Figure 4.6, this is known as a shearing stress. In the elastic region, the strain $\Delta y/\ell$ is proportional to the shearing stress F/A, where $F = mg$ and A is the cross-sectional beam area. We can, therefore, write the relation $F/A = G\Delta y/\ell$, where G is called the shear modulus.

This means that, for small angles, the load produces a torque $\tau_1 \approx F\ell = GA\Delta y$. Looking at small vibrations, because the force applied produces a torque about the pivot point at the support position, then the beam's response is a torque

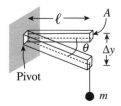

FIGURE 4.6   A beam under a shearing stress.

$\tau_2 = -\kappa \Delta y = I d^2\theta/dt^2$. For small oscillations we take $\Delta y \approx \ell\theta$ and equate the magnitude of both torques to get $\kappa = GA$ and

$$I\frac{d^2\theta}{dt^2} = -\kappa\ell\theta. \tag{4.7.1}$$

The oscillation frequency of this system is, therefore, $\omega = \sqrt{\kappa\ell/I} = \sqrt{GA\ell/I}$, where I is the moment of inertia of the beam and mass together about the pivot point.

■ 4.8 The Floating Sphere

When an air blower is pointed in the direction of a round object so as to make it float, as seen in Figure 4.7, it is possible to set up oscillations. These oscillations can be explained in terms of Bernoulli's equation for laminar, streamline motion of an incompressible fluid at constant pressure.

In Figure 4.7, suppose that the air leaves an air blower with speed v_0 and meets a sphere of mass m a distance y above the position of the air blower. If the sphere is held at equilibrium, then we can write the net force at the sphere's position as

$$F = -mg + cv, \tag{4.8.1}$$

where the lifting force due to air resistance is taken to be linear with speed, as in Chapter 2. For the present problem, Bernoulli's equation is

$$\frac{1}{2}\rho v_0^2 = \frac{1}{2}\rho v^2 + \rho g y; \tag{4.8.2}$$

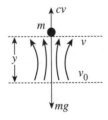

| FIGURE 4.7 A floating sphere.

that is, the air leaves the hose at speed v_0 and reaches speed v at the position of the mass after it has climbed a distance y. From here, we can solve for the velocity v at the sphere's position as

$$v = \sqrt{v_0^2 - 2gy} = v_0\sqrt{1 - \frac{2gy}{v_0^2}}. \tag{4.8.3}$$

For small oscillations, with $v_0^2 \gg 2gy$, we can expand the square root term to obtain, to first order in y,

$$v \approx v_0\left(1 - \frac{gy}{v_0^2}\right). \tag{4.8.4}$$

Substituting this expression into (4.8.1) and rearranging, we obtain the equation of motion

$$\frac{d^2y}{dt^2} + \frac{cg}{mv_0}(y - y_e) = 0, \tag{4.8.5}$$

where the equilibrium position $y_e = v_0(v_0/g - m/c)$ is the point where the net force (4.8.1) equals zero. The harmonic motion oscillatory frequency of the floating sphere is, therefore, $\omega = \sqrt{cg/mv_0}$.

■ 4.9 The Wire Under Tension

In Figure 4.8, a massless wire is shown under tension T with a mass m fastened to it at distance a from the lower support. If the mass is plucked through a small distance x, it is possible to obtain harmonic motion.

To see this notice that, for small displacements, the tension does not change and below we let $T_a = T_b \equiv T$. The x components of the tensions add to give

$$-T_b\sin(\theta) - T_a\sin(\phi) = m\frac{d^2x}{dt^2}, \tag{4.9.1}$$

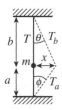

I FIGURE 4.8 A wire supported at both ends under tension.

which, according to the geometry of Figure 4.8, we can write as

$$m\frac{d^2x}{dt^2} = -T\left(\frac{1}{\sqrt{x^2 + a^2}} + \frac{1}{\sqrt{x^2 + b^2}}\right)x. \tag{4.9.2}$$

If we expand $1/\sqrt{1 + y} \sim 1 - y/2$, then an approximation can be made in the form

$$m\frac{d^2x}{dt^2} \approx -T\left[\frac{x}{a} + \frac{x}{b} - \frac{1}{2}\left(\frac{1}{a^3} + \frac{1}{b^3}\right)x^3\right], \tag{4.9.3}$$

which for small x becomes

$$m\frac{d^2x}{dt^2} + T\left(\frac{1}{a} + \frac{1}{b}\right)x = 0. \tag{4.9.4}$$

Thus the frequency of oscillation is

$$\omega = \sqrt{\frac{T}{m}\left(\frac{1}{a} + \frac{1}{b}\right)}. \tag{4.9.5}$$

Now, let's consider the special case of a taut wire of mass $m = \mu L$, where μ is the linear density and L is the total length. Imagine next that this mass is concentrated at the wire's center of mass so that $a = b = L/2$. This is like a mass m located at the center of a massless wire of length L whose ends are held fixed. Then, in this special case, (4.9.5) gives

$$\omega \rightarrow \sqrt{\frac{T}{m}\left(\frac{2}{L} + \frac{2}{L}\right)} = \sqrt{\frac{4T}{\mu L^2}}. \tag{4.9.6}$$

This leads to the frequency

$$f = \omega/2\pi = \frac{1}{\pi L}\sqrt{\frac{T}{\mu}} = \frac{v}{\lambda}, \tag{4.9.7}$$

which means that the associated waves move with a speed of $\sqrt{T/\mu}$ and a wavelength of πL.

■ 4.10   The RLC Circuit

In electricity and magnetism, the common application of the RLC circuit shown in Figure 4.9 is an important application of simple harmonic motion. Here, the time-dependent behavior of the charge as a function of time is sought.

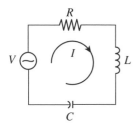

| FIGURE 4.9 An RLC circuit as an example of a driven harmonic oscillator.

The equation of motion for the charge $q(t)$ is easily obtained by applying Kirchhoff's loop rule for the voltage as follows:

$$V_0 \sin(\omega t + \phi) - IR - L\frac{dI}{dt} - \frac{q}{C} = 0 \tag{4.10.1}$$

due to the voltages associated with the source, the resistor, the inductor, and the capacitor. The current travels in a clockwise manner through the circuit, as indicated by the direction of the arrow in Figure 4.9. Since $I = dq/dt$, the differential equation can be rewritten as

$$\frac{d^2q}{dt^2} + \frac{R}{L}\frac{dq}{dt} + \frac{q}{LC} = \frac{V_0}{L}\sin(\omega t + \phi). \tag{4.10.2}$$

The result of the driven harmonic oscillator of Chapter 3 can be used to obtain the solution to our current problem. First we notice that the sine term on the right-hand side of (4.10.2) is equivalent to $\sin(\omega t + \phi) = imaginary[e^{i(\omega t + \phi)}]$, so that we can solve (4.10.2) with this complex exponential replacing the sine function, and assuming a solution for q in the form $q = Ae^{i(\omega t + \delta)}$. However, because we have solved a similar differential equation in Chapter 3, we can adapt the equations of Section 3.5 to our current problem if we make the replacements

$$m \rightarrow L, \quad c \rightarrow R, \quad k \rightarrow \frac{1}{C}, \quad F_0 \rightarrow V_0, \quad \text{and} \quad x \rightarrow q. \tag{4.10.3}$$

As mentioned, since the right-hand side of (4.10.2) involves the imaginary part of $e^{i(\omega t + \phi)}$, our steady state solution must be of the form

$$q = \text{Im}(Ae^{i(\omega t + \delta)}) = A\sin(\omega t + \delta). \tag{4.10.4}$$

Also we have the following definitions:

$$\delta = \theta - \tan^{-1}\left(\frac{2\gamma\omega}{\omega_0^2 - \omega^2}\right), \quad A = \frac{V_0/L}{\sqrt{(2\gamma\omega)^2 + (\omega_0^2 - \omega^2)^2}}, \quad \omega_0 = \sqrt{\frac{1}{LC}}, \quad \gamma = \frac{R}{2L}.$$

(4.10.5)

We can see, therefore, that the whole problem can be understood in terms of the driven harmonic oscillator of Chapter 3.

■ 4.11 The Hanging Spring-Mass System and the Spring-Mass Correction

If a mass hangs at the end of a spring, the equilibrium position about which the mass oscillates is not the same as the standard spring equilibrium position in the absence of gravity. In the present situation, a mass extends the spring to a new equilibrium position given by $y_e = \ell + mg/k$, as shown in Figure 4.10(a), where ℓ is the length of the spring.

The equation of motion is slightly different from that discussed in Chapter 3. The differential equation is

$$\frac{d^2y}{dt^2} + \omega^2(y - y_e) = 0,$$

(4.11.1)

but $\omega = \sqrt{k/m}$ as before, and the oscillation takes place about the new equilibrium position. If , for example, at $t = 0$ the mass is at equilibrium and the speed is v_0, then a solution of this equation can be written in the form

$$y = y_e + A\sin\omega t,$$

(4.11.2)

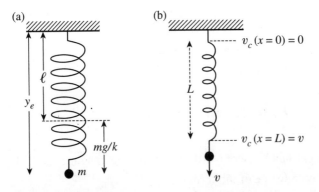

FIGURE 4.10 (a) A mass under gravity at the end of a spring. (b) The spring loops carry kinetic energy when the system oscillates.

as measured from the spring support shown in Figure 4.10(a). Here, A is the maximum amplitude of the oscillation away from equilibrium. We note that $v = dy/dt = A\omega\cos\omega t$, so that the maximum velocity (that is, the velocity it carries when it passes by the equilibrium position) is

$$v_{\max} = A\omega = A\sqrt{k/m} = v_0, \qquad (4.11.3)$$

because we have also said the speed at the equilibrium position is v_0. Considering only the effect due to gravity, we can look at the energy of the system, when the mass m is just at the end of the unstretched spring; that is, rather than considering oscillations about the equilibrium just discussed, let's ask, what is the lowest position it can reach when released from position ℓ from rest ($v_i = 0, y_i = \ell$)? We consider the support as the zero reference position. By energy conservation we have

$$E_{top} = E_{bottom} \quad \Rightarrow \quad -mg\ell = \frac{k}{2}(y_{bot} - \ell)^2 - mgy_{bot} \qquad (4.11.4)$$

because the speed is zero at the bottom. From here we see that

$$\left[\frac{k}{2}(y_{bot} - \ell) - mg\right](y_{bot} - \ell) = 0.$$

There are two positions where the speed is zero. One is the starting point at ℓ, and the other is the one we seek:

$$y_{bot} = \ell + 2\frac{mg}{k} = y_e + \frac{mg}{k}. \qquad (4.11.5)$$

Comparing this result to (4.11.2), we see that because this is equivalent to the value of y at $t = \tau/4 = \pi/2\omega$, then we also have that

$$A = \frac{mg}{k}. \qquad (4.11.6)$$

Putting this into (4.11.3) we see that

$$v_{\max} = \frac{mg}{k}\sqrt{k/m} = g\sqrt{m/k} = g/\omega. \qquad (4.11.7)$$

The results (4.11.6 and 4.11.7) depend only on gravity, the mass, and the spring constant because the initial conditions are such that at $t = 0$, $v_i = 0$ at $y_i = \ell$, when the spring is in the unstretched position.

To see that this result makes sense, let's suppose that the mass once again is dropped from position ℓ beginning at rest. Let's find the maximum velocity as well as the position at which it occurs. The energy conservation now gives $-mg\ell = k(y - \ell)^2/2 - mgy + mv^2/2$, or

$$mg(y - \ell) = \frac{k}{2}(y - \ell)^2 + \frac{1}{2}mv^2. \tag{4.11.8}$$

Taking the derivative of both sides with respect to y, we get

$$mg = k(y - \ell) + mv\frac{dv}{dy}. \tag{4.11.9}$$

Setting the derivative to zero and solving for y, we get the point where the velocity is a maximum, i.e., $y_e = \ell + \dfrac{mg}{k}$, which is the equilibrium position as shown in Figure 4.10(a), as it should be. Substituting this result back into (4.11.8) for y and solving for the speed at this position we get $v = v_{max} = g/\omega$, in agreement with (4.11.7).

As a further illustration, suppose that in contrast to the preceding case, the mass is instead at the equilibrium position when it is given an initial velocity of v_0 upward. How high does the mass climb? By energy conservation

$$\frac{1}{2}k(y_e - \ell)^2 - mgy_e + \frac{1}{2}mv_0^2 = -mgy + \frac{1}{2}k(y - \ell)^2, \tag{4.11.10}$$

because at the highest position the speed is zero, and we use the support as our point of reference position for the gravitational potential energy. Putting $y_e = \ell + \dfrac{mg}{k}$ into (4.11.10) and reorganizing we get the quadratic equation

$$(y - \ell)^2 - 2\frac{mg}{k}(y - \ell) - \frac{m}{k}v_0^2 + \left(\frac{mg}{k}\right)^2 = 0. \tag{4.11.11}$$

Solving for y we find

$$y = \ell + \frac{mg}{k} \pm v_0\sqrt{m/k}, \tag{4.11.12}$$

where the minus sign corresponds to the lowest position and the positive sign to the highest position where the speed is zero. This indicates that the maximum amplitude of oscillation about the equilibrium position is $v_0\sqrt{m/k}$. This is consistent with (4.11.2 and 4.11.3), since $v(0) = A\omega\cos(\omega t = 0) = A\omega = v_0$ in the present case, so that $A = v_0/\omega = v_0\sqrt{m/k}$ as well; that is, $v_{max} = v_0$.

Finally, in most of our discussion regarding springs, we have used massless springs. It is possible to include a correction due to the spring's mass. This is usually done by adding one third of the spring's mass to the hanging mass. Thus, in the present case, $\omega \to \sqrt{k/(m + m_s/3)}$, where m_s is the spring mass, which is assumed to be uniformly distributed along the spring. The idea that one should add one third the mass of the spring to the total mass can be justified from the point of view of energy. Consider that while the hanging mass is vibrating, the spring is stretched, and at a particular instant, the farthest spring turn moves at speed v, which is the same speed at which the hanging mass moves, as shown in Figure 4.10(b). At the same time, the closest turn to the pivot point can be thought to be at rest. Thus, each spring coil's speed can be approximated by

$$v_c(x) = (x/L)v, \qquad\qquad (4.11.13)$$

where L is the length of the spring. If the spring mass is assumed to be uniformly distributed, then we can let $\lambda = m_s/L$ be the spring's linear density. The total kinetic energy contribution due to the spring at this instant is

$$KE_s = \frac{1}{2}\int_0^L \lambda[v_c(x)]^2 dx = \frac{\lambda}{2}\int_0^L (vx/L)^2 dx = \frac{\lambda v^2 L}{2\ 3} = \frac{1}{2}\left(\frac{m_s}{3}\right)v^2, \qquad (4.11.14)$$

because each coil contribution to the energy has to be added. This suggests that the spring contribution can be approximately taken into account if one simply adds one third the spring's mass to the hanging mass. In (4.11.14), notice that if one assumes, as we have done before, that the spring is massless, then the spring contribution to the total energy is zero.

■ 4.12   Springs in Series and in Parallel

When springs are organized in a series configuration as shown in Figure 4.11(a) or in a parallel arrangement as in Figure 4.11(b), the springs act as a single spring with an effective spring constant.

(a)   (b)

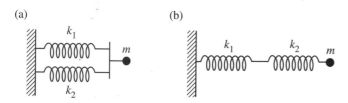

I FIGURE 4.11   (a) Springs in parallel. (b) Springs in series.

Parallel Configuration

When the mass m in Figure 4.11(a) is displaced by amount x, each spring is displaced by corresponding amounts $x_1 = x_2 = x$. The net restoring force applied by the springs is the sum of each spring contribution, i.e.,

$$F = -k_1 x_1 - k_2 x_2 = -(k_1 + k_2)x = -k_{eff}x, \qquad (4.12.1a)$$

so that the result is that the two springs act as a single spring with an effective spring constant

$$k_{eff} = k_1 + k_2. \qquad (4.12.1b)$$

Series Configuration

For this case, let the springs be organized in a series arrangement as in Figure 4.11(b). If a force F is applied so as to displace the mass m by amount x, then the first spring is displaced by amount x_1, and the second spring by amount x_2. The total displacement is given by

$$x = x_1 + x_2 = F\left(\frac{1}{k_1} + \frac{1}{k_2}\right) = \frac{F}{k_{eff}}, \qquad (4.12.2a)$$

which indicates that the two springs behave as a single spring with an effective spring constant whose value is

$$\frac{1}{k_{eff}} = \left(\frac{1}{k_1} + \frac{1}{k_2}\right), \qquad (4.12.2b)$$

with the understanding that Hooke's law still applies, $F = -k_{eff}x$.

Variation of the Parallel Configuration

A system that behaves similarly to the parallel configuration is shown in Figure 4.12. If the mass m is displaced by amount x in either direction, the effect is to have a reaction force from each spring in the same direction. If x is positive, then the first spring is stretched while the second spring is compressed by the same amount, yet both tend

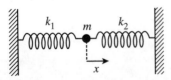

I FIGURE 4.12   A variation of a parallel spring-mass system.

to push the mass to the left. Similarly, if x is negative both springs will act so as to push the mass toward the right. Thus, in such case the net force on the mass m is

$$F = -(k_1 + k_2)x = -k_{eff}x, \tag{4.12.3}$$

with $k_{eff} = k_1 + k_2$, as in the parallel case.

■ 4.13   Interacting Spring-Mass Systems

In this section we consider systems of masses connected through springs. The possible configurations considered are shown in Figure 4.13. In the case of Figure 4.13(a), two springs with constants k_1 and k_2, while separately attached to supports, each has a mass m_1 and m_2 at its end. In the other case, Figure 4.13(b), the two end walls have been removed and there is no need for the end springs. In both cases, the masses themselves are interacting through a third spring of constant k_0.

Referring to Figure 4.13(a), notice that if m_1 moves by amount x_1, and m_2 moves by amount x_2 as shown, the first spring will pull m_1 back with a force proportional to x_1 and the middle spring will pull the opposite way with a force proportional to the difference of x_1 and x_2. A similar situation happens for m_2: the right spring pushes back with a force proportional to x_2 and the middle spring pulls back with a force proportional to the difference of x_2 and x_1. Thus, the equations of motion are

$$m_1 \frac{d^2 x_1}{dt^2} = -k_1 x_1 - k_0(x_1 - x_2) \quad \text{and} \quad m_2 \frac{d^2 x_2}{dt^2} = -k_2 x_2 - k_0(x_2 - x_1). \tag{4.13.1}$$

This is referred to as a coupled system of equations because the behavior of one mass affects the other and vice versa. Next, we will consider three cases of the preceding two equations. In the first case, the walls are removed and the springs k_1 and k_2 are no longer needed, which leaves a two-mass system interacting through a single spring as in Figure 4.13(b). In the second case, that of Figure 4.13(a), the coupled system of equations is solved assuming $m_1 = m_2 = m$, $k_1 = k_2 = k \neq k_0$. In the third case, we look at a particular limit of the equations for which $k_1 = k_2 = k_0 = k$ and

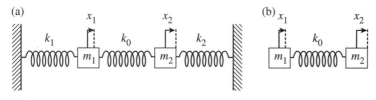

(a)

x_1 x_2

k_1 k_0 k_2

m_1 m_2

(b) x_1 x_2

k_0

m_1 m_2

I FIGURE 4.13   Interaction spring-mass system (a) with walls and (b) without walls.

$m_1 = m_2 = m$. The case of the masses and springs being all different is left for the problems at the end of the chapter.

Case 1: No Walls—Single Mode ($k_1 = k_2 = 0$)

In this case (see Figure 4.13b) Equations (4.13.1) can be written as

$$m_1 \frac{d^2 x_1}{dt^2} = k_0(x_2 - x_1) \quad \text{and} \quad m_2 \frac{d^2 x_2}{dt^2} = -k_0(x_2 - x_1), \tag{4.13.2}$$

so that multiplying the second by m_1 and the first by m_2 and then subtracting the result, we get

$$m_1 m_2 \frac{d^2 x_2}{dt^2} - m_1 m_2 \frac{d^2 x_1}{dt^2} = -m_1 k_0(x_2 - x_1) - m_2 k_0(x_2 - x_1), \tag{4.13.3}$$

or

$$\frac{d^2}{dt^2}(x_2 - x_1) = -\left(\frac{m_1 + m_2}{m_1 m_2}\right) k_0(x_2 - x_1). \tag{4.13.4}$$

If we define the relative coordinate

$$x_r = x_2 - x_1, \tag{4.13.5}$$

then (4.13.4) reduces to the equation

$$\frac{d^2 x_r}{dt^2} = -\frac{k_0}{\mu} x_r, \tag{4.13.6}$$

where μ is the reduced mass

$$\frac{1}{\mu} \equiv \frac{1}{m_1} + \frac{1}{m_2}. \tag{4.13.7}$$

From (4.13.6), we can readily see that the solution can be written in the form

$$x_r = A \sin \omega t + B \cos \omega t \tag{4.13.8}$$

where $\omega = \sqrt{k_0/\mu}$. Here, the initial relative position is $B = x_{r0} = x_{20} - x_{10}$, and the initial relative velocity is $\omega A = v_{r0} = v_{20} - v_{10}$, with v_{10} and v_{20} as the initial velocities of each mass, respectively.

According to Newton's first law of motion, in the absence of any external forces, the total momentum is constant as discussed in Chapter 1; the two-mass system's total momentum is

$$p_{total} = m_1 v_1 + m_2 v_2 = M v_{cm}, \tag{4.13.9}$$

where $M = m_1 + m_2$, and where v_{cm} (the center of mass velocity, which is also constant) is

$$v_{cm} = \frac{m_1 v_1 + m_2 v_2}{m_1 + m_2} = \frac{m_1 v_{10} + m_2 v_{20}}{m_1 + m_2}. \tag{4.13.10}$$

The center of mass velocity is therefore always known, and it in turn enables us to have knowledge of the center of mass coordinate

$$x_{cm}(t) - x_{cm0} = \frac{m_1 x_1 + m_2 x_2}{m_1 + m_2} = v_{cm} t, \tag{4.13.11}$$

where x_{cm0} is the center of mass at $t = 0$. Thus, for this two-mass system, from (4.13.5 and 4.13.11) we know the relative coordinate, $x_r(t)$, and from (4.13.11) we know the center of mass coordinate $x_{cm}(t)$ moves at constant velocity, indefinitely, because there are no walls to restrict its motion. Equations (4.13.5 and 4.13.11) represent two equations for the two unknowns x_1 and x_2. We can thus express the time-dependent positions x_1 and x_2 in terms of x_{cm} and x_r to obtain

$$x_1(t) = x_{cm}(t) - x_{cm0} - \frac{m_2}{m_1 + m_2} x_r(t) \text{ and } x_2(t) = x_{cm}(t) - x_{cm0} + \frac{m_1}{m_1 + m_2} x_r(t). \tag{4.13.12}$$

These are, therefore, the solutions to Equation (4.13.2). Figure 4.14 is a plot produced by the MATLAB script `inter_spr1.m`. The input parameters used are also shown in the figure. The amplitude for m_1 is greater than that of m_2 because m_1 is lighter. The motions take place about the center of mass whose behavior is a straight line, due to nonzero initial velocity values.

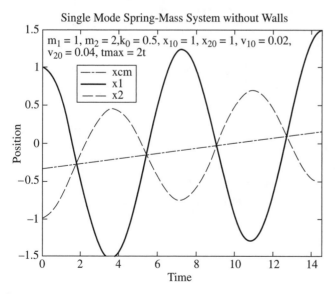

Single Mode Spring-Mass System without Walls

$m_1 = 1$, $m_2 = 2$, $k_0 = 0.5$, $x_{10} = 1$, $x_{20} = 1$, $v_{10} = 0.02$, $v_{20} = 0.04$, tmax = 2t

- - - xcm
—— x1
– – x2

I FIGURE 4.14   The coupled spring-mass system without walls with a single mode of vibrations.

SCRIPT

```
%inter_spr1.m
%plots the coordinate solutions for the single mode coupled spring-mass
%system without walls
clear;
m1=1;                                    %masses
m2=2;
k0=0.5;                                  %spring constant
x10=1;                                   %initial positions
x20=-1;
v10=0.02;                                %initial speeds
v20=0.04;
mu=m1*m2/(m1+m2);                        %reduced mass
xcm0=(m1*x10+m2*x20)/(m1+m2);            %initial center of mass
vcm=(m1*v10+m2*v20)/(m1+m2);            %center of mass speed
xr0=(x20-x10);                           %relative coordinate
vr0=(v20-v10);                           %relative speed
om=sqrt(k0/mu);                           %frequency
tau=2*pi/om;                             %period
A=vr0/om;                                %amplitudes
B=xr0;
tmax=2;
str=cat(2,'m_1=',num2str(m1),',m_2=',num2str(m2),',k_0=',num2str(k0),...
',x_{10}=',num2str(x10),',x_{20}=',num2str(x20),',v_{10}=',num2str(v10),...
',v_{20}=',num2str(v20),',tmax=',num2str(tmax),'\tau');
tmax=tmax*tau;
t=[0:tau/50:tmax];                       %plotting time range
xr=A*sin(om*t)+B*cos(om*t);              %solution
xcm=xcm0+vcm*t;                          %cm position vs time
x1=xcm-m2*xr/(m1+m2);                    %mass positions versus time
x2=xcm+m1*xr/(m1+m2);
plot(t,xcm,'k-.',t,x1,'b-',t,x2,'r-');
axis([0 tmax -1.5  1.5]);
text(0.25,1.35,str,'FontSize',11,'Color','black');
title('Single Mode Spring-Mass System Without Walls','FontSize',14)
h=legend('xcm','x1','x2',4); set(h,'FontSize',14)
ylabel('Position','FontSize',14);
xlabel('Time','FontSize',14);
```

Case 2: Full System—Bimodal ($m_1 = m_2 = m, k_1 = k_2 = k \neq k_0$)

We now turn our attention to the solution of the full set of Equations (4.13.1) for the system of Figure 4.13(a). In that process, you will get an opportunity to apply the eigenvalue–eigenvector technique to obtain the solutions for the positions of the masses as a function of time. We can write the equations in the matrix form

$$m\ddot{x} = -kx - k_0 Mx, \tag{4.13.13a}$$

where we have defined the matrices

$$m \equiv \begin{pmatrix} m & 0 \\ 0 & m \end{pmatrix}, \ x \equiv \begin{pmatrix} x_1 \\ x_2 \end{pmatrix}, \ k \equiv \begin{pmatrix} k & 0 \\ 0 & k \end{pmatrix}, \ M \equiv \begin{pmatrix} 1 & -1 \\ -1 & 1 \end{pmatrix}. \tag{4.13.13b}$$

Equations (4.13.13) can easily be solved if one can perform a rotational operation to decouple the system of equations. It is convenient to define the vector r and the matrix p such that

$$\begin{pmatrix} x_1 \\ x_2 \end{pmatrix} = \begin{pmatrix} p_{11} & p_{12} \\ p_{21} & p_{22} \end{pmatrix} \begin{pmatrix} r_1 \\ r_2 \end{pmatrix} \Rightarrow x = pr, \tag{4.13.14}$$

where the unknown elements $p_{11}, p_{12}, p_{21},$ and p_{22} of the p matrix are yet to be found and $r \equiv \begin{pmatrix} r_1 \\ r_2 \end{pmatrix}$. Putting this into (4.13.13) we get

$$mp\ddot{r} = -kpr - k_0 Mpr. \tag{4.13.15}$$

If we multiply by p^{-1} from the left, we obtain

$$mp^{-1}p\ddot{r} = -kp^{-1}pr - k_0 p^{-1}Mpr, \tag{4.13.16}$$

where we have used the fact that a diagonal matrix commutes with a nondiagonal matrix; for example, $p^{-1}m = mp^{-1}$. The idea is to find the matrix elements of p in such a way that the three-matrix product

$$p^{-1}Mp = \lambda, \tag{4.13.17}$$

where λ is a diagonal matrix whose elements are called eigenvalues and will be obtained next. In so doing, the equation for the vector r becomes

$$m\ddot{r} = -kr - k_0 \lambda r, \tag{4.13.18}$$

which represents a diagonal system of equations whose solutions can be carried out. Before we can obtain the solution to (4.13.18) we need the matrix λ, and before obtaining λ we need to obtain the matrix p. Equation (4.13.17) can be written as

$$Mp = p\lambda \Rightarrow (M - \lambda_n)p_n = 0, \tag{4.13.19}$$

where p_n is the nth column of the p matrix associated with the eigenvalue λ_n, as will be shown next. A nontrivial solution of this set of equations exist for nonzero elements of p, and the eigenvalues satisfy

$$M - \lambda = 0 = \begin{vmatrix} 1 - \lambda & -1 \\ -1 & 1 - \lambda \end{vmatrix} = 0, \tag{4.13.20}$$

or

$$(1 - \lambda)^2 - 1 = 0 \Rightarrow \lambda(\lambda - 2) = 0 \Rightarrow \lambda_{1,2} = 0, 2. \tag{4.13.21}$$

Putting $\lambda = \lambda_1 = 0$ into (4.13.19) we can get the first column of the matrix p; that is,

$$\begin{pmatrix} 1 - 0 & -1 \\ -1 & 1 - 0 \end{pmatrix} \begin{pmatrix} p_{11} \\ p_{21} \end{pmatrix} = 0 = \begin{pmatrix} p_{11} - p_{21} = 0 \\ -p_{11} + p_{21} = 0 \end{pmatrix} \Rightarrow p_{21} = p_{11}. \tag{4.13.22a}$$

Similarly, using $\lambda = \lambda_2 = 2$ into (4.13.19), we find, for the second column of p

$$\begin{pmatrix} 1 - 2 & -1 \\ -1 & 1 - 2 \end{pmatrix} \begin{pmatrix} p_{12} \\ p_{22} \end{pmatrix} = 0 = \begin{pmatrix} -p_{12} - p_{22} = 0 \\ -p_{12} - p_{22} = 0 \end{pmatrix} \Rightarrow p_{22} = -p_{12}. \tag{4.13.22b}$$

Each column of the matrix p represents an eigenvector corresponding to the given eigenvalue. It is generally accepted that the eigenvectors are to be normalized. For each column we write

$$\sum_i p_{ij}^2 = 1, \text{ or } p_{11}^2 + p_{21}^2 = 1, \text{ and } p_{12}^2 + p_{22}^2 = 1. \tag{4.13.23a}$$

With the use of (4.13.22) we find $p_{11} = \pm 1/\sqrt{2} = p_{21}$ for the first eigenvector and $p_{12} = \pm 1/\sqrt{2} = -p_{22}$ for the second eigenvector. Thus, for each eigenvalue we have obtained its corresponding eigenvector. Choosing the positive root, we have p and its inverse

$$p = \begin{pmatrix} 1/\sqrt{2} & 1/\sqrt{2} \\ 1/\sqrt{2} & -1/\sqrt{2} \end{pmatrix} \text{ and } p^{-1} = \begin{pmatrix} 1/\sqrt{2} & 1/\sqrt{2} \\ 1/\sqrt{2} & -1/\sqrt{2} \end{pmatrix}, \qquad (4.13.23b)$$

where we have used the inverse rule that $p^{-1} = Adj(p)/\det(p)$ (see Appendix B). One can easily verify that

$$p p^{-1} = \begin{pmatrix} 1 & 0 \\ 0 & 1 \end{pmatrix}, \qquad (4.13.24a)$$

and

$$p^{-1} M p = \begin{pmatrix} 1/\sqrt{2} & 1/\sqrt{2} \\ 1/\sqrt{2} & -1/\sqrt{2} \end{pmatrix} \begin{pmatrix} 1 & -1 \\ -1 & 1 \end{pmatrix} \begin{pmatrix} 1/\sqrt{2} & 1/\sqrt{2} \\ 1/\sqrt{2} & -1/\sqrt{2} \end{pmatrix} = \begin{pmatrix} 0 & 0 \\ 0 & 2 \end{pmatrix} = \begin{pmatrix} \lambda_1 & 0 \\ 0 & \lambda_2 \end{pmatrix} = \lambda, \qquad (4.13.24b)$$

as required in Equations (4.13.16, 4.13.17, and 4.13.21). The preceding eigenvalue–eigenvector results can be confirmed by running the MATLAB script eigen.m that follows. The values will appear in MATLAB's command line window. While the eigenvalues are the same, the eigenvectors can at most differ from ours by a negative sign depending on which sign is picked in (4.13.23a). This sign difference does not affect the motion studied here.

SCRIPT

```
%eigen.m
clear;
M={[1,-1];[-1,1]}   %input matrix M
[P,L]=eig(M)        %eigen vectors, P, and eigenvalues, L
PI=inv(P)           %inverse of P
P*PI                %the unit matrix should result
PI*M*P              %check eigenvectors. Should get eigenvalues back
```

We now go back to Equation (4.13.18) to write, with the use of (4.13.21),

$$m\frac{d^2 r_1}{dt^2} = -k r_1 - k_0 \lambda_1 r_1 \Rightarrow m\frac{d^2 r_1}{dt^2} + k r_1 = 0, \qquad (4.13.25a)$$

and

$$m\frac{d^2 r_2}{dt^2} = -k r_2 - k_0 \lambda_2 r_2 \Rightarrow m\frac{d^2 r_2}{dt^2} + (k + 2k_0) r_2 = 0. \qquad (4.13.25b)$$

For simplicity, we assume the solutions

$$r_1(t) = r_{10}\cos\omega_1 t \quad \text{and} \quad r_2(t) = r_{20}\cos\omega_2 t, \tag{4.13.26}$$

for each mode, with corresponding frequencies $\omega_1 \equiv \sqrt{k/m}$ and $\omega_2 \equiv \sqrt{(k + 2k_0)/m}$, where the initial values of r_1, and r_2 are related to the initial values of x_1 and x_2 by the inverse relation,

$$r = p^{-1}x \Rightarrow \begin{pmatrix} r_{10} \\ r_{20} \end{pmatrix} = \begin{pmatrix} 1/\sqrt{2} & 1/\sqrt{2} \\ 1/\sqrt{2} & -1/\sqrt{2} \end{pmatrix}\begin{pmatrix} x_{10} \\ x_{20} \end{pmatrix} = \begin{pmatrix} \dfrac{x_{10} + x_{20}}{\sqrt{2}} \\ \dfrac{x_{10} - x_{20}}{\sqrt{2}} \end{pmatrix}. \tag{4.13.27a}$$

To obtain the solutions to our initial problem, Equation (4.13.13), we use (4.13.14), to get

$$x = pr \Rightarrow \begin{pmatrix} x_1 \\ x_2 \end{pmatrix} = \begin{pmatrix} 1/\sqrt{2} & 1/\sqrt{2} \\ 1/\sqrt{2} & -1/\sqrt{2} \end{pmatrix}\begin{pmatrix} r_1 \\ r_2 \end{pmatrix} = \begin{pmatrix} \dfrac{r_1 + r_2}{\sqrt{2}} \\ \dfrac{r_1 - r_2}{\sqrt{2}} \end{pmatrix}. \tag{4.13.27b}$$

Thus, using Equations (4.13.26 and 4.13.27) we find the final solutions

$$x_1(t) = \left(\frac{x_{10} + x_{20}}{2}\right)\cos\omega_1 t + \left(\frac{x_{10} - x_{20}}{2}\right)\cos\omega_2 t \tag{4.13.28a}$$

and

$$x_2(t) = \left(\frac{x_{10} + x_{20}}{2}\right)\cos\omega_1 t - \left(\frac{x_{10} - x_{20}}{2}\right)\cos\omega_2 t. \tag{4.13.28b}$$

This indicates that there are two modes of vibration: the symmetric mode and the antisymmetric mode. Notice that if at $t = 0$, $x_{10} = x_{20}$, then $x_1(t) = x_2(t)$, which means that the masses vibrate in phase with the symmetric lower frequency (ω_1) *normal* mode. However, if at $t = 0$, $x_{20} = -x_{10}$ instead, then $x_2(t) = -x_1(t)$, which indicates that the masses vibrate out of phase at the antisymmetric higher frequency (ω_2) *normal* mode. Another characteristic of this system is that its motion develops beat-like behavior. This is easily seen if we rewrite (4.13.28a, b), with the use of trigonometric identities from Appendix C, as

$$x_1(t) = x_{10}\cos\overline{\omega}t\cos\omega_m t + x_{20}\sin\overline{\omega}t\sin\omega_m t, \tag{4.13.28c}$$

and

$$x_2(t) = x_{10}\sin\overline{\omega}t\sin\omega_m t + x_{20}\cos\overline{\omega}t\cos\omega_m t, \tag{4.13.28d}$$

where $\overline{\omega} = (\omega_1 + \omega_2)/2$ is the average frequency and $\omega_m = (\omega_2 - \omega_1)/2$ is the modulation frequency. This situation is similar to what is encountered in optics, where the beat frequency corresponds to twice the value of the modulation frequency because the square of the electric field displacement varies according to the beat frequency. Here, the modulation behavior of x_1 and x_2 is shown in Figure 4.15. The MATLAB script inter_spr2.m has been developed for this purpose. In order to see it, the values of x_1 and x_2 have to be other than equal or opposite. On the figure the parameters used are shown.

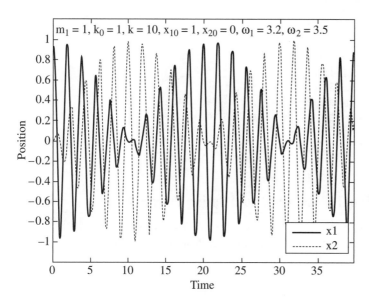

| FIGURE 4.15   The solution of the full coupled spring-mass bimodal system.

SCRIPT

```
%inter_spr2.m
%plots the coordinate solutions for the full bimodal coupled spring-mass
system
clear;
m=1;                            %masses
k0=1.0;  k=10.0;                %spring constants
x10=1;   x20=0;                 %initial positions
xs=(x10+x20)/2;
```

```
xd=(x10-x20)/2;
om1=sqrt(k/m);                   %frequencies
om2=sqrt((k+2*k0)/m);
om=min(om1,om2);                 %lowest frequency: for time range purposes
tau=2*pi/om; tmax=20*tau;
str=cat(2,'m=',num2str(m),', k_0=',num2str(k0),', k=',num2str(k),...
', x_{10}=',num2str(x10),', x_{20}=',num2str(x20),...
', \omega_1=',num2str(om1,2),', \omega_2=',num2str(om2,2));
t=[0:tau/50:tmax];               %plotting time range
x1=xs*cos(om1*t)+xd*cos(om2*t); %solutions
x2=xs*cos(om1*t)-xd*cos(om2*t);
% - can also write x1 and x2 as x3 and x4 if desired
%x3=x10*cos((om1+om2)*t/2).*cos((om2-om1)*t/2)...
%    +x20*sin((om1+om2)*t/2).*sin((om2-om1)*t/2);
%x4=x10*sin((om1+om2)*t/2).*sin((om2-om1)*t/2)...
%    +x20*cos((om1+om2)*t/2).*cos((om2-om1)*t/2);
%plot(t,x1,t,x2);
plot(t,x1,'b-',t,x2,'r:');
axis([0 tmax -1.2  1.2]);
text(0.5,1.05,str,'FontSize',12,'Color','black');
title('Coupled Spring-Mass Bimodal System','FontSize',14)
h=legend('x1','x2',4); set(h,'FontSize',14)
ylabel('Position','FontSize',14);
xlabel('Time','FontSize',14);
```

Case 3: Bimodal, Equal Mass, and Equal Spring Constants

In this case, taking $k_1 = k_2 = k_0 = k$, $m_1 = m_2 = m$, the solutions simplify in an interesting way. First, the frequencies of vibration from (4.13.26) simplify to

$$\omega_1 \equiv \sqrt{k/m} \quad \text{and} \quad \omega_2 \equiv \sqrt{3k/m} = \sqrt{3}\omega_1, \tag{4.13.29}$$

indicating that we still have two modes of vibration whose frequencies are related by the factor of $\sqrt{3}$. Second, the positions as a function of time, from (4.13.28a, b), can be more simply written as

$$x_1(t) = x_{cm0}\cos\omega_1 t + \frac{1}{2}x_{r0}\cos\sqrt{3}\omega_1 t \tag{4.13.30a}$$

and

$$x_2(t) = x_{cm0}\cos\omega_1 t - \frac{1}{2}x_{r0}\cos\sqrt{3}\omega_1 t, \tag{4.13.30b}$$

where $x_{cm0} = (x_{10} + x_{20})/2$, $x_{r0} = x_{10} - x_{20}$ are the initial values of the system's center of mass and relative coordinates. Finally, the case of the masses and springs being all different is left for the problems at the end of the chapter.

■ 4.14   The Method of Successive Approximations

As has been the case in this chapter, there are problems that under small displacements develop oscillatory motion. In such cases, a Taylor expansion reveals that a simple harmonic treatment can be carried out. While there are problems that require a full numerical scheme to obtain a solution, there are times when it is useful to be able to go beyond the simple harmonic approximation without resorting to a full numerical approach. For example, consider a problem for which the differential equation that needs to be solved is of the form

$$m\frac{d^2x}{dt^2} = f(x), \tag{4.14.1}$$

where if we expand $f(x)$ in a Taylor series expansion to third order, we get

$$f(x) = f_0(x_0) + f'(x_0)(x - x_0) + \frac{1}{2}f''(x_0)(x - x_0)^2 + \frac{1}{3!}f'''(x_0)(x - x_0)^3 + \cdots. \tag{4.14.2}$$

A system that develops oscillations does so about the equilibrium point x_0. Therefore, the first term is taken to be $f_0(x_0) = 0$; furthermore, if we define the effective spring constant in a similar way to Hooke's law for the spring force (say, $m\ddot{x} = -a_1 x$), then we can take $a_1 \equiv -f'(x_0)$ here. Further, we let $a_2 = f''(x_0)/2m$, and $a_3 = f'''(x_0)/3!m$; thus, (4.14.1) can be written in the form

$$\frac{d^2x}{dt^2} + \omega_0^2 x \approx a_2 x^2 + a_3 x^3 \tag{4.14.3}$$

to third order in x, where $x_0 \equiv 0$ and $\omega_0^2 = a_1/m$. The simplest approximation to this problem is when we take $a_2 = 0$ and $a_3 = 0$, in which case the right-hand side of (4.14.3) is zero and we have

$$x = A_1 \cos\omega_0 t \tag{4.14.4}$$

as a solution. Thus, the terms on the right-hand side of (4.14.3) correspond to deviations from linearity. In this section, we are interested in obtaining an approximation to (4.14.1) beyond the linear approximation and at least to include up to the third order in x. That is, we seek an approximate solution to (4.14.3) that is better than the linear approximation by taking a_2 and a_3 as nonzero values. The method of successive approximations does this as follows: guided by the linear solution (4.14.4), we guess that the approximate solution to (4.14.3) is of the form

$$x = A_0 + A_1 \cos \omega t + A_2 \cos 2\omega t + A_3 \cos 3\omega t, \tag{4.14.5}$$

where we have been careful to use ω and not ω_0 because it is expected that the frequency will deviate as well. The factors in the terms on the right in (4.14.5) are standard harmonics of $\cos \omega t$, guided by the powers of (4.14.3). Although we start with them, what we need to find are the unknowns A_0, A_2, A_3, and ω, with A_1 to be determined based on the initial conditions at $t = 0$. In order to find these unknowns, the idea is to substitute (4.14.5) into (4.14.3) and use the resulting relations to approximately determine these unknowns. To obtain better approximations, it is necessary to modify (4.14.5) further, based on the results of the substitution, until the solution would presumably no longer change. While doing so, if better approximations to (4.14.1) are desired, then more terms should be added to (4.14.3), and the process continues ad infinitum. Here, however, (4.14.5) will be a sufficient approximation to (4.14.3) for our purposes.

We proceed by obtaining the second derivative of x and write

$$\ddot{x} = -\omega^2 A_1 \cos \omega t - 4\omega^2 A_2 \cos 2\omega t - 9\omega^2 A_3 \cos 3\omega t. \tag{4.14.6}$$

We next consider the quadratic and cubic terms on the right of (4.14.3). With the use of (4.14.5) and the trigonometric identities (see Appendix C)

$$\cos^2 \omega t = (1 + \cos 2\omega t)/2, \qquad \cos^3 \omega t = (3 \cos \omega t + \cos 3\omega t)/4,$$

while keeping only the quadratic and cubic powers of the coefficients for x^2 and x^3 respectively, but discarding cross terms, in addition to neglecting harmonics higher than $3\omega t$ based on our assumed guess, we can write

$$x^2 \sim \frac{1}{2}(A_1^2 + A_2^2 + A_3^2) + \frac{A_1^2}{2} \cos 2\omega t + A_0^2 \tag{4.14.7a}$$

and

$$x^3 \sim \frac{3A_1^3}{4} \cos \omega t + \frac{3A_2^3}{4} \cos 2\omega t + \frac{1}{4}(A_1^3 + 3A_3^3) \cos 3\omega t + A_0^3. \tag{4.14.7b}$$

Putting (4.14.5–4.14.7) into (4.14.3) and rearranging we get

$$W_0 + W_1 A_1 \cos(\omega t) + W_2 A_2 \cos(2\omega t) + W_3 A_3 \cos(3\omega t) = 0, \tag{4.14.8a}$$

which in order to be true, each of the coefficients

$$W_0 = \omega_0^2 A_0 - a_2\left(A_0^2 + \frac{1}{2}[A_1^2 + A_2^2 + A_3^2]\right) - a_3 A_0^3,$$ (4.14.8b)

$$W_1 = -\omega^2 + \omega_0^2 - \frac{3a_3 A_1^2}{4},$$ (4.14.8c)

$$W_2 = A_2\omega_0^2 - 4\omega^2 A_2 - \frac{a_2}{2}A_1^2 - \frac{3a_3 A_2^3}{4},$$ (4.14.8d)

$$W_3 = A_3\omega_0^2 - 9\omega^2 A_3 - \frac{a_3}{4}(A_1^3 - 3A_3^3),$$ (4.14.8e)

must, therefore, vanish. Starting with $W_1 = 0$, we can get the oscillation frequency

$$\omega = \omega_0\sqrt{1 - 3a_3\left(\frac{A_1}{2\omega_0}\right)^2}.$$ (4.14.9a)

We next set $W_0 = 0$, and further ignoring the terms involving the powers A_0^3, A_2^2, A_3^2, in addition to the terms A_0^2 to avoid a quadratic equation, we obtain, to lowest order

$$A_0 = \frac{a_2 A_1^2}{2\omega_0^2}.$$ (4.14.9b)

Setting $W_2 = 0$, and further ignoring the A_2^3 term in order to avoid a cubic equation, and after substituting (4.14.9a) into it, we get

$$A_2 = \frac{a_2 A_1^2}{6(a_3 A_1^2 - \omega_0^2)}.$$ (4.14.9c)

Similarly, after setting $W_3 = 0$, ignoring the A_3^3 and substituting for ω, we obtain

$$A_3 = \frac{a_3 A_1^3}{(27a_3 A_1^2 - 32\omega_0^2)}.$$ (4.14.9d)

Thus, an approximate solution to (4.14.3) is (4.14.5) with the coefficients given by Equations (4.14.9). We finally note that, depending on the expansion of $f(x)$ in (4.14.1), if $a_2 \rightarrow 0$ in (4.14.3), then from the preceding coefficients $A_0 \rightarrow 0$ and $A_2 \rightarrow 0$, so that

$$x \rightarrow A_1\cos\omega t + A_3\cos 3\omega t,$$ (4.14.10)

with A_1 as determined by the initial condition for x at $t = 0$, and A_3 as given by (4.14.9d). If, however, $a_3 \to 0$ instead, then from the preceding equations we get that $\omega \to \omega_0$, $A_3 \to 0$, and A_2 simplifies so as to obtain the approximate solution

$$x \to A_1\cos\omega_0 t + \frac{a_2 A_1^2}{6\omega_0^2}(3 - \cos 2\omega_0 t). \tag{4.14.11}$$

Finally, if both a_2 and a_3 are zero, then to lowest order, we revert back to (4.14.4) as the solution.

■ 4.15 Beyond the Linear Approximation: Two Examples

The method of successive approximations solution of the previous section can readily be applied to two useful cases, namely, the pendulum and a model of a two-atom molecule potential considered earlier.

The Pendulum

In Section 4.2, we obtained the pendulum equation of motion

$$\frac{d^2\theta}{dt^2} + \omega_0^2\sin\theta = 0, \tag{4.15.1}$$

where the frequency $\omega_0 \equiv \sqrt{g/\ell}$ and the associated period is $\tau_0 = 2\pi/\omega_0 = 2\pi\sqrt{\ell/g}$ are standard characteristics of the $\sin\theta \sim \theta$ approximation for the simple pendulum. This simple pendulum has the solution $\theta = \theta_0\cos\omega_0 t$, as discussed in Section 4.2. If, however, the $\sin\theta$ function is expanded in a Taylor series and we keep terms to third order in θ, this equation becomes

$$\frac{d^2\theta}{dt^2} + \omega_0^2\theta = \frac{\omega_0^2\theta^3}{3!}. \tag{4.15.2}$$

An approximate solution of this equation can be written with the help of the previous section. The solution we seek is the case for which $a_2 \to 0$ in (4.14.3), so that as in (4.14.10) we can write for this nonlinear approximation

$$\theta = A_1\cos\omega t + A_3\cos 3\omega t, \tag{4.15.3a}$$

where A_3 is given in (4.14.9d) and A_1 is to be found in such a way that $\theta(t = 0) = \theta_0$. (This will be discussed shortly.) Further, from (4.14.3) and (4.15.2) we see that $a_3 = \omega_0^2/6$, so that (4.14.9a) gives

$$\omega = \omega_0\sqrt{1 - 3a_3\left(\frac{A_1}{2\omega_0}\right)^2} \to \omega_0\sqrt{1 - \frac{A_1^2}{8}}, \tag{4.15.3b}$$

or for the period

$$\tau_{nl} = \tau_0/\sqrt{1 - (A_1^2/8)}, \tag{4.15.3c}$$

where the subscript *nl* indicates the *nonlinear* approximation. This shows that the period is actually longer by a factor of $1/\sqrt{1 - (A_1^2/8)}$ compared to the simple pendulum. Thus, according to (4.5.3a and 4.5.3c), the period of the pendulum depends on the initial amplitude θ_0 in radians. If the amplitude is small, then τ_0 is a good approximation for the period. It is useful to compare τ versus θ_0 obtained this way to that obtained from the full numerical solution. However, Equation (4.15.1) can be conveniently solved numerically to obtain $\theta(t)$ but not $\tau(\theta)$. Thus, rather than using (4.15.1), it is best to look at the energy equation

$$E = mg\ell(1 - \cos\theta_0) = \frac{1}{2}I\omega^2 + mg\ell(1 - \cos\theta), \tag{4.15.4}$$

where the first term on the left is the initial potential energy, and the first and second terms on the right are the pendulum's kinetic and potential energies, respectively, with $I = m\ell^2$ and $\omega = d\theta/dt$ at any time later when the angle is θ. From here we can obtain an integral for the period using a formula similar to that of (2.4.12) in Chapter 2. That is,

$$\int_0^{\tau/4} dt = \frac{\tau}{4} = \int_0^{\theta_0} \frac{d\theta}{\sqrt{\dfrac{2[E - V(\theta)]}{I}}} = \sqrt{\frac{\ell}{2g}} \int_0^{\theta_0} \frac{d\theta}{\sqrt{\cos\theta - \cos\theta_0}}, \tag{4.15.5}$$

because the integral involves only one fourth of a full period's path. Although this integral can be carried out numerically in MATLAB, it is more convenient to use built-in functions. We notice that $\cos\theta = 1 - 2\sin^2\theta/2$; thus the preceding integral can be written as

$$\tau_f = 2\sqrt{\frac{\ell}{g}} \int_0^{\theta_0} \frac{d\theta}{\sin(\theta_0/2)\sqrt{1 - \left[\dfrac{\sin(\theta/2)}{\sin(\theta_0/2)}\right]^2}}, \tag{4.15.6}$$

where the subscript f indicates the *full* numerical solution. Thus, defining $\sin(\theta/2) = \sin(\theta_0/2)\sin\phi$, and since $\tau_0 = 2\pi\sqrt{\ell/g}$, one can write this integral as

$$\frac{\tau_f}{\tau_0} = \frac{2}{\pi} \int_0^{\pi/2} \frac{d\phi}{\sqrt{1 - \sin^2(\theta_0/2)\sin^2\phi}} = \frac{2}{\pi}K(k = \sin(\theta_0/2)), \tag{4.15.7}$$

where $K(k)$ is the well-known tabulated *complete elliptic integral of the first kind.*
For the nonlinear approximation, we need to find A_1 such that at $t = 0$ from (4.15.3a)

$$A_1 + A_3 = \theta_0 \quad \Rightarrow \quad f(A_1) \equiv A_1 + \frac{a_3 A_1^3}{(27 a_3 A_1^2 - 32 \omega_0^2)} - \theta_0 = 0. \quad (4.15.8)$$

This is a cubic equation and can be solved numerically using the Newton–Raphson iteration method for finding the zero of a function, in which

$$x_{i+1} = x_i - \frac{f(x_i)}{f'(x_i)}, \quad (4.15.9)$$

where $f(x)$ is given by (4.15.8), with x replacing A_1 and $f'(x)$ being the derivative, $1 + 3a_3 x^2/(27 a_3 x^2 - 32 \omega_0^2) - 54 a_3^2 x^4/(27 a_3 x^2 - 32 \omega_0^2)^2$. One starts with an initial value—say, $x_0 = \theta_0$—and the iteration proceeds forward until the Nth iteration when $x_{N+1} \approx x_N$, in which case $f(A_1 = x_N) \approx 0$. The MATLAB script pend0.m that follows uses this procedure for obtaining A_1 for a starting initial amplitude θ_0.

SCRIPT

```
%pend0.m
%This program shows the relationship between A1 and the initial angle, for
the
%non-linear approximation of the pendulum's sin(theta) term
clear;
w0=1;
a3=w0^2/6;
imax=10;
tol=1.e-5;
th0=0;
thmax=90;
N=25;
dth=(thmax-th0)/N;
for j=1:N,
  th(j)=th0+(j-1)*dth;
  x=th(j)-1;%initial guess
  xn=999;
  f=999;
```

```
i = 0;
while (abs(xn-x) >= tol) & (f ~= 0.0) & (i < imax)
    x=xn;
    f=x+a3*x^3/(27*a3*x^2-32*w0^2)-th(j);
    fp=1+3*a3*x^2/(27*a3*x^2-32*w0^2)...
        -54*a3^2*x^4/((27*a3*x^2-32*w0^2)^2);
    xn=x-f/fp;
    i = i + 1;
end
A1(j)=xn;
end
plot(th,A1)
xlabel('\theta_0 (Degrees)','FontSize',14)
ylabel('A_1 (Degrees)','FontSize',14)
title('Relationship between A_1 and \theta_0','FontSize',14)
```

Figure 4.16 shows the results obtained for A_1 versus θ_0. We notice that the relationship is very nearly linear with a slope of approximately unity.

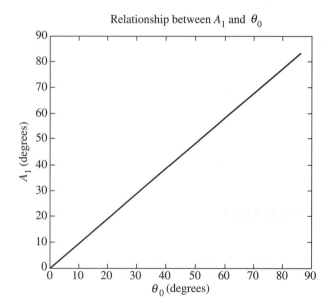

Relationship between A_1 and θ_0

FIGURE 4.16   The plot showing the linear relationship between the coefficient A_1 initial angle.

Thus, from here on we use the further approximation that in the nonlinear solution of the pendulum, Equation (4.15.3a),

$$A_1 \approx \theta_0, \qquad\qquad\qquad (4.15.10a)$$

and similarly, Equation (4.15.3c) becomes

$$\tau_{nl} \sim \tau_0 / \sqrt{1 - \theta_0^2/8}. \tag{4.15.10b}$$

The MATLAB script pend1.m that follows performs the comparison between Equations (4.15.7), (4.15.3c), and the standard simple harmonic oscillator (SHO) $\tau_0 = 2\pi\sqrt{\ell/g}$ value. Figure 4.17 contains the results.

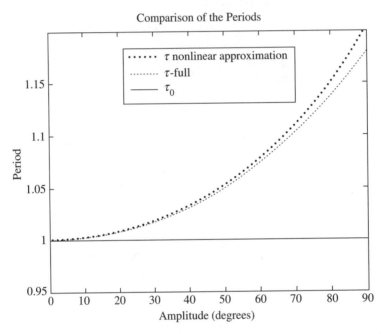

FIGURE 4.17   Comparison between the pendulum periods in the small angle approximation, the nonlinear approximation, and the full resolution.

SCRIPT

```
%pend1.m
%Program designed to compare the period of a pendulum with the next approximation
%beyond the linear simple pendulum formula. In the simple pendulum, the period is
%independent of amplitude
clear;
N=100;
thmax=90;% note at 180, the period is infinite - pendulum gets stuck
dth=thmax/N;
th=[0:dth:thmax];% in degrees
%MATLAB defines the complete elliptic integral of the first kind as:
% ellipke(m)=Integral of (1-m*sin^2(t)) on the interval 0 <= t <= pi/2
% Thus Matlab's m is actually our k^2, where k=sin(theta/2)
m=sin(th*2*pi/360/2).^2 ;
```

```
% Periods in units of simple pendulum tau=2*pi*sqrt(1/g)
y1=1./sqrt(1-(th*2*pi/360).^2/8);        %the nonlinear approximation
y2=2*ellipke(m)/pi;                      %full solution for the period versus amplitude
plot(th,y1,'b.',th,y2,'k-')
line([0,thmax],[1,1],'color','red'); %the simple harmonic oscilator period taken as one
axis([0 thmax 0.95 1.2]);
h=legend('\tau non-linear-approx','\tau-full','\tau_0',0);
set(h,'FontSize',14)
xlabel('Amplitude(Degrees)','FontSize',14)
ylabel('Period','FontSize',14)
title('Comparison of the Periods','FontSize',14)
```

Figure 4.17 shows that the approximation fares well compared to the full calculation. The deviation of the nonlinear approximation from the full calculation is even less than 2% at the highest angle shown in the figure of 90°. However, the standard SHO begins to deviate noticeably beyond about 15°.

It is useful to compare the $\theta(t)$ motion obtained from Equations (4.15.1), (4.15.3a), and the standard SHO $\theta = \theta_0 \cos(\omega_0 t)$. The MATLAB script pend2.m that follows does this. However, in order to solve the differential Equation (4.15.1) in MATLAB, it is necessary to express the differential equation as a system, i.e., let $d\theta_1/dt = \theta_2$ and $d\theta_2/dt = -\omega_0^2 \sin(\theta_1)$, and then θ_1 and θ_2 are treated as part of an array. The solution is carried out using MATLAB's built-in Runge–Kutta differential equation solver, which is much more accurate than the Euler method of Chapter 1. The right-hand sides of these two new differential equations are the derivatives of θ_1 and θ_2, respectively, and have to be incorporated in a second script, which we label pend2_der.m and whose listing is also shown. The main script pend2.m calls this second script at run time. Both scripts follow, and Figure 4.18 contains the results at the angle of 90°.

The figure shows that, even at the high angle of 90°, the difference between the nonlinear approximation and the full solution differ only slightly. There is a large discrepancy between these two and the standard SHO, however, as expected.

SCRIPT

```
%pend2.m
%program to plot the solutions of the harmonic oscillator using the simple, the
%nonlinear approximation, and the full sin(theta) term solution
clear;
cf=2*pi/360; %conversion factor from degrees to radians
w0=1; % let the frequency of the SHM be one
tau0=2*pi/w0; %period for the SHM
tmax=4*tau0; % maximum time
th=90; %initial amplitude in degrees
thr=th*cf;% initial angle in radians
```

```
ic1=[thr;0.0]; % initial conditions for angle and angular speed
%Use MATLAB's Runge-Kutta (4,5) formula
%[t,th2]=ode45('pend2_der',[0.0,tmax],ic1,[],w0);%numerical solution (default
tolerances)
opt=odeset('AbsTol',1.e-7,'RelTol',1.e-4);        %user set Tolerances
[t,th2]=ode45('pend2_der',[0.0,tmax],ic1,opt,w0);%numerical solution
%the nonlinear approximation
om=w0*sqrt(1-thr^2/8);
a3=w0^2/6;
A1=thr;
A3=a3*A1^3/(27*a3*A1^2-32*w0^2);
th1=thr*cos(om*t)+A3*cos(3*om*t);
%the SHO case
th0=thr*cos(w0*t);
plot(t,th0/cf,'r:',t,th1/cf,'k-.',t,th2(:,1)/cf,'b-');%Amplitude in degrees
h=legend('Standard SHO','Nonlinear Approx','Full Solution',1);
set(h,'FontSize',12)
axis([0 max(t) -th th*(1+0.4)]);
str=cat(2,'\theta_0=',num2str(th,3));
text(5,th*(1+0.2),str,'FontSize',12);
xlabel('Time (sec)','FontSize',14)
ylabel('Amplitude (degrees)','FontSize',14)
title('Comparison of Solutions','FontSize',14)
```

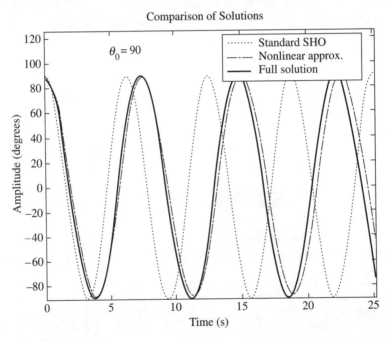

FIGURE 4.18   Comparison between the pendulum solutions in the small angle approximation, the nonlinear approximation, and the full solution.

```
function derivs = pend2_der( t, x, flag,w0)
% pend2_der: returns the derivatives for the pendulum's full solution
% The function pen2_der describes the equations of motion for a
% pendulum. The parameter w0, is part of the input
% Entries in the vector of dependent  variables are:
% x(1)-position, x(2)-angular velocity
derivs = [ x(2); -w0^2*sin(x(1))];
```

The Two-Atom Molecule

Example 3.2 used the two-atom molecular potential $V(x) = \left(\dfrac{A}{x^3} - \dfrac{B}{x^2} \right)$. As before,

we take $A = u_0 a_0^3$ and $B = u_0 a_0^2$, where u_0 is a unit of molecular energy and a_0 is a unit

of molecular distance. The force associated with this potential is

$F(x) = -V'(x) = \dfrac{3A}{x^4} - \dfrac{2B}{x^3}$ and   the   stable   equilibrium   point   was

found to be $x_b = \dfrac{3A}{2B} = \dfrac{3}{2}a_0$. If we expand the force to second order in a Taylor series

about this equilibrium point we get

$$F(x) \approx F(x_b) + F'(x_b)(x - x_b) + \frac{1}{2}F''(x_b)(x - x_b)^2, \tag{4.15.11}$$

where one can see that $F(x_b) = 0$, and

$$F'(x_b) = \left(-\frac{12A}{x^5} + \frac{6B}{x^4} \right)\bigg|_{x=x_b} = -\frac{32u_0}{81\,a_0^2}, \tag{4.15.12a}$$

and

$$F''(x_b) = \left(\frac{60A}{x^6} - \frac{24B}{x^5} \right)\bigg|_{x=x_b} = \frac{512u_0}{243\,a_0^3}, \tag{4.15.12b}$$

so that using these, we can write the differential equation for Newton's second law
corresponding to the force of (4.15.11) in the form

$$\frac{d^2x}{dt^2} + \omega_0^2(x - x_b) = \frac{c}{2m}(x - x_b)^2, \tag{4.15.13}$$

where $\omega_0 = \sqrt{k/m}$ is the simple harmonic oscillator frequency with $k \equiv \dfrac{32u_0}{81a_0^2}$ and

$c \equiv \dfrac{512u_0}{243a_0^3}$. An approximate solution to the equation can be obtained by setting $a_3 \to 0$ in (4.14.3), so that as in (4.14.11) we can write a nonlinear approximate solution in the form

$$x = x_b + A_1 \cos\omega_0 t + \frac{a_2 A_1^2}{6\omega_0^2}(3 - \cos 2\omega_0 t), \qquad (4.15.14a)$$

where by comparing (4.15.13) with (4.14.3) we identify the coefficient $a_2 = c/2m$. Here A_1 is to be found by requiring that at $t = 0$, $x - x_b = x_i$, the initial position, as measured from the equilibrium position. Here this corresponds to setting

$$A_1 + \frac{2a_2 A_1^2}{6\omega_0^2} = x_i \Rightarrow A_1 = \frac{3\omega_0^2}{2a_2}\left[-1 \pm \sqrt{1 + \frac{4a_2 x_i}{3\omega_0^2}}\right]. \qquad (4.15.14b)$$

The full differential equation corresponding to $F = ma$ of the problem is

$$m\frac{d^2 x_f}{dt^2} = \frac{3A}{x_f^4} - \frac{2B}{x_f^3}, \qquad (4.5.15a)$$

where the subscript f indicates the *full* numerical solution. If, once again, we use the preceding units, with distance $x_f = \bar{x}_f a_0$, time $t = \bar{t}\tau_0$, where $\tau_0 = 2\pi/\omega_0 = 2\pi\sqrt{81ma_0^2/32u_0}$, so that we can write the full differential equation in dimensionless units, we get

$$\frac{d^2 \bar{x}_f}{d\bar{t}^2} = \frac{81\pi^2}{8}\left(\frac{3}{\bar{x}_f^4} - \frac{2}{\bar{x}_f^3}\right). \qquad (4.15.15b)$$

We notice that the right-hand side vanishes at the dimensionless equilibrium point $\bar{x}_b = 3/2$, as expected. In MATLAB this equation is transformed to a system of two differential equations; that is, with $\bar{x}_1 = \bar{x}_f$

$$d\bar{x}_1/d\bar{t} = \bar{x}_2, \quad \text{and} \quad \frac{d\bar{x}_2}{d\bar{t}} = \frac{81\pi^2}{8}\left(\frac{3}{\bar{x}_1^4} - \frac{2}{\bar{x}_1^3}\right), \qquad (4.15.16)$$

where we seek $\bar{x}_1(\bar{t})$. In order to compare this full numerical solution to the nonlinear approximation, we write (4.15.14a) in dimensionless units as

$$\bar{x}_{nl} = \bar{x}_b + \bar{A}_1 \cos 2\pi \bar{t} + \frac{4\bar{A}_1^2}{9}(3 - \cos 4\pi \bar{t}), \tag{4.15.17a}$$

where the subscript *nl* refers to the *nonlinear* approximation. Also, in our units, we have that $\omega_0^2/a_2 = 3a_0/8$, $x_i = \bar{x}_i a_0$ so that $A_1 = \bar{A}_1 a_0$, with

$$\bar{A}_1 = \frac{9}{16}\left(-1 + \sqrt{1 + \frac{32}{9}\bar{x}_i}\right), \tag{4.15.17b}$$

where we have chosen the positive root. This guarantees that at $t = 0$, our initial condition, $\bar{x}_{nl} - \bar{x}_b = \bar{x}_i$, is obeyed.

Finally, the linear SHO solution is obtained simply by ignoring the term on the right of (4.15.13) to obtain $x = x_b + x_i \cos(\omega_0 t)$, which in dimensionless units is

$$\bar{x}_l = \bar{x}_b + \bar{x}_i \cos 2\pi \bar{t}, \tag{4.15.18}$$

where the subscript *l* refers to the *linear* approximation. We are now ready to compare the three solutions. The MATLAB script molec.m, along with the auxiliary file molec_der.m, does it, as listed in the following script. The results are shown on Figure 4.19.

SCRIPT

```
%molec.m
%program to plot the solutions of the two atom molecular potential model
%using the simple, the nonlinear approximation, and the full solutions
clear;
tmax=2; % maximum time in units of tau0
xb=3/2; %equilibrium position
xi=0.2; %initial position measured from equilibrium
ic1=[xb+xi;0.0]; % initial conditions: position, speed
%Use MATLAB's Runge-Kutta (4,5) formula
%[t,x2]=ode45('molec_der',[0.0,tmax],ic1,[]);%numeric soln (default tolerances)
opt=odeset('AbsTol',1.e-7,'RelTol',1.e-4);   %user set Tolerances
[t,x2]=ode45('molec_der',[0.0,tmax],ic1,opt);%numerical solution
%the non0linear approximation uses A1 for amplitude
% A1 is such that at t=0 x1-xb=xi
A1=9*(-1+sqrt(1+32*xi/9))/16;
x1=xb+A1*cos(2*pi*t)+4*A1^2*(3-cos(4*pi*t))/9;
%the SHM case
x0=xb+xi*cos(2*pi*t);
plot(t,x0,'r:',t,x1,'k-.',t,x2(:,1),'b-');%position versus time
```

```
h=legend('SHO','Nonlinear Approx','Full Solution',1); set(h,'FontSize',12)
line([0,max(t)],[xb,xb],'color','black');%the equilibrium position
text(0.8,xb-0.01,'Equilibrium','FontSize',14);
str=cat(2,'x_i=',num2str(xi,3));
text(0.2,xb+xi+0.025,str,'FontSize',14);
xlabel('Time (\tau_0)','FontSize',14)
ylabel('Position (a_0)','FontSize',14)
title('Comparison of Solutions','FontSize',14)
```

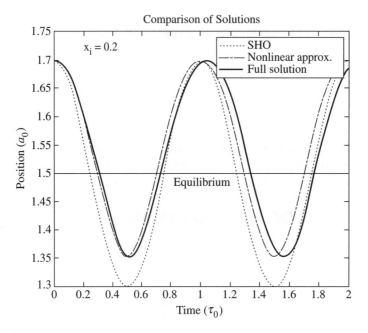

FIGURE 4.19   Comparison between a simple harmonic model of a molecule solution (dots), the nonlinear approximation (dash-dot), and the full numerical solution (solid line).

FUNCTION

```
function derivs = molec_der( t, x, flag)
% molec_der: returns the derivatives for the two atom molecule
% model's full solution
% Entries in the vector of dependent variables are:
% x(1)-position, x(2)-velocity
derivs = [ x(2); 81*pi^2*(3./(x(1).^4)-2./(x(1).^3))/8];
```

In the figure, the three solutions are shown for an initial $\bar{x}_i = 0.2$ in units of a_0. The equilibrium position $\bar{x}_b$, about which the motion occurs, is shown as a straight line. We notice that although the nonlinear approximation x_{nl} is an improvement over the SHO version, x_l, after the two periods shown, the deviation from the full numerical solution, x_f, is clearly visible. One could improve on the approximation by including the next order term in the expansion (4.15.11 and 4.15.13); however, the effort required renders the full numerical solution more valuable. The larger the value of the initial amplitude, x_i, is away from the equilibrium position, the higher is the order of approximation needed for reasonable results. The SHO approximation is only good for very small x_i. The MATLAB program can be used to investigate this aspect of the calculation.

■ Chapter 4 Problems

4.1 Prove by substitution that Equation (4.2.5) is a solution to Equation (4.2.4).

4.2 Prove by substitution that Equation (4.2.7) is a solution to Equation (4.2.6).

4.3 A 0.3 m-long adjustable wrench with a mass of 0.5 kg is pivoted at one end and allowed to swing as a pendulum. Its center of mass lies at 0.2 m from the pivoted end, and its oscillation period is 0.9 s. A 50-g nut is placed at its free end and allowed to swing. What is the period of the wrench–nut system?

4.4 For the hanging spring-mass system of Section 4.11, if at time $t = 0$ the mass is held at $y = 0.25$ cm below the equilibrium position, with zero initial velocity,

a. what is the maximum amplitude of the oscillation away from equilibrium after it is released? A

b. what is the equilibrium position? y_e

c. Write the particular solution for the displacement as a function of time.

d. What is the maximum velocity possible?

Assume $m = 0.2$ kg, with a spring length of 0.05 m.

4.5 Consider a floating sphere that experiences a lifting drag force proportional to the square of the air's speed. Assuming small oscillations, obtain an expression for the oscillatory frequency of the sphere.

4.6 Show that Equations (4.13.12) satisfy the differential Equations (4.13.2).

4.7 Use molec.m in order to investigate at what value of x_i is the SHO a good enough approximation to the two-atom molecule potential. Also, find the value of x_i beyond which the nonlinear approximation becomes less useful.

4.8 The coupled-spring problem of Figure 4.13(a) has differential equations (4.13.1). If we assume that the spring constants are equal; that is, that $k_1 = k_2 = k_0 = k$, and that the masses are also equal, $m_1 = m_2 = m$, rewrite the equations. Show that if you first subtract these resulting equations, you get one new differential equation. On the other hand, if you add them instead of subtracting them, you get a second new differential equation. These two new resulting differential equations can easily be solved for two new variables, $x_1 - x_2$ and $x_1 + x_2$. Give their solution and identify the vibrational frequencies with their corresponding formula.

■ Additional Problems

4.9 An aluminum rod 4.8 cm in diameter has a horizontal projection 5.3 m from a supporting wall. A 1200-kg mass hangs from its free end. Given that the shear modulus of aluminum is 1×10^{10} N/m^2 and neglecting the rod's mass, find the vertical deflection and the vibrational frequency of the rod.

4.10 Consider a simple pendulum (of length 0.25 m) that is displaced by 50° from the vertical and is let go.

   a. What would be an approximate value of its period according to the method of successive approximations?

   b. What is the percent error between this nearly correct value and the value normally used but only appropriate for small displacements?

4.11 The differential equation of a damped simple pendulum is given by
$$\frac{d^2\theta}{dt^2} + \frac{c}{m}\frac{d\theta}{dt} + \frac{g}{\ell}\theta = 0$$

   a. Explain how this differential equation is obtained.

   b. Write down a solution of this differential equation if at time $t = 0$ the pendulum is displaced by 15°.

   c. The pendulum is let go and after one full swing it returns to a position at the 5° mark. Obtain the pendulum's damping coefficient, c, given that the length ℓ is 0.5 m long and that a mass of 0.25 kg hangs from it.

4.12 A thick styrofoam raft of 1 m in length by 1 m in width is seen floating in the sea. The raft holds a tank full of fresh water. The raft appears to have a bobbing period of 1.5 s. If you ignore the masses of the raft and the container, approximately how many gallons of water does the container hold? (1 gal = 3.78 liter, 1 liter = 1000 cc)

4.13 Consider a mass attached to the spring configuration given in Figure 4.20. Each spring has the spring constant shown. Obtain a formula for the oscillation frequency observed if the mass develops simple harmonic motion. Give the frequency in terms of the spring constants k_1, k_2, k_3, and the mass m.

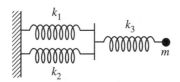

I FIGURE 4.20

4.14 The problem involving two masses and three springs of Figure 4.13 gave rise to the Equation (4.13.1).

a. For the case when the masses as well as the springs are all different, show that the equations can be written as $\ddot{x} = -Mx$, where

$$x = \begin{pmatrix} x_1 \\ x_2 \end{pmatrix}, \ M = \begin{pmatrix} \dfrac{k_0 + k_1}{m_1} & -\dfrac{k_0}{m_1} \\ \dfrac{k_0}{m_2} & \dfrac{k_0 + k_2}{m_2} \end{pmatrix}.$$

b. This system of equations can be solved by letting $x = pr$, where $r = \begin{pmatrix} r_1 \\ r_2 \end{pmatrix}$ and p is the eigenvector matrix, to obtain $\ddot{r} = -\lambda r$, where λ is the diagonal eigenvalue matrix. Both p and λ are obtained in such a way that $\lambda = p^{-1}Mp$, so that the determinant, $M - \lambda = 0$, and the columns of p satisfy $(M - \lambda_n)p_n = 0$ in a similar process to that of Section 4.13. Thus, possible solutions to the equations $\ddot{r} = -\lambda r$ are $r_1 = r_{10}\cos\omega_1 t$

and $r_2 = r_{20}\cos\omega_2 t$,   with $\omega_1 = \sqrt{\lambda_1}$, $\omega_2 = \sqrt{\lambda_2}$,   and   where

$\begin{pmatrix} r_{10} \\ r_{20} \end{pmatrix} = p^{-1}\begin{pmatrix} x_{10} \\ x_{20} \end{pmatrix}$. Obtain the eigenvalues λ_1 and λ_2 as well as the eigenvector matrix p.

c. The final solutions $x_1(t)$, $x_2(t)$ are obtained from the original definition, $x = pr$, where p and r are from part (b). Obtain the final solutions and give a plot of them using the parameters, $m_1 = 0.5$, $m_2 = 1$, $k_0 = 1$, $k_1 = 10$, $k_2 = 10$, $x_{10} = 1$, and $x_{20} = 0$. Bearing in mind the results of Section 4.13, what are your observations?

4.15 The nonlinear solution approximation to the pendulum problem up to third order in θ is given by $\theta = A_1\cos\omega t + A_3\cos 3\omega t$, where A_1 is such that at $t = 0$, $\theta = \theta_0$.

a. Show that this results in a cubic equation whose solution agrees closely with the numerical result of Figure 4.16.

b. From the cubic equation, comment on what happens to A_1 for small θ_0.

5 Vectors and Differential Calculus

■ 5.1 Introduction

A vector is a mathematical concept used to describe physical quantities. A vector is characterized by magnitude and direction. In Figure 5.1, **A** is the vector whose magnitude is denoted by $A \equiv |\mathbf{A}|$, which is its length, and whose direction is determined by the angle θ measured counterclockwise from the positive x-axis. A scalar is another mathematical concept used to describe physical quantities that only have magnitude. In the case of the vector **A**, its magnitude A is a scalar quantity. Examples of vectors include displacement, velocity, acceleration, and force, among others. Examples of scalar quantities are distance, speed, energy, temperature, and so on.

A vector can be multiplied by the magnitude of a scalar quantity with the result that a new vector is obtained, i.e., $s\mathbf{A} = \mathbf{B}$ whose magnitude is such that $sA = B$ but the direction remains the same. One example of this is Newton's second law, $\mathbf{F} = m\mathbf{a}$, in which the acceleration has the same direction as the force, but the magnitudes are different. From this we can also see that the units of the scalar play an important role. In the case of Newton's second law, the scalar quantity of mass has units of kg, which gives the unit of force, the Newton, its meaning as $\text{kg} \cdot \text{m/s}^2$. Vectors play an important role in physics, and this makes them indispensable.

The class of vectors, whose magnitude is unity, is known as unit vectors. For example, it is possible to create a unit vector out of the preceding vector **A** simply by dividing by its magnitude. The result is a vector whose magnitude will be unity, but with the same direction. This unit vector is written as $\hat{A} \equiv \mathbf{A}/A$. For this reason, it is

I FIGURE 5.1

119

a standard and convenient practice to have unit vectors in the Cartesian coordinate system, i.e., $\hat{i}, \hat{j}, \hat{k}$, corresponding to the x, y, and z directions, respectively. These three unit vectors are very useful in handling vectors in general. For example, the two-dimensional vector **A** in Figure 5.1 can be written in terms of its x and y components if we know their scalar values, A_x and A_y, as shown in Figure 5.2.

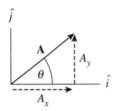

I FIGURE 5.2

Thus, we can write

$$\mathbf{A} = A_x\hat{i} + A_y\hat{j}. \tag{5.1.1}$$

Since

$$A_x = A\cos\theta \text{ and } A_y = A\sin\theta, \tag{5.1.2a}$$

then, by squaring these and adding them, and using $\sin^2\theta + \cos^2\theta = 1$, we see that the magnitude of **A** is

$$A = \sqrt{A_x^2 + A_y^2}, \tag{5.1.2b}$$

which is the Pythagorean theorem. Similarly, by taking the ratio of the y component to the x component, followed by the inverse tangent, the direction can be obtained as

$$\theta = \arctan\left(\frac{A_y}{A_x}\right). \tag{5.1.2c}$$

EXAMPLE 5.1

A motorcycle rider, after traveling for 5 km at a certain angle north of east (*NE*), runs out of gas. The rider is able to walk in a straight line back toward the road, and perpendicular to it, for a distance of 4 km. (a) How far is the rider from the initial starting position? (b) What was the initial direction of travel with respect to the road? (c) Write a unit vector corresponding to this direction.

Solution

(a) Let the perpendicular distance traveled be A_y = 4 km and let A = 5 km. By the Pythagorean theorem, the rider is at $A_x = \sqrt{A^2 - A_y^2} = \sqrt{5^2 - 4^2} =$ 3 km from the initial takeoff position.

(b) The direction of travel is $\theta = \arctan\left(\dfrac{4}{3}\right) = 53.13°\ NE$.

(c) A unit vector in this direction is $\hat{A} = (3\text{ km}\hat{i} + 4\text{ km}\hat{j})/5\text{ km} = 0.6\hat{i} + 0.8\hat{j}$.

Vectors are useful in representing equations. For example, in the case of $\mathbf{F} = m\mathbf{a}$, it means that if we write $\mathbf{F} = F_x\hat{i} + F_y\hat{j} + F_z\hat{k}$ and $\mathbf{a} = a_x\hat{i} + a_y\hat{j} + a_z\hat{k}$, then because the left side must equal to the right side of the equation, we must have like components on the left equal to like components on the right, or

$$a_x = \frac{F_x}{m},\ a_y = \frac{F_y}{m},\text{ and } a_z = \frac{F_z}{m}. \tag{5.1.3}$$

■ 5.2   Vector Addition and Subtraction

Vectors can be added, as in $\mathbf{C} = \mathbf{A} + \mathbf{B}$, or subtracted, as in $\mathbf{D} = \mathbf{A} + (-\mathbf{B})$, with subtraction being a special form of vector addition. Both of these examples are depicted in Figure 5.3.

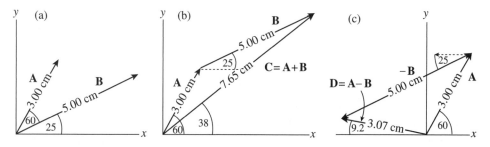

| FIGURE 5.3

Referring to the positive x direction as east and the positive y direction as north, Figure 5.3(a) shows two vectors $\mathbf{A}$ = 3 cm at $60°\ NE$ and $\mathbf{B}$ = 5 cm at $25°\ NE$ starting at the origin. These two vectors can be added graphically in an appropriate scale as in Figure 5.3(b), to estimate $\mathbf{C} = \mathbf{A} + \mathbf{B} \approx 7.65$ cm at $38°\,NE$. They also can be subtracted graphically, as shown in Figure 5.3(c), with the approximate result $\mathbf{D} = \mathbf{A} - \mathbf{B} \approx 3.07$ cm at $9.2°\,NW$. The graphical results depend on how precisely one reads the graph using a ruler and a protractor. However, $\mathbf{A}$ and $\mathbf{B}$ can be more accurately and conveniently added and subtracted analytically in terms of components. That is, in units of cm, let's write

$$\mathbf{A} = A_x\hat{i} + A_y\hat{j} = 3\cos(60)\hat{i} + 3\sin(60)\hat{j} = 1.50\hat{i} + 2.60\hat{j}, \tag{5.2.1a}$$

and similarly

$$\mathbf{B} = B_x\hat{i} + B_y\hat{j} = 5\cos(25)\hat{i} + 6\sin(25)\hat{j} = 4.53\hat{i} + 2.11\hat{j}. \tag{5.2.1b}$$

We find the vector $\mathbf{C} = \mathbf{A} + \mathbf{B} = (A_x + B_x)\hat{i} + (A_y + B_y)\hat{j}$ to get

$$\mathbf{C} = (1.50 + 4.53)\hat{i} + (2.60 + 2.11)\hat{j} = 6.03\hat{i} + 4.71\hat{j}. \tag{5.2.2a}$$

The magnitude and direction of $\mathbf{C}$ are therefore as in Section 5.1,

$$C = \sqrt{6.03^2 + 4.71^2} = 7.65 \text{ cm and } \theta = \tan^{-1}(4.71/6.03) = 37.99°\,NE. \tag{5.2.2b}$$

Referring to Figure 5.3(b), we see that the graphical method closely agrees with these accurate results. Similarly, we can find $\mathbf{D} = \mathbf{A} - \mathbf{B} = (A_x - B_x)\hat{i} + (A_y - B_y)\hat{j}$, to obtain

$$\mathbf{D} = (-3.03\hat{i} + 0.49\hat{j}),\ D = 3.07 \text{ cm, and } \theta = \tan^{-1}\!\left(\frac{0.49}{3.03}\right) = 9.19°\,NW, \tag{5.2.3}$$

where the angle has been calculated in the second quadrant for convenience. Once again, the graphical results shown in Figure 5.3(c) are the estimated values. In the analytical values given, we have employed the understanding that angles measured from the $+x$ direction are referred to as being NE, and those angles measured from the $-x$ direction are referred to as being NW, and so on. Finally, three-dimensional vectors of the form $\mathbf{V} = V_x\hat{i} + V_y\hat{j} + V_z\hat{k}$ can be plotted in Cartesian coordinates, as shown in Figure 5.4.

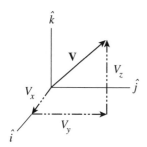

We will come back to discuss vectors in systems other than the Cartesian coordinate system in a later section. We also note that MATLAB can draw vectors, as shown in Appendix A.

■ 5.3 Vector Multiplication

Here we discuss the dot product and the cross product. Both of these operations are very useful in carrying out a description of the physical world. The result of a dot product between two vectors is a scalar. The result of the cross product between two vectors is another vector.

Dot Product

The dot product or scalar product of two vectors, **A** and **B**, is defined as

$$\mathbf{A} \cdot \mathbf{B} = A_x B_x + A_y B_y + A_z B_z = \mathbf{B} \cdot \mathbf{A}. \tag{5.3.1}$$

Based on this, suppose a vector lies along the $+x$ axis, $\mathbf{A} = A_x\hat{i}$, and another vector has two components, one along the $+x$ axis and the other along the $+y$ axis, $\mathbf{B} = B_x\hat{i} + B_y\hat{j}$. By (5.3.1) their dot product is $A_x B_x = AB\cos\theta$, so that one can think of this dot product as the projection of **B** onto **A**. We can, therefore, write in general that the dot product between two vectors is also given by

$$\mathbf{A} \cdot \mathbf{B} = AB\cos\theta, \tag{5.3.2}$$

where the angle θ is the angle between the two vectors, as shown in Figure 5.5.

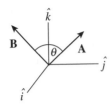

FIGURE 5.5

Equations (5.3.1 and 5.3.2) also define the orthogonality of the Cartesian coordinate unit vectors $\hat{i}$, $\hat{j}$, and $\hat{k}$. The dot product of two unit vectors that are perpendicular to one another is zero; in contrast the dot product between two unit vectors that are parallel to each other is unity, so we can write, $\hat{i} \cdot \hat{i} = 1$, $\hat{i} \cdot \hat{j} = 0$, $\hat{i} \cdot \hat{k} = 0$, $\hat{j} \cdot \hat{i} = 0$, $\hat{j} \cdot \hat{j} = 1$, $\hat{j} \cdot \hat{k} = 0$, $\hat{k} \cdot \hat{i} = 0$, $\hat{k} \cdot \hat{j} = 0$, and $\hat{k} \cdot \hat{k} = 1$. By applying this multiplication rule to the vectors **A** and **B** on the left-hand side of (5.3.2), it can be seen that it is quite consistent with (5.3.1). Both results—Equations (5.3.1) and (5.3.2)—are scalars and are equivalent. They can, therefore, be used to obtain the angle between the two vectors as

$$\theta = \cos^{-1}\left(\frac{\mathbf{A} \cdot \mathbf{B}}{AB}\right). \tag{5.3.3}$$

EXAMPLE 5.2

Although a particle in space is subject to the force $\mathbf{F} = (2\hat{i} - 3\hat{j} + \hat{k})$N, it is seen to be displaced by $\mathbf{r} = (-\hat{i} + 3\hat{j} + 4\hat{k})$ m. (a) What is the work done by the force on the particle? (b) What is the angle between the applied force and the displacement?

Solution

(a) Work is defined by the dot product between the force vector and the displacement vector. We find $W = \mathbf{F} \cdot \mathbf{r} = -2 - 9 + 4 = -7$ J. The force performs seven Joules of negative work.

(b) To obtain the angle, we note that the magnitudes of the force and the displacement are $F = \sqrt{2^2 + 3^2 + 1^2}\,\text{N} = \sqrt{14}\,\text{N}$, $r = \sqrt{26}\,\text{m}$, thus

$$\theta = \cos^{-1}\left(\frac{-7}{\sqrt{14}\sqrt{26}}\right) = 111.52°.$$

Finally, note that MATLAB can perform the dot product of vectors, as shown in Appendix A.

Cross Product

The cross product or vector product of two vectors, **A** and **B**, is defined in such a way that the mathematical operation produces a vector that is perpendicular to both vectors **A** and **B**. Thus we define the vector product as

$$\mathbf{C} = \mathbf{A} \times \mathbf{B} = (A_x\hat{i} + A_y\hat{j} + A_z\hat{k}) \times (B_x\hat{i} + B_y\hat{j} + B_z\hat{k})$$

$$= (A_yB_z - A_zB_y)\hat{i} - (A_xB_z - A_zB_x)\hat{j} + (A_xB_y - A_yB_x)\hat{k} \qquad (5.3.4)$$

$$\equiv \begin{vmatrix} \hat{i} & \hat{j} & \hat{k} \\ A_x & A_y & A_z \\ B_x & B_y & B_z \end{vmatrix} = C_x\hat{i} + C_y\hat{j} + C_z\hat{k},$$

where **C** is the vector that results from the operation, and the last line represents the fact that the cross product can be thought of as the determinant of the given matrix. By convention, this prescription obeys the right-hand rule (RHR), as shown in Figure 5.6.

(a)   (b)

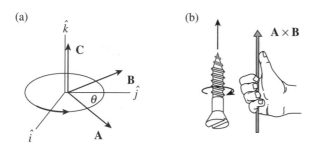

❙ FIGURE 5.6 Rotation according to the right-hand rule (RHR).

For example, if vectors **A** and **B** lie on the *x-y* plane, as shown in Figure 5.6(a), then if **A** is crossed into **B** according to the direction of the planar circle, the result is the vector **C**, which lies in the *z* direction. If the planar circle is thought of as a rotation, then the direction of rotation and the result of the cross product describe a kind of right-handed screw, as shown in Figure 5.6(b).

According to the preceding definition, two vectors that are parallel to each other have zero cross product, and the unit vectors in such system satisfy (5.3.4) as follows:
$$\hat{i} \times \hat{i} = 0, \hat{i} \times \hat{j} = \hat{k}, \hat{i} \times \hat{k} = -\hat{j}, \hat{j} \times \hat{i} = -\hat{k}, \hat{j} \times \hat{j} = 0, \hat{j} \times \hat{k} = \hat{i}, \hat{k} \times \hat{i} = \hat{j},$$
$$\hat{k} \times \hat{j} = -\hat{i}, \text{ and } \hat{k} \times \hat{k} = 0.$$ If we take the dot product of the vector **C** in (5.3.4) with itself, we get

$$
\begin{aligned}
\mathbf{C} \cdot \mathbf{C} &= (A_y B_z - A_z B_y)^2 + (A_x B_z - A_z B_x)^2 + (A_x B_y - A_y B_x)^2 \\
&= A_y^2 B_z^2 + A_z^2 B_y^2 - 2A_y B_z A_z B_y + A_x^2 B_z^2 + A_z^2 B_x^2 - 2A_x B_z A_z B_x \\
&\quad + A_x^2 B_y^2 + A_y^2 B_x^2 - 2A_x B_y A_y B_x \\
&= (A_x^2 + A_y^2 + A_z^2)(B_x^2 + B_y^2 + B_z^2) - (A_x^2 B_x^2 + A_y^2 B_y^2 + A_z^2 B_z^2) - 2(A_x B_x A_y B_y \\
&\quad + A_x B_x A_z B_z + A_y B_y A_z B_z) \\
&= (A_x^2 + A_y^2 + A_z^2)(B_x^2 + B_y^2 + B_z^2) - (A_x B_x + A_y B_y + A_z B_z)^2 \\
&= (\mathbf{A} \cdot \mathbf{A})(\mathbf{B} \cdot \mathbf{B}) - (\mathbf{A} \cdot \mathbf{B})^2 \\
&= A^2 B^2 - A^2 B^2 \cos^2\theta = A^2 B^2 (1 - \cos^2\theta) = A^2 B^2 \sin\theta,
\end{aligned}
$$

which leads to

$$C = |\mathbf{A} \times \mathbf{B}| \equiv AB \sin(\theta) = \sqrt{C_x^2 + C_y^2 + C_z^2}. \tag{5.3.5b}$$

We also notice that built into the definition (5.3.4) is the fact that

$$\mathbf{A} \times \mathbf{B} = -\mathbf{B} \times \mathbf{A}, \tag{5.3.6}$$

indicating that the cross product is not commutative. Finally, notice that Appendix A contains an example of a cross product using MATLAB.

EXAMPLE 5.3

A force of 40 N is applied to the top edge of a 0.25 m radius wheel at 30° *NE* and parallel to its plane, as shown in Figure 5.7. If the wheel is free to rotate about an axis passing through its center and perpendicular to its plane, find the torque experienced by the wheel.

Solution

We assume that the wheel lies on the *x-y* plane and write the force as $\mathbf{F} = (40\cos 30\hat{i} + 40\sin 30\hat{j})$ N $= (34.64\hat{i} + 20\hat{j})$ N. The moment arm can be written as $\mathbf{r} = 0.5\hat{j}$ m. The torque about the center of the wheel is found as, $\boldsymbol{\tau} = \mathbf{r} \times \mathbf{F} = (0.5\hat{j}\,\text{m}) \times (34.64\hat{i} + 20\hat{j})$ N $= -17.32$ N $\cdot$ m$\hat{k}$. Thus the wheel will experience an angular acceleration into the plane of this paper, i.e., in a clockwise sense. Notice that while another way to get the direction of the torque can be seen by applying the RHR, its magnitude can also be obtained by writing $\tau = rF\sin(\theta_{rF}) = (0.5\,\text{m})(\sqrt{34.64^2 + 20^2}\;\text{N})\sin(90 - 30) = 17.32$ N $\cdot$ m, because the angle between the moment arm and the force is $\theta_{rF} = 60°$, as seen in Figure 5.7.

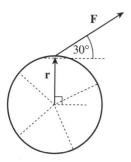

I FIGURE 5.7

EXAMPLE 5.4

The parallelepiped shown in Figure 5.8 contains base sides of magnitudes 3.22 m and 2.41 m at 50° from each other and form a parallelogram. The upright solid is

3.60 m high. Find its volume.

Solution

The volume of any parallelepiped can be obtained by the expression $V = (\mathbf{A} \times \mathbf{B}) \cdot \mathbf{C}$. If we pick $\mathbf{A}$ and $\mathbf{B}$ for the base vectors, then the area of the bottom parallelogram is a vector obtained through $\mathbf{A} \times \mathbf{B} = (3.22\text{ m})(2.41\text{ m})\sin(50°)\hat{k} = 5.94\text{ m}^2\hat{k}$, by the RHR. Letting $\mathbf{C} = 3.6\text{ m}\hat{k}$, the volume obtained after performing the dot product is 21.40 m^3.

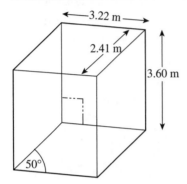

| FIGURE 5.8

■ 5.4 Coordinate System Transformations

A general vector $\mathbf{V}$ can be expressed in terms of components whose magnitudes are determined according to the coordinate system from which the vector is viewed. For example, consider Figure 5.9, where we illustrate two coordinate systems S and S'.

Here, the unit vectors $\hat{i}$ and $\hat{j}$ coincide with the coordinates x and y of the S frame, respectively, and similarly for the S' frame. The S' frame of reference is rotated by an angle θ with respect to the S frame. We have that $\mathbf{V} = V_x\hat{i} + V_y\hat{j}$ in the S frame, and $\mathbf{V} = V'_x\hat{i}' + V'_y\hat{j}'$ in S' frame. The question that arises is, what are the components of $\mathbf{V}$ as seen in one frame in terms of the components of $\mathbf{V}$ as seen in the other frame? A convenient way to answer this question is to notice that in the S frame, $V_x = \hat{i} \cdot \mathbf{V}$ and $V_y = \hat{j} \cdot \mathbf{V}$; however, since $\mathbf{V}$ is the same whether it is seen from either frame of reference, then we also have that

$$V_x = \hat{i} \cdot \mathbf{V} = \hat{i} \cdot (V'_x\hat{i}' + V'_y\hat{j}') = V'_x\hat{i} \cdot \hat{i}' + V'_y\hat{i} \cdot \hat{j}' \tag{5.4.1a}$$

and

$$V_y = \hat{j} \cdot \mathbf{V} = \hat{j} \cdot (V'_x\hat{i}' + V'_y\hat{j}') = V'_x\hat{j} \cdot \hat{i}' + V'_y\hat{j} \cdot \hat{j}'. \tag{5.4.1b}$$

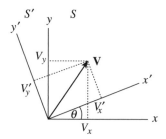

| FIGURE 5.9 The S' coordinate system rotated with respect to the S frame.

We can write this in matrix form as

$$\begin{pmatrix} V_x \\ V_y \end{pmatrix} = \begin{pmatrix} \hat{i} \cdot \hat{i}' & \hat{i} \cdot \hat{j}' \\ \hat{j} \cdot \hat{i}' & \hat{j} \cdot \hat{j}' \end{pmatrix} \begin{pmatrix} V_x' \\ V_y' \end{pmatrix}.$$

(5.4.2)

With the help of Figure 5.9 and the dot product rule applied to these unit vectors, we can obtain the matrix elements, i.e.,

$$\hat{i} \cdot \hat{i}' = \cos\theta_{x,x'} = \cos\theta, \quad \hat{i} \cdot \hat{j}' = \cos\theta_{x,y'} = \cos(90° + \theta) = -\sin\theta$$
$$\hat{j} \cdot \hat{i}' = \cos\theta_{y,x'} = \cos(90° - \theta) = \sin\theta, \quad \hat{j} \cdot \hat{j}' = \cos\theta_{y,y'} = \cos\theta$$
, (5.4.3)

so that (5.4.2) becomes

$$\begin{pmatrix} V_x \\ V_y \end{pmatrix} = \begin{pmatrix} \cos\theta & -\sin\theta \\ \sin\theta & \cos\theta \end{pmatrix} \begin{pmatrix} V_x' \\ V_y' \end{pmatrix}$$

(5.4.4a)

for the S frame components in terms of those of the S' frame. This equation can also be inverted to obtain the S' frame components in terms of those in the S frame

$$\begin{pmatrix} V_x' \\ V_y' \end{pmatrix} = \begin{pmatrix} \cos\theta & \sin\theta \\ -\sin\theta & \cos\theta \end{pmatrix} \begin{pmatrix} V_x \\ V_y \end{pmatrix}.$$

(5.4.4b)

This process can be extended to three-dimensional vectors, as shown in Figure 5.10, to obtain

$$\begin{pmatrix} V_x' \\ V_y' \\ V_z' \end{pmatrix} = \begin{pmatrix} \hat{i}' \cdot \hat{i} & \hat{i}' \cdot \hat{j} & \hat{i}' \cdot \hat{k} \\ \hat{j}' \cdot \hat{i} & \hat{j}' \cdot \hat{j} & \hat{j}' \cdot \hat{k} \\ \hat{k}' \cdot \hat{i} & \hat{k}' \cdot \hat{j} & \hat{k}' \cdot \hat{k} \end{pmatrix} \begin{pmatrix} V_x \\ V_y \\ V_z \end{pmatrix},$$

(5.4.5)

where the unit vectors $\hat{i}, \hat{j}, \hat{k}$ $(\hat{i}', \hat{j}', \hat{k}')$ have corresponding directions x, y, z (x', y', z') in their respective S (S') frame of reference and with the various angles between the axes to be found.

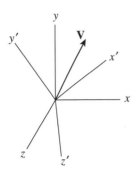

❙ FIGURE 5.10

EXAMPLE 5.5

The magnetic field vector seen in the frame of an observer is given by $\mathbf{B} = (5\hat{i} + 3\hat{j})T$. What is the magnetic field vector seen by a second observer in a frame that is rotated 30° from the first observer's frame?

Solution

The components of the rotated reference frame vectors are given by (5.4.4b) as

$$
\begin{pmatrix} V_x' \\ V_y' \end{pmatrix} = \begin{pmatrix} \cos 30° & \sin 30° \\ -\sin 30° & \cos 30° \end{pmatrix}\begin{pmatrix} 5 \\ 3 \end{pmatrix} = \begin{pmatrix} \sqrt{3}/2 & 1/2 \\ -1/2 & \sqrt{3}/2 \end{pmatrix}\begin{pmatrix} 5 \\ 3 \end{pmatrix} = \begin{pmatrix} \dfrac{5\sqrt{3}+3}{2} \\ \dfrac{-5+3\sqrt{3}}{2} \end{pmatrix},
$$

so that $\mathbf{B} = \left(\dfrac{5\sqrt{3}+3}{2}\hat{i}' + \dfrac{3\sqrt{3}-5}{2}\hat{j}' \right)T.$

■ 5.5   Vector Derivatives, Relative Displacement, and Velocity

Derivative of a Vector

The time derivative of the position vector $\mathbf{r}$ is defined as

$$
\frac{d\mathbf{r}}{dt} = \lim_{\Delta t \to 0} \frac{\mathbf{r}(t + \Delta t) - \mathbf{r}(t)}{\Delta t} \equiv \mathbf{v}, \tag{5.5.1}
$$

where the velocity $\mathbf{v}$ can also be represented as $\dot{\mathbf{r}}$ (pronounced *r-dot*). Thus, writing $\mathbf{r} = x\hat{i} + y\hat{j} + z\hat{k}$, we see that

$$
\mathbf{v} = \dot{\mathbf{r}} = \frac{d}{dt}\left(x\hat{i}\right) + \frac{d}{dt}\left(y\hat{j}\right) + \frac{d}{dt}\left(z\hat{k}\right) = \dot{x}\hat{i} + \dot{y}\hat{j} + \dot{z}\hat{k} + x\dot{\hat{i}} + y\dot{\hat{j}} + z\dot{\hat{k}}. \tag{5.5.2}
$$

However, because the coordinate system is presently not changing, we set $\dot{\hat{i}} = \dot{\hat{j}} = \dot{\hat{k}} = 0$, so that

$$\mathbf{v} = \dot{x}\hat{i} + \dot{y}\hat{j} + \dot{z}\hat{k}, \text{ and } v = \sqrt{\dot{x}^2 + \dot{y}^2 + \dot{z}^2}. \tag{5.5.3}$$

In a similar way the acceleration is written as

$$\mathbf{a} = \dot{\mathbf{v}} = \ddot{\mathbf{r}} = \ddot{x}\hat{i} + \ddot{y}\hat{j} + \ddot{z}\hat{k}, \tag{5.5.4}$$

where the second derivative with respect to time has been represented as $\ddot{\mathbf{r}} \equiv d^2\mathbf{r}/dt^2$, pronounced *r-double-dot*.

Relative Displacement and Velocity

Consider points P_1 and P_2 located at positions $\mathbf{r}_1$ and $\mathbf{r}_2$ from the origin, respectively, as shown in Figures 5.11(a) and (b).

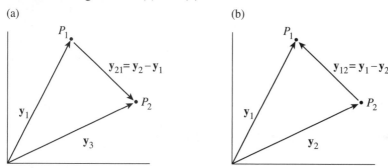

(a)

(b)

I FIGURE 5.11   Relative displacement of (a) P_2 with respect to P_1, and (b) P_1 with respect to P_2.

The relative displacement of P_2 with respect to P_1 is given by

$$\mathbf{r}_{21} = \mathbf{r}_2 - \mathbf{r}_1, \tag{5.5.5a}$$

and, similarly, the relative displacement of P_1 with respect to P_2 is

$$\mathbf{r}_{12} = \mathbf{r}_1 - \mathbf{r}_2, \tag{5.5.5b}$$

where one notices that $\mathbf{r}_{12} = -\mathbf{r}_{21}$, as shown in Figure 5.11. With $\mathbf{v}_1 = \dot{\mathbf{r}}_1$ and $\mathbf{v}_2 = \dot{\mathbf{r}}_2$, as the velocities of points P_1 and P_2, respectively, the relative velocity of P_2 with respect to P_1 is

$$\mathbf{v}_{21} = \dot{\mathbf{r}}_{21} = \dot{\mathbf{r}}_2 - \dot{\mathbf{r}}_1, \tag{5.5.6}$$

and the relative acceleration of P_2 with respect to P_1 is

$$\mathbf{a}_{21} = \dot{\mathbf{v}}_{21} = \ddot{\mathbf{r}}_{21} = \ddot{\mathbf{r}}_2 - \ddot{\mathbf{r}}_1. \tag{5.5.7}$$

EXAMPLE 5.6

Two particles have position vectors given by $\mathbf{r}_1 = (3t\hat{i} - 5t^3\hat{j} + 2\sin(t)\hat{k})$ m and $\mathbf{r}_2 = [(5t^2 - 6)\hat{i} + 7t\hat{j} + 2\cos(t)\hat{k}]$ m. Find (a) the relative displacement, (b) the relative velocity, and (c) the relative acceleration of second particle with respect to the first at the instant when $t = 4$ s.

Solution
The relative displacement is

$$\mathbf{r}_{21}(t) = \mathbf{r}_2 - \mathbf{r}_1 = [(5t^2 - 6 - 3t)\hat{i} + (7t + 5t^3)\hat{j} + 2(\cos(t) - \sin(t))\hat{k}] \text{ m.}$$

The relative velocity is

$$\mathbf{v}_{21}(t) = \dot{\mathbf{r}}_{21}(t) = [(10t - 3)\hat{i} + (7 + 15t^2)\hat{j} + 2(-\sin(t) - \cos(t))\hat{k}] \frac{\text{m}}{\text{s}}.$$

The relative acceleration is

$$\mathbf{a}_{21}(t) = \dot{\mathbf{r}}_{21}(t) = [10\hat{i} + 30t\hat{j} + 2(-\cos(t) + \sin(t))\hat{k}] \frac{\text{m}}{\text{s}}.$$

Thus at $t = 4$ s,

$$\mathbf{r}_{21}(4) = [62\hat{i} + 348\hat{j} + 0.206\hat{k}, \mathbf{v}_{21}(4) = 37\hat{i} + 247\hat{j} + 2.820\hat{k}] \text{ m,}$$

$$\mathbf{v}_{21}(4) = [37\hat{i} + 247\hat{j} + 2.820\hat{k}] \frac{\text{m}}{\text{s}}, \text{ and}$$

$$\mathbf{a}_{21}(4) = [10\hat{i} + 120\hat{j} - 0.206\hat{k}] \frac{\text{m}}{\text{s}^2}.$$

■ 5.6   Gradient, Divergence, and Curl

The gradient of a scalar S, ∇S, divergence, $\nabla \cdot \mathbf{F}$, and curl, $\nabla \times \mathbf{F}$, of a vector $\mathbf{F}$ are differential operators whose role is to enable the proper expression of physical properties that occur in nature. By definition the gradient operates on a scalar function of the spatial coordinates, although the divergence and curl operate on quantities that have vector properties.

The Gradient

The gradient of a scalar function makes use of the *del* operator, which in Cartesian coordinates has the form

$$\nabla = \frac{\partial}{\partial x}\hat{i} + \frac{\partial}{\partial y}\hat{j} + \frac{\partial}{\partial z}\hat{k}. \tag{5.6.1}$$

The gradient, as defined, operates on scalar functions, say $f(x, y, z)$, whose gradient is

$$\nabla f \equiv \frac{\partial f}{\partial x}\hat{i} + \frac{\partial f}{\partial y}\hat{j} + \frac{\partial f}{\partial z}\hat{k}, \tag{5.6.2}$$

and the result is a vector.

EXAMPLE 5.7

Given the function $f = xe^{-\mathbf{r}\cdot\mathbf{r}}$, (a) obtain its gradient as a function of x, y, and z, and (b) evaluate the function, the gradient, and the magnitude of the gradient at the point $P = (-1, 1.5, 0.5)$. (c) Generate suitable plots for the function and the gradient.

....

Solution

(a) Since $\mathbf{r} \cdot \mathbf{r} = r^2 = x^2 + y^2 + z^2$, then

$$\nabla f = \hat{i}\frac{\partial}{\partial x}xe^{-(x^2+y^2+z^2)} + \hat{j}\frac{\partial}{\partial y}xe^{-(x^2+y^2+z^2)} + \hat{k}\frac{\partial}{\partial z}xe^{-(x^2+y^2+z^2)}$$

$$= [(1 - 2x^2)\hat{i} - 2xy\hat{j} - 2xz\hat{k}]e^{-r^2}.$$

(b) The function value at P is $f(-1, 1.5, 0.5) = -e^{-1-2.25-0.25} = -0.030$, and $\nabla f(-1, 1.5, 0.5) = -0.030\hat{i} + 0.090\hat{j} + 0.030\hat{k}$, and the magnitude is $|\nabla f|(-1, 1.5, 0.5) = \sqrt{0.030^2 + 0.090^2 + 0.030^2} = 0.100$.

(c) Figures 5.12 (a, b) contain plots of the function versus x, y with $z = 0.4$.

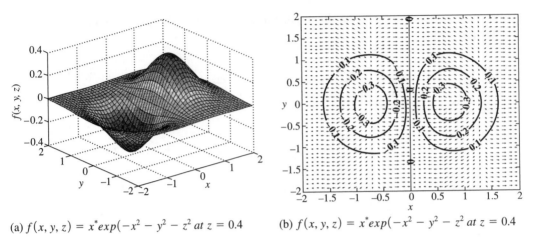

(a) $f(x, y, z) = x^* exp(-x^2 - y^2 - z^2$ at $z = 0.4$ (b) $f(x, y, z) = x^* exp(-x^2 - y^2 - z^2$ at $z = 0.4$

I FIGURE 5.12   (a) Surface plot versus x, y. (b) Gradient and contour plot from Example 5.7.

In Figure 5.12(a) the function changes sign along x as expected. The gradient is shown along with a contour plot for visualization purposes in Figure 5.12(b). The contour plot is useful in visualizing where the values of the function are highest, because the contour density is higher in those regions. The contours themselves are lines of constant function values, as shown. Also, notice that the arrows, indicating the gradient directions, cross perpendicularly to the contour lines. This is an important property of gradients.

The figures were created by the MATLAB script gradient_ex.m that follows. The three-dimensional plots are carried out at a set value of $z = 0.4$. We notice that from the analytic expression in part (a) of the problem, at the point $(x, y, z) = (0, 0, 0.4)$, the x component is the only part of the gradient that survives; that is, $\nabla f(0, 0, 0.4) \rightarrow (\nabla f)_x|_{x=0, y=0, z=0.4} = (1 - 2(0)^2)e^{-(0^2 + 0^2 + 0.4^2)} = 0.85$, and is the reason why the gradient's arrow points in the x direction in Figure 5.12(b) at this point.

If after running the script in MATLAB we interact with the command line, we can find the array index values that correspond to the $(x, y, z) = (0, 0, 0.4)$ point. For example, typing (0+xmax)/xs+1 and then pressing Enter gives the array value 21 corresponding for $x = 0$, and similarly for $y = 0$. For $z = 0.4$, the array index value is obtained by typing (0.4+zmax)/zs+1, which gives the array index value 25. Then, interactively typing dfx(21,21,25), which is the numerical x component of the gradient, in the command line, we get 0.8437, which is reasonably close to the value we obtained analytically. For more accurate results, one needs to decrease the size of the gradient grid variables as seen in the script listed below, albeit at the cost of computational time.

SCRIPT

```
%gradient_ex.m
%This script plots and evaluates the function f=x*exp(-(x^2+y^2+z^2)) and its
% gradient
%The plots are done versus x,y at a certain z value
warning off; %suppress unwanted warnings by plotter if needed
clear;
vmax=2.0;
xmax=vmax; ymax=vmax; zmax=vmax; % x,y,z limits
vs=0.1;
xs=vs; ys=vs; zs=vs; % step size
N=2*vmax/vs; % number of points to be plotted is NxNxN
dv=0.1;
dx=dv; dy=dv; dz=dv; % used in the gradient
m=round(N/2+5); zm=-zmax+(m-1)*zs; % value of z at which we plot f(x,y,z)
[x,y,z]=meshgrid(-xmax:xs:xmax,-ymax:ys:ymax,-zmax:zs:zmax);
f=x.*exp(-(x.^2+y.^2+z.^2)); % the desired function
[dfx,dfy,dfz] = gradient(f,dx,dy,dz);%gradient of f(x,y,z)
%mesh(x(:,:,m),y(:,:,m),f(:,:,m)) % can do a mesh if desired
surf(x(:,:,m),y(:,:,m),f(:,:,m)) % surface plot
xlabel('x','FontSize',14)
ylabel('y','FontSize',14)
zlabel('f(x,y,z)','FontSize',14)
str=cat(2,'f(x,y,z)=x*exp(-x^2-y^2-z^2) at ','z=',num2str(zm,3));
title(str,'FontSize',14)
figure
%contour(x(:,:,m),y(:,:,m),f(:,:,m),20)%contour plot 20 line case
[C,h] = contour(x(:,:,m),y(:,:,m),f(:,:,m));% generate contour plot
clabel(C,h,'FontSize',12)% add contour labels
hold on
quiver(x(:,:,m),y(:,:,m),dfx(:,:,m),dfy(:,:,m))%draw the gradient as arrows at x
xlabel('x','FontSize',14)
ylabel('y','FontSize',14)
str=cat(2,'f(x,y,z)=x*exp(-x^2-y^2-z^2) at ','z=',num2str(zm,3));
title(str,'FontSize',14)
```

In the preceding example we alluded to the fact that the gradient lies perpendicularly to the constant function value contours. It turns out that if, in general, we consider a surface of constant value—that is, $\phi(x, y, z) = c$, as shown in Figure 5.13—and letting $\mathbf{r}(t)$ be a vector to the point $P(\mathbf{r})$, which lies on the surface, then from the figure we see that $\Delta\mathbf{r} = \mathbf{r}(t + \Delta t) - \mathbf{r}(t)$, where $\mathbf{r}(t + \Delta t)$ is the position vector Δt later. In the limit as $\Delta t \to 0, \Delta\mathbf{r} \to d\mathbf{r}$, which lies tangent to the surface of constant value.

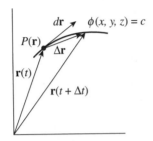

| FIGURE 5.13   Point P on a constant value surface.

Since $\mathbf{r} = x\hat{i} + y\hat{j} + z\hat{k}$, then we also have that $d\mathbf{r} = dx\hat{i} + dy\hat{j} + dz\hat{k}$. Furthermore, notice that for our constant surface,

$$d\phi = \frac{\partial\phi}{\partial x}dx + \frac{\partial\phi}{\partial y}dy + \frac{\partial\phi}{\partial z}dz = \left(\frac{\partial\phi}{\partial x}\hat{i} + \frac{\partial\phi}{\partial y}\hat{j} + \frac{\partial\phi}{\partial z}\hat{k}\right) \cdot (dx\hat{i} + dy\hat{j} + dz\hat{k}) = \nabla\phi \cdot d\mathbf{r} = 0.$$

Since, as we have shown, $d\mathbf{r}$ lies tangent to the surface, then the gradient $\nabla\phi$ must always be a vector that lies perpendicular to the surface of constant value. Its magnitude gives the rate of maximum change of ϕ along that direction.

Finally, the gradient is useful in obtaining the force from the potential energy function. In a previous chapter, in one dimension, we had that if the potential is given, $f(x) = -\partial V(x)/\partial x$, or if the force is known, $V(x) = -\int_x f(x')\,dx'$. In general, the force is a vector and the potential is a scalar; thus, with the use of the gradient, we can generalize this and write in three dimensions

$$\mathbf{F}(\mathbf{r}) \equiv -\nabla V, \tag{5.6.3a}$$

for the force, if the potential is known, and

$$V(r) = -\int_r \mathbf{F}(\mathbf{r}) \cdot d\mathbf{r}, \tag{5.6.3b}$$

for the potential if the force is given instead.

The Divergence

The divergence of a vector **F**, in Cartesian coordinates, has the form

$$\nabla \cdot \mathbf{F} = \left(\frac{\partial}{\partial x}\hat{i} + \frac{\partial}{\partial y}\hat{j} + \frac{\partial}{\partial z}\hat{k}\right) \cdot (F_x\hat{i} + F_y\hat{j} + F_z\hat{k}) = \frac{\partial F_x}{\partial x} + \frac{\partial F_y}{\partial y} + \frac{\partial F_z}{\partial z}. \quad (5.6.4)$$

As can be seen, while the divergence is defined for vectors, its result is a scalar quantity.

EXAMPLE 5.8

Given the vector function **r**, (a) obtain its divergence. (b) Evaluate the function and its divergence at the point $P = (0.5, -0.5, 0.6)$. (c) Generate suitable plots for the function and the divergence.

Solution

(a) $\nabla \cdot \mathbf{r} = \left(\frac{\partial}{\partial x}\hat{i} + \frac{\partial}{\partial y}\hat{j} + \frac{\partial}{\partial z}\hat{k}\right) \cdot (x\hat{i} + y\hat{j} + z\hat{k}) = 3.$

(b) $\mathbf{r}|_{(0.5, -0.5, 0.6)} = 0.5\hat{i} - 0.5\hat{j} + 0.6\hat{k}.$ Because the divergence of this vector is a constant, the result is equal to 3.

(c) The plots generated with the MATLAB script divergence.m are shown in Figures 5.14(a) and (b).

In Figure 5.14(a) the length of the arrows is a measure of the magnitude of **r**, and the arrows' directions tend to spread apart more as one moves farther from the origin. This is in fact a property of a vector with a nonzero divergence. The divergence measures the amount of flow out of a region of space. Figure 5.14b indicates the constant nature of the particular result for $\nabla \cdot \mathbf{r}$ and so happens to yield a constant value. The script divergence.m that follows is capable of performing similar plots for different vector functions. One needs to uncomment the needed example as well as to comment the unneeded lines by removing the "%" symbols or placing them, respectively. Other examples can be performed as desired by just adding new lines of code.

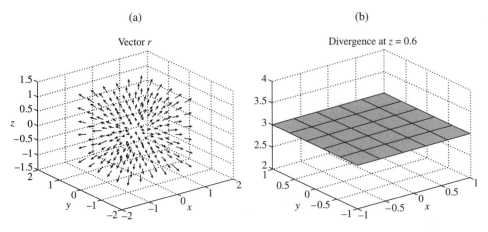

FIGURE 5.14   (a) Vector function and (b) divergence from Example 5.8.

SCRIPT

```
%divergence_ex.m
%This script plots and evaluates the function r=x i + y j + z k
%and its divergence. Other functions are possible as well.
warning off; %suppress unwanted warnings by plotter, if needed
clear; vmin=-1; vmax=1;
xmax=vmax; ymax=vmax; zmax=vmax; % x,y,z upper limits
xmin=vmin; ymin=vmin; zmin=vmin;
vs=0.4; xs=vs; ys=vs; zs=vs; % step size
N=(vmax-vmin)/vs; % number of points to be plotted is NxNxN
m=round(N); zm=zmin+(m-1)*zs; % z at which the vector function is plotted
%[x,y,z]=meshgrid(|minusns|xmax:xs:xmax,-ymax:ys:ymax,-zmax:zs:zmax);
[x,y,z]=meshgrid(xmin:xs:xmax,ymin:ys:ymax,zmin:zs:zmax);
% The desired vector is F = fx i + fy j + fz k. Uncomment appropriate example
% Example 1
fx=x; fy=y; fz=z;str1='Vector r';
% Example 2
%fx=x.*z; fy=-y.^2; fz=2*x.^2.*y;str1='Vector (x*z,-y^2,2*x^2*y)';
% Example 3
% fx=x.^3; fy=y; fz=z;str1='Vector (x^3,y,z)';
% Example 4
% fx=x.^2.*y; fy=y.^2.*z; fz=z.^2.*x;str1='Vector (x^2*y,y^2*z,z^2*x)';
%quiver3(x,y,z,fx,fy,fz,2); %uses scaling
quiver3(x,y,z,fx,fy,fz); %draws arrows in three dimensions
title (str1,'FontSize',14);
xlabel('x','FontSize',14); ylabel('y','FontSize',14);
zlabel('z','FontSize',14)
```

```
div = divergence(x,y,z,fx,fy,fz);%divergence of F
figure
surf(x(:,:,m),y(:,:,m),div(:,:,m)) % surface plot
str2=cat(2,'Divergence at ','z=',num2str(zm,3));
title(str2,'FontSize',14)
xlabel('x','FontSize',14), ylabel('y','FontSize',14)
```

Gauss' Theorem or Divergence Theorem

Associated with the divergence properties of a vector, the divergence theorem, or Gauss' theorem, states that *the integral of the divergence of a vector field over a volume equals the integral of the vector field across the surface bounding the volume.* Stated mathematically, we have

$$\iiint_V \boldsymbol{\nabla} \cdot \mathbf{F}\,dV = \iint_S \mathbf{F} \cdot d\mathbf{S}, \qquad (5.6.5)$$

where the surface element can be written as $d\mathbf{S} = \hat{n}\,dS$, with $\hat{n}$ a unit vector normal to the surface. A simple way to view the theorem is as follows: Anything that flows out of a volume must cross the area that encloses the volume. This theorem is very useful, as demonstrated in Figures 5.15(a) and (b).

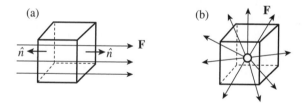

I FIGURE 5.15   Examples of a function with (a) zero divergence and (b) nonzero divergence.

In Figure 5.15(a), we see that because $\mathbf{F}$ is constant and crosses perpendicularly to the faces of the cube as shown in the figure, the same amount enters and leaves the cube, so that $\int \hat{n} \cdot \mathbf{F}\,dA = -FA + FA = 0$, where A is the area of each face of the cube whose unit vectors on each side are opposite in direction. There is no contribution from faces that are perpendicular to the force. By the divergence theorem this implies that $\int \boldsymbol{\nabla} \cdot \mathbf{F}\,dV = 0$, or $\boldsymbol{\nabla} \cdot \mathbf{F} = 0$. Because the function does not have a divergence, there is no net flow within the cube region. The situation is different in

Figure 5.15(b). If we define the flux $\Phi \equiv \iint\limits_{S} \mathbf{F} \cdot d\mathbf{S}$; that is, the flux is the number of field lines passing through an area, then one can see that a flux actually emanates from the inside of the cube. Thus we can conclude that $\int \mathbf{\nabla} \cdot \mathbf{F} dV = \int \hat{n} \cdot \mathbf{F} dA \neq 0$; therefore, we must have that $\mathbf{\nabla} \cdot \mathbf{F} \neq 0$.

EXAMPLE 5.9

Apply the divergence theorem to the vector function of Example 5.8; that is, $\mathbf{F} = \mathbf{r}$, over a sphere of radius R.

Solution

Referring to Figure 5.16, and using the divergence result from Example 5.8, we see that the left side of (5.6.5) becomes

$$\iiint\limits_{V} \mathbf{\nabla} \cdot \mathbf{r} dV = 3 \int\limits_{radius=R} dx\,dy\,dz = 3V_{sphere} = 3\left(\frac{4}{3}\pi R^3\right) = 4\pi R^3.$$

The right side of (5.6.5) can be carried out if we write $d\mathbf{S} = \hat{n} dS$, and $\mathbf{r}$ points in the same direction, so $\mathbf{r} = R\hat{n}$, whose magnitude is constant over the spherical surface S. Because the integral is to be carried out over the spherical surface S bounding the volume $V(radius = R)$, we have

$$\iint\limits_{S} \mathbf{r} \cdot d\mathbf{S} = R \int\limits_{radius=R} \hat{n} \cdot \hat{n} dS = RS_{sphere} = R(4\pi R^2) = 4\pi R^3$$

and Gauss' theorem is satisfied.

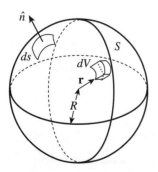

| FIGURE 5.16   Surface and volume elements in spherical coordinates.

One common application of the divergence occurs in electricity, where the divergence of the electric field has a nonzero value when charges are present in a certain region of space, so that $\nabla \cdot \mathbf{E} = \rho/\varepsilon_0$, where $\mathbf{E}$ is the electric field, ρ is the volume charge density, and ε_0 is the permittivity of free space. If the divergence of the electric field, for example, has a nonzero value, then there must be a charge density present that is responsible for the existence of a divergenceless electric field. We can thus think of the charges (associated with ρ) distributed within the surface as the source of that field.

The Curl

The curl of a vector $\mathbf{F}$, in Cartesian coordinates, has the form

$$\nabla \times \mathbf{F} \equiv \begin{vmatrix} \hat{i} & \hat{j} & \hat{k} \\ \dfrac{\partial}{\partial x} & \dfrac{\partial}{\partial y} & \dfrac{\partial}{\partial z} \\ F_x & F_y & F_z \end{vmatrix} = \left(\frac{\partial F_z}{\partial y} - \frac{\partial F_y}{\partial z} \right)\hat{i} - \left(\frac{\partial F_z}{\partial x} - \frac{\partial F_x}{\partial z} \right)\hat{j} + \left(\frac{\partial F_y}{\partial x} - \frac{\partial F_x}{\partial y} \right)\hat{k}. \quad (5.6.6)$$

As with the divergence, the curl is defined for vector-valued functions of the spatial coordinates, and the result is another vector-valued function.

EXAMPLE 5.10

Given the vector function $\mathbf{F} = -y\hat{i} + x\hat{j}$, (a) obtain its curl, and (b) plot the function and its curl.

Solution

(a) Using (5.6.6) we find $\nabla \times \mathbf{F} = 0\hat{i} - 0\hat{j} + (\partial x/\partial x - \partial(-y)/\partial y)\hat{k} = 2\hat{k}$.
(b) Figures 5.17(a), (b), and (c) contain useful plots.

These were generated using the MATLAB script curl_ex.m that follows. Figure 5.17(a) is the function itself where the arrows show the direction and their lengths are proportional to the magnitude. Figure 5.17(b) shows a two-dimensional plot of $\mathbf{F}(x, y, z = z_m)$, where $z_m = -0.1$ and is indicated by the arrows. The two-dimensional surface plot shown in the figure is the curl average velocity, $\omega_z \equiv \dfrac{1}{2}\left(\dfrac{\partial F_y}{\partial x} - \dfrac{\partial F_x}{\partial y} \right)$, which is generated by the MATLAB script automatically

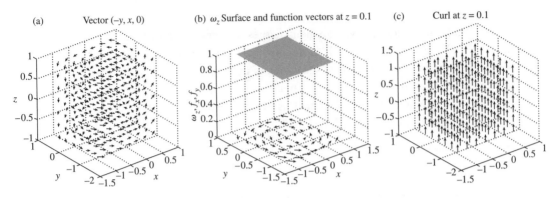

FIGURE 5.17   (a) Vector function, (b) two-dimensional plot, and (c) the curl from Example 5.10.

while performing the curl operation. As can be seen, this is one half the value of the z-component of the curl. It is useful in visualizing rotational properties associated with the function **F**. The function, for example, could describe an electromagnetic field or a velocity field that could have rotational properties. In the present case, because the result of the curl is a constant with a magnitude of two units, the value of ω_z is unity. Other examples included in the script have interesting behaviors worth exploring. To do so, as explained before, comment and uncomment appropriate lines. Figure 5.17(c) shows the actual curl in three dimensions.

The arrows point in the z-direction and are all of the same size, as expected in this case. Furthermore, the curl agrees with the right-hand rule for the given function; that is, we have a rotation in the x-y plane and clockwise as viewed from the bottom, the curl is perpendicular to both, in the z-direction. The script does the plots in the range determined by the parameters vmin and vmax in steps of vs, which can all be changed as desired.

SCRIPT

```
%curl_ex.m
%This script plots and evaluates the function v = -y i + x j
%and its curl. Other functions are possible as well.
warning off; %suppress unwanted warnings by plotter, if needed
clear; vmin=-1; vmax=1;
xmax=vmax; ymax=vmax; zmax=vmax; % x,y,z upper limits
```

```
xmin=vmin; ymin=vmin; zmin=vmin;
vs=0.3; xs=vs; ys=vs; zs=vs; % step size
N=(vmax-vmin)/vs; % number of points to be plotted is NxNxN
m=round(N/2+1); zm=zmin+(m-1)*zs; % z at which the vector function is plotted
%[x,y,z]=meshgrid(|minusns|xmax:xs:xmax,-ymax:ys:ymax,-zmax:zs:zmax);
[x,y,z]=meshgrid(xmin:xs:xmax,ymin:ys:ymax,zmin:zs:zmax);
% The desired vector is F = fx i + fy j + fz k. Uncomment appropriate example
% Example 1
  fx=-y;fy=x;fz=0.*z;str1='Vector (|minusns|y,x,0)';
% Example 2
% fx=x.*z;fy=-y.^2;fz=2*x.^2.*y;str1='Vector (x*z,-y^2,2x^2*y)';
% Example 3
% fx=y.^2;fy=(2*x.*y+z.^2);fz=2*y.*z;str1='Vector (y^2,(2*x*y+z^2),2*y*z)';
% Example 4
% fx=x.*y;fy=y.*z;fz=z.*x;str1='Vector (xy,yz,zx)';
% Example 5
% fx=x.^2.*y;fy=y.^2.*z;fz=z.^2.*x;str1='Vector (x^2*y,y^2*z,z^2*x)';
% quiver3(x,y,z,fx,fy,fz,2); %uses scaling
 quiver3(x,y,z,fx,fy,fz); %draws arrows in three dimensions
 title (str1,'FontSize',14)
 xlabel('x','FontSize',14), ylabel('y','FontSize',14)
 zlabel('z','FontSize',14)
 [curlx,curly,curlz,cav] = curl(x,y,z,fx,fy,fz);%Curl and angular velocity in rad/sec of F
figure
% surf(x(:,:,m),y(:,:,m),cav(:,:,m)) % surface plot of the angular velocity
%===============================
 x2=x(:,:,m); y2=y(:,:,m); fx2=fx(:,:,m); fy2=fy(:,:,m);
 cav2 = curl(x2,y2,fx2,fy2);% curl's average angular velocity in one plane of the volume
 surf(x2,y2,cav2) % surface plot of the angular velocity
 shading interp
 hold on;
 quiver(x2,y2,fx2,fy2);%plots the function vectors at z=zm
 str2=cat(2,'\omega_z surface and function vectors at ','z=',num2str(zm,3));
 str4=cat(2,'\omega_z, f_x, f_y, ');
 title (str2,'FontSize',14)
 xlabel('x','FontSize',14), ylabel('y','FontSize',14)
 zlabel(str4,'FontSize',14)
 hold off
%===============================
figure
quiver3(x,y,z,curlx,curly,curlz); %draws curl's result as arrows in three dimensions
str3=cat(2,'Curl at ','z=',num2str(zm,3));
title (str3,'FontSize',14)
xlabel('x','FontSize',14), ylabel('y','FontSize',14)
zlabel('z','FontSize',14)
% %subplot(2,2,4) % Lines below prove equality to cav2 = curl(x2,y2,fx2,fy2)/2
% figure
```

```
% cav3=0.5*curlz(:,:,m);% same as curl's average angular velocity about the axis
% surf(x2,y2,cav3);%plots the velocity vectors
% shading interp
% hold on;
% quiver(x2,y2,fx2,fy2);%plots the velocity vectors
% hold off
```

Stokes' Theorem or the Fundamental Theorem for Curls

The Fundamental Theorem for Curls, which is also known as Stokes' theorem, states that *the integral of the normal component of the curl of a vector field over a surface equals the integral of the vector along a closed contour bounding the surface.* Stated mathematically, we have

$$\iint_S \mathbf{\nabla} \times \mathbf{F} \cdot d\mathbf{S} = \oint_C \mathbf{F} \cdot d\mathbf{l}, \tag{5.6.7}$$

where the surface area element can be written as $d\mathbf{S} = \hat{n}\,dS$, with $\hat{n}$ a unit vector normal to the surface. The vector field, according to the right-hand rule, determines the line integral's sense of direction. This theorem can be intuitively understood by referring to Figures 5.18(a) and (b).

In Figure 5.18(a) the vector function $\mathbf{F}$ is illustrated. The contributions of the component of its curl, $\mathbf{\nabla} \times \mathbf{F}$, along the surface's $\hat{n}$ direction, are equivalent to a bounding surface contour along a path in whose direction lie all the contributions from the vector function $\mathbf{F}$. The contour is generally drawn around the surface of interest, but it corresponds to the equality shown in the figure. Figure 5.18(b) shows that with the sense of direction of the curl along $d\mathbf{S}$, then, according to the right-hand rule, this helps to determine the direction of the contour path, $d\mathbf{l}$, that bounds the surface S.

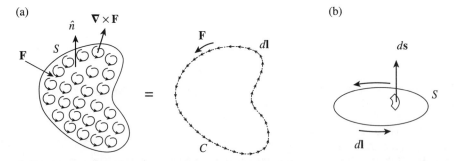

FIGURE 5.18 Pictorial representation of Stokes' theorem.

EXAMPLE 5.11

Apply Stokes' theorem to the vector function of Example 5.10; that is, $\mathbf{F} = -y\hat{i} + x\hat{j}$, over a square of sides a and parallel to the x-y plane.

Solution

Referring to Figure 5.19, and using the curl result from Example 5.10, we see that the left side of (5.6.7) is $\iint_S \boldsymbol{\nabla} \times \mathbf{F} \cdot d\mathbf{S} = 2\int_0^a dx \int_0^a dy = 2a^2$, where we have used $d\mathbf{S} = dx\,dy\hat{k}$ and $\hat{n} = \hat{k}$ as in the figure. The right side of (5.6.7), with $d\mathbf{l} = d\mathbf{l}_1 + d\mathbf{l}_2 + d\mathbf{l}_3 + d\mathbf{l}_4$, and setting $d\mathbf{l}_1 = dx\hat{i}$, $d\mathbf{l}_2 = dy\hat{j}$, $d\mathbf{l}_3 = -dx\hat{i}$, and $d\mathbf{l}_4 = -dy\hat{j}$, becomes

$$\oint_C \mathbf{F} \cdot d\mathbf{l} = \int (-y\hat{i} + x\hat{j}) \cdot (dx\hat{i}|_{y=0} + dy\hat{j}|_{x=a} - dx\hat{i}|_{y=a} - dy\hat{j}|_{x=0})$$

$$= -\int_0^a y|_{y=0}dx + \int_0^a y|_{y=a}dx + \int_0^a x|_{x=a}dy - \int_0^a x|_{x=0}dy = 0 + a^2 + a^2 - 0$$

$$= 2a^2,$$

where, because the directions were taken care of by $d\mathbf{l}$, the limits are on $[0, a]$.

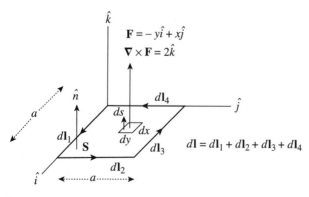

$$\mathbf{F} = -y\hat{i} + x\hat{j}$$
$$\boldsymbol{\nabla} \times \mathbf{F} = 2\hat{k}$$
$$d\mathbf{l} = d\mathbf{l}_1 + d\mathbf{l}_2 + d\mathbf{l}_3 + d\mathbf{l}_4$$

| FIGURE 5.19

◼ Chapter 5 Problems

5.1 Once Jack enters the Surelife Insurance Company where he works, he walks 10 m north to the elevator, which takes him up 100 m to the desired floors, and then he walks 25 m east right to his office. Write a unit vector corresponding to his displacement vector.

5.2 A bug initially performs a flight according to the vector $\mathbf{A} = 5\hat{i} + 2\hat{j} - \hat{k}$, and later according to $\mathbf{B} = 3\hat{i} + 7\hat{j} + 3\hat{k}$. Find the bug's total displacement.

5.3 At a particular time in a beetle's flight path, its position is given by the vector $\mathbf{A} = 5\hat{i} + 2\hat{j} - \hat{k}$ in arbitrary units of distance. Similarly, a second beetle's position is given by $\mathbf{B} = 3\hat{i} + 7\hat{j} + 3\hat{k}$. If the first beetle wanted to move toward the second beetle, how far would it need to go and in what direction?

5.4 A certain force is given by $\mathbf{F} = K\dfrac{x\hat{i} + y\hat{j}}{(x^2 + y^2)^{3/2}}$. Find an expression for the amount of work required to move an object from very far away to a point a distance a from the origin, assuming the motion takes place along the x-axis.

5.5 At a particular time of observation, a 10,000-kg asteroid, located at 0.75 AU from the sun 35° NE from it, is seen to be moving at a speed of 50 km/s in a 160° NE direction. If the motion takes place on a plane, what is the asteroid's angular momentum with respect to the sun, in units of $AU \cdot \text{kg} \cdot \text{km/s}$? Here, assume the sun is fixed and located at the origin of its own coordinate system, and take east and north as the $+x$ and $+y$ directions.

5.6 a. Show that the triple vector product $\mathbf{C} \cdot (\mathbf{A} \times \mathbf{B})$ can be written in the form of a determinant of a 3 × 3 matrix.

b. Show that $\mathbf{A} \cdot (\mathbf{B} \times \mathbf{C}) = \mathbf{B} \cdot (\mathbf{C} \times \mathbf{A}) = \mathbf{C} \cdot (\mathbf{A} \times \mathbf{B})$.

5.7 Two particles are located each at the end of its own spring. The time-dependent position of each particle is given by $\mathbf{r}_1(t) = \hat{i}R\cos(\omega t)$, $\mathbf{r}_2(t) = \hat{j}R\sin(\omega t)$, for the first and second particles, respectively, where $R = 20$ cm and $\omega = 3$ rad/s. Obtain the displacement, velocity, and acceleration of the first particle with respect to the second at $t = 6$ s.

5.8 Run the MATLAB script `gradient_ex.m`. While interacting with the MATLAB command line, show that the script results agree, within a reasonable tolerance, with the analytic result of Example 5.7 for the function, gradient, and magnitude at the point $P = (-1, 1.5, 0.5)$.

5.9 Given the vector function $\mathbf{F} = 5x^2\hat{i} + 3y^2\hat{j}$, verify Gauss' theorem on a cube of sides a with origin at, $x = 0$, $y = 0$, $z = 0$, i.e., a cube corner and oriented according to the standard Cartesian axes.

5.10 A charge q is located in free space. Use Gauss' theorem on a sphere of radius r to show that the magnitude of the electric field from a point charge is given by $\mathbf{E} = \hat{r}kq^2/r^2$, where $k = 1/4\pi\varepsilon_0$.

5.11 Check Stokes' theorem for the vector function $\mathbf{V} = (2xz + 3y^2)\hat{j} + 4yz^2\hat{k}$ on a square surface with sides of value equal to unity in arbitrary distance units. The square is oriented parallel to the y-z Cartesian plane with a corner at the origin.

5.12 Given the function $\mathbf{f} = x^2zy\hat{i}$, verify Stokes' theorem on a face of a parallelepiped oriented in the x-z plane and displaced by $y = 1/2$. The height along the z-direction is 3 units, and the width along the x direction is 2 units.

5.13 Show that the triple vector cross product obeys the relationship $\mathbf{A} \times (\mathbf{B} \times \mathbf{C}) = \mathbf{B}(\mathbf{A} \cdot \mathbf{C}) - \mathbf{C}(\mathbf{A} \cdot \mathbf{B})$, also known as the BAC-CAB rule.

5.14 The three-dimensional Dirac delta function is defined as $\int \delta^3(r)d^3r \equiv 1$, where d^3r is a three-dimensional volume element of a given coordinate system. For example, in Cartesian coordinates

$$\int \delta^3(r)d^3r = \int \delta(x)\delta(y)\delta(z)dxdydz = \int \delta(x)dx \int \delta(y)dy \int \delta(y)dz = 1.$$

Working with spherical coordinates, show that $\nabla \cdot \dfrac{\hat{r}}{r^2} = 4\pi\delta^3(r)$. (*Hint*: Use Gauss' theorem.)

5.15 Newton's universal law of gravitation for the force between masses M and m separated by distance r is given by $\mathbf{F}_g = -\dfrac{GMm}{r^2}\hat{r}$, where G is the universal gravitational constant. Show the gravitational force flux associated with this force, that is, $\Phi = \displaystyle\int \boldsymbol{\nabla} \cdot \mathbf{F}_g dV$, in a spherical geometry, is equal to $-4\pi GMm$. (*Hint:* Use the Dirac delta function.)

6 Motion in Two and Three Dimensions

■ **6.1 Introduction**

It is a wonderful experience to be able to understand and analyze equations of motion in one dimension. Many applications in physics, however, involve more than one dimension. For example, projectile motion is a three-dimensional problem when one considers aiming a projectile at a certain angle while taking into account forces due to air resistance and the projectile's spinning motion. There are problems that can be treated strictly in two dimensions as, for example, the motion of planets around the sun, though it is best to work in polar coordinates in such a case. Other interesting applications involve the motion of charged particles in electromagnetic fields, which in general involve more than one dimension. It is the experience of solving problems in one dimension that enables us to go beyond and study more realistic situations, but it is the need to explain realistic dynamic behavior, which takes place in dimensions other than one, that propels us forward to develop the formulation we need in order understand such motion beyond that of one dimension. In this chapter vector forces that depend on time, displacement, and velocity are considered. Newton's second law of motion is now expressed in vector form

$$\mathbf{F}(t, \mathbf{r}, \dot{\mathbf{r}}) = m\mathbf{a} \tag{6.1.1}$$

and leads to motion that involves more than one dimension. In general, the force and the acceleration are conveniently expressed in component form

$$\mathbf{F}(t, \mathbf{r}, \dot{\mathbf{r}}) = f_x(t, \mathbf{r}, \dot{\mathbf{r}})\hat{i} + f_y(t, \mathbf{r}, \dot{\mathbf{r}})\hat{j} + f_z(t, \mathbf{r}, \dot{\mathbf{r}})\hat{k},$$

$$\mathbf{a} = \ddot{x}\hat{i} + \ddot{y}\hat{j} + \ddot{z}\hat{k}. \tag{6.1.2}$$

With these, (6.1.1) leads to three differential equations

$$m\ddot{x} = f_x(t, x, y, z, \dot{x}, \dot{y}, \dot{z}),$$

$$m\ddot{y} = f_y(t, x, y, z, \dot{x}, \dot{y}, \dot{z}), \text{ and}$$

$$m\ddot{z} = f_z(t, x, y, z, \dot{x}, \dot{y}, \dot{z}). \tag{6.1.3}$$

Although these equations are extremely complicated in their most general form due to the possible coupling of the coordinates, it is convenient to consider special cases of the motion. Such special cases lead to simplifications of the equations, which under particular conditions can be solved numerically or even analytically.

■ 6.2   Uncoupled or Separable Forces

In many physical problems it is possible to write the force in such a way that its component along x depends only on x, and likewise for the other components. In this case (6.1.3) becomes

$$m\ddot{x} = f_x(t, x, \dot{x}), \quad m\ddot{y} = f_y(t, y, \dot{y}), \quad m\ddot{z} = f_z(t, z, \dot{z}). \tag{6.2.1}$$

Because this equation expresses the dynamics of a particle's coordinate independent of other coordinates, we refer to this motion as uncoupled motion. The forces are thus separable and make it possible to solve for each particle's coordinate as a function of time independently of the other coordinates. Here we consider two examples for which these equations apply, i.e., free fall and motion under the acceleration due to gravity in the presence of air resistance.

Projectile Motion Without Air Resistance

The direction of the force due to gravity near Earth's surface will be taken along the negative z direction, with zero initial position and velocity in the y direction so that, in what follows, the motion will be restricted to the zx-plane, as shown in Figure 6.1.

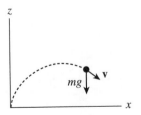

I FIGURE 6.1 Projectile motion described in two dimensions.

Thus, ignoring the y-direction, the rest of the forces in (6.2.1) are $f_x = 0, f_z = -mg$ and the solutions for x and z as a function of time are well known from previous chapters.

$$m\ddot{x} = 0 \quad \Rightarrow \quad x = x_i + v_{ix}t, \quad v_x = v_{ix},$$
$$m\ddot{z} = -mg \quad \Rightarrow \quad z = z_i + v_{iz}t - gt^2/2, \quad v_z = v_{iz} - gt. \tag{6.2.2}$$

These equations can be written in a vector notation as

$$\mathbf{r} = \mathbf{r}_i + \mathbf{v}_i t - \frac{1}{2}gt^2\hat{k}, \quad \mathbf{v} = \mathbf{v}_i - gt\hat{k},$$
$$\mathbf{r}_0 = x_i\hat{i} + z_i\hat{k}, \quad \mathbf{v}_i = v_{ix}\hat{i} + v_{iz}\hat{k}. \tag{6.2.3}$$

We also take $x_i \equiv 0$ and $z_i \equiv 0$, and then it is possible to obtain an equation for z as a function of x. This is accomplished by solving the first of (6.2.2) for time to get $t = x/v_{ix}$. Putting this into the z-component of (6.2.2) we get

$$z = v_{iz}\left(\frac{x}{v_{ix}}\right) - \frac{1}{2}g\left(\frac{x}{v_{ix}}\right)^2, \tag{6.2.4}$$

which can be thought of as a quadratic equation for x with solutions $x/v_{ix} = v_{iz}/g \pm \sqrt{(v_{iz}/g)^2 - 2z/g}$. Here, the initial height and displacement at $t = 0$ can be incorporated by making the replacement $z \to z - z_i$ and $x \to x - x_i$ if desired. After rearranging this, we obtain

$$\left(x - \frac{v_{iz}v_{ix}}{g}\right)^2 = -\frac{2v_{ix}^2}{g}\left(z - \frac{v_{iz}^2}{2g}\right). \tag{6.2.5}$$

Recall that the equation of a parabola with a vertex at (x_0, z_0) and focal point F at $(x_0, z_0 + e)$ is given by the equation

$$(x - x_0)^2 = 4e(z - z_0), \tag{6.2.6a}$$

where e is a number that stands for eccentricity. The parabola is concave down (up) if $e < 0$ ($e > 0$). Figure 6.2 shows a series of parabolas for $e < 0$. The lower values of e have higher curvatures.

The focal points corresponding to each e are shown to move farther away as the curvature decreases. The figure was created with the MATLAB code parabola.m given in the following script.

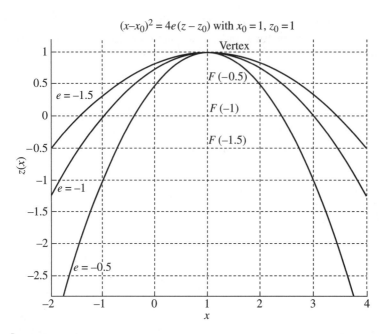

I FIGURE 6.2 Parabolas of different eccentricities.

```
%parabola.m
%plots parabolas (x-x0)^2=4e(z-z0), whose curvatures=1/2e,
%(x0,z0) is the vertex, and (x0,z0+e) is the focal point position
clear;
x0=1; z0=1; %parameters
xmin=-3; xmax=3; xs=0.1; % x-range
x=[xmin+x0:xs:xmax+x0]; % x array
hold on; zt=0; zb=0; % useful in plotting
for e=-0.5:-0.5:-1.5, %vary e
z=z0+(x-x0).^2/4/e;
plot(x,z,x0,z0,'r.',x0,z0+e,'rx');
str1=cat(2,'F (',num2str(e),')');% focal point
str2=cat(2,'e=',num2str(e));
text(x0*(1+0.2),z0*(1+0.1)+e,str1,'FontSize',12,'Color','red');
text(x(4),z(4),str2,'FontSize',10,'Color','red');
if abs(max(z)) > abs(zt); zt=max(z); end
if abs(min(z)) > abs(zb); zb=min(z); end
end
str3=cat(2,'(x-x_0)^2=4e(z-z_0)',' with x_0=',num2str(x0),...
```

```
   ',z_0=',num2str(z0));
%title('Plot of (x-x_0)^2=4e(z-z_0)')
title(str3,'FontSize',10); ylabel('z(x)','FontSize',14)
xlabel('x','FontSize',14); grid on;
axis([xmin+x0 xmax+x0 zb*(1-0.2) zt*(1+0.2)]);
text(x0*(1+0.2),z0*(1+0.1),'Vertex','FontSize',12,'Color','red');
```

Equation (6.2.6a) is equivalent to an equation of the form

$$z = ax^2 + bx + c, \tag{6.2.6b}$$

where

$$a = 1/4e, b = -x_0/2e, \text{ and } c = z_0 + x_0^2/4e. \tag{6.2.6c}$$

From (6.2.6b), since the derivative $dz/dx = 2ax + b$ is a line with slope corresponding to the preceding parabola of curvature $2a$, then through (6.2.6c) we see that the curvature of the parabola in (6.2.6a) is $1/2e$. By comparing (6.2.6a) and (6.2.5) one identifies that free-fall motion in the absence of air resistance is a concave down parabola with vertex (x_{half}, z_{max}), where $x_{half} \equiv x_0$ is the horizontal midpoint of the motion and $z_{max} \equiv z_0$ is the maximum height. These are given by

$$x_{half} = \frac{v_{iz}v_{ix}}{g} = \frac{v_i^2 \sin 2\theta}{2g}, z_{max} = \frac{v_{iz}^2}{2g}, \tag{6.2.7}$$

where we have used Figure 6.3 to obtain the components of the velocity and where, from Appendix C, the identity $\sin 2\theta = 2\cos\theta\sin\theta$ was also used. The parabolic curvature is given by $1/2e = -g/v_{ix}^2$, which is affected not only by g and v_i but also by the initial launching angle, θ.

| FIGURE 6.3

From (6.2.5) we also see that there are two values of x where $z = 0$. One is the origin, $x = 0$, and the other is when x reaches its maximum value $R \equiv x_{max}$; that is, the range. This is given by

$$R = \frac{2v_{iz}v_{ix}}{g} = \frac{v_i^2 \sin 2\theta}{g}. \qquad (6.2.8)$$

This can also be seen to be the case from the symmetry properties of the parabola about the midpoint so that $R = 2x_{half}$. These projectile motion properties are shown in Figure 6.4.

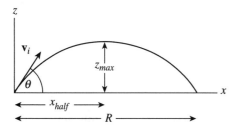

| FIGURE 6.4   The maximum height and the range associated with projectile motion in two dimensions.

EXAMPLE 6.1

(a) Given that a projectile is launched at an initial speed v_i, at what angle should the projectile be aimed in order to achieve the maximum range? (b) Obtain plots of $z(x)$ for various initial angles using suitable values for v_i with $z_i = 0 = x_i$, $g = 9.8$ m/s^2, and explain your observations.

....

Solution

(a) The maximum range is obtained by setting $dR/d\theta = 2v_i^2 \cos 2\theta/g = 0$, which gives $\theta = \pi/4$.

(b) Figure 6.5 shows the plots of $z(x)$ for various initial angles with $v_i = 5.0$ m/s and $g = 9.8$ m/s^2. We notice that not only does the curvature change with the angle, as expected, but also it is confirmed that the angle of $45°$ gives the largest range.

Finally, one notices that complementary launching angles $(\theta, \pi/2 - \theta)$ have the same range. Table 6.1 shows some examples of the term $\sin 2\theta = 2\cos\theta \sin\theta =$

$2\cos\theta\cos(\pi/2 - \theta)$ of (6.2.8) responsible for this behavior. Here this term is evaluated for four launching angles (1–4) and their complements (4–7). The $\sin 2\theta$ term has the same value for the launching angle tests 1, 7; 2, 6; 3, 5; test 4 is its own complement. This symmetry property of projectile motion occurs only in the absence of air resistance. Such property is expected to disappear when the effects due to friction are included.

Although complementary angles share the same range, projectiles launched at higher angles have a longer time of flight. The MATLAB script projectile.m used to calculate the curves is shown in Figure 6.5.

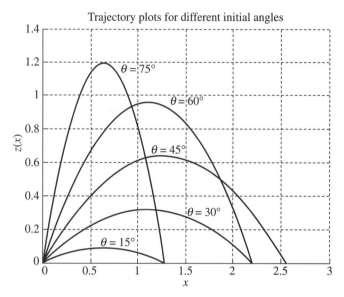

FIGURE 6.5 Example 6.1 projectile motion versus distance for various initial launching angles.

Table 6.1   Projectile Motion Symmetry Between the Angle and Its Complement.

Test launching Angle: θ	Complement: $\pi/2 - \theta$	$\sin 2\theta = \cos(\theta)\cos(\pi/2)$
(1) 0	$\pi/2$	$\cos(0)\cos(\pi/2) = 0$
(2) $\pi/12 = \pi/2 - 5\pi/12$	$5\pi/12$	$\cos(\pi/12)\cos(5\pi/12) = 0.25$
(3) $\pi/6 = \pi/2 - \pi/3$	$\pi/3$	$\cos(\pi/6)\cos(\pi/3) = 0.433$
(4) $\pi/4 = \pi/2 - \pi/4$	$\pi/4$	$\cos(\pi/4)\cos(\pi/4) = 0.5$
(5) $\pi/3 = \pi/2 - \pi/6$	$\pi/6$	$\cos(\pi/3)\cos(\pi/6) = 0.433$
(6) $5\pi/12 = \pi/2 - \pi/12$	$\pi/12$	$\cos(5\pi/12)\cos(\pi/12) = 0.25$
(7) $\pi/2$	0	$\cos(\pi/2)\cos(0) = 0$

SCRIPT

```
%projectile.m - without air resistance
%plots the projectile trajectory z(x)=viz*x/vix-0.5*g*(x/v0x)^2
clear;
vi=5; zi=0; xi=0; g=9.8;%initial: speed, positions, gravity
xmin=xi;xs=0.005; thi=15; ths=15; thm=75;% x, angle ranges
hold on;
for th=thi:ths:thm;
    vix=vi*cos(th*2*pi/360);% initial speed components
    viz=vi*sin(th*2*pi/360);
    xmax=2*vix*viz/g + xi;
    N=(xmax-xmin)/xs+1; M=round(N/2);
    x=[xmin:xs:xmax];
    z=zi+viz*(x-xi)/vix-0.5*g*((x-xi)/vix).^2;
    plot(x,z);
str1=cat(2,'\theta =',num2str(th),'^o');
text(x(M+25),z(M+25)+0.025,str1,'FontSize',10,'Color','red');
end
str3=cat(2,'Trajectory plots for different initial angle');
title(str3,'FontSize',14); ylabel('z(x)','FontSize',14)
xlabel('x','FontSize',14); grid on;
```

Projectile Motion With Air Resistance

Turning our attention to the air resistance problem, recall from Sections 1.3 and 2.5 that there are various models for air resistance. Here we consider the case of $\mathbf{R} = -C\mathbf{v}$. The vector equation of motion takes the form

$$\mathbf{F}(t, \mathbf{r}, \dot{\mathbf{r}}) = -mg\hat{k} - C\mathbf{v} = m\ddot{\mathbf{r}}, \tag{6.2.9a}$$

or

$$m\ddot{x} = -Cv_x, \quad m\ddot{y} = -Cv_y, \quad m\ddot{z} = -mg - Cv_z. \tag{6.2.9b}$$

From Section 2.5, the solution to the third part of (6.2.9b) is already known

$$z(t) = -\frac{mg}{C}t - \frac{m}{C}\left(\frac{mg}{C} + v_{iz}\right)\left(e^{-\frac{C}{m}t} - 1\right), \quad v_z = \left(\frac{mg}{C} + v_{iz}\right)e^{-\frac{C}{m}t} - \frac{mg}{C}, \tag{6.2.10}$$

where we have taken $z_i \equiv 0$. The first part of (6.2.9b) can be solved

$$m\frac{dv_x}{dt} = -Cv_x, \text{ or } \int_{v_{ix}}^{v_x} \frac{dv_x}{v_x} = -\frac{C}{m}\int_0^t dt \Rightarrow v_x = v_{ix}e^{-\frac{C}{m}t}, \tag{6.2.11a}$$

from which, by applying $x = x_i + \int_0^t v_x dt$, one finds

$$x = x_i + \frac{m}{C} v_{ix} \left(1 - e^{-\frac{C}{m} t} \right). \tag{6.2.11b}$$

The second of (6.2.9b) has a similar solution; however, we again let $v_{iy} = 0$, and $y_i = 0$, so that the motion takes place in the xz-plane. In a similar way to the previous section, we seek an equation for $z(x)$. Solving (6.2.11b) for t and rearranging, find

$$t = \frac{m}{C} \ln \left[\frac{m v_{ix}}{m v_{ix} - C(x - x_i)} \right], \tag{6.2.12}$$

which when substituted into the first part of (6.2.10), obtain after rearranging

$$z = \left(\frac{mg}{C} + v_{iz} \right) \frac{(x - x_i)}{v_{ix}} + \frac{m^2 g}{C^2} \ln \left(1 - \frac{C(x - x_i)}{m v_{ix}} \right). \tag{6.2.13}$$

This equation is to be contrasted to the result without air resistance, that is, equation (6.2.4). To see how air resistance affects the shape of $z(x)$, it is useful to consider the behavior of (6.2.13) for small C. We first take $x_i = 0$ and use the expansion $\ln(1 + \varepsilon) = \varepsilon - \varepsilon^2/2 + \varepsilon^3/3 - \cdots$, so that expanding the natural log term to third order, we have

$$z \approx \left(\frac{mg}{C} + v_{iz} \right) \frac{x}{v_{ix}} + \frac{m^2 g}{C^2} \left(-\frac{Cx}{m v_{ix}} - \frac{C^2 x^2}{2 m^2 v_{ix}^2} - \frac{C^3 x^3}{3 m^3 v_{ix}^3} \right),$$

which simplifies to

$$z \approx \frac{v_{iz}}{v_{ix}} x - \frac{1}{2} \frac{g}{v_{ix}^2} x^2 - \frac{1}{3} \frac{Cg}{m v_{ix}^3} x^3. \tag{6.2.14}$$

The first two terms of this equation correspond to the parabola associated with projectile motion without air resistance. The third term, which disappears in the limit as $C \to 0$, represents the first term in the deviation from parabolic motion and the associated symmetry in the motion discussed earlier.

Strictly speaking, from (6.2.11b), we see that as $t \to \infty$, the maximum horizontal position reached by a projectile in the present case is $x \to x_i + \frac{m}{C} v_{ix}$. This is true provided the object does not reach the ground first. However, if the projectile reaches the ground before too long, it is of interest to know when it does so and how far it gets. We can answer both of these questions if the first of (6.2.10) is set to zero

$$z\big(t\big) = 0 = \left[-\frac{mg}{C}t - \frac{m}{C}\left(\frac{m^2}{C} + v_{iz}\right)\left(e^{-\frac{C}{m}t} - 1\right)\right]_{t-t_{max}} \tag{6.2.15}$$

from which we can obtain the time at which the projectile touches the ground as

$$t_{max} = \left(\frac{m}{C} + \frac{v_{iz}}{g}\right)\left(1 - e^{-\frac{C}{m}t_{max}}\right). \tag{6.2.16}$$

The range of the projectile when under the influence of air resistance is obtained by substituting this result into (6.2.11b), to get (taking $x_i = 0$)

$$x_{max} = \frac{m}{C}v_{ix}\left(1 - e^{-\frac{C}{m}t_{max}}\right). \tag{6.2.17}$$

Going back to (6.2.16) notice that it is a transcendental equation for t_{max}, which does not have an analytic solution and must therefore be solved numerically. However, it is possible to obtain an approximate result for low air resistance. Using the expansion $e^{-\varepsilon} \sim 1 - \varepsilon + \varepsilon^2/2$ to second order in ε, then for small C in (6.2.16) we obtain to second order in time

$$t_{max} \approx t_{max} - \frac{C}{2\,m}t_{max}^2 + \frac{v_{iz}C}{mg}t_{max} - \frac{v_{iz}C^2}{2\,m^2g}t_{max}^2,$$

from which, after simplifying and solving for t_{max}, we get

$$t_{max} \sim \frac{2v_{iz}}{g}\frac{1}{\left(1 + \dfrac{Cv_{iz}}{mg}\right)} \sim \frac{2v_{iz}}{g}\left(1 - \frac{Cv_{iz}}{mg}\right), \tag{6.2.18}$$

where we've used the Taylor expansion $1/(1 + \varepsilon) \sim 1 - \varepsilon$ to first order in ε. In the limit as $C \to 0$, we have $t_{max} \sim 2v_{iz}/g$, which is the expected result for the projectile's time of flight in the absence of air resistance. To obtain an approximate result for x_{max}

in (6.2.17), we can expand the exponential term to first order in C and simplify to get $x_{max} \sim v_{ix} t_{max}$. Finally, substituting (6.2.18) into this we find

$$x_{max} \sim \frac{2 v_{ix} v_{iz}}{g} - \frac{2 C v_{ix} v_{iz}^2}{m g^2}, \tag{6.2.19}$$

the first order in C. In the limit as $C \to 0$, we see that $x_{max} \to 2 v_{ix} v_{iz}/g$, which is the standard result for the range of a projectile in the absence of air resistance equation (6.2.8). We can thus conclude that air resistance shortens both the range and the round trip time of a projectile. Further, the maximum range of a projectile in the presence of air resistance is no longer 45°.

EXAMPLE 6.2

Referring to the z direction as north and the x direction as east, a projectile is launched near Earth's surface with an initial speed of 25 m/s at 45° NE. (a) Find the time of flight for the projectile under two conditions: with and without air resistance. (b) Find the range of the projectile under both conditions. (c) Obtain a plot where both motions are compared and explain your observations. For the air resistance case, assume a drag coefficient value of 0.5 kg/s and a mass $m = 1$ kg.

....

Solution

(a) For the case without air resistance, the time of flight is $t_R = \dfrac{2 v_{iz}}{g} = \dfrac{2(25\,\text{m/s})\sin(45)}{9.8\,\text{m/s}^2} =$

3.60 s. For when there is air resistance, one could apply (6.2.18) to estimate the value; however, the drag coefficient is not small enough to use the approximation (see Problem 6.4). Instead we need to solve Equation (6.2.16). To do this we could employ the brute

force iteration process, $t_{i+1} = \left(\dfrac{m}{C} + \dfrac{v_{iz}}{g} \right)\left(1 - e^{-\frac{C}{m} t_j} \right)$, where i is a counter, t_{i+1} is a new guess, with t_i a previous guess. Since i starts at zero, we must provide the initial guess $t_{i=0}$. However, this process is slow. A popular method, which works pretty well here, is the Newton-Raphson technique we used in Section 4.15, according to which $t_{i+1} = t_i - f(t_i)/f'(t_i)$, where here $f(t_i)$ is the function whose zero we seek. In the present example this is

$$f(t_i) = \left(\frac{m}{C} + \frac{v_{iz}}{g} \right) \left(1 - e^{-\frac{C}{m}t_j} \right) - t_i,$$

and its derivative is

$$f'(t_i) = \left(\frac{m}{C} + \frac{v_{iz}}{g} \right) \frac{C}{m} e^{-\frac{C}{m}t_j} - 1,$$

As the process works, after a few iterations (say, N) the function $f(t_{i=N})$ should be near zero with some tolerance ($\sim 1*10^{-4}$) and $t_{i=N}$ should be a reasonable approximation to the true time at which the projectile reaches the ground. The MATLAB script projectile2.m incorporates this to obtain a value of $t_R = 2.92$ s.

(b) The range without air resistance is $R = v_i^2 \sin(2\theta)/g = (25)^2 \sin(90)/9.8 = 63.78$ m. When there is air resistance we have, using the given value of t_R for air resistance,

$$x_{max} = (m/C)v_{ix}[1 - \exp(-Ct_{max}/m)]$$

$$= (1/0.5)(25 \cdot \cos 45)[1 - \exp(-0.5 \cdot 2.92/1)]$$

$$= 27.15 \text{ m}.$$

(c) Figure 6.6 shows the trajectory plots of the motion with and without air resistance. We notice that with air resistance the range is much shorter than the free case, and the trajectory is no longer symmetric. Furthermore, the script projectile2.m can be used to investigate how the maximum range actually increases as the angle gets lower than 45° (see Problem 6.5).

Trajectory Plots without Air Resistance at $\theta = 45°$, $c = 0.5$ kg/s, $vi = 25$ m/s

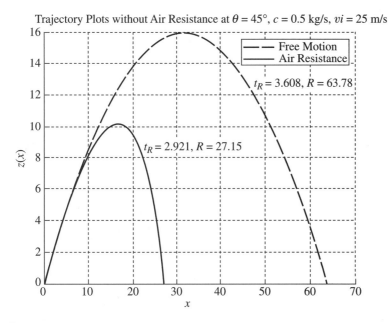

FIGURE 6.6 Two-dimensional projectile motion comparing free fall and motion with air resistance from Example 6.2.

The MATLAB script `projectile2.m` can be used to study various aspects of projectile motion under air resistance by changing the input parameters.

SCRIPT

```
%projectile2.m - motion with air resistance
%plots the projectile trajectory:
%z(x)=(m*g/c+viz)*(x-xi)/vix+m^2*g*log(1-c*(x-xi)/m/vix)/c^2
clear;
vi=25;zi=0;xi=0;th=45;g=9.8;%initial: speed, positions, angle, gravity
c=0.5; m=1; % Drag Coeff and mass
vix=vi*cos(th*2*pi/360); viz=vi*sin(th*2*pi/360);%initial vi components
xmin=xi;xs=0.005;xmax=2*vix*viz/g;% Range without air resistance
x=[xmin:xs:xmax];
z=zi+viz*(x-xi)/vix-0.5*g*((x-xi)/vix).^2;% z(x) without air resistance
tmax=2*viz/g; %no air resistance round trip time for initial guess
%fprintf('Free Motion: t_R %f sec, R= %f m\n',tmax,xmax)
stra=cat(2,'t_R=',num2str(tmax,4),',R=',num2str(xmax,4));
%======= Newton Raphson Method for time to land
Nt=0; tdiff=1;
while Nt < 5 & tdiff > 0.0001, % iteration conditions
    Nt=Nt+1;                    % counter
    f=(m/c+viz/g)*(1-exp(-c*tmax/m))-tmax;% function whose zero we seek
```

```
    fp=c*(m/c+viz/g)*exp(-c*tmax/m)/m-1;  % derivative
    tdiff=f/fp;tmax=tmax-tdiff;% tmax becomes the new guess
end
% =============================
xmaxf=m*vix*(1-exp(-c*tmax/m))/c;%Range with air resistance
xf=[xmin:xs:xmaxf];
zf=(m*g/c+viz)*(xf-xi)/vix+m^2*g*log(1-c*(xf-xi)/m/vix)/c^2;
plot(x,z,'b-',xf,zf,'r');
str0=cat(2,'\theta =',num2str(th),'^o',',c=',...
    num2str(c),'kg/s', ',vi=',num2str(vi),'m/s');
str1=cat(2,'Trajectory plots w/o Air Resistance',' at ',str0);
title(str1,'FontSize',13); ylabel('z(x)','FontSize',14)
xlabel('x','FontSize',14); grid on;
h=legend('Free Motion','Air Resistance',1); set(h,'FontSize',12)
%fprintf('With Air Resistance: t_R %f sec, R= %f m\n',tmax,xmaxf)
strb=cat(2,'t_R=',num2str(tmax,4),',R=',num2str(xmaxf,4));
Ma=round(4*length(x)/5); Mb=round(4*length(xf)/5);
text(x(Ma)*(1+0.01),z(Ma),stra,'FontSize',9,'Color','blue');
text(xf(Mb)*(1+0.02),zf(Mb),strb,'FontSize',9,'Color','red');
```

■ 6.3 Potential Energy Function

The potential energy function and its relation to a vector force were briefly described in Chapter 5. We now look into this relationship in more detail, especially in the context of conservative and nonconservative forces.

We recall that a force is the negative of the gradient of the potential energy function,

$$\mathbf{F}(\mathbf{r}) \equiv - \boldsymbol{\nabla} V(\mathbf{r}), \tag{6.3.1}$$

and vice versa. The potential energy function is equal to the negative of the integral of the force dotted with the path

$$V(\mathbf{r}) = - \int_{\mathbf{r}_0}^{r} \mathbf{F}(\mathbf{r}') \cdot d\mathbf{r}'. \tag{6.3.2}$$

The question arises as to whether Equations (6.3.1 and 6.3.2) are always true. It turns out that we must make a distinction between conservative and nonconservative forces. As we will see, for conservative forces, Equations (6.3.1 and 6.3.2) are always true. Equation (6.3.2) is a "line integral," and we will consider two important cases of its application.

Case 1: Conservative Force

We refer to $\mathbf{F}(\mathbf{r})$ in (6.3.2) as a conservative force whenever the line integral is independent of the path. To illustrate this point, suppose $\mathbf{F}(\mathbf{r}) = A_0 x \hat{i}$. Let's find the potential associated with this force as we move from the origin to point P on the square of sides ℓ shown in Figure 6.7.

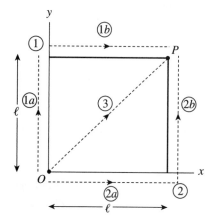

I FIGURE 6.7 Different paths taken in a line integral on a square.

While there are an infinite number of paths to choose from to go from point O to point P, only three are shown in the illustration. Each of the paths 1 and 2 has sub-paths labeled $1a$, $1b$ and $2a$, $2b$, respectively. If we choose path 1, the line integral (6.3.2) is written as

$$V_1 = - \int_{\substack{path\ 1a, \\ y = \ell,\, dx = 0}} \mathbf{F} \cdot d\mathbf{r} \quad \int_{\substack{path\ 1b, \\ y = \ell,\, dy = 0}} \mathbf{F} \cdot d\mathbf{r}$$

$$= -A_0 \int_{x=0,\,y=0}^{x=0,\,y=\ell} 0\hat{i} \cdot \hat{j}\, dy - A_0 \int_{x=0,\,y=\ell}^{x=\ell,\,y=\ell} x\hat{i} \cdot \hat{i}\, dx - A_0 \left.\frac{x^2}{2}\right|_{x=0}^{x=\ell} = -A_0 \frac{\ell^2}{2}, \tag{6.3.3}$$

where $d\mathbf{r} = dx\hat{i} + dy\hat{j}$ has been used. Path 2 can be worked out in a similar way (Problem 6.6) to obtain the same answer; thus we find that $V_2 = V_1$. Path 3 is obtained as follows:

$$V_3 = -\int_{path\ 3} \mathbf{F} \cdot d\mathbf{r} = -A_0 \int_{x=0,\,y=0}^{x=\ell,\,y=\ell} x\hat{i} \cdot \left(\hat{i}\, dx + \hat{j}\, dy\right) = -A_0 \left.\frac{x^2}{2}\right|_{x=0}^{x=\ell} = -A_0 \frac{\ell^2}{2}. \tag{6.3.4}$$

Regardless of what path we take to perform the line integral from point O to point P on the above square of Figure 6.7, the potential works out to be the same. This is an example of a conservative force. Whenever the potential is obtained from (6.3.2), where $\mathbf{F}(\mathbf{r}')$ is a conservative force, we say that the integral is independent of the path chosen. In fact, in all such cases the integral depends only on the end points. For example, if in the present case of $\mathbf{F}(\mathbf{r}) = Ax\hat{i}$, we perform the integral (6.3.2) strictly without having a particular path in mind, the result works out to be that of (6.3.4) where, after the integration has been carried out, the result is evaluated at the end points. That is,

$$V = -\int \mathbf{F} \cdot d\mathbf{r} = -A_0 \int_{x=0,\,y=0}^{x=\ell,\,y=\ell} x\hat{i} \cdot (\hat{i}\,dx + \hat{j}\,dy) = -A_0 \frac{x^2}{2}\Bigg|_{x=0}^{x=\ell} = -A_0 \frac{\ell^2}{2}, \quad (6.3.5)$$

where no reference was made to a particular path, yet the answer is the same as before.

Case 2: Nonconservative Force

Consider next an example for which $\mathbf{F}(\mathbf{r}) = Axy\hat{i} + By\hat{j}$. Again, we wish to obtain the potential energy function on the square of Figure 6.7 in going from point O to point P. Choosing the first path we write

$$V_1 = \int_{\substack{path\ 1a,\\x=0,\,dx=0}} \mathbf{F} \cdot d\mathbf{r} - \int_{\substack{path\ 1b,\\y=\ell,\,dy=o}} \mathbf{F} \cdot d\mathbf{r}$$

$$= -B \int_{x=0,\,y=0}^{x=0,\,y=\ell} \left(y\hat{j}\right) \cdot \hat{j}\,dy - A \int_{x=0,\,y=\ell}^{x=\ell,\,y=\ell} xy\hat{i} \cdot \hat{i}\,dx \qquad (6.3.6)$$

$$= -B \frac{y^2}{2}\Bigg|_{y=0}^{y=\ell} - Ay\frac{x^2}{2}\Bigg|_{\substack{x=0\\y=\ell}}^{\substack{x=\ell\\y=\ell}} = -\frac{\ell^2}{2}\left(B + A\ell\right).$$

We next perform the line integral using path 2. We have

$$V_2 = -\int_{\substack{path\ 2a,\\y=0,\,dy=0}} \mathbf{F} \cdot d\mathbf{r} - \int_{\substack{path\ 2b,\\x=\ell,\,dx=0}} \mathbf{F} \cdot d\mathbf{r}$$

$$= -\int_{x=0,\,y=0}^{x=\ell,\,y=0} \left(0\right) \cdot \hat{i}\,dx = \int_{x=\ell,\,y=0}^{x=\ell,\,y=\ell} \left(Axy\,\hat{i} + By\,\hat{j}\right) \cdot \hat{j}\,dy = -\frac{By^2}{2}\Bigg|_{y=0}^{y=\ell} = -\frac{B\ell^2}{2}, \qquad (6.3.7)$$

which is already different from the result from path 1 in (6.3.6). This indicates that the line integral depends on the path chosen; therefore, there is not a unique potential energy function for the given force. Thus, whenever the potential energy obtained from (6.3.2) depends on the path taken, then $\mathbf{F}(\mathbf{r})$ is not conservative; in such event, $V(\mathbf{r})$ is not uniquely defined.

We refer to potential energy associated with conservative forces as mechanical energy. As such, mechanical energy can be defined if and only if the work done in moving a body from an initial point $\mathbf{r}_i$ to a final point $\mathbf{r}_f$ is such that

$$W = -V_{fi} = \int_{\mathbf{r}_i}^{\mathbf{r}_f} \mathbf{F} \cdot d\mathbf{r}, \qquad (6.3.8)$$

is path independent, i.e., it depends only on the end points, $\mathbf{r}_i$ and $\mathbf{r}_f$. This can only be if the force $\mathbf{F}(\mathbf{r})$ is conservative. Another important question arises: How can we determine whether a given force is conservative or not? Certainly, performing line integrals for many different paths is no guarantee that a force can surely be found to be conservative, because there are an infinite number of paths. Fortunately, this question can be answered quite definitively through the application of Stokes' theorem discussed in Chapter 5. Recall that according to this theorem

$$\iint_S \mathbf{\nabla} \times \mathbf{F} \cdot d\mathbf{S} = \oint_C \mathbf{F} \cdot d\mathbf{l}, \qquad (6.3.9)$$

so that if a force is conservative, then

$$\oint_C \mathbf{F} \cdot d\mathbf{r} = \int_A^B \mathbf{F} \cdot d\mathbf{r} + \int_B^A \mathbf{F} \cdot d\mathbf{r} = -V_{BA} - V_{AB} = W_{BA} + W_{AB} = 0,$$

or $W_{AB} = -W_{BA}$; that is, the work done in going from point A to point B is the negative of that done on returning to the initial point A. As shown in Figure 6.8, the force is conservative and the path taken does not matter.

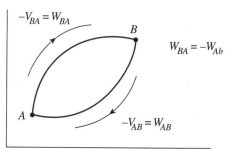

FIGURE 6.8 The initial and final points of a closed line integral are equal. The net work done by a conservative force is zero.

The integral depends only on the end points. Because after the round trip, the end point is the same as the beginning point, then the line integral about a closed path is zero. From (6.3.9) it follows that, for a conservative force

$$\iint_S \nabla \times \mathbf{F}_{\text{conservative}} \cdot d\mathbf{S} = 0 \quad \Rightarrow \quad \nabla \times \mathbf{F}_{\text{conservative}} = 0 \tag{6.3.10}$$

everywhere. This provides a recipe to test whether a force is conservative or not.

EXAMPLE 6.3

Apply the conservative force test criteria (6.3.10) to the forces $\mathbf{F}_1 = A_0 x \hat{i}$, and $\mathbf{F}_2 = Axy\hat{i} + By\hat{j}$.

....

Solution

$$\nabla \times \mathbf{F}_1 = \begin{vmatrix} \hat{i} & \hat{j} & \hat{k} \\ \dfrac{\partial}{\partial x} & \dfrac{\partial}{\partial y} & \dfrac{\partial}{\partial z} \\ A_0 x & 0 & 0 \end{vmatrix} = \hat{i}(0) - \hat{j}\left(-\dfrac{\partial}{\partial z}A_0 x\right) + \hat{k}\left(-\dfrac{\partial}{\partial y}A_0 x\right) = 0, \tag{6.3.11a}$$

which shows that $\mathbf{F}_1$ is a conservative force and therefore has a potential function associated with it, as shown in case 1. Repeating this process for $\mathbf{F}_2$, we get

$$\nabla \times \mathbf{F}_2 = \begin{vmatrix} \hat{i} & \hat{j} & \hat{k} \\ \dfrac{\partial}{\partial x} & \dfrac{\partial}{\partial y} & \dfrac{\partial}{\partial z} \\ Axy & By & 0 \end{vmatrix} = \hat{i}\left(-\dfrac{\partial}{\partial z}By\right) - \hat{j}\left(-\dfrac{\partial}{\partial z}Axy\right) + \hat{k}\left(\dfrac{\partial}{\partial x}By - \dfrac{\partial}{\partial y}Axy\right) = -Ax\hat{k},$$

$$\tag{6.3.11b}$$

which shows the nonconservative character of this force, as discussed in case 2.

We can conclude by saying that, for a given force $\mathbf{F}$, if $\nabla \times \mathbf{F} = 0$, then there is a potential energy function for which $\mathbf{F}(\mathbf{r}) = -\nabla V(\mathbf{r})$ and $V(\mathbf{r}) = -\displaystyle\int_{\mathbf{r}_i}^{r} \mathbf{F}(\mathbf{r}) \cdot d\mathbf{r}$.

These equations have a dual purpose. If the potential energy function is known, then the gradient of the potential allows us to obtain the force components as $F_x = -\partial V/\partial x$, $F_y = -\partial V/\partial y$, $F_z = -\partial V/\partial z$. If, however, the force is known instead, then the potential can be found from the given force, as shown in Example 6.4.

EXAMPLE 6.4

Given the force $\mathbf{F} = ax\hat{i} + by\hat{j} + cz\hat{k}$, obtain its related potential energy function.

Solution

First, notice that the curl of this vector function is zero. Using the components of the force we can write

$$V(x, y, z) = -\int \mathbf{F} \cdot d\mathbf{r} = -\int (ax\,dx + by\,dy + cz\,dz) = -\frac{1}{2}(ax^2 + by^2 + cz^2) + D$$

$$(6.3.12)$$

where D is a constant of integration; i.e., $V(0, 0, 0) = D$.

■ 6.4 Two-Dimensional Motion of a Charged Particle in an Electromagnetic Field

A charged particle, q, in the presence of an electric field, $\mathbf{E}$, experiences an electric force given by

$$\mathbf{F}_e = q\mathbf{E}(\mathbf{r}), \tag{6.4.1}$$

as shown in Figure 6.9(a). A charge moving with velocity $\mathbf{v}$ in the presence of a magnetic field, $\mathbf{B}$, experiences a magnetic force

$$\mathbf{F}_m(\mathbf{r}) = q\mathbf{v} \times \mathbf{B}, \tag{6.4.2}$$

as illustrated in Figure 6.9(b).

If both fields are present and the charge is moving, then by the principle of superposition, we can add the forces to get the so-called Lorentz force

$$\mathbf{F} = \mathbf{F}_e + \mathbf{F}_m = q(\mathbf{E}(\mathbf{r}) + \mathbf{v} \times \mathbf{B}). \tag{6.4.3}$$

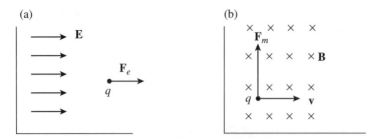

FIGURE 6.9 Charged particle in (a) an electric field and (b) a magnetic field.

At this point, it is useful to review some important aspects of electromagnetic theory. For an electrostatic field $\mathbf{E}$ with conservative property, $\boldsymbol{\nabla} \times \mathbf{E} = 0$ or $\boldsymbol{\nabla} \times \mathbf{F}_e = 0$. This means that there is an *electrostatic potential energy* such that

$$\mathbf{F}_e = -\boldsymbol{\nabla} V_e(\mathbf{r}) \text{ and } V_e(\mathbf{r}) = -\int \mathbf{F}_e \cdot d\mathbf{r}, \tag{6.4.4}$$

or equivalently an *electrostatic potential* ($\varphi = V_e/q$)

$$\mathbf{E} = -\boldsymbol{\nabla}\varphi(\mathbf{r}) \text{ and } \varphi(\mathbf{r}) = -\int \mathbf{E} \cdot d\mathbf{r}. \tag{6.4.5}$$

From (6.4.2) we see that the magnetic force is always perpendicular to the velocity of the charged particle. Then, since $d\mathbf{r} \propto \mathbf{v}$, we see that

$$\int \mathbf{F}_m \cdot d\mathbf{r} = 0; \tag{6.4.6}$$

thus the magnetic force does no work. Next we work with some examples of the resulting motion of the charged particle in the presence of electric and magnetic fields.

EXAMPLE 6.5

Obtain the time-dependent position and velocity of a charged particle q with mass m in the presence of an electric field $\mathbf{E} = E_0\hat{k}$.

....

Solution

The electric force on the charged particle is $\mathbf{F} = qE_0\hat{k}$, which implies a constant acceleration, $\mathbf{a} = qE_0\hat{k}/m$. The position and velocity of the particle as a function of time can be written as in the case of projectile motion by replacing the gravitational acceleration with the current one. We obtain

$$x = x_0 + v_{0x}t, \quad v_x = v_{0x}, \quad y = y_0 + v_{0y}t, \quad v_y = v_{0y}.$$
$$z = z_{0z} + v_{0z}t + \frac{q}{2m}E_0t^2, \quad v_z = v_{0z} + \frac{q}{m}E_0t$$

(6.4.7)

EXAMPLE 6.6

Obtain the time-dependent position and velocity of a moving charged particle q with mass m in the presence of a magnetic field $\mathbf{B} = B_0\hat{k}$.

....

Solution

The magnetic force on the charged particle is

$$\mathbf{F} = q\mathbf{v} \times \mathbf{B} = qB_0(v_x\hat{i} + v_y\hat{j} + v_z\hat{k}) \times \hat{k} = qB_0(\hat{i}v_y - \hat{j}v_x) = m\mathbf{a} = ma_x\hat{i} + ma_y\hat{j}.$$

(6.4.8)

From here, we need to separate the x, y, and z motions as

$$a_x = \frac{qB_0}{m}v_y, \quad a_y = -\frac{qB_0}{m}v_x, \quad a_z = 0,$$

(6.4.9a)

or

$$\ddot{x} = \frac{qB_0}{m}\dot{y}, \quad \ddot{y} = -\frac{qB_0}{m}\dot{x}, \quad \ddot{z} = 0.$$

(6.4.9b)

We notice that the differential equations represent a coupled system of equations. The x and y motions depend on each other. Because there is no acceleration along the z direction, we let $z_0 = 0$ and $v_{0z} = 0$. The motion takes place on the xy-plane. A convenient way to solve this system of equations is to use complex variables. Let $r = x + iy$, where $x \equiv \text{Re}(r)$, and $y \equiv \text{Im}(r)$ for real and imaginary parts of r, respectively, and similarly for the derivatives. From these and (6.4.9), we see that

$$\ddot{r} = \frac{qB_0}{m}(\dot{y} - i\dot{x}) = -i\frac{qB_0}{m}(i\dot{y} + \dot{x}) = -i\frac{qB_0}{m}\dot{r},$$

(6.4.10a)

from which follows a solvable differential equation for $\dot{r}$

$$\frac{d\dot{r}}{dt} = -i\frac{qB_0}{m}\dot{r} \quad \Rightarrow \int \frac{d\dot{r}}{\dot{r}} = -i\frac{qB_0}{m}\int dt \quad \Rightarrow \dot{r} = A_1 e^{-i\frac{qB_0}{m}t}.$$

(6.4.10b)

Integrating once more with respect to time, we get

$$r = \int \dot{r} \, dt = A_1 \int e^{-i\frac{qB_0}{m}t} \, dt = A_2 \, e^{-i\frac{qB_0}{m}t} + A_3, \tag{6.4.11}$$

where we have defined $A_2 \equiv -mA_1/iqB_0$. Since A_2 and A_3 are unknown, for convenience we let them take the forms $A_2 = ce^{-i\varphi}$, $A_3 = a + ib$, where φ, a, and b are constants that depend on the initial conditions. With these definitions and the use of Euler's formula $e^{i\theta} = \cos(\theta) + i\sin(\theta)$, we obtain from (6.4.11)

$$r = c\left[\cos\left(\frac{qB_0}{m}t + \varphi\right) - i\sin\left(\frac{qB_0}{m}t + \varphi\right) \right] + a + ib, \tag{6.4.12}$$

from which, by separating the real and imaginary parts, we find

$$x(t) = \mathrm{Re}(r) = c\cos(\omega_c t + \varphi) + a, \quad y(t) = \mathrm{Im}(r) = -c\sin(\omega_c t + \varphi) + b, \tag{6.4.13}$$

where we have used the cyclotron frequency $\omega_c \equiv qB_0/m$ or period $\tau = 2\pi/\omega_c = 2\pi m/(qB_0)$. The frequency of the particle's oscillation turns out to be independent of the particle's speed. Two more aspects of this result are worth noting. One is that (6.4.13) can be written as

$$(x - a)^2 + (y - b)^2 = c^2, \tag{6.4.14}$$

which indicates that the motion is a circle of radius $R = c$ centered at $(x = a, y = b)$. Another is that, from (6.4.13), we can obtain the radius of the circle as follows,

$$\dot{x}^2 + \dot{y}^2 = R^2\omega_c^2 = v^2 \quad \Rightarrow \quad R = v/\omega_c = mv/qB. \tag{6.4.15}$$

If in the last part of (6.4.15) we multiply both sides by v and rearrange the result, this also means that the particle, while performing circular motion, is subject to a centripetal force $mv^2/R = qvB$, whose source in the magnetic field and whose direction is toward the center of the circle, as shown in Figure 6.10.

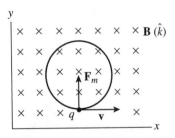

I FIGURE 6.10   The charged particle from Example 6.6 moving in a magnetic field develops circular motion.

In the figure, the magnetic field lies along the positive z-axis ($\hat{k} = \hat{i} \times \hat{j}$), and the motion takes place according to the right-hand rule. The force points toward the center of the circle. In this example, the initial velocity along the z direction was chosen to be zero. However, if this had not been the case, the circle would develop into a helix along the z coordinate. The circular shape would still be visible in the xy-plane. Finally, the components of the velocity of the particle as a function of time are obtained from the time rate of change of (6.4.13), i.e.,

$$\dot{x}(t) = -c\omega_c \sin(\omega_c t + \varphi), \quad \dot{y}(t) = -c\omega_c \cos(\omega_c t + \varphi). \qquad (6.4.16)$$

EXAMPLE 6.7

Obtain the time-dependent position and velocity of a moving charged particle q with mass m in the presence of a magnetic field $\mathbf{B} = B_0 \hat{k}$ and an electric field $\mathbf{E} = E_0 \hat{i}$. Obtain a plot of the resulting motion as a function of time with appropriate values of the electric and magnetic fields.

....

Solution

In addition to the force of (6.4.8), we now include the electric field to get

$$\mathbf{F} = qE_0\hat{i} + q\mathbf{v} \times \mathbf{B} = q(E_0 + B_0 v_y)\hat{i} - qB_0 v_x \hat{j} = m(\ddot{x}\hat{i} + \ddot{y}\hat{j} + \ddot{z}\hat{k}), \qquad (6.4.17a)$$

from which, separating components, we have

$$\ddot{x} = \frac{q}{m}E_0 + \frac{q}{m}B_0\dot{y}, \quad \ddot{y} = -\frac{qB_0}{m}\dot{x}, \quad \ddot{z} = 0. \qquad (6.4.17b)$$

Similar to Example 6.6, we use the complex variable $r = x + iy$, and we see that, except for the first constant term due to the electric field, the equations are similar to (6.4.9b). The constant simply displaces the real part, so that we add qE_0/m to our previous equation (6.4.10a) for $\ddot{r}$ to get

$$\ddot{r} = \frac{qE_0}{m} - i\frac{qB_0}{m}\dot{r}. \qquad (6.4.18)$$

This is a differential equation for $\dot{r}(t)$ of the form

$$\frac{d\dot{r}}{dt} = -i\frac{qB_0}{m}\left(\frac{iE_0}{B_0} + \dot{r}\right) \Rightarrow \int \frac{d\dot{r}}{\dfrac{iE_0}{B_0} + \dot{r}} = -i\frac{qB_0}{m}\int_0^t dt, \qquad (6.4.19)$$

whose integration can be carried out to obtain

$$\dot{r} = A_1 e^{-i\frac{qB_0}{m}t} - \frac{iE_0}{B_0}, \tag{6.4.20}$$

where A_1 is a constant of integration. Performing the integral once more, we have

$$r = \int_0^t \dot{r} \, dt = A_2 e^{-i\frac{qB_0}{m}t} - \frac{iE_0}{B_0}t + A_3, \tag{6.4.21}$$

where we have again used the definition of A_2 from Example 6.6. Once more, we redefine the constants as $A_2 = Ae^{-i\varphi}$, and $A_3 = a + ib$, followed by the use of Euler's formula and a separation of real and imaginary parts, as before, to get

$$x(t) = A\cos(\omega_c t + \varphi) + a, \quad y(t) = -A\sin(\omega_c t + \varphi) - \frac{E_0}{B_0}t + b. \tag{6.4.22}$$

Notice that if in these equations we let $E_0 \to 0$, the result is identical to that of (6.4.13). Finally, the velocity components are given by the time derivative of (6.4.22) to get

$$\dot{x}(t) = -\omega_c A\sin(\omega_c t + \varphi), \quad \dot{y}(t) = -\omega_c A\cos(\omega_c t + \varphi) - \frac{E_0}{B_0}. \tag{6.4.23}$$

We can get an understanding of the resulting motion if we rewrite (6.4.22) as

$$(x - a)^2 + \left(y + \frac{E_0}{B_0}t - b\right)^2 = A^2, \tag{6.4.24}$$

which corresponds to a circle with radius $R = A$ centered at $(x = a, y = b - E_0 t/B_0)$; that is, the center of the circle moves along the $-y$ direction with speed E_0/B_0. The shape of the trajectory is known as a cycloid. The higher the electric field is, the faster the circle moves. In order to produce a plot, we need to obtain the constants A, a, b, and φ, based on the initial conditions. If at time $t = 0$, $x(0) = x_0$, $y(0) = y_0$, $\dot{x}(0) = \dot{x}_0$, and $\dot{y}(0) = \dot{y}_0$, then we can solve for these constants from (6.4.22, 23) to get

$$A = \pm\frac{\sqrt{\dot{x}_0^2 + (\dot{y}_0 + E_0/B_0)^2}}{\omega_c}, \quad \varphi = \tan^{-1}\left(\frac{\dot{x}_0}{\dot{y}_0 + E_0/B_0}\right) \tag{6.4.25a}$$

and

$$a = x_0 - A\cos(\varphi), \qquad b = y_0 + A\sin(\varphi). \tag{6.4.25b}$$

The plot of this motion for a proton, created using the MATLAB script `cycloid2d.m`, is illustrated in Figure 6.11. In the figure, the magnetic field points out of the plane of the drawing, and the electric field is in the positive x direction. The particle moves in the negative y direction, but oscillates in the x direction. By varying the strength of the fields or the initial speeds, the particular shape of the loops can be changed. The `cycloid2d.m` script for a proton in an electric field of 5.5×10^{-8} V/m and a magnetic field of 3.13×10^{-8} T follows.

Motion of a charge in $E(+x$, or east), and $B(+z$, out of view) fields

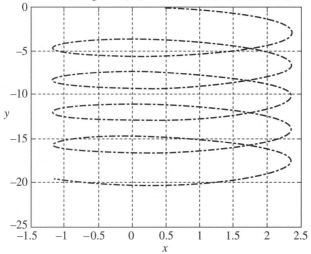

FIGURE 6.11   The charged particle from Example 6.7 moving in the presence of electric and magnetic fields. The motion developed has the shape of a cycloid.

SCRIPT

```
%cycloid2d.m - plots the motion of a charged particle in the presence of
%electric (+x direction) and magnetic (+z direction) fields.
clear;
tmax=10;
q=1.6e-19; m=1.67e-27; %proton charge(Coulombs), proton mass (kg)
B0=3.13e-8; E0=5.5e-8; %B and E fields in Tesla & volts/meter
```

```
v0x=5; v0y=0.0; %initial x,y speeds in m/s
x0=0.0; y0=0.0; %initial x,y position (m)
f=atan(v0x/(v0y+E0/B0)); %the angle fi
wc=q*B0/m;                %cyclotron frequency
A=-sqrt(v0x^2+(v0y+E0/B0)^2)/wc;%(pick - sign for convenience) constant A
a=x0-A*cos(f); b=y0+A*sin(f);   %constants based on init conditions
t=[0:0.01:tmax];
x=A*cos(wc*t+f)+a;
y=-A*sin(wc*t+f)-E0*t/B0+b;
plot(x,y,'b-.','LineWidth',1.5)
str=cat(2,'Motion of a charge in E(+x-east), and B(+z-out of paper) fields');
title(str,'FontSize',13), ylabel('y','FontSize',14)
xlabel('x','FontSize',14); grid on;
```

In Example 6.7, if the charge particle's velocity is picked as $\mathbf{v} = \dot{y}\hat{j}$, with $\dot{x}_0 = 0 = \dot{z}_0$, the force on the charge is now

$$\mathbf{F} = q(E_0 + B_0\dot{y})\hat{i} = m\mathbf{a} \Rightarrow \ddot{x} = \frac{q}{m}E_0 + \frac{q}{m}B_0\dot{y}, \quad \ddot{y} = 0, \quad \ddot{z} = 0, \qquad (6.4.26)$$

rather than (6.4.17). We notice that in this equation, the velocity in the y direction is arbitrary. Suppose that we choose $\dot{y} = -E_0/B_0$, then from (6.4.26) we find that $\ddot{x} = 0$; that is, the electric and magnetic forces cancel. A beam of charges with this velocity passes undeflected and straight through the fields. This is shown in Figure 6.12. In order for this to happen, of course, the velocity must be carefully tuned. Thus, a device operating this way is a velocity selector. This principle may be used in the creation of monoenergetic beams. In fact, J. J. Thomson's discovery of the electron used this concept in his measurement of the electron's charge-to-mass ratio in 1897. Before that, electrons were known to be a type of radiation called cathode rays.

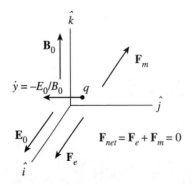

FIGURE 6.12   A charged particle with the proper velocity passes undeflected in an electromagnetic field.

■ 6.5 Three-Dimensional Motion of a Charged Particle in an Electromagnetic Field

It is very useful to investigate the motion of a charged particle in a three-dimensional electromagnetic field and to experiment with the parameters. This is possible if the equations of motion for the problem are solved numerically in three dimensions (3D). For the general 3D situation we have

$$\mathbf{F} = q\mathbf{v} \times \mathbf{B} + q\mathbf{E} = m\mathbf{a}, \tag{6.5.1}$$

which leads to the three differential equations

$$\frac{d^2x}{dt^2} = q(v_y B_z - v_z B_y + E_x)/m, \quad \frac{d^2y}{dt^2} = q(v_z B_x - v_x B_z + E_y)/m,$$

$$\text{and } \frac{d^2z}{dt^2} = q(v_x B_y - v_y B_x + E_z)/m \tag{6.5.2}$$

for the particle's coordinates as a function of time. As can be seen, this is a multicoupled system of equations for x, y, and z that is suitable for a numerical investigation. The MATLAB scripts cycloid3d.m and cycloid3d_der.m solve this system of equations numerically. The process to solve the equations works by transforming each second-order equation to a first-order pair of equations. We therefore define the variables according to an array notation as follows: $x \to r(1)$, $\dot{x} \to r(2)$, $y \to r(3)$, $\dot{y} \to r(4)$, and $z \to r(5)$, $\dot{z} \to r(6)$. Thus, the three second-order equations in (6.5.2) become six first-order equations given by

$$\frac{dr(1)}{dt} = r(2), \quad \frac{dr(2)}{dt} = q[r(4)B(3) - r(6)B(2) + E(1)]/m, \tag{6.5.3a}$$

$$\frac{dr(3)}{dt} = r(4), \quad \frac{dr(4)}{dt} = q[r(6)B(1) - r(2)B(3) + E(2)]/m, \tag{6.5.3b}$$

$$\frac{dr(5)}{dt} = r(6), \quad \frac{dr(6)}{dt} = q[r(2)B(2) - r(4)B(1) + E(3)]/m, \tag{6.5.3c}$$

where we have also defined arrays for the electric, $E = (E_x, E_y, E_z) = E(1, 2, 3)$, and magnetic, $B = (B_x, B_y, B_z) = B(1, 2, 3)$, field components. The main script, cycloid3d.m, defines the initial conditions and makes a call to MATLAB's simplest differential equation solver, ode23. Among various parameters, the solver accepts the script cycloid2d_der.m as input. The script cycloid3d_der.m's primary function is to perform the evaluation of the right-hand side of each of the Equations (6.5.3), given that the main script has defined their initial values.

Charge in General E & B Fields—3D

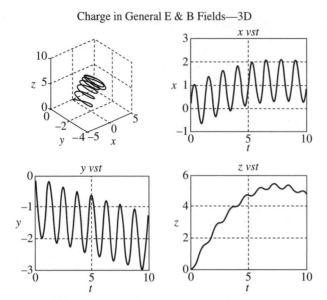

I FIGURE 6.13   A charged particle moving in the presence of a three-dimensional electromagnetic field.

The results of this scheme are shown in Figure 6.13, with electric and magnetic field components chosen to be $E = [0.5 \times 10^{-8}, 1 \times 10^{-9}, -3 \times 10^{-9}]$ and $B = [1 \times 10^{-8}, -1 \times 10^{-9}, 5.13 \times 10^{-8}]$ in units of volts/meter and Tesla, respectively.

The figure shows four plots. The upper left-hand corner is the three-dimensional plot of the dependent variable solutions, $x(t)$, $y(t)$, and $z(t)$, to the differential equations, each on its corresponding axis. The rest are two-dimensional plots of the individual variables $x(t)$, $y(t)$, and $z(t)$ versus time, each of which changes in a nontrivial way. Their behavior depends on the actual field values used as well as on the initial conditions. The figure, therefore, shows the interesting behavior that's made possible through the investigation of the multicoupled system of equations. The charge and mass used are those of a proton. These parameters can also be changed within the script. If the electric and magnetic field values as well as the initial conditions are picked to match those used in the analytic formulas whose plots are shown in Figure 6.11, the results will agree, as expected.

SCRIPT

```
%cycloid3d.m
%solves the motion of a charged particle in an electromagnetic
%field in 3 dimensions. Here, x(1)=x, x(2)=vx, x(3)=y, x(4)=vy,
%x(5)=z, x(6)=vz
clear;
q=1.6e-19; m=1.67e-27; qm=q/m; %charge, mass, & ratio
```

```
t0=0.0;tmax=10.0;        %time range
%———————-
%Use the next 3 lines to reproduce the 2D motion case (use plot command below)
%B=[0.0;0.0;3.13e-8];  %Bx,By,Bz
%E=[5.5e-8;0.0;0.0];   %Ex,Ey,Ez
%ic=[0.0;5.0;0.0;0.0;0.0;0.0];%x0,vx0,y0,vx0,z0,vz0 - init conditions
%———————-
B=[1.e-8;-1.e-9;5.13e-8];    %Bx,By,Bz
E=[0.5e-8;1.e-9;-3.e-9];       %Ex,Ey,Ez
ic=[0.0;5.0;0.0;0.0;0.0;0.5];%x0,vx0,y0,vy0,z0,vz0 - init conditions
[t,r]=ode23('cycloid3d_der',[t0 tmax],ic,[],qm,B,E); %simple ode solver
%[t,r]=ode45('cycloid3d_der',[t0 tmax],ic,[],qm,B,E);% use this for a better solver
%plot(r(:,1),r(:,3),'LineWidth',1.0,'Color','black'); % 2D motion case
subplot(2,2,1)
plot3(r(:,1),r(:,3),r(:,5),'r-','LineWidth',2)%use plot3 in 3D motion
view(320,30)% can change the view angle
str=cat(2,'Charge in general E&B fields - 3D');
title(str,'FontSize',14); xlabel('x','FontSize',11);
ylabel('y','FontSize',11); zlabel('z','FontSize',11); grid on;
axis square
subplot(2,2,2);plot(t,r(:,1));title('x vs t','FontSize',11);
xlabel('t','FontSize',11); ylabel('x','FontSize',12); grid on;
subplot(2,2,3);plot(t,r(:,3));title('y vs t','FontSize',11);
xlabel('t','FontSize',11); ylabel('y','FontSize',12); grid on;
subplot(2,2,4);plot(t,r(:,5));title('z vs t','FontSize',11);
xlabel('t','FontSize',11); ylabel('z','FontSize',11); grid on;
```

FUNCTION

```
% cycloid3d_der.m
function derivs = cycloid3d_der( t, r, flag, qm,B,E )
% Cycloid3d_der: returns the derivatives needed by cycloid3d
% for a charged particle in an electromagnetic field
% arrays are used as follows:
% B=[Bx;By;Bz], E=[Ex;Ey;Ez] - magnetic, electric fields
% r(1)=x, r(2)=dx/dt=vx
% r(3)=y, r(4)=dy/dt=vy
% r(5)=z, r(6)=dz/dt=vz
% qm - charge to mass ratio
%derivs calculates the derivatives of r. For example
%dx/dt=vx->r(2), dvx/dt=q*(vy*Bz-vz*By+Ex)/m->dr(2)/dt
derivs = [ r(2); qm*(r(4)*B(3)-r(6)*B(2)+E(1)); ...
           r(4); qm*(r(6)*B(1)-r(2)*B(3)+E(2)); ...
           r(6); qm*(r(2)*B(2)-r(4)*B(1)+E(3))];
```

■ Chapter 6 Problems

6.1 Derive the equation for a parabola $(x - x_0)^2 = 4e(z - z_0)$ and show its equivalency to $z = ax^2 + bx + c$ with the coefficients given in Equation (6.2.6).

6.2 The range of a projectile moving in the xz-plane, in the absence of air resistance, is given by the formula $R = 2v_{0z}v_{0x}/g$, where v_{0z} and v_{0x} are the initial velocity components. According to this formula, complementary angles (for example, $10°$ and $80°$, or $15°$ and $75°$, and so on) have the same range. The time to hit the ground is, however, different. Find a simple relationship between the time of flight for complementary angles. That is, if you know the time of flight at a given angle, then, by such relationship, you can obtain the time of flight of the complementary angle.

6.3 A more accurate approximation than (6.2.19) can be obtained for the range of a projectile in the presence of air resistance if Equation (6.2.14) is set to zero and the positive root for x is expanded to first order in the drag coefficient. Find the expression for the projectile range obtained this way.

6.4 Using mass, velocity, and any other needed information from Example 6.2, estimate the value of C beyond which the air resistance round-trip time formula $t_R \sim 2v_{0z}(1 - Cv_{0z}/mg)/g$ is no longer a good approximation within a 3% tolerance level.

6.5 In Example 6.2 in the case of air resistance, $C = 0.5$ kg/s. Using the same values for the rest of the parameters, investigate the approximate angle at which the maximum range occurs.

6.6 In the case of the conservative force $\mathbf{F}(\mathbf{r}) = Ax\hat{i}$, prove that the result of performing the line integral $V(\mathbf{r}) = -\int_o^P \mathbf{F}(\mathbf{r}') \cdot d\mathbf{r}'$ on the square of Figure 6.7 using path 2 is equal to $V_2 = -A\ell^2/2$.

6.7 Use the nonconservative force $\mathbf{F}(\mathbf{r}) = xy\hat{i} + y\hat{j}$ to perform a line integral for the potential on the square of Figure 6.7, using path 3. Explain your answer.

6.8 Assume a general potential energy function $V = V(x, y, z)$, and show that if $\mathbf{F} = -\nabla V$, then, in fact, $\nabla \times \mathbf{F} = 0$ actually follows.

6.9 Given $V(r) = Ae^{-r^2}$, find the associated force $\mathbf{F}(\mathbf{r})$.

6.10 Verify Equation (6.2.12).

6.11 Obtain Equation (6.2.18) starting from Equation (6.2.16).

6.12 Modify `projectile.m` to one that incorporates a charged particle moving in an electric field instead. Test the script with an electron ($q = -1.6 \times 10^{-19}$C, $m = 9.1 \times 10^{-31}$ kg) as well as with a positron ($q = 1.6 \times 10^{-19}$C, $m = 9.1 \times 10^{-31}$ kg.) (*Hint:* See Example 6.1, and try an electric field that produces an acceleration equivalent to that of the value of gravity $\{g = 9.8$ m/s$^2\}$.)

6.13 Consider a charged particle, q, with mass m in the presence of a magnetic field $\mathbf{B} = B_0\hat{k}$.

a. Show that the acceleration experienced by the charge obeys the form $\ddot{\mathbf{r}} = -D\mathbf{r}$, where D is a constant of proportionality.

b. Find the constant of proportionality and explain the result.

■ Additional Problems

6.14 A particle of charge $+q$ and mass m is made to overcome its weight and floats due to the presence of a constant electric field of 10 N/C in magnitude that points in the direction shown. A constant 0.1 T magnetic field is also present and points in the same direction as the electric field. The charge has a velocity with a magnitude of 0.35 m/s and makes a circular orbit, as shown in Figure 6.14. Find the mass of the particle and the radius of the orbit if $q = 6\ \mu$C.

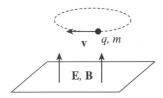

I FIGURE 6.14

6.15 In an experiment similar to that performed by J. J. Thomson in 1897, electrons are emitted from a hot filament and travel through a region where they are accelerated by an applied potential V, as shown in Figure 6.15. Once they leave the accelerating region at a speed v_0, the electrons enter a parallel plate capacitor of length L where a constant field E deflects them by an amount y.

a. What is the speed v_0 of the electrons?

b. Obtain an expression for the deflection y in terms of e, v_0, E, L, and m_e.

c. If we wished to ensure the electrons are not deflected within the capacitor, we can add a constant magnetic field there. Obtain an expression for the magnitude of this magnetic field in terms of v_0 and E. In what direction should the magnetic field be oriented?

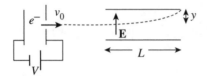

| FIGURE 6.15

6.16 An approximate expression for the maximum range of a projectile traveling in the presence of linear air resistance, for small C, is given by (6.2.19) as

$$x_{max} \sim \frac{2v_{ix}v_{iz}}{g} - \frac{2Cv_{ix}v_{iz}^2}{mg^2}.$$ Show that if this expression is maximized with

respect to the launching angle and if this angle is not far from $45°$, an approximate expression for the angle of maximum range in the presence of air resistance results in the form

$$\sin 2\theta \approx -\frac{Cv_0}{2mg} + \sqrt{1 + \frac{Cv_0}{2mg}\left(\frac{Cv_0}{2mg} + \frac{\sqrt{8}}{4}\right)}.$$

7 | Systems of Coordinates

■ 7.1 Introduction

Thus far we have studied problems in more than one dimension, but we restricted our-selves mainly to motion in Cartesian coordinates. There are many applications in physics, however, that are more conveniently studied in their natural coordinates. For example, the rotation of a point on a rolling wheel and the motion of a planet in orbit around the sun are more naturally understood in polar coordinates. The fact that Earth spins, for example, makes the motion associated with the weather patterns complicat-ed. We need to learn to deal with accelerated coordinate systems to understand at least some observed behaviors due to winds, or even a rock falling over the spinning Earth. We will find that noninertial coordinate systems find it convenient to incorporate, for example, forces that are natural from an inertial reference frame, but are fictitious when viewed from an accelerated coordinate system, such as the notorious Coriolis and centrifugal forces. In this chapter, coordinate systems other than Cartesian are considered. In particular, transformations to plane, cylindrical, and spherical polar coordinates are made. Following this construction, the motion of a body in moving coordinate systems will be studied.

■ 7.2 Plane Polar Coordinates

This system of coordinates is convenient in situations of plane polar symmetry. This system is basically a two-dimensional one, similar to the Cartesian system, as can be seen in Figure 7.1.

From Figure 7.1(a) we know that the Cartesian coordinates x and y are related to the plane polar coordinates r and θ through the equations $x = r\cos\theta$ and $y = r\sin\theta$. In fact, the point P's location in Cartesian coordinates is $\mathbf{r} = x\hat{i} + y\hat{j}$, where $\hat{i}$ and $\hat{j}$ are the unit vectors; this can also be expressed in polar coordinates as

$$\mathbf{r} = r\hat{r}. \tag{7.2.1}$$

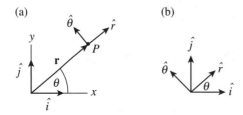

FIGURE 7.1   Plane polar coordinates.

The polar $\hat{r}$ and $\hat{\theta}$ unit vectors depend on the angle θ. They can be expressed in terms of the unit vectors of the Cartesian xy-plane through the coordinate system transformations method discussed in Chapter 5, Section 4. That is, we write

$$\begin{pmatrix} \hat{r} \\ \hat{\theta} \end{pmatrix} = \begin{pmatrix} \hat{r} \cdot \hat{i} & \hat{r} \cdot \hat{j} \\ \hat{\theta} \cdot \hat{i} & \hat{\theta} \cdot \hat{j} \end{pmatrix} \begin{pmatrix} \hat{i} \\ \hat{j} \end{pmatrix}, \tag{7.2.2a}$$

or for the case of the Cartesian unit vector in terms of the polar unit vectors

$$\begin{pmatrix} \hat{i} \\ \hat{j} \end{pmatrix} = \begin{pmatrix} \hat{i} \cdot \hat{r} & \hat{i} \cdot \hat{\theta} \\ \hat{j} \cdot \hat{r} & \hat{j} \cdot \hat{\theta} \end{pmatrix} \begin{pmatrix} \hat{r} \\ \hat{\theta} \end{pmatrix}. \tag{7.2.2b}$$

From Figure 7.1(b) we see that $\hat{r} \cdot \hat{i} = \cos\theta$, $\hat{r} \cdot \hat{j} = \cos(90 - \theta)$, $\hat{\theta} \cdot \hat{i} = \cos(90 + \theta)$, and $\hat{\theta} \cdot \hat{j} = \cos\theta$, so that, from (7.2.2a) we get

$$\hat{r} = \hat{i}\cos\theta + \hat{j}\sin\theta, \hat{\theta} = -\hat{i}\sin\theta + \hat{j}\cos\theta, \tag{7.2.3a}$$

where the identities $\cos(90 + \theta) = -\sin\theta$ and $\cos(90 - \theta) = \sin\theta$ have been used. In a similar way the Cartesian unit vectors can be expressed in terms of the polar unit vectors (see Problem 7.2),

$$\hat{i} = \hat{r}\cos\theta - \hat{\theta}\sin\theta, \hat{j} = \hat{r}\sin\theta + \hat{\theta}\cos\theta. \tag{7.2.3b}$$

The polar coordinate velocity is obtained by taking the time derivative of (7.2.1)

$$\mathbf{v} = \frac{d}{dt}\mathbf{r} = \frac{d}{dt}(r\hat{r}) = \dot{r}\hat{r} + r\frac{d\hat{r}}{dt}. \tag{7.2.4}$$

We first notice that

$$\frac{d\hat{r}}{dt} = \frac{d\hat{r}}{d\theta}\dot{\theta}, \quad \text{and} \quad \frac{d\hat{\theta}}{dt} = \frac{d\hat{\theta}}{d\theta}\dot{\theta}, \tag{7.2.5}$$

so that with

$$\frac{d\hat{r}}{d\theta} = \frac{d}{d\theta}(\hat{i}\cos\theta + \hat{j}\sin\theta) = -\hat{i}\sin\theta + \hat{j}\cos\theta = \hat{\theta} \tag{7.2.6a}$$

and

$$\frac{d\hat{\theta}}{d\theta} = \frac{d}{d\theta}(-\hat{i}\sin\theta + \hat{j}\cos\theta) = -\hat{i}\cos\theta - \hat{j}\sin\theta = -\hat{r} \tag{7.2.6b}$$

we find that

$$\frac{d\hat{r}}{dt} = \dot{\theta}\hat{\theta} \quad \text{and} \quad \frac{d\hat{\theta}}{dt} = -\dot{\theta}\hat{r}. \tag{7.2.7}$$

Putting the first of these into (7.2.4), the final expression for the velocity in polar coordinates becomes

$$\mathbf{v} = \dot{\mathbf{r}} = \dot{r}\hat{r} + r\dot{\theta}\hat{\theta}. \tag{7.2.8}$$

The polar acceleration is

$$\mathbf{a} = \frac{d\mathbf{v}}{dt} = \ddot{\mathbf{r}} = \frac{d}{dt}(\dot{r}\hat{r} + r\dot{\theta}\hat{\theta}) = \ddot{r}\hat{r} + \dot{r}\frac{d\hat{r}}{dt} + \dot{r}\dot{\theta}\hat{\theta} + r\ddot{\theta}\hat{\theta} + r\dot{\theta}\frac{d\hat{\theta}}{dt}. \tag{7.2.9}$$

Substituting (7.2.7) into this, we obtain $\mathbf{a} = \ddot{r}\hat{r} + \dot{r}\dot{\theta}\hat{\theta} + \dot{r}\dot{\theta}\hat{\theta} + r\ddot{\theta}\hat{\theta} - r\dot{\theta}^2\hat{r}$, or

$$\mathbf{a} = \ddot{\mathbf{r}} = (\ddot{r} - r\dot{\theta}^2)\hat{r} + (r\ddot{\theta} + 2\dot{r}\dot{\theta})\hat{\theta}. \tag{7.2.10}$$

The term $r\dot{\theta}^2$ is due to centripetal acceleration from the angular motion. Its direction is toward the center of rotation, as expected. The term $2\dot{r}\dot{\theta}$, due to simultaneous radial and angular speeds, is sometimes referred to as the Coriolis acceleration, where

$\dot{\theta}$ is the angular frequency (ω). The terms $\ddot{r}$ and $r\ddot{\theta}$ correspond to the magnitudes of the radial (a_r) and tangential (a_t) acceleration terms, respectively, where $\ddot{\theta}$ is the angular acceleration (α).

EXAMPLE 7.1

A particle of mass m is held by a massless rod in circular motion of radius R, as shown in Figure 7.2. The particle is whirled around in a vertical loop about an axis passing through its center and perpendicular to the orbital plane shown. Assuming there is no friction, (a) find the particle's position, velocity, acceleration, and the tension on the rod as a function of time in the absence of gravity. Let the initial rotational frequency ω_0 be constant, because gravity is assumed absent ($g = 0$). (b) What are the particle's position, velocity, acceleration, and the tension on the rod as a function of time if gravity is taken into account, assuming the mass starts with an initial angular speed ω_0?

Solution

(a) Because the circle's radius is fixed, we let $\dot{r} = \ddot{r} = 0$, and $r = R$. The angular velocity is constant, so that $\dot{\theta} = \omega = \omega_0$, $\ddot{\theta} = 0$, and using (7.2.1, 7.2.8, 7.2.10), we get $\mathbf{r} = R\hat{r}$, $\mathbf{v} = -R\omega_0\hat{\theta}$, and $\mathbf{a} = -R\omega_0^2\hat{r}$. The negative sign in the velocity is due to clockwise rotation, and the negative sign in the acceleration shows its centripetal direction. The tension on the rod is obtained by Newton's second law, $m\mathbf{a} = -mR\omega_0^2\hat{r} = \sum \mathbf{F} = \mathbf{T}$ and points toward the center.

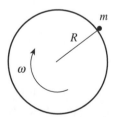

FIGURE 7.2   Example 7.1 for a particle at the end of a massless rod.

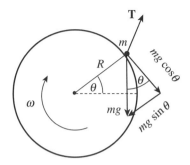

| FIGURE 7.3 Example 7.1 for a particle at the end of a massless rod and its forces.

(b) If gravity is present, referring to Figure 7.3, the mass rotates with angular frequency ω. It has radial and tangential components of acceleration due to the force of gravity acting on it, which depend on the angle θ. The tension **T** is the general reaction force associated with the rod.

Because the circle's radius is fixed, we set $\dot{r} = \ddot{r} = 0$, and $r = R$. The tangential acceleration magnitude is

$$a_t = R\alpha = g\cos\theta,$$

which gives

$$\ddot{\theta} = \alpha = g\cos\theta/R.$$

This shows that θ depends on time in a nontrivial manner. We next let

$$\dot{\theta} = \omega(t) = \omega_0 + \int_0^t \alpha\, dt = \omega_0 + \frac{g}{R}\int_0^t \cos\left(\theta(t')\right) dt',$$

or

$$\dot{\theta} = \dot{\theta}(t) = \omega_0 + f(t)g/R$$

where we have defined

$$f(t) = \int\limits_0^t \cos\left(\theta(t')\right) dt'.$$

Again, using (7.2.1, 7.2.8), we have for the position $\mathbf{r} = R\hat{r}$ and velocity $\mathbf{v} = -r\dot{\theta}\hat{\theta} = -[R\omega_0 + gf(t)]\hat{\theta}$. Notice that at $t = 0$ $f(t = 0) = 0$ and the velocity simplifies to that of part (a) as expected. The acceleration is obtained using (7.2.10) as $\mathbf{a} = -r\dot{\theta}^2\hat{r} - r\ddot{\theta}\hat{\theta} = -[(R\omega_0 + gf(t))^2/R]\hat{r} - g\cos\theta\,\hat{\theta}$. The first term is due to a centripetal acceleration, with two tangential velocity contributions, and points toward the center of the circle. The second term is due to the tangential acceleration that's a function of angle with a negative sign due to the clockwise direction of the rotation. The acceleration simplifies to that of part (a) in the limit as $t \to 0$ and $g \to 0$. Finally, by Newton's second law

$$m\mathbf{a} = -m\{[(R\omega_0 + gf(t))^2/R]\,\hat{r} + g\cos\theta\,\hat{\theta}\} = \sum \mathbf{F} = \mathbf{T} - mg\sin\theta\,\hat{r},$$

which gives $\mathbf{T} = m\{[g\sin\theta - (R\omega_0 + gf(t))^2/R]\,\hat{r} - g\cos\theta\,\hat{\theta}\}$.

■ 7.3 Cylindrical Polar Coordinates

Cylindrical coordinates are useful for situations that involve cylindrical geometry, as shown in Figure 7.4(a). The unit vectors of interest are also shown in Figure 7.4(b).

Here, the cylindrical coordinate unit vectors are $\hat{\rho}$, $\hat{\phi}$, and $\hat{k}$ and form an orthogonal coordinate system. The $\hat{\rho}$ and $\hat{\phi}$ unit vectors are exactly analogous to the $\hat{r}$ and $\hat{\theta}$ of the plane polar coordinates, respectively, and the $\hat{k}$ coincides with its Cartesian version. Thus, in this coordinate system, the position vector for point P of Figure 7.4(b) is

$$\mathbf{r} = \rho\hat{\rho} + z\hat{k}, \tag{7.3.1}$$

because it represents a projection onto the xy-plane with the additional component in the z-direction, and as before, $x = \rho\cos\phi$ and $y = \rho\sin\phi$. In analogy to the plane polar coordinate, therefore, we see that

$$\mathbf{v} = \dot{\mathbf{r}} = \dot{\rho}\hat{\rho} + \rho\dot{\phi}\hat{\phi} + \dot{z}\hat{k}, \tag{7.3.2}$$

(a)   (b)

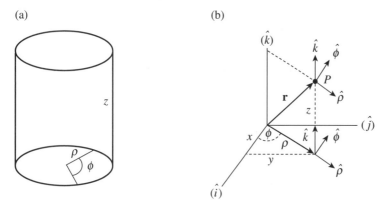

FIGURE 7.4 Cylindrical coordinates.

where we have used $d\hat{k}/dt = 0$, because the $\hat{k}$ direction is constant. Similarly, the acceleration is

$$\mathbf{a} = \ddot{\mathbf{r}} = (\ddot{\rho} - \rho\dot{\phi}^2)\hat{\rho} + (\rho\ddot{\phi} + 2\dot{\rho}\dot{\phi})\hat{\phi} + \ddot{z}\hat{k}. \tag{7.3.3}$$

EXAMPLE 7.2

Consider a mass m attached to a wheel or radius R that's rotating in the xy-plane, in a direction according to the right-hand rule (thumb toward z-axis), with a constant angular frequency ω whose vector lies along the z-axis. If while rotating the wheel is released and free falls near Earth's surface, write the expression for the mass' position, velocity, and acceleration as a function of time.

Solution
The radius of the wheel is $\rho = R$, and because this is constant, then $\dot{\rho} = 0 = \ddot{\rho}$. In addition, because the wheel is free falling, $\ddot{z} = -g$, $\dot{z} = v_{0z} - gt$, and $z = z_0 + v_{0z}t - \frac{1}{2}gt^2$. With $v_{0z} \equiv 0$, $\dot{\phi} = \omega$, $\ddot{\phi} = 0$, we have from (7.3.1 and 7.3.2)

for the position $\mathbf{r} = R\hat{\rho} + \left(z_0 - \frac{1}{2}gt^2\right)\hat{k}$, and for the velocity $\dot{\mathbf{r}} = R\omega\hat{\phi} - gt\hat{k}$.

Similarly, from (7.3.3) the acceleration is, $\ddot{\mathbf{r}} = -R\omega^2\hat{\rho} - g\hat{k}$.

■ 7.4 Spherical Polar Coordinates

Spherical coordinates are used in problems that involve spherical symmetry. The three mutually perpendicular unit vectors used are $\hat{r}$, $\hat{\theta}$, and $\hat{\phi}$, which are displayed in Figure 7.5(a). The relationship between the spherical component and the Cartesian coordinates is shown in Figure 7.5(b).

(a)   (b)

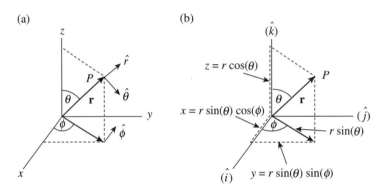

| FIGURE 7.5 Spherical coordinates.

The $\hat{\theta}$ or polar direction is a clockwise rotation in the direction of increasing θ, and the $\hat{\phi}$ or azimuthal direction is a right-hand rule rotation (thumb toward the z-axis) in the direction of increasing ϕ. As a matter of convention, the range of the polar angle is $0 < \theta < \pi$, and that of the azimuth angle is $0 < \phi < 2\pi$.

While the projections of $\hat{r}$ onto the x-, y-, and z-axes are identical to those of $\mathbf{r}$ that are shown in Figure 7.5(b), we notice that the unit vector $\hat{\theta}$ makes an angle θ with respect to the xy-plane as well as an angle ϕ with respect to the x-axis. Similarly, the unit vector $\hat{\phi}$, although located on the xy-plane, makes an angle ϕ with respect to the y-axis. Both of these situations are shown in Figure 7.6.

(a)   (b)

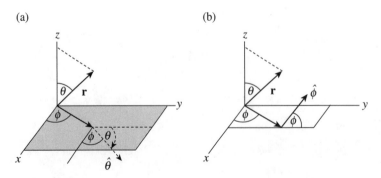

| FIGURE 7.6 Unit vectors in spherical coordinates.

The relationship between the spherical unit vector and the Cartesian ones can be written in matrix form, as was done with the plane polar coordinate case of Section 7.2. We have

$$
\begin{pmatrix} \hat{r} \\ \hat{\theta} \\ \hat{\phi} \end{pmatrix} = \begin{pmatrix} \hat{r}\cdot\hat{i} & \hat{r}\cdot\hat{j} & \hat{r}\cdot\hat{k} \\ \hat{\theta}\cdot\hat{i} & \hat{\theta}\cdot\hat{j} & \hat{\theta}\cdot\hat{k} \\ \hat{\phi}\cdot\hat{i} & \hat{\phi}\cdot\hat{j} & \hat{\phi}\cdot\hat{k} \end{pmatrix} \begin{pmatrix} \hat{i} \\ \hat{j} \\ \hat{k} \end{pmatrix},
\tag{7.4.1}
$$

Using Figures 7.5 and 7.6 and the above understanding, we have that the projections of $\hat{r}$ onto the Cartesian directions are $\hat{r}\cdot\hat{i} = \sin\theta\cos\phi$, $\hat{r}\cdot\hat{j} = \sin\theta\sin\phi$, $\hat{r}\cdot\hat{k} = \cos\theta$. Similarly, for $\hat{\theta}$ and $\hat{\phi}$ we obtain $\hat{\theta}\cdot\hat{i} = \cos\theta\cos\phi$, $\hat{\theta}\cdot\hat{j} = \cos\theta\sin\phi$, $\hat{\theta}\cdot\hat{k} = -\sin\theta$; $\hat{\phi}\cdot\hat{i} = -\sin\phi$, $\hat{\phi}\cdot\hat{j} = \cos\phi$, and $\hat{\phi}\cdot\hat{k} = 0$. Thus, from (7.4.1) we obtain

$$
\begin{aligned}
\hat{r} &= \sin\theta\cos\phi\,\hat{i} + \sin\theta\sin\phi\,\hat{j} + \cos\theta\,\hat{k} \\
\hat{\theta} &= \cos\theta\cos\phi\,\hat{i} + \cos\theta\sin\phi\,\hat{j} - \sin\theta\,\hat{k} \\
\hat{\phi} &= -\sin\phi\,\hat{i} + \cos\phi\,\hat{j} + 0\hat{k}.
\end{aligned}
\tag{7.4.2}
$$

In spherical coordinates, the position of the point P of Figure 7.5(a) is written as

$$
\mathbf{r} = r\hat{r},
\tag{7.4.3}
$$

where $\hat{r}$ is shown in (7.4.2). To obtain the velocity and acceleration in spherical coordinates, we need to obtain the time derivatives of the unit vectors. We'll use

$$
\frac{d\hat{r}}{dt} = \frac{d\hat{r}}{d\theta}\dot{\theta} + \frac{d\hat{r}}{d\phi}\dot{\phi}, \quad \frac{d\hat{\theta}}{dt} = \frac{d\hat{\theta}}{d\theta}\dot{\theta} + \frac{d\hat{\theta}}{d\phi}\dot{\phi}, \text{ and } \frac{d\hat{\phi}}{dt} = \frac{d\hat{\phi}}{d\phi}\dot{\phi}.
\tag{7.4.4}
$$

With (7.4.2) into (7.4.4), we get

$$
\frac{d\hat{r}}{dt} = [\cos\theta\cos\phi\,\dot{\theta} - \sin\theta\sin\phi\,\dot{\phi}]\hat{i} + [\cos\theta\sin\phi\,\dot{\theta} + \sin\theta\cos\phi\,\dot{\phi}]\hat{j} - \sin\theta\,\dot{\theta}\,\hat{k}
$$

or

$$
\frac{d\hat{r}}{dt} = \dot{\theta}\hat{\theta} + \sin\theta\,\dot{\phi}\hat{\phi}.
\tag{7.4.5a}
$$

Similarly, we find

$$
\frac{d\hat{\theta}}{dt} = [-\sin\theta\cos\phi\,\dot{\theta} - \cos\theta\sin\phi\,\dot{\phi}]\hat{i} + [-\sin\theta\sin\phi\,\dot{\theta} + \cos\theta\cos\phi\,\dot{\phi}]\hat{j} - \cos\theta\,\dot{\theta}\,\hat{k}
$$

or

$$\frac{d\hat{\theta}}{dt} = \hat{\phi}\dot{\phi}\cos\theta - \hat{r}\dot{\theta}, \tag{7.4.5b}$$

and

$$\frac{d\hat{\phi}}{dt} = (-\cos\phi\,\hat{i} - \sin\phi\,\hat{j})\,\dot{\phi} = (-\sin\theta\,\hat{r} - \cos\theta\,\hat{\theta})\dot{\phi}. \tag{7.4.5c}$$

The spherical coordinate velocity is obtained by taking the time derivative of (7.4.3); that is, $\dot{\mathbf{r}} = d\mathbf{r}/dt = \dot{r}\hat{r} + rd\hat{r}/dt$, which with (7.4.5a) gives

$$\mathbf{v} = \dot{\mathbf{r}} = \dot{r}\hat{r} + r\dot{\theta}\,\hat{\theta} + r\dot{\phi}\sin\theta\,\hat{\phi}. \tag{7.4.6}$$

The acceleration is the time derivative of this expression; that is,

$$\mathbf{a} = \ddot{\mathbf{r}} = \ddot{r}\hat{r} + \dot{r}\frac{d\hat{r}}{dt} + \dot{r}\dot{\theta}\hat{\theta} + r\ddot{\theta}\hat{\theta} + r\dot{\theta}\frac{d\hat{\theta}}{dt} + \dot{r}\dot{\phi}\sin\theta\,\hat{\phi} + r\ddot{\phi}\sin\theta\hat{\phi} + r\dot{\phi}\frac{d\sin\theta}{dt}\hat{\phi} + r\dot{\phi}\sin\theta\frac{d\hat{\phi}}{dt}, \tag{7.4.7a}$$

which with the use of (7.4.5) becomes

$$\mathbf{a} = [\ddot{r} - r\dot{\theta}^2 - r\dot{\phi}^2\sin^2\theta]\hat{r} + [2\dot{r}\dot{\theta} + r\ddot{\theta} - r\dot{\phi}^2\sin\theta\cos\theta]\hat{\theta}$$
$$+ [2\dot{r}\dot{\phi}\sin\theta + 2r\dot{\theta}\dot{\phi}\cos\theta + r\ddot{\phi}\sin\theta]\hat{\phi}. \tag{7.4.7b}$$

EXAMPLE 7.3

An Earth satellite of mass m is on an equatorial orbit with equatorial (azimuthal) angular speed ω. Let its distance from the center of Earth be b and its latitude angle α be a constant. (a) Obtain the satellite's position, velocity, and acceleration. (b) If the force of gravity on the satellite due to Earth (M_E) is given by $\mathbf{F}_g = -GM_E m\hat{r}/r^2$, write Newton's second law for the satellite in orbit and obtain the angular velocity and force needed to maintain the satellite in a constant latitude angle.

....

Solution

(a) Assuming Earth is a sphere, we let the spherical polar angle be $\theta = 90° - \alpha$. We have: $r = b$, $\dot{\phi} = \omega$, $\dot{\theta} = 0 = \ddot{\theta} = \ddot{\phi}$, and $\dot{r} = 0 = \ddot{r}$. Thus, $\mathbf{r} = b\hat{r}$, $\mathbf{v} = b\omega \sin(90° - \alpha)\hat{\phi}$, and

$$\mathbf{a} = -b\omega^2 \sin^2(90° - \alpha)\hat{r} - b\omega^2 \sin(90° - \alpha)\cos(90° - \alpha)\hat{\theta}.$$

Or, using identities from Appendix C, with $\sin(90° - \alpha) = \cos\alpha$, and $\cos(90° - \alpha) = \sin\alpha$, get $\mathbf{a} = -b\omega^2 \cos^2\alpha\,\hat{r} - (1/2)b\omega^2 \sin(2\alpha)\hat{\theta}$. At the equator, $\alpha = 0°$, and $\mathbf{v} \to b\omega\,\hat{\phi}$, a tangential speed, although $\mathbf{a} \to -b\omega^2\,\hat{r}$, a centripetal acceleration. At the North Pole the situation is slightly complicated. The ω needed to maintain a balanced orbit depends on the angle α, as discussed in the second part of this example.

(b) Newton's second law for the satellite is

$$m\mathbf{a} = -mb\omega^2 \cos^2\alpha\,\hat{r} - (1/2)mb\omega^2 \sin(2\alpha)\hat{\theta} = \sum \mathbf{F} = F_g\hat{r} + F_\theta\hat{\theta},$$

where $F_g = -GM_E m/b^2$ and F_θ is the polar component of the total force needed to maintain a constant latitude angle α. First, by equating radial components, we see $mb\omega^2 \cos^2\alpha = GM_E m/b^2$ or $\omega = \sqrt{GM_E/(b^3 \cos^2\alpha)}$ for the azimuthal angular velocity. Second, by equating the polar components, we get $F_\theta = -(1/2)mb\omega^2 \sin(2\alpha)$, which with the above ω becomes $F_\theta = -(1/2)GM_E m \sin(2\alpha)/(b^2 \cos^2\alpha) = -GM_E m \tan\alpha/b^2$. At the equator, $\omega = \sqrt{GM_E/b^3}$, $F_\theta = 0$, and the satellite is maintained in orbit by the radial force component alone, i.e., F_g. At a higher latitude angle, ω increases and so does F_θ until at the North Pole it is impossible to maintain the equatorial orbit since both ω and F_θ become infinitely large. Finally, with both F_g and F_θ known, the total force needed to maintain the satellite in orbit, in a constant latitude angle, is $\mathbf{F} = F_g\hat{r} + F_\theta\hat{\theta}$.

■ 7.5   Moving Coordinate Systems

When a body experiences motion, the motion itself takes place with respect to a coordinate system. The question arises as to whether the coordinate system in reference to which motion takes place is itself accelerating, is at rest, or is moving at constant velocity. Thus we can have an *inertial* coordinate system that is defined to be one in which Newton's first law of motion applies. That is, it is one that, although it could be moving at constant velocity, it does not experience an acceleration. We can also have a *noninertial* or accelerated coordinate system in which Newton's first law of motion does not apply; rather, because the reference frame is accelerated, inertial or fictitious forces arise, as will be discussed shortly. Here, we consider translational motion of bodies with respect to reference frames. One frame is the inertial reference frame or the S frame of reference with coordinates (x, y, z), and the other is a non-inertial reference frame or the S' frame of reference with primed coordinates (x', y', z'). In Figure 7.7 we consider a particle of mass m located at point P whose position is $\mathbf{r}'$ with respect to the S' frame.

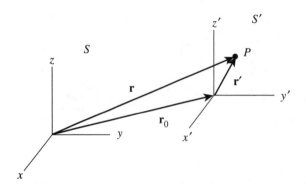

I FIGURE 7.7   The S' frame moving with respect to the inertial reference frame S.

We see that the S' frame is located at position $\mathbf{r}_0$ with respect to the S frame. Thus, we can find the particle's position with respect to the S frame as

$$\mathbf{r} = \mathbf{r}_0 + \mathbf{r}'. \tag{7.5.1}$$

The particle's velocity as seen from the S frame is obtained by taking the time derivative of this, or

$$\mathbf{v} = \mathbf{v}_0 + \mathbf{v}', \tag{7.5.2}$$

where $\mathbf{v}_0$ is the velocity of the S' frame with respect to the S frame, and $\mathbf{v}'$ is the velocity of the particle with respect to the S' frame. Taking the time derivative of (7.5.2), we have that the particle's acceleration with respect to the S frame is the sum of the accelerations of the S' frame with respect to the S frame, and the acceleration of the particle with respect to the S' frame; that is,

$$\mathbf{a} = \mathbf{a}_0 + \mathbf{a}'. \tag{7.5.3}$$

Because the S frame is an inertial frame, Newton's second law of motion holds, so that we can write

$$f = m\mathbf{a}, \tag{7.5.4}$$

which is what an observer in the S frame sees. From (7.5.3), an observer in the S' frame will see

$$f' = m\mathbf{a}' = m\mathbf{a} - m\mathbf{a}_0. \tag{7.5.5a}$$

The term $m\mathbf{a}_0$ stems from the fact that the S' frame is an accelerated coordinate system. Thus, the S' frame observer has to add an extra term, the so-called *fictitious force*, $f_{fict} \equiv - m\mathbf{a}_0$, in order to find agreement with the inertial frame of reference observer. In the event that the S' frame moves at constant velocity, $\mathbf{a}_0 \to 0$ and $f' = f$, without mention being made to the fictitious force. Newton's second law, in this case, becomes identical in each frame. The point of this discussion is that when one describes the motion of a body from the point of view of an accelerated coordinate system, one needs to account for the fictitious force due to such acceleration. Therefore, the noninertial frame of reference observer must write

$$f' = f + f_{fict} \tag{7.5.5b}$$

in order to agree with the motion seen by the inertial frame of reference observer. This way, Newton's second law is applicable to both reference frames.

EXAMPLE 7.4

Consider a person of mass m, at rest, at a location on Earth's equator and unaware of Earth's rotation. (a) Find the weight of the person as seen by an Earth observer. (b) Find the weight of the person as seen from an inertial frame of reference far away in space. (c) In light of the fact that Earth rotates about its own axes, what must be done for these results to agree?

Solution

(a) The Earth observer will write the equilibrium equation

$$\sum \mathbf{F}_{ext} = m\mathbf{a}' = 0 = f', \tag{7.5.6a}$$

because this observer is not aware that the motion takes place in a noninertial frame of reference. We then have $\mathbf{N} - mg\hat{k} = 0$, where the normal force results in $\mathbf{N} = mg\hat{k}$; that is, the person's weight is $w = mg$.

(b) An observer from the inertial position in space sees that Earth rotates, so that the equation of motion is

$$\sum \mathbf{F}_{ext} = m\mathbf{a} = \mathbf{f}, \tag{7.5.6b}$$

where $\mathbf{a}$ is the *centripetal* acceleration, directed toward Earth's center, associated with rotational motion present in the inertial frame, or

$$\mathbf{N} - mg\hat{k} = f = -m(v^2/R_E)\hat{k}, \tag{7.5.6c}$$

and leads to $\mathbf{N} = (mg - mv^2/R_E)\hat{k}$, so that the weight of the person is more appropriately $w = |\mathbf{N}| = mg - mv^2/R_E$.

(c) In order for the two observers to agree, the noninertial frame of reference observer must add the fictitious force f_{fict}. From (7.5.5b) and (7.5.6a) we can write $f' = f + f_{fict} = 0$ or $f_{fict} = -f$, and using f from (7.5.6c) we see that $f_{fict} = m(v^2/R_E)\hat{k}$. Thus the noninertial frame of reference observer must now add this so-called *centrifugal* force $m(v^2/R_E)\hat{k}$ to the force analysis. Equation (7.5.6a) is to be replaced with

$$f' = f + f_{fict} = \sum \mathbf{F}_{ext} + m(v^2/R_E)\hat{k}, \tag{7.5.6d}$$

but because the $f' = 0$, then $\sum \mathbf{F}_{ext} = -f_{fict}$, with the left-hand side of this equation being identical to that of part (a). From here, the analysis to obtain the person's weight leads to $\mathbf{N} - mg\hat{k} = -m(v^2/R_E)\hat{k}$, which is identical to that of part (b), and both observers agree.

■ 7.6   Rotating Coordinate Systems

Consider a S' frame rotating with respect to a fixed frame S whose origins coincide, as shown in Figure 7.8.

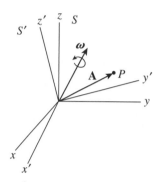

| FIGURE 7.8   The S' frame rotating with respect to the inertial S frame of reference.

In the figure, $\boldsymbol{\omega}$ is the angular rotational frequency vector according to the right-hand rule and $\mathbf{A}$ is a vector whose origin coincides with the coordinate systems. Let $\mathbf{A}$ represent a vector that refers to a physical property as viewed from the fixed coordinate system S. According to the S frame,

$$\mathbf{A} = A_x\hat{i} + A_y\hat{j} + A_z\hat{k}. \tag{7.6.1}$$

Let $\mathbf{A}'$ be the same vector as seen by the rotating S' frame; thus,

$$\mathbf{A}' = A'_x\hat{i}' + A'_y\hat{j}' + A'_z\hat{k}', \tag{7.6.2}$$

where $\hat{i}'$, $\hat{j}'$, and $\hat{k}'$ are unit vectors that belong to the S' coordinate system, because the magnitudes of the components of both vectors as well as the unit vector directions are different from those of the S frame. However, $\mathbf{A}$ does not change, because it is fixed; therefore, we must have that

$$\mathbf{A} = \mathbf{A}'. \tag{7.6.3}$$

Next, let's take the time derivative of this vector, and we have

$$\hat{i}\frac{dA_x}{dt} + \hat{j}\frac{dA_y}{dt} + \hat{k}\frac{dA_z}{dt} = \hat{i}'\frac{dA'_x}{dt} + \hat{j}'\frac{dA'_y}{dt} + \hat{k}'\frac{dA'_z}{dt} + A'_x\frac{d\hat{i}'}{dt} + A'_y\frac{d\hat{j}'}{dt} + A'_z\frac{d\hat{k}'}{dt};$$
$$\tag{7.6.4}$$

that is,

$$\dot{\mathbf{A}} = \dot{\mathbf{A}}' + A_x'\frac{d\hat{i}'}{dt} + A_y'\frac{d\hat{j}'}{dt} + A_z'\frac{d\hat{k}'}{dt}. \tag{7.6.5}$$

The term on the left is what we normally expect to be the time derivative of the left side of (7.6.3). The first term on the right is the contribution due to the change in magnitude, which would still be present even if the S' frame were not rotating. The rest of the terms on the right are the contribution due to the fact that S' is rotating.

We need to find the derivative of the unit vectors in (7.6.5). Consider Figure 7.9, where we imagine the rotating three-dimensional coordinate system for which the only unit vector of interest is the $\hat{i}'$ unit vector. Let θ be the angle between $\boldsymbol{\omega}$ and $\hat{i}'$.

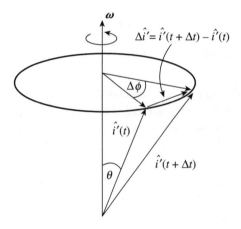

| FIGURE 7.9   The rate of change of a unit vector in the S' frame due to the frame's rotation.

From the figure, we see that $|\Delta\hat{i}'| = |\hat{i}'(t)\sin(\theta)|\Delta\phi$ with $\Delta\phi = |\boldsymbol{\omega}|\Delta t$, so that, according to the right-hand rule, we can write $\Delta\hat{i}' = \hat{i}'(t + \Delta t) - \hat{i}'(t) = (\boldsymbol{\omega} \times \hat{i}')\Delta t$, or

$$\frac{d\hat{i}'}{dt} = \lim_{\Delta t \to 0}\frac{\Delta\hat{i}'}{\Delta t} = \boldsymbol{\omega} \times \hat{i}'. \tag{7.6.6}$$

Similarly, we can repeat the above process for $\hat{j}'$ and $\hat{k}'$ to get $d\hat{j}'/dt = \boldsymbol{\omega} \times \hat{j}'$ and $d\hat{k}'/dt = \boldsymbol{\omega} \times \hat{k}'$. Putting these results together into (7.6.5), we find

$$\dot{\mathbf{A}} = \dot{\mathbf{A}}' + \boldsymbol{\omega} \times (A_x'\hat{i}' + A_y'\hat{j}' + A_z'\hat{k}') = \dot{\mathbf{A}}' + \boldsymbol{\omega} \times \mathbf{A}', \tag{7.6.7a}$$

or

$$\left(\frac{d\mathbf{A}}{dt}\right)_{fixed} = \left(\frac{d\mathbf{A}}{dt}\right)_{rot} + \boldsymbol{\omega} \times \mathbf{A}', \qquad (7.6.7b)$$

where $(d\mathbf{A}/dt)_{rot} \equiv \dot{\mathbf{A}}'$, i.e., the time derivative of $\mathbf{A}$ as measured by the S' frame. Equations (7.6.7) express the time derivative in the fixed frame and equals to what would be found in the rotating frame plus the extra term $\boldsymbol{\omega} \times \mathbf{A}'$. Notice that for a scalar quantity, this second term vanishes and the time rate of change of the scalar would be identical in both frames of reference.

We next apply this result to a position vector. If the S' frame is rotating only, then the position vector is

$$\mathbf{r} = \mathbf{r}'. \qquad (7.6.8)$$

The velocity is

$$\mathbf{v} = \frac{d\mathbf{r}}{dt} = \left(\frac{d\mathbf{r}}{dt}\right)_{rot} + \boldsymbol{\omega} \times \mathbf{r}' = \mathbf{v}' + \boldsymbol{\omega} \times \mathbf{r}', \qquad (7.6.9)$$

where $\mathbf{v}'$ is the velocity as seen by the rotating frame. The acceleration is

$$\mathbf{a} = \left(\frac{d}{dt}\right)\Bigg|_{fixed} (\mathbf{v}' + \boldsymbol{\omega} \times \mathbf{r}') = \left(\frac{d\mathbf{v}'}{dt}\right)\Bigg|_{fixed} + \left[\left(\frac{d\boldsymbol{\omega}}{dt}\right)\Bigg|_{fixed}\right] \times \mathbf{r}' + \boldsymbol{\omega} \times \left(\frac{d\mathbf{r}'}{dt}\right)\Bigg|_{fixed}.$$

However, $(d\boldsymbol{\omega}/dt)_{fixed} = (d\boldsymbol{\omega}/dt)_{rot} + \boldsymbol{\omega} \times \boldsymbol{\omega}'$, and $\boldsymbol{\omega} \times \boldsymbol{\omega}' = 0$ since by (7.6.3) $\boldsymbol{\omega} = \boldsymbol{\omega}'$, so that $(d\boldsymbol{\omega}/dt)_{fixed} = (d\boldsymbol{\omega}/dt)_{rot}$. This indicates that

$$\mathbf{a} = \left(\frac{d\mathbf{v}'}{dt}\right)\Bigg|_{rot} + \boldsymbol{\omega} \times \mathbf{v}' + \left[\left(\frac{d\boldsymbol{\omega}}{dt}\right)\Bigg|_{rot}\right] \times \mathbf{r}' + \boldsymbol{\omega} \times \left\{\left(\frac{d\mathbf{r}'}{dt}\right)\Bigg|_{rot} + \boldsymbol{\omega} \times \mathbf{r}'\right\}$$

or

$$\mathbf{a} = \left(\frac{d}{dt}\right)_{rot} (\mathbf{v}' + \boldsymbol{\omega} \times \mathbf{r}') + \boldsymbol{\omega} \times (\mathbf{v}' + \boldsymbol{\omega} \times \mathbf{r}')$$

$$= \dot{\mathbf{v}}' + \boldsymbol{\omega} \times \dot{\mathbf{r}}' + \dot{\boldsymbol{\omega}} \times \mathbf{r}' + \boldsymbol{\omega} \times \mathbf{v}' + \boldsymbol{\omega} \times (\boldsymbol{\omega} \times \mathbf{r}'),$$

which becomes

$$\mathbf{a} = \mathbf{a}' + 2\boldsymbol{\omega} \times \mathbf{v}' + \dot{\boldsymbol{\omega}} \times \mathbf{r}' + \boldsymbol{\omega} \times (\boldsymbol{\omega} \times \mathbf{r}'), \tag{7.6.10}$$

with $\mathbf{a}'$ the acceleration seen by the rotating frame.

More generally, to include a possible translational motion of the S' frame whose initial location, speed, and acceleration are $\mathbf{r}_0$, $\mathbf{v}_0$ and $\mathbf{a}_0$, respectively, we write

$$\mathbf{r} = \mathbf{r}_0 + \mathbf{r}', \tag{7.6.11a}$$

$$\mathbf{v} = \mathbf{v}_0 + \mathbf{v}' + \boldsymbol{\omega} \times \mathbf{r}', \tag{7.6.11b}$$

and

$$\mathbf{a} = \mathbf{a}_0 + \mathbf{a}' + 2\boldsymbol{\omega} \times \mathbf{v}' + \dot{\boldsymbol{\omega}} \times \mathbf{r}' + \boldsymbol{\omega} \times (\boldsymbol{\omega} \times \mathbf{r}'). \tag{7.6.11c}$$

Thus if the motion of a body is known as seen from a S' frame, in addition to the quantities $\mathbf{r}_0$, $\mathbf{v}_0$, $\mathbf{a}_0$, and $\boldsymbol{\omega}$, then using Equations (7.6.11) we can find the body's motion as seen in the S frame. Because $\mathbf{f} = m\mathbf{a}$ holds in the S frame and not in the S' frame, then $\mathbf{r}$, $\mathbf{v}$, and $\mathbf{a}$ are the ones to be understood and determined. To find f' we use (7.6.11c) and write for a body of mass m,

$$f' = m\mathbf{a}' = m\mathbf{a} - m\mathbf{a}_0 - 2m\boldsymbol{\omega} \times \mathbf{v}' - m\dot{\boldsymbol{\omega}} \times \mathbf{r}' - m\boldsymbol{\omega} \times (\boldsymbol{\omega} \times \mathbf{r}'), \tag{7.6.12}$$

that is,

$$f' = f + f_{fict}, \tag{7.6.13}$$

where in this general case

$$f_{fict} = -m\mathbf{a}_0 - 2m\boldsymbol{\omega} \times \mathbf{v}' - m\dot{\boldsymbol{\omega}} \times \mathbf{r}' - m\boldsymbol{\omega} \times (\boldsymbol{\omega} \times \mathbf{r}'). \tag{7.6.14}$$

The first term on the right is due to relative acceleration. The second is due to the Coriolis force, which is always perpendicular to both $\boldsymbol{\omega}$ and $\mathbf{v}'$. The third term involves the angular acceleration and represents a transverse force that is always perpendicular to both the radius vector $\mathbf{r}'$ and the angular acceleration $\dot{\boldsymbol{\omega}}$. The last term is the centrifugal force.

EXAMPLE 7.5

In Figure 7.10, a wheel of radius b that rotates with angular frequency ω_0 and whose center moves with a linear velocity v_0 is shown. Find the instantaneous location of a point P (at the top of the wheel), its velocity, and acceleration in the S and S' reference frames. Assume that the S' reference frame's center, located on the wheel, coincides with the wheel as the wheel rotates around it; that is, the S' frame keeps its axes aligned with the S frame, which is at rest, but its origin does follow the wheel's center. Also, assume there is no slipping.

....

Solution

As a preliminary discussion, it is worth noting that the wheel of radius b rotates with angular speed ω_0, travels a distance $d = 2\pi b$ in time $t = \tau = 2\pi/\omega_0$, so that its center of mass speed, if there is no slipping, is $\mathbf{v}_{cm} = (d/\tau)\hat{j} = b\omega_0\hat{j} \equiv v_0\hat{j}$. A point P at the top edge of the wheel, as in the case of Figure 7.10, has a velocity $\mathbf{v} = \mathbf{v}_{cm} + b\omega_0\hat{j} = 2v_0\hat{j}$.

Next, referring to Figure 7.10, and looking at the primed coordinate system, at this instant in time, $\mathbf{r}' = b\hat{k}$, $\dot{\mathbf{r}}' = v_0\hat{j} = b\omega_0\hat{j}$, and $\ddot{\mathbf{r}}' = -(v_0^2/b)\hat{k}$, because the wheel is not accelerating, but point P moves around the S' frame. Next, according to the S frame, using Equation (7.6.11a), we have $\mathbf{r} = v_0t\hat{j} + 2b\hat{k}$ for

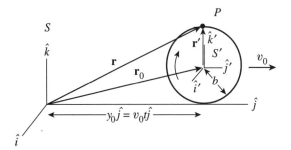

I FIGURE 7.10   Example 7.5 for a rotating wheel.

the position. For the velocity, because the S' coordinate system moves at $\mathbf{v}_0 = v_0\hat{j}$ but does not rotate ($\boldsymbol{\omega} = 0$), from (7.6.11b) we get $\mathbf{v} = 2v_0\hat{j}$, which agrees with our previous discussion. Finally, for the acceleration, with $\mathbf{a}_0 = 0$, we get from (7.6.11c) and the above results, $\mathbf{a} = -(v_0^2/b)\hat{k}$. A variation of this problem is to let the S' frame rotate with the wheel, but the results do not change.

EXAMPLE 7.6

In Figure 7.11, a wheel of radius R and mass M, while fixed at position h, is made to rotate by a falling mass m through a rope wrapped around the wheel as shown. The S' coordinate system is painted on the wheel. At the instant m is let go, the S' coordinate system's orientation coincides with that of the S coordinate system. Find the instantaneous position, velocity, and acceleration of point P as seen from the S frame of reference.

Solution

First, one needs to find the acceleration produced by the falling mass m. With the tension on the rope T, we can write $T - mg = -ma$ for the mass and $TR = I\alpha$ for the torque on the wheel, where α is the angular acceleration and I is the wheel's moment of inertia. The first gives $T = m(g - a)$, which when substituted into the second, with $\alpha = a/R$ and $I = MR^2$, we can solve for a to obtain $a = g^*$, where $g^* \equiv mg/[M + m]$. With this, at this instant, we are ready to identify the following: $\mathbf{r}_0 = b\hat{j} + h\hat{k}$, $\mathbf{r}' = R\hat{j}$. Because S' is rotating with the

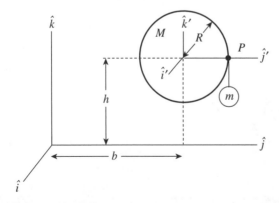

FIGURE 7.11   Example 7.6 for a falling mass that causes the rotation of the wheel through a rope.

wheel we also have $\mathbf{v}' = 0 = \mathbf{a}'$, $\boldsymbol{\omega} = -\alpha t\hat{i} = -(g^*/R)t\hat{i}$, and $\dot{\boldsymbol{\omega}} = -(g^*/R)\hat{i}$, for clockwise rotation. The wheel does not have any translational motion, so that $\mathbf{v}_0 = 0 = \mathbf{a}_0$. Finally, using (7.6.11), we find the position $\mathbf{r} = (b + R)\hat{j} + h\hat{k}$, the velocity $\mathbf{v} = 0 + 0 + [-(g^*/R)t\hat{i}] \times [R\hat{j}] = -g^*t\hat{k}$, and the acceleration

$$\mathbf{a} = 0 + 0 + 2[-(g^*/R)t\hat{i}] \times 0 + [-(g^*/R)\hat{i}] \times [R\hat{j}]$$

$$+ [-(g^*/R)t\hat{i}] \times ([-(g^*/R)t\hat{i}] \times [R\hat{j}])$$

or $\mathbf{a} = -g^*\hat{k} + [-(g^*/R)t\hat{i}] \times (-g^*t\hat{k}) = -[(g^*t)^2/R]\hat{j} - g^*\hat{k}$. That is, point P has two components of acceleration, one due to centripetal and the other due to gravity. The centripetal component increases with time as the mass m continues to fall.

■ 7.7   Applications to the Rotating Earth

Earth is a noninertial system by virtue of its rotation about its own axis, its translational motion around the sun, its motion associated with the movement of the sun about the center of the Milky Way, as well as any other motion involving galactic interactions. In what follows we consider Earth's spin about its own axis only.

Static Case: Object in S′ Frame at Rest

As an example, consider an object at rest on the surface of Earth. Let us find the normal force acting on the mass as a function of the latitude angle. We will let the S' frame be located at the center of Earth but rotating with it at a position $\mathbf{r}_0$ from the S frame, as shown in Figure 7.12(a). At the instant shown, the S and S' frames' unit vectors coincide. The radius of Earth is $R = 6.37 \times 10^6$ m, and its angular velocity is written as $\boldsymbol{\omega} = \omega_0\hat{k}$ with $\dot{\boldsymbol{\omega}} = 0$, where because Earth does a revolution per day, we have $\omega_0 = (2\pi \text{ rad}/86400 \text{ s}) = 7.272 \times 10^{-5}$ rad/s.

Since the length of $\mathbf{r}_0$ is arbitrary, we take it to be zero. Because Earth's only assumed motion is that of its rotation, we have $\mathbf{v}_0 = 0 = \mathbf{a}_0$. In Figure 7.12(a), θ is the latitude angle, so that $\mathbf{r}' = R\cos\theta\,\hat{j} + R\sin\theta\,\hat{k}$, because the mass is at rest at $\mathbf{v}' = 0$ and $\mathbf{a}' = 0$. From (7.6.11) for the position we have

$$\mathbf{r} = \mathbf{r}_0 + \mathbf{r}' = R\cos\theta\,\hat{j} + R\sin\theta\,\hat{k}. \tag{7.7.1}$$

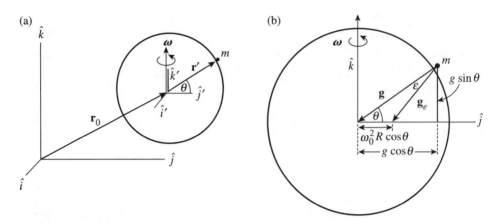

FIGURE 7.12 (a) An object at rest on the surface of a rotating Earth. (b) Due to Earth's rotation, the object experiences a net acceleration at the plumb line angle with respect to the direction of the true gravitational force.

For the velocity, we have

$$\mathbf{v} = 0 + 0 + (\omega_0 \hat{k}) \times (R\cos\theta\,\hat{j} + R\sin\theta\,\hat{k}) = -\omega_0 R\cos\theta\,\hat{i}, \qquad (7.7.2)$$

and for the acceleration we have

$$\mathbf{a} = 0 + 0 + 2(\omega_0\hat{k}) \times (0) + 0 \times \mathbf{r}' + (\omega_0\hat{k}) \times (\omega_0\hat{k} \times (R\cos\theta\,\hat{j} + R\sin\theta\,\hat{k})),$$

or

$$\mathbf{a} = -\omega_0^2 R\cos\theta\,\hat{j}. \qquad (7.7.3)$$

With this acceleration, we can write Newton's second law as

$$\sum \mathbf{F} = \mathbf{N} + m\mathbf{g} = -m\omega_0^2 R\cos\theta\,\hat{j}. \qquad (7.7.4)$$

At the point of the mass, at the angle θ, $\mathbf{g} = -g\cos\theta\,\hat{j} - g\sin\theta\,\hat{k}$, where the magnitude of Earth's acceleration at its surface is taken as $g = GM/R^2$. With $G = 6.67 \times 10^{-11}\,\dfrac{\text{N}\cdot\text{m}^2}{\text{kg}^2}$, $M = 5.98 \times 10^{24}$ kg, and R as before, we have $g \approx 9.83\,\text{m/s}^2$. From (7.7.4) we then have for the sought normal force

$$\mathbf{N} = m(g - \omega_0^2 R)\cos\theta\,\hat{j} + mg\sin\theta\,\hat{k} \equiv -m\mathbf{g}_e, \qquad (7.7.5)$$

where we have defined the effective value of Earth's gravitational acceleration $\mathbf{g}_e = -(g - \omega_0^2 R)\cos\theta\,\hat{j} - g\sin\theta\,\hat{k}$. At the equator, where $\theta = 0°$, the y component reaches its maximum value of $g = 9.83 \text{ m/s}^2$ decreased by $(7.272 \times 10^{-5} \text{rad/s})^2 (6.37 \times 10^6 \text{ m}) \approx 0.0337 \text{ m/s}^2$ due to centripetal acceleration; that is, $g_e(\theta = 0) = (g - \omega_0^2 R) = (9.83 - 0.0337) \approx 9.8 \text{ m/s}^2$. This is only about 0.34% change from the value of g. At the North Pole, where $\theta = 90°$, the z component reaches the expected maximum value of 9.83 m/s^2. Figure 7.12(b) shows the vectors $\mathbf{g}$ and $\mathbf{g}_e$ and their components. A plumb line on Earth will point along $\mathbf{g}_e$, and the plumb line angle ε between $\mathbf{g}$ and $\mathbf{g}_e$ measures the deviation between the two. From Figure 7.12(b), and using the law of sines, we see that $\dfrac{\sin \varepsilon}{\omega_0^2 R \cos \theta} = \dfrac{\sin \theta}{g_e(\theta)}$, so

$\varepsilon = \sin^{-1} \left(\dfrac{\omega_0^2 R \cos \theta \sin \theta}{g_e(\theta)} \right)$ or $\varepsilon = \sin^{-1} \left(\dfrac{\omega_0^2 R \sin 2\theta}{2 g_e(\theta)} \right)$. This deviation angle depends on the latitude angle θ; therefore, ε is zero at the North Pole as well as at the equator.

At a latitude of $\theta = 34°$, the magnitude of $\mathbf{g}_e$ is

$$g_e(\theta = 34°) = \sqrt{((g - \omega_0^2 R) \cos \theta)^2 + (g \sin \theta)^2}\Big|_{\theta = 34°}$$

$$= \sqrt{((9.83 - 0.0337) \cos 34°)^2 + (9.83 \sin 34°)^2} \approx 9.81 \text{ m/s}^2,$$

and so $\varepsilon(\theta = 34°) = \sin^{-1} \left(\dfrac{0.0337 \sin 68°}{2 \cdot 9.81} \right) = 1.59 \times 10^{-3} \text{rad} \approx 0.091°$.

Dynamic Case: Motion Along Earth's Radius

We next consider a situation where an object is free-falling near Earth's surface. Let's say that we are interested in the motion of the object from the point of view of the rotating Earth. In this case, we would be interested in the behavior of $\mathbf{r}'(t)$. First, looking at Figure 7.13(a), we notice that the S frame is located at the center of Earth and

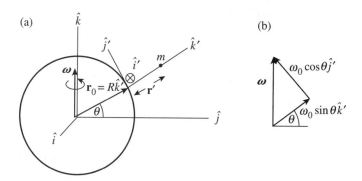

FIGURE 7.13   (a) A mass falling over the spinning Earth. (b) Components of Earth's rotation in the rotating frame at the latitude of interest.

fixed in space and that the S' frame is now located at a point on Earth's surface at a latitude angle θ.

The direction coordinates of the S' frame are such that $\hat{i}'$ is into the plane of the paper and defines a right-handed coordinate system as shown. As far as the S frame goes, $\boldsymbol{\omega}$ points in the $\hat{k}$ direction, but as seen from the S' frame, it has components along the $\hat{j}'$ and $\hat{k}'$ directions as shown in Figure 7.13(b), or $\boldsymbol{\omega} = \omega_0 \cos\theta \hat{j}' + \omega_0 \sin\theta \hat{k}'$, where $\omega_0 = 7.272 \times 10^{-5}$ rad/s as calculated before but with $\dot{\boldsymbol{\omega}} = 0$. From Figure 7.13(a), we have from (7.6.11a) that $\mathbf{r} = R\hat{k}' + \mathbf{r}'$. Because the only motion of the S' frame is its rotation, then $\mathbf{v}_0 = 0$, and from (7.6.11b) $\mathbf{v} = \mathbf{v}' + (\omega_0 \cos\theta \hat{j}' + \omega_0 \sin\theta \hat{k}') \times \mathbf{r}'$, which we write for completeness. What we seek, however, is the equation of motion of $\mathbf{r}'$. This is obtained from (7.6.11c), which, since $\mathbf{a}_0 = 0 = \dot{\boldsymbol{\omega}}$, we write it as

$$\mathbf{a} = \ddot{\mathbf{r}}' + 2\boldsymbol{\omega} \times \dot{\mathbf{r}}' + \boldsymbol{\omega} \times (\boldsymbol{\omega} \times \mathbf{r}'), \tag{7.7.6}$$

where we have used $\mathbf{a}' = \ddot{\mathbf{r}}'$ and $\mathbf{v}' = \dot{\mathbf{r}}'$. Since for an object in free-fall the inertial frame sees that the falling object experiences a gravitational acceleration that points toward the center of Earth ($\mathbf{a} = -g\hat{k}'$), then we can write an equation of motion for $\mathbf{r}'$ as follows:

$$\ddot{\mathbf{r}}' = -g\hat{k}' - 2\boldsymbol{\omega} \times \dot{\mathbf{r}}' - \boldsymbol{\omega} \times (\boldsymbol{\omega} \times \mathbf{r}'). \tag{7.7.7a}$$

Further, because the last term involves a quadratic factor of ω_0, this term is expected to be very small near Earth's surface and can be neglected. This leaves only two forces present: the force due to gravity and the Coriolis force. If we also write $\mathbf{r}' = x'\hat{i}' + y'\hat{j}' + z'\hat{k}'$ and perform the cross product,

$$\boldsymbol{\omega} \times \dot{\mathbf{r}}' = \begin{vmatrix} \hat{i}' & \hat{j}' & \hat{k}' \\ 0 & \omega_0\cos\theta & \omega_0\sin\theta \\ \dot{x}' & \dot{y}' & \dot{z}' \end{vmatrix} = (\dot{z}'\omega_0\cos\theta - \dot{y}'\omega_0\sin\theta)\hat{i}' + \dot{x}'\omega_0\sin\theta\hat{j}' - \dot{x}'\omega_0\cos\theta\hat{k}'.$$

$$\tag{7.7.7b}$$

Equation (7.7.6) yields the three coupled equations of motion of the falling object as seen by the rotating Earth-based frame

$$\ddot{x}' = -2\dot{z}'\omega_0\cos\theta + 2\dot{y}'\omega_0\sin\theta$$

$$\ddot{y}' = -2\dot{x}'\omega_0\sin\theta \tag{7.7.8}$$

$$\ddot{z}' = -g + 2\dot{x}'\omega_0\cos\theta.$$

A standard use of these equations is to integrate both sides once over time to obtain

$$\dot{x}' = -2\omega_0(z'\cos\theta - y'\sin\theta) + \dot{x}'_0$$
$$\dot{y}' = -2x'\omega_0\sin\theta + \dot{y}'_0 \qquad\qquad (7.7.9)$$
$$\dot{z}' = -gt + 2x'\omega_0\cos\theta + \dot{z}'_0$$

From these, if we put $\dot{y}'$ and $\dot{z}'$ into the equation for $\ddot{x}'$, we get

$$\ddot{x}' = -2\omega_0\dot{z}'_0\cos\theta + 2\omega_0 gt\cos\theta - 4\omega_0^2 x'\cos^2\theta - 4\omega_0^2 x'\sin^2\theta + 2\omega_0\dot{y}'_0\sin\theta,$$

which, if the ω_0^2 terms are neglected due to their small contribution, becomes

$$\ddot{x}' \approx 2\omega_0 gt\cos\theta - 2\omega_0(\dot{z}'_0\cos\theta - \dot{y}'_0\sin\theta).$$

Integrating this twice over time gives

$$x' = \omega_0 gt^3\cos\theta/3 - \omega_0(\dot{z}'_0\cos\theta - \dot{y}'_0\sin\theta)t^2 + \dot{x}'_0 t + x'_0. \qquad (7.7.10)$$

Putting this result for x' into $\dot{y}'$ and $\dot{z}'$ of (7.7.9) and neglecting the ω_0^2 followed by an integration over time, we get

$$y' \approx -\dot{x}'_0 t^2\omega_0\sin\theta - 2x'_0\omega_0 t\sin\theta + \dot{y}'_0 t + y'_0, \qquad (7.7.11)$$

and

$$z' \approx -gt^2/2 + (\dot{x}'_0\omega_0 t^2 + 2x'_0 t\omega_0)\cos\theta + \dot{z}'_0 t + z'_0. \qquad (7.7.12)$$

Let us now consider the mass initially at rest, at a height $z'_0 = h$, and where $x'_0 = y'_0 = \dot{x}'_0 = \dot{y}'_0 = \dot{z}'_0 = 0$. If the object is let go, where does it land? Because when it lands it is at $z' = 0$, then from (7.7.12), the time taken to reach ground is $t_g = \sqrt{2h/g}$, which when put into (7.7.10) yields

$$x' = \omega_0 g(2h/g)^{3/2}\cos\theta/3. \qquad (7.7.13)$$

This indicates the mass lands east of its initial position. The Coriolis force is responsible for this deviation. One way to think of the Coriolis force is through a time rate of change of angular momentum. A mass performing circular motion has angular momentum

$$L' = r' \times p' = mr' \times v' = mr' \times (\omega \times r') = m[\omega(r' \cdot r') - r'(r' \cdot \omega)],$$
$$(7.7.14)$$

where we have used the cross product rule $A \times (B \times C) = B(A \cdot C) - C(A \cdot B)$ and the fact that $v' = \omega \times r'$. While not generally true, if for the moment we let r' and dr'/dt be perpendicular to ω, which is constant, then we get the torque

$$\tau' = \frac{d}{dt}L' = 2m\omega\left(r' \cdot \frac{dr'}{dt}\right) = 2mr' \times (\omega \times v').$$
$$(7.7.15)$$

This torque can be identified with a force; that is, $\tau' = r' \times (-F'_{Cor})$, where we see that

$$F'_{Cor} = -2m\omega \times v'$$
$$(7.7.16)$$

is the fictitious Coriolis force term of (7.6.14). Thus, the Coriolis force arises due to the torque needed in order to move an object along a radius. It is a sidewise force applied at a distance r' from the center of rotation. This force is actually a fictitious force in the case of the rotating Earth, because no deflection will be seen (still taking $x'_0 = y'_0 = \dot{x}'_0 = \dot{y}'_0 = \dot{z}'_0 = 0$) by the inertial reference frame observer; however, because Earth does rotate, the noninertial observer will see such deflection. In the case of a merry-go-round, though, an observer on the rotating platform will have to apply a sideways force in order to move radially outward. In fact, the effect will be visible to an inertial observer outside the platform in the form of a curved path followed by the noninertial observer.

Dynamic Case: Motion Along Earth's Surface

Referring to Figure 7.13, we now pick v' along the surface of Earth, so $v' = v'_{0x}\hat{i}' + v'_{0y}\hat{j}'$. Then using the third term, i.e., the Coriolis force term from (7.6.12), we have

$$F_{Cor} = -2m\omega \times v' = -2m(\omega_0\cos\theta\,\hat{j}' + \omega_0\sin\theta\,\hat{k}') \times (v'_{0x}\hat{i}' + v'_{0y}\hat{j}')$$
$$= -2m\omega_0(-v'_{0y}\sin\theta\,\hat{i}' + v'_{0x}\sin\theta\,\hat{j}' - v'_{0x}\cos\theta\,\hat{k}') = F_H + F_V,$$
$$(7.7.17)$$

where $\mathbf{F}_H$ and $\mathbf{F}_V$ are the horizontal and vertical components of the force, respectively. The horizontal component

$$\mathbf{F}_H = 2m\omega_0\sin\theta\,(v'_{0y}\hat{i}' - v'_{0x}\hat{j}')\qquad(7.7.18)$$

is particularly interesting because it is always perpendicular to the above surface $\mathbf{v}'$, since $\mathbf{F}_H \cdot \mathbf{v}' = 0$. Its magnitude is

$$F_H = 2m\omega_0 v'\sin\theta,\qquad(7.7.19)$$

where $v' = \sqrt{v'^2_{0x} + v'^2_{0y}}$; that is, its magnitude is independent of the direction of the surface $\mathbf{v}'$. Thus, in the absence of other horizontal forces, it would produce circular motion, clockwise in the northern hemisphere and counterclockwise in the southern hemisphere, due to the $\sin\theta$ contribution, as shown in Figure 7.14(a).

Air flow on Earth is strongly influenced by this Coriolis effect. At the equator, where $\theta = 0$, unless there is some residual angular momentum, the Coriolis force is negligible and weather systems have less probability of formation. However, the dynamics of weather systems have other effects that this example does not encompass, as Figure 7.14(b) shows.

For example, considering a low-pressure region in the northern hemisphere as shown in Figure 7.14(b), pressure gradients cause air masses to flow into the low-pressure region as indicated by the dashed arrows. The Coriolis force causes a sideways

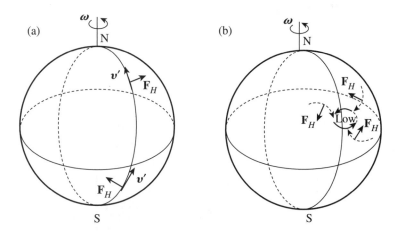

FIGURE 7.14   (a) Coriolis forces perpendicular to the motion along the surface. (b) Air flow on Earth's surface affected by Coriolis and pressure forces.

deflection of wind flow, and the net result is a counterclockwise rotation of the winds within isobars (lines of constant pressure). In contrast, the opposite happens in a high-pressure region with a net effect that causes a clockwise rotation. Both of these situations are reversed in the southern hemisphere.

■ 7.8   The Foucault Pendulum

The Foucault pendulum refers to a pendulum system designed by Jean Foucault (1819–1868) in order to demonstrate Earth's rotation. The pendulum is ideally held from a nearly frictionless support, and care is taken that no torques are exerted there. It is also useful to have a large mass, in order for gravitational and Coriolis forces to be much greater than friction or drag forces. A long plumb line has a large period and helps to minimize friction losses. The pendulum, once set into motion, swings on a plane. Due to the rotation of Earth under it, the pendulum's plane of motion is seen to precess with a period that depends on the latitude. Figure 7.15(a) shows the pendulum at a latitude angle θ, and Figure 7.15(b) shows the pendulum as it swings with the appropriate tension components in each direction.

The equation of motion for the Foucault pendulum is as in (7.7.6), but now $m\mathbf{a} = \mathbf{T} + m\mathbf{g}$, where $\mathbf{g} = -g\hat{k}'$, and neglecting the small $\boldsymbol{\omega} \times (\boldsymbol{\omega} \times \mathbf{r}')$ term from the start, to write

$$\ddot{\mathbf{r}}' = \mathbf{T}/m - g\hat{k}' - 2\boldsymbol{\omega} \times \dot{\mathbf{r}}'. \tag{7.8.1}$$

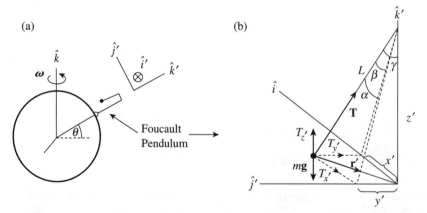

I FIGURE 7.15   (a) The Foucault pendulum and (b) the forces on it.

With $\mathbf{T} = T_x \hat{i}' + T_y \hat{j}' + T_z \hat{k}'$, and as before, $\mathbf{r}' = x'\hat{i}' + y'\hat{j}' + z'\hat{k}'$, $\boldsymbol{\omega} = \omega_0 \cos\theta \, \hat{j}' + \omega_0 \sin\theta \, \hat{k}'$, the cross product has already been carried out in (7.7.7), to get for each acceleration component

$$\ddot{x}' = -2\dot{z}'\omega_0 \cos\theta + 2\dot{y}'\omega_0 \sin\theta + T_{x'}/m$$

$$\ddot{y}' = -2\dot{x}'\omega_0 \sin\theta + T_{y'}/m \qquad (7.8.2)$$

$$\ddot{z}' = -g + 2\dot{x}'\omega_0 \cos\theta + T_{z'}/m.$$

We will consider only the x'-y' motion, so we set $\ddot{z}' = 0$. Also from Figure 7.15(b)

$$T_{x'} = -T \sin\alpha = -T\cos\beta = -Tx'/L, T_{y'} = -T\sin\beta = -T\cos\alpha = -Ty'/L,$$

and $T_{z'} = T\cos\gamma = Tz'/L$; furthermore, for small displacements (small γ) from the vertical, $T_{z'} \sim mg \sim T$. Thus, we have two coupled differential equations for the Foucault pendulum in the form

$$\ddot{x}' = 2\dot{y}'\omega_0 \sin\theta - gx'/L$$

$$\ddot{y}' = -2\dot{x}'\omega_0 \sin\theta - gy'/L. \qquad (7.8.3)$$

We wish to solve these equations for $x'(t)$ and $y'(t)$. First, in (7.8.3) let us label what in fact is Earth's latitude-dependent precessional frequency (which is also the $\hat{k}'$ component of $\boldsymbol{\omega}$) as $\omega_1 = \omega_0 \sin\theta$, and the Foucault pendulum swinging frequency as $\omega_f = \sqrt{g/L}$. Also, in order to convert the two coupled variable differential equations to one, it is useful to define the complex variable $r = x' + iy'$, where $x' = real(r)$, $y' = imag(r)$. Thus, multiplying the second of (7.8.3) by i and adding the result to the first and rearranging to get

$$\ddot{r} = -2i\omega_1 \dot{r} - (\omega_2^2 - \omega_1^2)r, \qquad (7.8.4)$$

where we have defined, $\omega_2 \equiv \sqrt{\omega_1^2 + \omega_f^2}$. This is now a standard differential equation that can be solved for $r(t)$ if we write it in the form

$$[D^2 + 2i\omega_1 D + (\omega_2^2 - \omega_1^2)]r = 0, \qquad (7.8.5)$$

where as usual $D \equiv d/dt$. The quantity in brackets has roots

$$D = -i\omega_1 \pm i\omega_2, \qquad (7.8.6)$$

which commonly suggests that there are two possible solutions to (7.8.4) of the form

$$r(t) = \{Ae^{-i(\omega_1 + \omega_2)t}, \, Be^{-i(\omega_1 - \omega_2)t}\}. \tag{7.8.7}$$

For the present problem, it suffices to take $A = B \equiv r_0/2$ and make a linear combination of these solutions to write

$$r(t) = r_0(e^{-i(\omega_1 + \omega_2)t} + e^{-i(\omega_1 - \omega_2)t})/2 = r_0 e^{-i\omega_1 t}\cos\omega_2 t. \tag{7.8.8}$$

Finally, letting the initial positions at time $t = 0$ be $r_0 = x'_0 + iy'_0$, we can separate the real and imaginary parts of (7.8.8) to obtain $x'(t)$ and $y'(t)$, respectively, as

$$x'(t) = x'_0\cos\omega_1 t\cos\omega_2 t + y'_0\sin\omega_1 t\cos\omega_2 t$$
$$y'(t) = -x'_0\sin\omega_1 t\cos\omega_2 t + y'_0\cos\omega_1 t\cos\omega_2 t. \tag{7.8.9}$$

EXAMPLE 7.7

Suppose a Foucault pendulum is located at 34°N latitude. (a) What is its precessional period? (b) If at time $t = 0$, the pendulum is displaced by 1 m in the x-direction, use the formulas in (7.8.9) to obtain a plot of the resulting motion during a 24-hour period. To make matters simple and illustrative, assume the period of the pendulum is 1 hour.

Solution
Figure 7.16 contains the plot sought for a 24-hour period at the 34° latitude.

(a) Earth rotates once every 24 hours, which gives $\omega_0 = 2\pi/24$ rad/hr, so that at 34°, $\tau_1 = 2\pi/\omega_1 = 24/\sin(34°)$ hr $= 42.9$ hr.

(b) The MATLAB script foucault.m was used for this purpose. A listing of the script follows. Since the pendulum has a period τ_f of 1 hour, its length is given by $L = (\tau_f/2\pi)^2 g$. The script plots the first 12 hours as blue and the second 12 hours as red. At $t = 0$, the motion in the figure starts at $x = 1$ and proceeds clockwise. Because this pendulum makes a full swing every hour, and since the precessional period is 42.9 hours at the 34°N latitude, the pendulum does not retrace itself fully, as seen in the figure by the colored swings. The color appears on the screen when the script is run. It will take 18.9 (42.9 − 24) more hours to do so; thus only about three blue swings were retraced by red swings in the figure. At the North Pole, because Earth rotates once every 24 hours, that will be

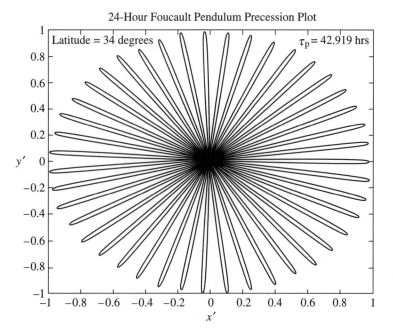

I FIGURE 7.16   Equations (7.8.9) for a Foucault pendulum with a 24-hour period.

the precessional period of the pendulum. There will be 24 distinct swings of the pendulum, with 12 blue and 12 red. In the figure one must be careful to notice that there are twice as many half swings as there are swings.

Finally, because Earth rotates under the pendulum, the solution of the differential equation of motion indicates that there will be a slight deviation in the period of the pendulum. The actual oscillatory frequency will be $\omega_2 \equiv \sqrt{\omega_1^2 + \omega_f^2}$, rather than ω_f. However, usually $\omega_1 \ll \omega_f$, then $\omega_2 \sim \omega_f$, and the deviation is very small.

SCRIPT

```
%foucault.m incorporates the solution to the Foucault pendulum formulas
clear;
lth=34;              % latitude angle in degrees
th=lth*2*pi/360;     % convert to radians
g=9.8;               % acceleration due to gravity
tf=3600;             % pendulum's period is one hour for convenience
```

```
L=(tf/2/pi)^2*g;       % pendulum's length
wf=2*pi/tf;            % Foucault pendulum frequency same as sqrt(g/L)
w0=2*pi/24/3600;       % Earth's rotational frequency
w1=w0*sin(th);         % Precession frequency
w2=sqrt(w1^2+wf^2);    % roughly pendulum frequency
x0=1.0; y0=0.0;        % Initial conditions
tp=abs(2*pi/w1);       % precession period
tmax=24*3600;          % 24-hour time limit
ts=tmax/1000;          % time step in seconds
t=[0:ts:tmax]; n=length(t); nh=round(n/2);
x=x0*cos(w1*t).*cos(w2*t)+y0*cos(w1*t).*sin(w2*t);
y=-x0*sin(w1*t).*cos(w2*t)+y0*cos(w1*t).*cos(w2*t);
str1=cat(2,'Latitude =',num2str(lth),' Degrees');
str2=cat(2,'\tau_p =',num2str(tp/3600),' hrs');
plot(x(1:nh),y(1:nh),'b') %first 12 hours as blue
hold on; plot(x(nh:n),y(nh:n),'r') %2nd 12 hours as red
title('24 Hour Foucault Pendulum Precession Plot','FontSize',14)
ylabel('y\prime','FontSize',14); xlabel('x\prime','FontSize',14);
text(min(x)*(1-0.05),max(y)*(1-0.1),str1,'FontSize',10,'Color','blue');
text(max(x)*(1-0.45),max(y)*(1-0.1),str2,'FontSize',10,'Color','blue');
```

■ Chapter 7 Problems

7.1 Using the expressions leading to the result of (7.2.3a), show that equations (7.2.3b) can be obtained from (7.2.2b).

7.2 Show that the polar coordinate version of **r**, Equation (7.2.1), is consistent with the standard Cartesian coordinate expression $\mathbf{r} = x\,\hat{i} + y\,\hat{j}$.

7.3 Consider Example 7.5 with the S' coordinate system painted on the wheel so that it rotates with the wheel. Obtain the instantaneous location of a point P (at the top of the wheel), its velocity, and its acceleration in the S and S' reference frames.

7.4 In Example 7.6, imagine that the S' reference frame is fixed in space instead of rotating with the wheel, but with its origin pinned to coincide with the wheel's. Let the direction vectors point in the same direction as the S frame. Obtain the expressions for the instantaneous position, velocity, and acceleration of point P as seen from the S frame of reference.

7.5 A one-gram bug is found on the surface of a spinning record. The record rotates at 45 rpm. If the bug moves at 2 cm/s radially outward, it requires a constant force to do so. What is the value of this force?

7.6 Derive Equations (7.8.9) that are solutions to the Foucault pendulum differential Equations (7.8.3).

7.7 A Foucault pendulum is built at the North Pole. If at time $t = 0$ the pendulum is displaced by 1 m in the x-direction, use the formulas in (7.8.9) to obtain a plot of the resulting motion during a 24-hour period. As in Example 7.7, assume the period of the pendulum is 1 hour. Also, what is the precessional period of the pendulum?

7.8 Derive Equations (a) 7.4.5a, (b) 7.4.5b, and (c) 7.4.5c.

7.9 Derive Equation (7.4.7a).

7.10 Derive Equation (7.4.7b).

■ Additional Problems

7.11 What is the plumb line angle corresponding to a north latitude of 45°?

7.12 At a 45° north latitude a 1800-kg helicopter is flying due east at 90 mi/hr. What is the magnitude and direction of the Coriolis force on the helicopter?

7.13 A rock is dropped from a height of 50 m above Earth's surface. If the rock is dropped from rest at a north latitude of 30°, what is the rock's lateral displacement (eastward) due to the Coriolis effect?

7.14 At a south latitude of 40° a Foucault pendulum is built and pegs are set up 2 m from below the suspension point 10° apart. How long does a visitor have to wait to see a peg hit just after one was knocked over?

7.15 A realistic Foucault pendulum is built at 35° north latitude. If at time $t = 0$, the pendulum is displaced by 1 m in the x-direction, use the formulas in (7.8.9) to obtain a plot of the resulting motion during its first 50 swings. Let the pendulum length be 30 m. In your answer include the precessional period, the pendulum period, and the angle covered by the precessing plane during the first 50 swings.

8 Central Forces

■ 8.1 Introduction

In classical mechanics we refer to a central force as the force a body exerts on another body with the force directed along the line connecting the centers of the bodies. A central force points along the radius, either toward or away from a center point, which is the origin or force center, and whose magnitude depends solely on the distance from the origin. Figure 8.1 shows an example of a general central force. In the figure, a mass m moves along the shown path. The mass is acted on by the force $\mathbf{F}(r)$, whose source is located at the origin. This force is a central force if it is directed along the radius vector $\mathbf{r} = r\hat{r}$.

The central force shown in the figure can thus be written as

$$\mathbf{F}(r) = F(r)\hat{r} = \frac{F(r)}{r}\mathbf{r}. \tag{8.1.1}$$

Central forces play a significant role in nature. Some examples of central forces are electrostatic, atomic, and molecular, as well as gravitational. In this chapter we will investigate the role of central forces and their characteristics. Applications are made to planetary motion, including Kepler's laws of motion. The chapter culminates with Newton's universal law of gravitation.

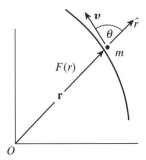

I FIGURE 8.1

■ 8.2   Central Forces and Potential Energy

A central force is of particular interest in mechanics because, as we will see, there is a well-defined potential energy function associated with it. In Chapter 6, we learned that a force leads to a potential energy function through

$$\mathbf{F}(r) = -\nabla V(r), \tag{8.2.1}$$

if the force is conservative; that is, $\nabla \times \mathbf{F}(r) = 0$. This is, in fact, what happens in the case of a central force. Writing (8.1.1) as

$$\mathbf{F}(r) = \frac{F(r)}{r}\mathbf{r} = \frac{F(r)}{r}(x\hat{i} + y\hat{j} + z\hat{k}), \tag{8.2.2}$$

and taking its curl, gives

$$\nabla \times \mathbf{F} = \begin{vmatrix} \hat{i} & \hat{j} & \hat{k} \\ \frac{\partial}{\partial x} & \frac{\partial}{\partial y} & \frac{\partial}{\partial z} \\ \frac{x}{r}F(r) & \frac{y}{r}F(r) & \frac{z}{r}F(r) \end{vmatrix} = \begin{aligned} &\hat{i}\left(\frac{\partial}{\partial y}\frac{z}{r}F(r) - \frac{\partial}{\partial z}\frac{y}{r}F(r)\right) - \hat{j}\left(\frac{\partial}{\partial x}\frac{z}{r}F(r) - \frac{\partial}{\partial z}\frac{x}{r}F(r)\right) \\ &+\hat{k}\left(\frac{\partial}{\partial x}\frac{y}{r}F(r) - \frac{\partial}{\partial y}\frac{x}{r}F(r)\right) \end{aligned}.$$

$$\tag{8.2.3}$$

Notice, however, that in the first expression in parentheses on the right,

$$\frac{\partial}{\partial y}\frac{z}{r}F(r) = z\frac{\partial}{\partial y}\frac{F(r)}{r} = z\frac{\partial}{\partial r}\left[\frac{F(r)}{r}\right]\frac{\partial r}{\partial y}, \text{ and similarly, } \frac{\partial}{\partial z}\frac{y}{r}F(r) = y\frac{\partial}{\partial r}\left[\frac{F(r)}{r}\right]\frac{\partial r}{\partial z}.$$

Also, since $r = \sqrt{x^2 + y^2 + z^2}$, $\frac{\partial r}{\partial y} = \frac{y}{r}$, and similarly $\frac{\partial r}{\partial z} = \frac{z}{r}$, so that this first term vanishes. In a similar way one can show that each of the remaining terms also vanishes. We can conclude that all central forces are conservative, or

$$\nabla \times \mathbf{F}(r) = 0 \Rightarrow \mathbf{F}(r) = -\nabla V(r), \tag{8.2.4}$$

where $V(r)$ is the potential energy function associated with the central force. If we write ∇ in spherical coordinates

$$\nabla = \frac{\partial}{\partial r}\hat{r} + \frac{1}{r}\frac{\partial}{\partial \theta}\hat{\theta} + \frac{1}{r\sin\theta}\frac{\partial}{\partial \phi}\hat{\phi} \tag{8.2.5}$$

then since $\mathbf{F}(r)$ is central—that is, it does not depend on θ or ϕ—we can ignore the last two terms of the del operator, so that the right-hand side (RHS) of (8.2.4) becomes

$$\nabla V(r) \rightarrow \frac{\partial V}{\partial r}\hat{r} = -F(r)\hat{r} \quad \Rightarrow \quad V(r) = -\int_{r_0}^{r} F(r)\,dr. \tag{8.2.6}$$

In conclusion, this result says that the potential associated with a central force can be found by performing the integral of the force with respect to the variable r.

■ 8.3   Angular Momentum of a Central Force System

First, recall that Newton's second law for a particle of linear momentum $\mathbf{p} = m\mathbf{v}$ can be written as the time derivative of the momentum

$$\mathbf{F} = \frac{d\mathbf{p}}{dt}. \tag{8.3.1}$$

Referring to Figure 8.1, by definition, the angular momentum about the origin is obtained by

$$\mathbf{L} = \mathbf{r} \times \mathbf{p}, \tag{8.3.2}$$

where $\mathbf{r}$ is the moment arm. Taking the time derivative of this, we see that $d\mathbf{L}/dt = (d\mathbf{r}/dt) \times \mathbf{p} + \mathbf{r} \times (d\mathbf{p}/dt)$. The first term is just $m\mathbf{v} \times \mathbf{v}$ and thus vanishes, because the velocity is parallel to itself. We then find that

$$\frac{d\mathbf{L}}{dt} = \mathbf{r} \times (d\mathbf{p}/dt) = \mathbf{r} \times \mathbf{F} = \boldsymbol{\tau}; \tag{8.3.3}$$

that is, the time rate of change of a body's angular momentum is the torque about the origin exerted on that body by the force. The significance of (8.3.3) regarding central forces is that since $\mathbf{F} = F(r)\hat{r}$, and $\mathbf{r} \propto \hat{r}$, then the cross product

$$\boldsymbol{\tau} = \mathbf{r} \times \mathbf{F} = 0; \tag{8.3.4}$$

thus, central forces produce no torques. Consequently, looking at the left-hand side of (8.3.3), the implication is that

$$\frac{d\mathbf{L}}{dt} = 0, \tag{8.3.5a}$$

or

$$\mathbf{L} = const; \tag{8.3.5b}$$

that is, the angular momentum is a constant of the motion and therefore conserved under the action of a central force. Further insight into this result is obtained from (8.3.2); because the angular momentum magnitude and direction do not change, with its direction perpendicular to both $\mathbf{r}$ and $\mathbf{p}$, a body's motion under a central force is confined to a plane. Under these circumstances, only two dimensions are needed to describe the motion under a central force. The motion can be described in plane polar coordinates whose variables are **r** and **θ** as in Chapter 7. Thus, we can write Newton's second law, with the use of (7.2.10), as

$$\mathbf{F} = m\mathbf{a} = F(r)\hat{r} = m[(\ddot{r} - r\dot{\theta}^2)\hat{r} + (r\ddot{\theta} + 2\dot{r}\dot{\theta})\theta], \tag{8.3.6}$$

which by equating terms gives

$$F(r) = m(\ddot{r} - r\dot{\theta}^2) \tag{8.3.7a}$$

and

$$0 = (r\ddot{\theta} + 2\dot{r}\dot{\theta}). \tag{8.3.7b}$$

These two equations form a coupled system of differential equations for r and θ. It is generally noticed that the term

$$\frac{d}{dt}(mr^2\dot{\theta}) = 2mr\dot{r}\dot{\theta} + mr^2\ddot{\theta} = mr(2\dot{r}\dot{\theta} + r\ddot{\theta}) = 0, \tag{8.3.8}$$

due to (8.3.7b). We next use cylindrical coordinates and let $\mathbf{r} = r\hat{r}$, $\mathbf{p} = m\mathbf{v} = m(v_r\hat{r} + v_\theta\theta)$. Here v_r is the radial component of the velocity ($\dot{r} = v\cos\theta$, $v = \sqrt{v_r^2 + v_\theta^2}$, θ the angle between **r** and **v** as in Figure 8.1) and v_θ the tangential component of the velocity ($v_\theta = v\sin\theta = r\dot{\theta}$), respectively. The angular momentum then becomes $\mathbf{L} = \mathbf{r} \times \mathbf{p} = mv_\theta r\hat{k}$. The magnitude of the angular momentum is $L = mv_\theta r = mr^2\dot{\theta} = I\dot{\theta}$, where I is the moment of inertia. This means that (8.3.8) is consistent with

$$L = mr^2\dot{\theta} = const, \tag{8.3.9}$$

as expected from (8.3.5). Further, from here

$$\dot{\theta} = L/mr^2, \tag{8.3.10}$$

which when substituted into (8.3.7a) gives

$$m\ddot{r} = F(r) + \frac{L^2}{mr^3} \equiv F_{eff}, \tag{8.3.11}$$

where we have defined the result as an effective force F_{eff}. Since $F(r)$ is the real force, then the L^2/mr^3 term is the fictitious force associated with the central force motion problem. This fictitious force term is the centrifugal force discussed in Chapter 7. Related to this effective force, there is an effective potential

$$V_{eff} = -\int_{r_0}^{r} F_{eff}(r)dr = -\int_{r_0}^{r} F(r)dr - \frac{L^2}{m}\int_{r_0}^{r} \frac{1}{r^3}dr. \tag{8.3.12}$$

The first integral on the right is the real force standard potential of (8.2.6), and if we pick the lower limit as $r_0 = \infty$ and perform the integration, we get

$$V_{eff} = V(r) + \frac{L^2}{2mr^2}. \tag{8.3.13}$$

This represents the potential energy associated with bringing a body of mass m from very far away to position r in the presence of a central force.

EXAMPLE 8.1

Consider a body of mass m moving under the action of a central force in the form $f(r) = -ar^p$. If at time $t = 0$ the body's initial position, tangential velocity, and angle are 1 m, 6 m/s, and 0 rad, obtain plots of $r(t)$, $v(t)$, $\theta(t)$, and $r(\theta)$. Assume $m = 1$ kg, $p = 1$, and $a = 108$ N/m.

Solution

Since $p = 1$, the given central force is a harmonic type—that is, $f(r) = -ar$—and from (8.2.6) the potential is $V(r) = ar^2/2$, using a zero value for the lower limit of the integral. The bulk of this problem consists in solving the differential Equations (8.3.10 and 8.3.11) for $r(t)$ and $\theta(t)$. This can be done numerically in MATLAB. The results are shown in Figures 8.2.

The upper-left figure shows $r(t)$ and $\dot{r}(t)$ and they both appear to oscillate as a function of time. The shape of the oscillations are not exactly harmonic, though, as can be seen; however, because there are two minimum and two maximum positions of r in one revolution (see the lower-right figure, also), the revolution period is four times a single minimum-to-maximum time leg. Because there are about three min-max legs in about 0.5 seconds, each leg is about 0.15 seconds long. A revolution is, therefore, about 0.6 seconds long. The upper-right figure is $\theta(t)$ and although it does increase in time, it is periodic. Combining the results of the top two figures, we obtain a Cartesian plot of $r(\theta)$, which is shown in the lower-left figure. The radius begins at the initial value at $t = 0$, increases to a maximum value, and then decreases, as the angle increases, down to a minimum value, and finally increases back up to a maximum value. The motion repeats with a half revolution period of about 0.3 seconds in correlation with the upper-left figure. A very useful polar plot of $r(\theta)$ is that shown on the lower right. We see the motion is elliptical in shape. At $t = 0$ the radius has a value of $r = 1$ m when $\theta = 0$ rad,

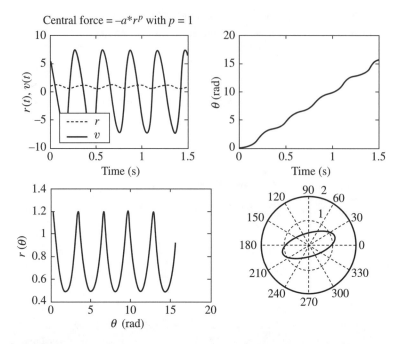

FIGURE 8.2   Example 8.1 of the numeric solution of Equations (8.3.10 and 8.3.11) for the motion of a particle under a central force.

but because the initial velocity is nonzero, the particle can continue to increase its orbital distance up to a maximum radius and then decreases as described in the lower-left figure.

In order to solve the differential equations of this problem using MATLAB, we have defined the three-dimensional vector w, where $w[1] \equiv r$, $w[2] \equiv \dot{r}$, and $w[3] \equiv \theta$. Thus, Equations (8.3.10 and 8.3.11) become three-coupled differential equations for the vector w as follows:

$$\frac{dw[1]}{dt} = w[2], \frac{dw[2]}{dt} = \ddot{r} = -\frac{aw[1]^p}{m} + \frac{L^2}{m^2 w[1]^3}, \text{ and } \frac{dw[3]}{dt} = \frac{L}{mw[1]^2}.$$

$$(8.3.14)$$

The main MATLAB script central.m that follows uses the numerical Runge–Kutta differential equation solver. The derivatives are placed in the function central_der.m, which is called by the main script. The main program performs the plot commands based on the preceding definitions. Both scripts follow. The script is made in such a general way that it can be modified with little effort to repeat the calculation for a force with a different value of p. In the present problem, the initial parameters used, a, p, r_0, v_0, and θ_0, have respective values 108 N/m, 1, 1 m, 6 m/s, and 0 rad. The calculations are carried out for a time $t_{max} = 1.5$ s, which is longer than the observed revolution period. Finally, it is noted that the script is designed to work with a general force $F(r) = -ar^p$ with the values of a and p as input parameters.

SCRIPT

```
%central.m
%program to plot the solution of a body of mass m under the action of
%a central force of the form -a*r^p
clear;
a=108.0;              %force strength in N/m^p
p=1;                  %power of r
tmax=1.5;             %maximum time in seconds
r0=1; v0=6.0; th0=0;  %initial position(m), tangential velocity(m/s), angle(rad)
m=1;                  %object's mass in kg
L=m*v0*r0;            %angular momentum, v0=tangential velocity
ic1=[r0;v0;th0];      %initial conditions: position, v0, angle
%Use MATLAB's Runge-Kutta (4,5) formula (uncomment/comment as needed)
%opt=odeset('AbsTol',1.e-7,'RelTol',1.e-4);    %user set Tolerances
%[t,w]=ode45('central_der',[0.0,tmax],ic1,opt,a,L,m,p);%with set tolerance
[t,w]=ode45('central_der',[0.0,tmax],ic1,[],a,L,m,p);%with default tolerance
str=cat(2,'Central Force=-a*r^p',' with p=',num2str(p,3));
subplot(2,2,1)
plot(t,w(:,1),'b:',t,w(:,2),'r-')
h=legend('r','v',0); set(h,'FontSize',12)
xlabel('time (sec)','FontSize',14);
ylabel('r(t), v(t)','FontSize',14);title(str,'FontSize',14)
subplot(2,2,2)
plot(t,w(:,3))
xlabel('time (sec)','FontSize',14),ylabel('\theta (rad)','FontSize',14)
subplot(2,2,3)
plot(w(:,3),w(:,1))
xlabel('\theta (rad)','FontSize',14);ylabel('r(\theta)','FontSize',14)
subplot(2,2,4)
polar(w(:,3),w(:,1)) %plot r(theta) in polar coordinates
```

FUNCTION

```
%central_der.m: returns the derivatives for -a*r^p central force
function derivs = central_der( t, w, flag,a,L,m,p)
% a=force strength, L=angular momentum, m=mass, p=power of r
% Entries in the vector of dependent variables are:
% w(1)-position(t), w(2)-velocity(t), w(3)-theta(t)
derivs = [ w(2);-(a/m)*w(1).^p+L^2/m^2./w(1).^3;L/m./w(1).^2];
```

■ 8.4 Total Energy and Central Forces

We now look at the total energy of a system under the action of a central force. The total energy of a system is conserved in the absence of any dissipative forces. The total energy E is the scalar sum of the kinetic and potential energies T and V, respectively; i.e.,

$$E = T + V. \tag{8.4.1}$$

The kinetic energy has translational ($m\dot{r}^2/2$) as well as rotational ($mr^2\dot{\theta}^2/2$) contributions. The potential energy is just $V(r)$. The total energy becomes

$$E = \frac{1}{2}m\dot{r}^2 + \frac{1}{2}mr^2\dot{\theta}^2 + V(r) = \frac{1}{2}m\dot{r}^2 + \frac{1}{2}\frac{L^2}{mr^2} + V(r), \tag{8.4.2}$$

where (8.3.10) has been used. We notice that the energy can also be understood as the sum of the translational kinetic energy and the effective potential energy; that is,

$$E = \frac{1}{2}m\dot{r}^2 + V_{eff}(r). \tag{8.4.3}$$

Since the energy is a constant, Equation (8.4.2) can be solved for $\dot{r}$ to get

$$\dot{r} = \pm\sqrt{\left\{\frac{2}{m}\left[E - V(r) - \frac{1}{2}\frac{L^2}{mr^2}\right]\right\}} = \frac{dr}{dt}. \tag{8.4.4}$$

The radius is a positive quantity that changes from an initial value of, say, r_i, to a final value r, so that (8.4.4) can be rewritten to obtain the time required for this change as

$$\int_0^t dt = \int_{r_i}^r \frac{dr}{\sqrt{\left\{\frac{2}{m}\left[E - V(r) - \frac{1}{2}\frac{L^2}{mr^2}\right]\right\}}}. \tag{8.4.5}$$

This equation has the analytic form of a position-dependent force problem as discussed in Chapter 2, Section 4. It can be used to obtain $t(r)$, which can be inverted to get $r(t)$.

We note that (8.4.5) is the time of travel from r_i to r, which depending on $V(r)$ may not, in general, necessarily involve a periodic orbit; in the case of a body orbiting another, this equation can be used to obtain the period of the orbit.

EXAMPLE 8.2

Consider the body discussed in Example 8.1 that moves under the action of a central force in the form $f(r) = -ar^p$ where $p = 1$ and $a = 108$ N/m. Obtain the period of its orbit. As in Example 8.1, at time $t = 0$, the body's initial position, tangential velocity, and angle are 1 m, 6 m/s, and 0 rad, and that the body's mass $m = 1$ kg.

Solution

We notice from Figure 8.2 that the orbit passes twice through its maximum and minimum positions. So its full orbital period can be obtained if we take the time between the minimum and the maximum position and multiply by a factor of four. In (8.4.5) we then take the lower limit of the motion as $r_i \to r_{min}$ and the upper limit as $r \to r_{max}$, to write for the revolution period

$$\tau = 4 \int_0^{\tau/4} dt = 4 \int_{r_{min}}^{r_{max}} \frac{dr}{\sqrt{\left\{ \frac{2}{m} \left[E - V(r) - \frac{1}{2} \frac{L^2}{mr^2} \right] \right\}}}. \tag{8.4.6}$$

For a general value of p, the potential has the form $V(r) = a \int r^p dr$, which gives

$$V(r) = \begin{cases} a r^{(p+1)}/(p+1) & \text{if } p \neq -1 \\ a \ln(r) & \text{if } p = -1 \end{cases}. \tag{8.4.7}$$

The initial conditions allow us to find the angular momentum $L = m r_0 v_0$, where $v_0 = v_\theta$ is the tangential velocity, and the energy

$$E = 0.5 m v_r^2 + 0.5 L^2 / m r_0^2 + V(r_0), \tag{8.4.8}$$

where r_0 is the given initial position and is not necessarily equal to r_{min} or r_{max}, and where $v_r = \dot{r}$ is the radial velocity component. Here, we need to estimate the value of v_r. Since angular momentum is constant then we make the approximation that $L = m r_0 v_0 = m r_{min} v \Rightarrow v = r_0 v_0 / r_{min}$, where $v = \sqrt{v_r^2 + v_0^2}$, and solving this for v_r we get $v_r = v_0 \sqrt{(r_0/r_{min})^2 - 1}$. Since r_{min} is not known exactly, we use the estimated value obtained from Figure 8.2 or by typing min(w(:,1)) within

MATLAB after running the script central.m, with the potential parameter value of $p = 1$, to get $r_{\min} \approx 0.4813$ m. This value is not as good as value found below (in regards to the integral [8.4.6]), but short of performing a long iterative calculation, it will be fine here. Although the integral (8.4.6) can't be done analytically for a general value of p, it can be performed numerically if we know the min-max limit values of r. To find these values, one notices that the quantity under the square root, in the integral (8.4.6), i.e.,

$$E - V(r) - \frac{1}{2}\frac{L^2}{mr^2},\qquad(8.4.9)$$

has to be positive in order for the period to be *real*. If r is too small—that is, less than $r_{\min}$—this quantity becomes negative, due to the angular momentum term. If r is too large—that is, greater than $r_{\max}$—the quantity becomes negative due to the potential term. If (8.4.9) becomes negative, the integrand of (8.4.6) becomes imaginary. Therefore, we need to seek the values of $r_{\min}$ and $r_{\max}$ for which the integrand is real before performing the integration. Figure 8.3 shows the solution to this problem. First, notice that Figure 8.3(a) shows the difference between the three values of r associated with this problem. Figure 8.3(b) shows the integrand of (8.4.6), and the shaded area is the value of the integral between $r_{\min}$ and $r_{\max}$. The orbital period is four times the area so that we obtain a value of $\tau = 0.598$ s. This value agrees with our expectations from Example 8.1. Figure 8.3(a) is generated using the MATLAB script central2.m (not listed but included with the

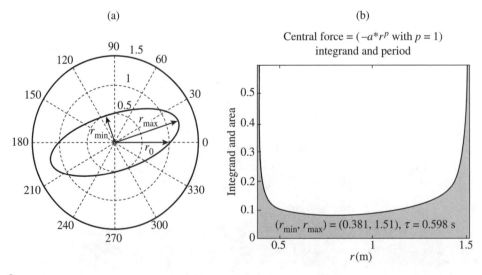

(a)

(b)

I FIGURE 8.3 Example 8.2 for the orbital period of a particle under the central force of Example 8.1.

text's website), which is similar to central.m but uses the parameter tmax=0.6 to confirm that a full orbit takes place in about this period of time.

The MATLAB script orbit_period.m has been used to perform the calculation of the integral (8.4.6) and to generate Figure 8.3(b). Using the initial conditions of the problem, the script searches for the values of r_{min} and r_{max} as well. As discussed before, the value of $r_{min} = 0.381$ m actually found is different from the value we used to estimate the energy E in (8.4.8), but it would require an iterative process to get a better answer, which we forgo for now. To perform the search for these values of r, the script seeks the turning points where the integrand of (8.4.6) is no longer *real*, as discussed earlier in connection to (8.4.9). The listing of the script orbit_period.m follows. Finally, the initial parameters used are as in Example 8.1. Also, the script is designed to work with a general force $F(r) = -ar^p$ with the values of a and p as input parameters.

SCRIPT

```
%orbit_period.m
%Script made to obtain the time it takes to go from rmin to rmax
%in an orbit due to a force of the form -a*r^p
clear;
a=108.0;              %force strength in N/m^p
p=1;                  %power of r
r0=1; v0=6.0; th0=0; %initial position(m),tangential velocity(m/s), angle(rad)
%ra is an approximate value of rmin needed to estimate vr in the next line
ra=0.4813; vr=v0*sqrt((r0/ra)^2-1); %vr=radial velocity component needed in E
m=1;                  %object's mass in kg
L=m*v0*r0;            %angular momentum, v0=tangential velocity
% Below E is the constant energy, I is the integrand definition
if p ==-1
    E=0.5*m*vr^2+L^2/m/r0^2/2+a*log(r0);
    I = inline('1./sqrt(2*(E-a*log(r)-L^2/m./r.^2/2)/m)',...
        'r','a','p','E','m','L');
else
    E=0.5*m*vr^2+L^2/m/r0^2/2+a*r0^(p+1)/(p+1);
    I = inline('1./sqrt(2*(E-a*r.^(p+1)/(p+1)-L^2/m./r.^2/2)/m)',...
        'r','a','p','E','m','L');
end
% search the min, max turning points
st=0.0002; rlim=10*r0; % search step size and upper limit of search
for r=[r0:-st:0]
    if real(I(r,a,p,E,m,L)) <= 0, rmin=r+st; break;
    end
```

```
end
for  r=[r0:st:rlim]
    if  real(I(r,a,p,E,m,L))  <=0,   rmax=r-st;  break;
    end
end
T = 4*quad(I,rmin,rmax,[],[],a,p,E,m,L);%period is 4 times area
%r=[rmin:0.01:rmax]; plot(r,I(r,a,p,E,m,L))%integrand plot
r=[rmin:0.01:rmax];
area(r,I(r,a,p,E,m,L),'FaceColor',[0.7 0.8 0.9])%area plot
str1=cat(2,'Central Force (-a*r^p',', p=',num2str(p,3),')',...
    ' Integrand & Period');
str2=cat(2,'(rmin,rmax)=(',num2str(rmin,3),...
    ',',num2str(rmax,3),'),',' \tau = ',num2str(T,3),' sec');
axis([rmin rmax 0 0.6]);xlabel('r(m)','FontSize',14);
ylabel('Integrand & Area','FontSize',14);title(str1,'FontSize',14)
text(0.6,0.1,str2,'FontSize',14)
```

■ 8.5 The Equation for $r(\theta)$

In the previous two sections it has been demonstrated that it is possible to solve the equations of motion numerically for $r(t)$, $\theta(t)$, $t(r)$, and indirectly for $r(\theta)$. It is possible, however, to obtain a direct relationship for r with θ as the independent variable. To this end, it is useful to define

$$u = u(r) \equiv \frac{1}{r}, \tag{8.5.1}$$

so that (8.3.10) becomes

$$\dot{\theta} = \frac{L}{m}u^2. \tag{8.5.2}$$

Using (8.5.1) $r = 1/u$, we have for the time derivative of r and assuming a θ dependence, $du/dt = (du/d\theta)(d\theta/dt)$, so that

$$\dot{r} = -u^{-2}\frac{du}{dt} = -u^{-2}\dot{\theta}\frac{du}{d\theta}, \tag{8.5.3}$$

which when (8.5.2) is substituted into it, we get

$$\dot{r} = -\frac{L}{m}\frac{du}{d\theta}. \tag{8.5.4}$$

Similarly,

$$\ddot{r} = \frac{d\dot{r}}{dt} = \frac{d}{dt}\left(-\frac{L}{m}\frac{du}{d\theta}\right) = -\frac{L}{m}\dot{\theta}\frac{d^2u}{d\theta^2} = -\frac{L^2}{m^2}u^2\frac{d^2u}{d\theta^2}. \tag{8.5.5}$$

The equation of motion for $r(t)$ is given by (8.3.11), which using Equations (8.5.1, 8.5.2, and 8.5.5) becomes

$$F\left(\frac{1}{u}\right) = -\frac{L^2}{m}u^2\frac{d^2u}{d\theta^2} - \frac{L^2}{m}u^3, \tag{8.5.6}$$

and after rearranging this, we obtain an equation for $r(\theta)$ as

$$\frac{d^2u}{d\theta^2} + u = -\frac{m}{L^2}\frac{1}{u^2}F\left(\frac{1}{u}\right). \tag{8.5.7}$$

EXAMPLE 8.3

Referring to Example 8.1, consider a body of mass $m = 1$ kg moving under the action of a central force in the form $f(r) = -ar^p$, with $p = 1$ and $a = 108$ N/m. For the same initial conditions at $t = 0$, that is, the body's initial position, tangential velocity and angle are 1 m, 6 m/s, and 0 rad. Compare the $r(\theta)$ orbit obtained using (8.5.7) to that obtained in Example 8.1, where (8.3.10 and 8.3.11) were used instead. Comment on your results.

Solution

Solving (8.5.7) numerically can be done in MATLAB. The force is written as $f(1/u) = -a(1/u)^p$, and we define the two-dimensional array vector v, where $v[1] \equiv u$ and $v[2] = du/d\theta \equiv u'$. Thus, Equation (8.5.7) becomes a system of two coupled differential equations for the vector v as follows:

$$\frac{dv[1]}{d\theta} = v[2] \tag{8.5.8a}$$

and

$$\frac{dv[2]}{d\theta} = \frac{d^2u}{d\theta^2} = -u - \frac{m}{L^2}\frac{1}{u^2}F\left(\frac{1}{u}\right) = -v[1] - \frac{m}{L^2}\frac{1}{v[1]^2}\left(-\frac{a}{v[1]^p}\right). \tag{8.5.8b}$$

We are to solve for $u(\theta)$ followed by its reciprocal at the end of the calculation to get $r(\theta)$. The problem's initial conditions are expressed in terms of r_0 (position) and v_0 (the tangential velocity), so that $u_0 = 1/r_0$, and (8.5.4) is used to get the initial value of $(du/d\theta)_0$ as $u'_0 = -(m/L)v_0$. The results of both methods are shown in Figure 8.4.

Figure 8.4 shows $r(\theta)$ in Cartesian as well as in polar coordinates for completeness. The blue lines correspond to the results of using Equation (8.5.7) for

Force = $-a*r^p$ with $p = 1$, Two Methods Compared

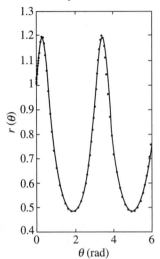

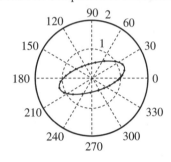

Polar Plot Comparison of Two Methods

FIGURE 8.4   Example 8.3: Solution of a central force problem using Equations (8.3.10 and 8.3.11) compared to the solution using Equation (8.5.7).

$u(\theta)$ as outlined earlier in this section, and the dots (red in MATLAB) correspond to Equations (8.3.10 and 8.3.11) for $r(t)$ and $\theta(t)$. In summary, both methods give identical results, as can be seen, but (8.5.7) allows the calculation to take less numerical effort than using (8.3.10 and 8.3.11) and is a more direct approach to the orbit $r(\theta)$.

The main MATLAB script `centralu.m` that follows employs the numerical Runge–Kutta differential equation solver. It makes use of the MATLAB function `centralu_der.m`, which contains the derivatives, Equation (8.5.8), and which is listed below as well. To make a comparison with the earlier method of Equations (8.3.10 and 8.3.11), the script also makes further use of `central_der.m`, which was listed in Example 8.1. When run, the script generated will be in color.

SCRIPT

```
%centralu.m
%program to plot the solution of a body of mass m under the action of
%a central force of the form -a*r^p where r=1/u
clear;
a=108.0;              %force strength in N/m^p
p=1;                  %power of r
r0=1; v0=6.0; hmin=0;%initial position(m), tangential velocity(m/s), angle(rad)
```

```
m=1;                    %object's mass in kg
L=m*v0*r0;              %angular momentum, v0=tangential velocity
u0=1/r0;                %initial value of u
uv0=-m*v0/L;            %initial value of the time derivative of u
hmax=2*pi;             %maximum angle
%————
%r(theta) as obtained from the u(theta) diff. eq.
ic1=[u0;uv0];           %initial conditions: position, initial u-dot
opt1=odeset('AbsTol',1.e-7,'RelTol',1.e-4);%user set Tolerances
[h,v]=ode45('centralu_der',[hmin,hmax],ic1,opt1,a,L,m,p);%with default tolerance
r=1./v(:,1);
%————
% r(theta) as obtained from the r(theta) diff. eq.
ic2=[r0;v0;hmin]; tmax=0.6;% init conds. and orbit period for tmax
[t,w]=ode45('central_der',[0.0,tmax],ic2,[],a,L,m,p);%with default tolerance
%————
%r(theta) comparison plots for the two methods
str=cat(2,'Central Force=-a*r^p',' with p=',num2str(p,3),', Two Methods Compared');
subplot(1,2,1); plot(h,r,'b-'); hold on
plot(w(:,3),w(:,1),'r.','MarkerSize',5)
xlabel('\theta (rad)','FontSize',14);ylabel('r(\theta)','FontSize',14)
title(str,'FontSize',14), axis([0 6 0.4 1.3])
%plot r(theta) in polar coordinates
subplot(1,2,2);polar(h,r,'b-'); hold on
m=length(w(:,3));polar(w(1:3:m,3),w(1:3:m,1),'r.');%plot every 3rd point
title('Polar Plot Comparison of Two Methods','FontSize',12)
```

FUNCTION

```
%centralu_der.m: returns the derivatives for
%-a*r^p central force, where r=1/u
function derivs = centralu_der( h, v, flag,a,L,m,p)
% a=force strength, L=angular momentum, m=mass, p=power of r
% Entries in the vector of dependent variables are:
% v(1) is u(h), v(2) is u'(t), here h=angle, u is a function of angle
derivs = [v(2);-v(1)-((m/L^2)./v(1)^2).*(-a./v(1)^p)];
```

EXAMPLE 8.4

Using the orbit Equation (8.5.7), obtain the equation of the shape of a body's orbital path $r(\theta)$ in the absence of any force and plot it. Assume an initial condition such that the body is very far away at $\theta = 0$ and that at $\theta = 90°$, $r = 1.5\ AU$. Note that an *AU* is an astronomical unit of distance that corresponds to a value of 1.496×10^{11} m, and that it is a measure of the average Earth–sun distance.

Solution

In the absence of a force, (8.5.7) becomes

$$\frac{d^2u}{d\theta^2} + u = 0,$$ (8.5.9a)

whose solution can be written as $u = A\sin\theta + B\cos\theta$. According to the initial condition, $r = \infty$ or $u = 1/r \to 0$ at $\theta = 0$, and we must have that the particular solution is $u = A\sin\theta$, or

$$r = C\csc\theta,$$ (8.5.9b)

where $C = 1.5AU$ to satisfy the second initial condition at $\theta = 90°$. The plots of the orbit in both Cartesian and polar coordinates are shown in Figure 8.5.

The Cartesian plot shows how the body is very far away for a small angle, reaches the closest approach at $\theta = 90°$, and moves far away as the angle reaches π. The polar plot is more informative in the sense that the motion is just a straight line motion as is expected when there are no forces present acting on a body according to Newton's first law. The code used for this exercise, `simple_orbit.m`, follows.

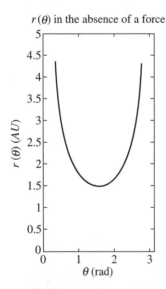

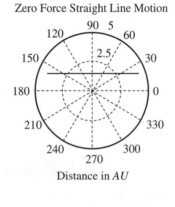

FIGURE 8.5 Example 8.4: Simple force analytic orbit solution of (8.5.7).

SCRIPT

```
%simple_orbit.m - plots the orbit for a zero force case,
u=1.5AU*sin(theta)=o/r
th=[0.35:0.01:pi-0.35];% angle range
r=1.5*csc(th);          % distance versus angle
subplot(1,2,1);plot(th,r);
axis([0 pi 0 5]),
xlabel('\theta (rad)','FontSize',14)
ylabel('r(\theta) (AU)','FontSize',14);
title('r(\theta) in the absence of a force','FontSize',14)
subplot(1,2,2);polar(th,r,'r')
title('Zero Force Straight Line Motion','FontSize',14)
xlabel('Distance in AU','FontSize',14)
```

■ 8.6   General Inverse Squared Force

A general force that varies as the inverse of the distance squared has broad applications in physics. One example is Newton's universal law of gravitation:

$$\mathbf{F}_g = -\frac{Gm_1m_2}{r^2}\hat{r}, \tag{8.6.1}$$

which expresses the gravitational force of attraction between two bodies, m_1 and m_2, separated by distance r, where $G = 6.673 \times 10^{-11}$ Nm²/kg². A second example is Coulomb's law:

$$\mathbf{F}_e = \frac{kq_1q_2}{r^2}\hat{r}, \tag{8.6.2}$$

which expresses the electrical interactive force between two charges, q_1 and q_2, separated by distance r, where $k = 8.988 \times 10^9$ Nm²/C². Both of these forces can be expressed in the form

$$\mathbf{F}(r) = F(r)\hat{r}, \text{ and } F(r) \equiv \frac{K}{r^2}, \tag{8.6.3a}$$

along with the corresponding potential

$$V(r) = -K\int\frac{1}{r^2}dr = \frac{K}{r}, \tag{8.6.3b}$$

where K is negative for attractive and positive for repulsive forces. The purpose of this section is to obtain a general equation for $r(\theta)$ that is a solution to (8.5.7) for the inverse distance squared force. Using $r = 1/u$, the force of interest is $F(r) \to Ku^2$, which is substituted into (8.5.7) to get

$$\frac{d^2u}{d\theta^2} + u = -\frac{m}{L^2}\frac{1}{u^2}Ku^2 = -\frac{mK}{L^2}. \tag{8.6.4}$$

This is a second-order inhomogeneous differential equation. It has homogeneous and particular solutions. The homogenous part is obtained when the RHS is zero, which suggests a harmonic type of solution. We can write the general homogeneous form

$$u_h = u_0 \cos(\theta - \theta_0), \tag{8.6.5}$$

where θ_0 is the value of θ at $t = 0$ when $u_h = u_0$. For the particular case, one can let the trial solution be a constant $u_p = C$, which when substituted back into (8.6.4) the result is $u_p = C = -mK/L^2$. By the superposition principle, the total solution is the sum of the particular and homogenous parts, so that we write

$$u = u_h + u_p = u_0 \cos(\theta - \theta_0) - \frac{mK}{L^2} \tag{8.6.6}$$

as a solution to (8.6.4) corresponding to the inverse distance squared force problem. Since $r = 1/u$, then

$$r(\theta) = 1/u(\theta) = \left\{ u_0 \cos\theta - \frac{mK}{L^2} \right\}^{-1}, \tag{8.6.7}$$

where we have taken $\theta_0 \equiv 0$. This equation can be rearranged to read

$$r = -\frac{L^2}{mK}\frac{1}{\left(1 - \frac{u_0 L^2}{mK}\cos\theta\right)} \equiv r_{\min}\left(\frac{1+e}{1+e\cos\theta}\right), \tag{8.6.8}$$

where the definitions

$$r_{\min} \equiv -\frac{L^2}{mK}\frac{1}{(1+e)} \quad \text{and} \quad e \equiv -\frac{u_0 L^2}{mK} \tag{8.6.9}$$

for the minimum radius and the eccentricity of the orbit have been made. The eccentricity is a quantity that is associated with an ellipse. In the range of $0 \le e \le 1$, the shape described by (8.6.8) can go from a circle ($e = 0$) to a parabola ($e = 1$), with an ellipse in between ($0 \le e \le 1$). Figure 8.6 contains a simple plot of (8.6.8) obtained using the ellipse.m MATLAB code with $r_{\min} = 2.5\ AU$, $r_{\max} = 10\ AU$, and $e = 0.6$, where the AU is an astronomical unit of distance defined in Example 8.4. The code follows. The semimajor axis is the average value of the minimum ($r_{\min}$) and maximum ($r_{\max}$) distances from the focal point; that is, $a \equiv (r_{\min} + r_{\max})/2$. The semimajor axis is sometimes referred to just as the average ellipse distance from the focal point. In Figure 8.6, the ellipse focus is located at the center of the polar diagram. An ellipse is characterized by two foci. The source of the inverse distance squared force is posi-

Ellipse with $(r_{min}, r_{max}) = (2.5, 10)\, AU, e = 0.6$

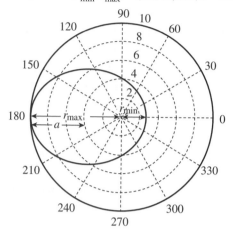

tioned at one of the focal points. Appendix C contains more details regarding plane geometry, from where we also find that $r_{min} = a(1 - e)$ and $r_{max} = a(1 + e)$.

SCRIPT

```
%ellipse.m - script to draw an ellipse with minimum radius rmim and
%eccentricity e.
clear;
rmin=2.5;
e=0.6;
th=[0:0.1:2*pi];
r=rmin*(1+e)./(1+e*cos(th));
rmax=rmin*(1+e)./(1+e*cos(pi));
polar(th,r)
str=cat(2,'Ellipse with (rmin, rmax) = (',num2str(rmin,3),',',...
    num2str(rmax,3),')AU, e = ',num2str(e,3));
title(str,'FontSize',14)
```

From (8.6.8) we learn that in the angle range $0 < \theta < \pi$, the radius range is $r_{min} < r < r_{max}$, which helps to define the maximum radius

$$r_{max} = r_{min}\left(\frac{1+e}{1-e}\right) = -\frac{L^2}{mK}\left(\frac{1}{1-e}\right). \tag{8.6.10}$$

The eccentricity of the elliptical orbit Earth ($m_E = 5.98 \times 10^{24}$ kg) makes around the sun ($m_S = 1.99 \times 10^{30}$ kg) is 0.017. A planet's closest approach to the sun (that is, its r_{min} distance) is known as *perihelion* distance. The farthest point on its orbit (that is, its r_{max} distance) is known as the *aphelion* distance. Earth's average distance to the sun is about 149.6×10^6 km, which is about one *AU* and is also Earth's ellipse's semimajor axis distance. The analog minimum and maximum distances in the case of the moon orbiting Earth are known as *perigee* and *apogee*, respectively. The moon $(m_m = 7.36 \times 10^{22}$ kg) orbits Earth with an orbital eccentricity value of 0.066 and its perigee distance is about 384,400 km. The radii r_{min} and r_{max} are also referred to as turning points. At these positions the radial speed $\dot{r} = 0$, so that from (8.4.2 and 8.4.3), the energy at these locations is given by

$$E = \frac{1}{2} \frac{L^2}{mr_m^2} + V(r_m) = \frac{1}{2} \frac{L^2}{mr_m^2} + \frac{K}{r_m} = V_{eff}(r_m), \tag{8.6.11}$$

where r_m can take the values of r_{min} and r_{max} and where (8.6.3) has been used. This equation can be rearranged to read

$$\frac{1}{r_m^2} + \frac{2mK}{L^2} \frac{1}{r_m} - \frac{2mE}{L^2} = 0, \tag{8.6.12}$$

which is a quadratic equation for $1/r_m$ in terms of the angular momentum and the energy, whose solutions are

$$\frac{1}{r_m} = \begin{cases} -\dfrac{mK}{L^2} + \sqrt{\left(\dfrac{mK}{L^2}\right)^2 + \dfrac{2mE}{L^2}} = \dfrac{1}{r_{min}} \\[2ex] -\dfrac{mK}{L^2} - \sqrt{\left(\dfrac{mK}{L^2}\right)^2 + \dfrac{2mE}{L^2}} = \dfrac{1}{r_{max}} \end{cases}, \tag{8.6.13}$$

where, for an attractive force $(K < 0)$, the positive (negative) root gives the larger (smaller) value of r.

We notice from Equations (8.6.13) and (8.6.9 and 8.6.10) that it is possible to obtain an equation for the eccentricity by equating corresponding radii. Defining

$$D \equiv \sqrt{\left(\frac{mK}{L^2}\right)^2 + \frac{2mE}{L^2}}, \tag{8.6.14}$$

we see that

$$\frac{1}{r_{min}} = -\frac{mK}{L^2}(1 + e) = -\frac{mK}{L^2}\left(1 - \frac{DL^2}{mK}\right), \tag{8.6.15a}$$

and

$$\frac{1}{r_{max}} = -\frac{mK}{L^2}(1 - e) = -\frac{mK}{L^2}\left(1 + \frac{DL^2}{mK}\right). \tag{8.6.15b}$$

Both of these indicate that we can write

$$e = -\frac{DL^2}{mK} \Rightarrow e^2 = D^2\left(\frac{L^2}{mK}\right)^2 = 1 + \frac{2EL^2}{mK} \Rightarrow e = \pm\sqrt{1 + \frac{2EL^2}{mK^2}}, \tag{8.6.16}$$

where we take the $e > 0$ root for a negative K (attractive force) and the $e < 0$ root for a positive K (repulsive force). Recalling (8.4.2) the energy is $E = [m\dot{r}^2 + mr^2\dot{\theta}^2 + 2V(r)]/2$, so that if $V(r)$ is negative, then E can be negative as well as positive. When $-mK^2/2L^2 \le E < 0$, which is the case of bound orbits, then $0 \le e < 1$, which involves elliptical motion. When $E = 0$, the motion is unbound and parabolic ($e = 1$). When $E > 0$, the motion is also unbound but hyperbolic ($e > 1$). Notice from here that the minimum energy required to have a circular motion is obtained when $e = 0$, that is, $E \to E_c$, where

$$E_c = -\frac{mK^2}{2L^2}, \tag{8.6.17}$$

in which case, from (8.6.15) $r_{max} = r_{min} \to r_c$, or

$$r_c \equiv -\frac{L^2}{mK}. \tag{8.6.18}$$

EXAMPLE 8.5

Give a plot of the individual terms in Equations (8.4.2 and 8.4.3) versus distance, including the effective potential of an inverse distance squared force law. Calculate the eccentricity, total energy, and angular momentum associated with the body's orbit. Show the positions of r_{min} and r_{max} as well as r_c and indicate where the total energy lies, including the energy minimum for a circular orbit. Use the following parameters for initial conditions: $m = 1$ kg, $r_i = 1$ m, $v_i = 5$ m/s, with the angle between $\mathbf{r}_i$ and $\mathbf{v}_i$ as $\theta_i = 75°$ and $K = -45$ Nm^2.

Solution
The positions r_m are illustrated in Figure 8.7 along the various parts of the total energy Equations (8.4.2 and 8.4.3) for an attractive potential ($K < 0$). The parameters used are posted on the figure, including the angular momentum, $L = mr_iv_\theta$ with $v_\theta = v_i\sin\theta$ the initial tangential velocity; the energy, $E = mv_r^2/2 + L^2/(2mr_i^2) + K/r_i$ with the radial velocity as $v_r = v_i\cos\theta_i$; and the eccentricity, $e = \sqrt{1 + 2EL^2/(mK^2)}$. Also calculated are $r_{min} = (-mK/L^2 + D)^{-1}$ and $r_{max} = (-mK/L^2 - D)^{-1}$, where $D = \sqrt{(mK/L^2)^2 + 2mE/L^2}$. Notice that if the total energy were to change, the r_m positions would change as well. When the total energy is exactly zero, the body has an infinite r_{max}, that is, the body is no longer bound to an orbit. This situation corresponds to the eccentricity, taking on a value of unity as can be seen from

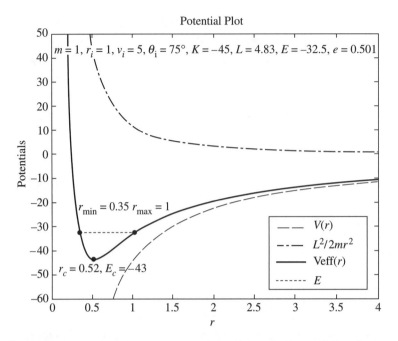

Potential Plot

$m = 1, r_i = 1, v_i = 5, \theta_i = 75°, K = -45, L = 4.83, E = -32.5, e = 0.501$

$r_{min} = 0.35 \; r_{max} = 1$

$r_c = 0.52, E_c = -43$

- - - - $V(r)$
- · - · - $L^2/2mr^2$
——— Veff(r)
········· E

FIGURE 8.7   Example 8.5 shows the force potential, the centrifugal, and the effective potentials as well as the total energy of a particle in an inverse square force.

(8.6.10), and the orbit becomes a parabola. The figure also shows the r_c, E_c values from (8.6.17 and 8.6.18) corresponding to the circular orbit possibility. Notice that the minimum of the effective potential corresponds to the minimum energy needed to have a circular orbit.

· This figure was created using the MATLAB script potential.m, which follows, for the parameters of the problem.

SCRIPT

```
%potential.m - script to plot an attractive potential, energy, etc.
%for a body under a central force
m=1; ri=1; vi=5; K=-45; thi=75; %inputs
th=thi*pi/180; vt=vi*sin(th); vr=vi*cos(th);%velocity components
L=m*ri*vt;                 %angular momentum, vt=tangential velocity
E=m*vr^2/2+L^2/m/ri^2/2+K/ri;  %E is constant
e=sqrt(1+2*E*L^2/m/K^2);   %eccentricity
r=[0.2:0.01:5];            %variable
```

```
V=K./r; CT=L^2./r.^2/2/m;        %V(r), and Ang. Momentum term
Veff=V+CT;                       %Effective potential
rmin=1/(-m*K/L^2+sqrt((m*K/L^2)^2+2*m*E/L^2));
rmax=1/(-m*K/L^2-sqrt((m*K/L^2)^2+2*m*E/L^2));
Ec=-m*K^2/L^2/2; rc=-L^2/m/K;
plot(r,V,'k-',r,CT,'-.',r,Veff,'r')
line([rmin;rmax],[E;E],'Color','b','LineStyle',':')
axis([0 4 -60 50])
h=legend('V(r)','L^2/2mr^2','Veff(r)','E',4);
set(h,'FontSize',14)
xlabel('r','FontSize',14); ylabel('Potentials','FontSize',14); hold on;
plot(rmin,E,'k.',rmax,E,'k.',rc,Ec,'k.');%dot rmin,rmax,rc positions
str1=cat(2,'r_{min}=',num2str(rmin,2));
str2=cat(2,'r_{max}=',num2str(rmax,2));
str3=cat(2,'r_c=',num2str(rc,2),', E_c=',num2str(Ec,3));
text(rmin*(1+0.1),E*(1-0.2),str1,'FontSize',10); %post rmin
text(rmax*(1-0.1),E*(1-0.2),str2,'FontSize',10); %post rmax
text(rc*(1-0.8),Ec*(1+0.1),str3,'FontSize',9);   %post rc,Ec
str4=cat(2,' m=',num2str(m),', r_i=',num2str(ri,2),', v_i=',num2str(vi,2),...
     ', \theta_i=',num2str(thi,3),'^0',', K=',num2str(K,3),', L=',num2str(L,3),...
     ', E=',num2str(E,3),', e=',num2str(e,3));
text(0.5,40,str4,'FontSize',11);            %post the parameters used
title('Potential Plot','FontSize',14)
```

■ 8.7   Kepler's Laws

Johannes Kepler was a German astronomer and professor of mathematics. He moved to Prague around the beginning of the 17th century to work with Tycho Brahe, who was also an astronomer and was famous for his naked-eye astronomical observations. Kepler studied Brahe's data on planetary motion and in the period between 1609 and 1619 announced his three laws of planetary motion. These laws are important because they became the groundwork of Isaac Newton's universal law of gravitation. The laws are stated as follows:

1. The orbits of the planets around the sun are ellipses with the sun at one focus.

2. A line joining a planet and the sun sweeps out equal areas in equal times.

3. The ratio between the square of a planet's period and the cube of its average distance (semimajor axis) is the same for all planets; that is, $\tau^2/a^3 = $ constant. In astronomical units of distance (Earth–sun separation) and units of years for time, the constant is unity.

A. The First Law

This is a result of the inverse distance squared force law, which we investigated in the previous section. For an orbiting body to have circular motion, the body must have the exact minimum energy E_c of (8.6.17), and any small amount above this value will result in an elliptical orbit, whose eccentricity will deviate from zero.

B. The Second Law

This is depicted in Figure 8.8(a) which, as we will see, is a result of angular momentum conservation. Suppose that a planet initially located at point a takes time τ_{ab} to get to point b; then the area swept in this period of time is A_{ab}. During a different time period on its orbit around the sun, suppose the planet takes τ_{cd} to travel from point c to point d; then the area swept during this time is A_{cd}. According to Kepler's second law, if these two time periods are the same, that is, if $\tau_{ab} = \tau_{cd}$, then the swept areas are also the same, or $A_{cd} = A_{ab}$. This is actually possible because the planets speed up when they are nearest to the sun, but slow down when they are farthest from it.

From Figure 8.8(b), we can write $\mathbf{r} = r\hat{r}$, and $d\mathbf{r} = dr\hat{r} + r\,d\theta\,\theta$, in polar coordinates, and since the area of the triangle shown is formally given by $dA = |\mathbf{r} \times d\mathbf{r}|/2$, then $dA = |(r\hat{r}) \times (dr\hat{r} + r\,d\theta\,\theta)|/2 = |r^2d\theta\,\hat{k}|/2 = r^2d\theta/2$. Therefore, we have

$$\frac{dA}{dt} = \frac{1}{2}r^2\frac{d\theta}{dt} = \frac{1}{2m}mr^2\dot{\theta} = \frac{L}{2m} = const, \tag{8.7.1}$$

which is Kepler's second law and, accordingly, this rate is constant. This is directly linked to the angular momentum conservation concept mentioned before. Notice also that if we substitute $d\mathbf{r} = \mathbf{v}dt = \mathbf{p}dt/m$ into $dA = |\mathbf{r} \times d\mathbf{r}|/2$, we get $dA = |\mathbf{r} \times \mathbf{p}|dt/2\,m \Rightarrow dA/dt = L/2\,m$.

(a)   (b)

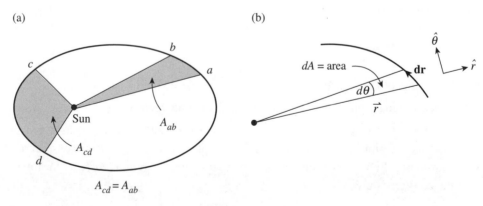

I FIGURE 8.8 (a) Kepler's second law and (b) the area under a general triangle.

C. The Third Law

Noticing that if we integrate (8.7.1) over time, between $t = 0$ and $t = \tau$, where τ is the period of revolution, we see that the total area swept by a planet is

$$A = \frac{L}{2m}\tau. \tag{8.7.2}$$

Since an ellipse of eccentricity e is characterized by the area

$$A_{ellipse} = \pi a^2 \sqrt{1 - e^2}. \tag{8.7.3}$$

using (8.6.15), the semimajor axis can be written as

$$a = \frac{r_{\min} + r_{\max}}{2} = \frac{1}{2}\left(-\frac{L^2}{mK}\frac{1}{(1+e)} - \frac{L^2}{mK}\frac{1}{(1-e)}\right) = -\frac{L^2}{mK}\frac{1}{(1-e^2)}, \tag{8.7.4a}$$

from which we obtain that

$$\sqrt{1 - e^2} = \frac{L}{\sqrt{m(-K)a}}. \tag{8.7.4b}$$

Substituting this expression into (8.7.3) results in

$$A_{ellipse} = \pi a^{3/2}\frac{L}{\sqrt{m(-K)}}.$$

We finally equate this expression with that of (8.7.2) to get

$$\pi a^{3/2}\frac{L}{\sqrt{m(-K)}} = \frac{L}{2m}\tau \quad \Rightarrow \quad \frac{\tau^2}{a^3} = \left(\frac{4\pi^2 m}{-K}\right) = \frac{4\pi^2}{Gm_S}, \tag{8.7.5}$$

which is Kepler's third law of planetary motion, where we have used (8.6.1) to identify $K = -Gm_S m$ for a planet of mass m orbiting the sun with mass m_s. The right-hand side is a constant that only depends on the sun's mass. Its value can be calculated as

$$\frac{4\pi^2}{Gm_S} = \frac{4\pi^2}{(6.673 \times 10^{-11}\ \text{Nm}^2/\text{kg}^2)(1.99 \times 10^{30}\ \text{kg})} = 2.973 \times 10^{-19}\frac{\text{s}^2}{\text{m}^3} = 1.008\frac{yr^2}{AU^3} = \frac{\tau^2}{a^3} \tag{8.7.6}$$

where we have used 3.156×10^7 s per year and 1.5×10^{11} m per AU. For our purposes and in these units, the constant in Kepler's third law, therefore, can numerically be taken equal to 1. In Table 8.1, the observed sidereal periods (periods as seen from stars) of the planets and their semimajor axis distances are shown in the first two columns. The last two columns contain the period squared and the cubed of the semimajor axis. The last two columns are very close to each other, with the biggest percent difference being about 0.15% for the case of Uranus.

| Table 8.1 | Data for the sidereal period of the planets and their semimajor axes. |

Planet	$\tau(yr)*$	$a(AU)*$	$\tau^2(yr^2)$	$a^3(AU^3)$
Mercury	0.241	0.387	0.058	0.058
Venus	0.615	0.723	0.378	0.378
Earth	1.000	1.000	1.000	1.000
Mars	1.881	1.524	3.538	3.540
Jupiter	11.86	5.203	140.660	140.852
Saturn	29.46	9.539	867.892	867.978
Uranus	84.01	19.191	7057.680	7067.939
Neptune	164.8	30.061	27159.040	27165.040
Pluto	248.5	39.529	61752.250	61765.720

* *Universe.* W. J. Kaufmann III, 4th ed. (W. H. Freeman Co., 1994)

In Figure 8.9, we plot τ^2 versus a^3 as indicated by the dots on the left plot. Since the behavior is very nearly linear, we then obtain a linear curve fit to the data as indicated by the dashed line. The circles around the dots provide a visual comparison of the data with the value predicted by the fit. The curve fit shows that a straight line with a slope of 0.9998 and an intercept of -1.08 works very well. This confirms Kepler's third law [Equation (8.7.6)] with very low error. The goodness of fit, as measured by the correlation factor r^2, has a value close to unity, which is excellent. We also plotted the theoretical relation $\tau = a^{3/2}$ in the figure on the right, shown as a red dashed line. For better visualization, the period is plotted using a logarithmic base 10 scale. The actual planets' period for a given semimajor axis of Table 8.1 was plotted as shown by the black dots named according to the corresponding planets. The theoretical relation is followed very closely.

Finally, the MATLAB script kepler3rd.m used to make these figures follows. (The script-generated figures appear in color.) It used parameters from Table 8.1.

SCRIPT

```
%kepler3rd.m - This script is designed to look at the solar system planets
%and to look at the tau versus a relationship.
clear;
planet=['Mercury ';'Venus   ';'Earth   ';'Mars    ';'Jupiter ';...
        'Saturn  ';'Uranus  ';'Neptune ';'Pluto   '];
p=[0.241;0.615;1.000;1.881;11.86;29.46;84.01;164.8;248.5];%period
a=[0.387;0.723;1.000;1.524;5.203;9.539;19.191;30.061;39.529];%semimajor axis
x=a.^3; y=p.^2;% define the variables
```

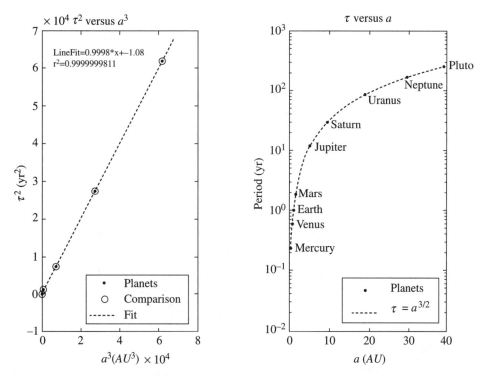

FIGURE 8.9 Planets data compared with the results of Kepler's third law for (a) the period squared, and (b) the period in a logarithmic scale.

```
[c,e]=polyfit(x,y,1);%Linear fit of y(x) can give coefficient b, error e(3)
[cp,ep]=polyfit(y,x,1);%Linear fit of x(y) needed to calculate r-squared
rr=c(1)*cp(1); %definition of r_squared (Bevington's Data Reduction...)
ybest1 = c(1).*x + c(2);%used to compare predicted and actual
x2=[min(x):(max(x)-min(x))/9:max(x)*(1+0.15)];%more points for the prediction
ybest2 = c(1).*x2 + c(2);                %general predicted
subplot(1,2,1);plot(x,y,'k.',x,ybest1,'o',x2,ybest2,'r:')
xlabel('a^3 (AU^3)','FontSize',14); ylabel('\tau^2 (yr^2)','FontSize',14)
title('\tau^2 Versus a^3','FontSize',14)
h=legend('Planets','Comparison',' Fit',4); set(h,'FontSize',12)
str1=cat(2,' Line Fit=',num2str(c(1),4),'*x + ',num2str(c(2),4));
str2=cat(2,' r^2=',num2str(rr,10));
text(x(6),y(9)*(1+0.05),str1,'FontSize',9);%post rmin
text(x(6),y(9),str2,'FontSize',9);         %post rmin
%— tau = a^{3/2} figure
xth=[0:0.1:max(a)];                        %variable for pth
```

```
pth=xth.^(3/2);                              %theoretical curve
subplot(1,2,2);semilogy(a,p,'k.',xth,pth,'r:')%use semilog for better viewing
for i=1:9
text(a(i),p(i),planet(i,:),'Color','b','FontSize',12);%The planets
end
str3=cat(2,'\tau = a^{3/2}');
h=legend('Planets',str3,4); set(h,'FontSize',14)
xlabel('a (AU)','FontSize',14); ylabel('Period (yr)','FontSize',14);
title('\tau versus a','FontSize',14)
```

EXAMPLE 8.6

Four moons of Jupiter (known as the Galilean moons) were discovered by Galileo, two of which are Ganymede and Europa. If Europa is found to have a sidereal period of 3.551 days and a mean distance of 670,900 km from the planet's center, what is Ganymede's period if it is about 1,070,000 km from the planet?

....

Solution
We employ Kepler's third law, using the subscripts G and E to signify Ganymede and Europa, respectively, and write

$$\frac{\tau_E^2}{a_E^3} = \frac{\tau_G^2}{a_G^3} \Rightarrow \tau_G = \sqrt{\frac{\tau_E^2}{a_E^3} a_G^3} = \sqrt{\frac{3.551^2}{670,900^3} 1,070,000^3} = 7.15 \text{ days.} \quad (8.7.7)$$

EXAMPLE 8.7

The universal law of gravitation was actually discovered by Sir Isaac Newton in 1665, many years after Kepler had died in 1630. Kepler's third law, however, can be used to deduce the inverse distance squared force law between the sun and the planets. Explain how this is possible.

....

Solution
If we assume that the distance a planet travels on its orbit around the sun is $2\pi a$, letting the planet's speed be v, then its period is $\tau = 2\pi a/v$, and Kepler's third law becomes

$$\frac{\left(\frac{2\pi a}{v}\right)^2}{a^3} = \left(\frac{4\pi^2 m}{-K}\right) \Rightarrow mv^2 = -\frac{K}{a}, \quad (8.7.8)$$

which if both sides are divided by a, we get

$$\frac{mv^2}{a} = -\frac{K}{a^2},$$ (8.7.9)

which indicates that the centripetal force needed to hold the planet in orbit is that of an inverse distance squared type. Of course, Newton's laws are an underlying concept here that was not available during Kepler's time.

■ 8.8   Orbit Transfers

A body moving in a circle around the sun has a centripetal force associated with it

$$F_c = -\frac{mv_c^2}{r_0},$$ (8.8.1)

where r_0 is the radius of a circular orbit and v_c is the circular orbital speed. This force is due to gravity, so that

$$\frac{K}{r_0^2} = -m\frac{v_c^2}{r_0} \quad \Rightarrow \quad K = -mr_0v_c^2.$$ (8.8.2)

If we next consider a body located at r_0 from the origin but moving at a speed that's greater than v_c but perpendicular to r_0, the body's orbit will deviate from a circular shape, in which case the perihelion from (8.6.9) is $r_{min} \to r_0$, and we have for the eccentricity

$$e = -\frac{L^2}{mKr_0} - 1.$$ (8.8.3)

Or, using (8.8.2) and $L = |m\mathbf{r}_0 \times \mathbf{v}_0| = mr_0v_0$, where v_0 is the tangential velocity, we also have

$$e = \frac{v_0^2}{v_c^2} - 1.$$ (8.8.4)

Notice that if $v_0 \to v_c$, then $e \to 0$ as expected for a circle. For speeds greater than v_c, the motion will have an aphelion and a perihelion distance associated with it. Also, notice that the case of $e \to 1$ corresponds to $v_0 \to v_e = \sqrt{2}v_c$, which implies that a body's escape kinetic energy ($mv_e^2/2$) is twice the circular orbit kinetic energy ($mv_c^2/2$), where v_e is the escape speed. With e given by (8.8.4), we can also rewrite the aphelion distance using (8.6.10) as

$$r_{max} = r_0\frac{v_0^2}{v_c^2[2 - (v_0/v_c)^2]}.$$ (8.8.5)

From this equation we see that if $v_0 \to v_c$, then $r_{max} \to r_0$, as expected. Also, notice that if $v_0 \to v_e = \sqrt{2}v_c$, then $r_{max} \to \infty$, which is consistent with the case when $e \to 1$ in (8.8.4), indicating that the body has an open orbit and will have escaped.

EXAMPLE 8.8

The Earth–sun average distance is about 1.496×10^{11} m and Earth's orbital eccentricity is 0.017. Obtain Earth's perihelion and aphelion distances, as well as its orbital speeds at those positions. Give a polar plot of Earth's orbit around the sun under these conditions.

....

Solution

We can obtain the perihelion distance from the equation for an ellipse, as discussed in Section 8.6; thus, $r_{min} \to r_0 = a(1 - e) = 1.496 \times 10^{11}(1 - 0.017) \approx 1.471 \times 10^{11}$ m. The perihelion speed is from (8.8.4), $v_0 = v_c\sqrt{1 + e}$, where from (8.8.2), $v_c = \sqrt{-K/m_E r_0} = \sqrt{Gm_E m_S/m_E r_0}$, to get

$$v_c \approx \sqrt{(6.67 \times 10^{11})(1.991 \times 10^{30})/1.471 \times 10^{11}} \approx 3.005 \times 10^4 \text{ m/s}$$

or 30.05 km/s. We then get $v_0 \approx 3.005 \times 10^4\sqrt{1 + 0.017} \approx 3.031 \times 10^4$ m/s or 30.31 km/s. The aphelion distance is from (8.8.5), $r_{max} \approx 1.471 \times 10^{11}(30.31/30.05)^2/(2 - (30.31/30.05)^2) \approx 1.521 \times 10^{11}$ m. It remains to find the orbital speed at r_{max}. Using the angular momentum conservation rule, which is equivalent to Kepler's second law, we see that $m_E r_0 v_0 = m_E r_{max} v_{rmax}$, or $v_{rmax} = r_0 v_0/r_{max} \approx (1.471/1.521)(30.31) \approx 29.29$ km/s. The final numbers shown are those obtained with the script earthorb.m, according to the accuracy of MATLAB, which also gives the full plot of Earth's orbit in Figure 8.10 according to Equation (8.6.8) in the form $r = r_0(1 + e)/(1 + e\cos\theta)$.

Earth Orbit $e = 0.017$, $(r_0, r_{max}) = (1.471\ e + 008, 1.521\ e + 008)$ km

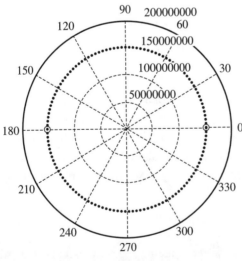

$(v_0, v_{rmax}) = (30.31, 29.29)$ km/s

FIGURE 8.10   Example 8.8 for Earth's orbit around the sun.

In the figure, the blue (red) circle corresponds to the perihelion (aphelion) position. Earth's orbit around the sun is very nearly circular due to the small value of its eccentricity, which in turn is due to its primordial value of v_0 being very close to the value v_c; in fact, the ratio $v_0/v_c = 1.0085$ is very nearly unity. One possible theory maintains that Earth was captured by the sun under such conditions billions of years ago. Earth is about 4.5 billion years old, and the sun is about 1.5 billion years older. The black dots correspond to the planet's path. Another (and perhaps more probable) theory regarding Earth's present location and orbit suggest it formed from solar material that was already in orbit around the sun billions of years ago. The script, earthorb.m, follows, according to the described calculations. The script-generated figure will contain the colors mentioned above.

SCRIPT

```
%Earthorb.m - script to draw Earth's ellipse with minimum radius rmim and
%eccentricity e given
clear;
m=5.98e24;        %Earth's mass in kg
G=6.67e-11;       %Universal gravitational constant
M=1.991e30;       %Sun's mass in kg
a=1.496e8;        %average Earth-sun distance in km, or semimajor axis
e=0.017;          %Earth's eccentricity
r0=a*(1-e);       %r0 in km, same as rmin
K=-G*M*m/1000^3;  %in km^3.kg/s^2
vc=sqrt(-K/r0/m);%speed in km/s (circular orbit speed)
v0=vc*sqrt(1+e);  %speed at r0
th=[0:0.05:2*pi];%range variable
r=r0*(1+e)./(1+e*cos(th));%orbit formula, with ro=rmin
rmax=r0*(v0/vc)^2/(2-(v0/vc)^2);%same as ro*(1+e)/(1-e)
v_rmax=r0*v0/rmax;%using angular momentum conservation
polar(th,r,'k.');hold on
polar(0,r0,'bo')
polar(pi,rmax,'ro')
str1=cat(2,'Earth Orbit: e = ',num2str(e,3),', (r_0, r_{max}) = (',...
    num2str(r0,4),',',num2str(rmax,4),')km');
title(str1,'FontSize',12)
str2=cat(2,' (v_0,v_{rmax})=(',num2str(v0,4),',',num2str(v_rmax,4),') km/s');
xlabel(str2,'FontSize',12)
```

EXAMPLE 8.9

A probe is launched at a distance of $r_0(r_{\min})$ equal to Earth's orbital position, as shown in Figure 8.11. What is the necessary speed so that the probe will make an elliptical orbit with a maximum radius ($r_{\max}$) equal to Mars' orbital radius?

....

Solution
We set $r_0 = R_E$, and $r_{\max} = R_M$, where R_E and R_M are Earth and Mars' orbital radii, respectively. We then use (8.8.5), with $v_c \to v_E$, that is, the speed of Earth in its orbit around the sun, to get $R_M = R_E(v_0^2/v_E^2)(1/[2 - (v_0/v_E)^2])$, and solve for v_0 to obtain

$$v_0 = v_E\sqrt{2(R_M/R_E)/[1 + (R_M/R_E)]}. \tag{8.8.6}$$

This v_0 will be the needed speed by the probe to achieve Mars' orbit.

EXAMPLE 8.10

Compare the orbital velocity of Mars to that of Earth around the sun.

....

Solution
We can set the centripetal force equal to the gravitational force. For Earth, we have that its orbital velocity is

$$-\frac{mv_E^2}{R_E} = -\frac{Gm_Sm}{R_E^2} \quad \Rightarrow \quad v_E = \sqrt{\frac{Gm_S}{R_E}}, \tag{8.8.7a}$$

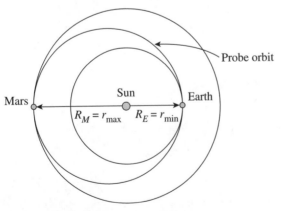

I FIGURE 8.11   Example 8.9 of an orbit transfer.

and similarly for Mars' orbital velocity

$$v_M = \sqrt{\frac{Gm_S}{R_M}}. \tag{8.8.7b}$$

Taking their ratio, we get

$$\frac{v_M}{v_E} = \sqrt{\frac{R_E}{R_M}}. \tag{8.8.7c}$$

This indicates that because $R_E < R_M$, then planets that are farther from the sun move more slowly.

EXAMPLE 8.11

When the probe of Example 8.9 reaches Mars, what will its speed be?

....

Solution
The probe loses energy in the transfer process. Let the transfer speed between Earth to Mars from (8.8.6) be

$$v_0 \to v_{EM} = v_E\sqrt{2(R_M/R_E)/[1 + (R_M/R_E)]}. \tag{8.8.8}$$

We next look at the initial and final energies. We have the initial energy at Earth's position

$$E_i = \frac{1}{2}mv_{EM}^2 + \frac{K}{R_E}, \tag{8.8.9}$$

and the final energy at arrival to Mars

$$E_f = \frac{1}{2}mv_{ME}^2 + \frac{K}{R_M}, \tag{8.8.10}$$

where m is the probe's mass and v_{ME} is the velocity of approach we seek. Equating these energies, we have

$$v_{ME}^2 = v_{EM}^2 - 2\frac{K}{mR_M} + 2\frac{K}{mR_E}. \tag{8.8.11}$$

However, using (8.8.2) we see that $K/mR_M = -Gm_S/R_M = -v_M^2$, and similarly $K/mR_E = -Gm_S/R_E = -v_E^2$, so that (8.8.11) becomes

$$v_{ME}^2 = v_{EM}^2 + 2v_M^2 - 2v_E^2. \tag{8.8.12}$$

We need v_{EM}. From (8.8.7c), we have that

$$v_E = v_M\sqrt{R_M/R_E},$$
(8.8.13)

which when substituted into (8.8.8) we get

$$v_{EM} = v_M\sqrt{\frac{2(R_M/R_E)R_M}{R_E + R_M}}.$$
(8.8.14)

Finally, substituting (8.8.13 and 8.8.14) into (8.8.12) and simplifying we get

$$v_{ME}^2 = 2v_M^2\left[\frac{R_E}{R_E + R_M}\right],$$
(8.8.15)

which gives the speed with which Mars is reached. Since $R_E < R_M$, then $R_E + R_M > 2R_E$, or $R_E/(R_E + R_M) < 1/2$; thus, $v_{ME}^2 < v_M^2$ or $v_{ME} < v_M$ so that the probe arrives with less speed than Mars' orbital speed. Mars will approach it from behind and overtake it.

Finally, it is useful to point out that, if in this example, the takeoff velocity, v_0, were to be less than v_c, then (8.8.4) results in $e < 0$, which means that in (8.6.10) the roles of $r_{\min}$ and $r_{\max}$ become reversed. In such case, $r_{\min}$ becomes the aphelion position and $r_{\max}$ becomes the perihelion position, from where the takeoff occurs. If we let R_M, where the subscript m corresponds to, say, Mercury, for which $R_M < R_E$, then $R_M + R_E < 2R_E$, and from (8.8.15) $v_{ME} > v_M$. This means that when transferring toward an inner planet, the planet will be overtaken because it will be moving more slowly than the probe when the probe reaches it.

■ 8.9 Oscillations about Circular Orbits

In Chapter 3, Section 3, we learned that for small oscillations, a potential energy function can be expanded in a Taylor series, from which, according to the idea of restoring forces (Hooke's law type), an effective spring constant can be associated with it and thereby an angular frequency

$$\omega_0 = \sqrt{\frac{k_{eff}}{m}},$$
(8.9.1)

where here, $k_{eff} \equiv V''_{eff}(r)|_{r=r_c}$; that is, the second derivative of the effective potential is evaluated at the circular orbit radius r_c. A general V_{eff} is given by (8.3.13), and we then have

$$V''_{eff}(r)\big|_{r=r_c} = \left[V''(r) + \frac{3L^2}{mr^4}\right]_{r=r_c} = -F'(r_c) + \frac{3L^2}{mr_c^4}, \tag{8.9.2}$$

since $F(r) = -V'(r)$. We are looking at oscillations about circular orbits. According to the theory of small oscillations, at the equilibrium position

$$V'_{eff}(r)\big|_{r=r_c} = V'(r_c) - \frac{2L^2}{2mr_c^3} = 0 = -F(r_c) - \frac{L^2}{mr_c^3} \Rightarrow F(r_c) = -\frac{L^2}{mr_c^3}, \tag{8.9.3}$$

and (8.9.2) can be written as

$$V''_{eff}(r)\big| = -F'(r_c) - 3\frac{F(r_c)}{r_c}. \tag{8.9.4}$$

Therefore, the oscillations about a circle will have associated with them the r_{min} and the r_{max} radii with frequency $\omega_0 = \sqrt{(-F'(r_c) - 3F(r_c)/r_c)/m}$, or period

$$\tau = \frac{2\pi}{\omega} = 2\pi\sqrt{\frac{m}{-F'(r_c) - 3\frac{F(r_c)}{r_c}}}. \tag{8.9.5}$$

It turns out that in order to have orbits that are stable, the frequency must be real so that $-F'(r_c) - 3F(r_c)/r_c$ must be positive. Additionally, in order to have closed orbits, the angle traveled in half this period, that is, $\theta_{\tau/2} = \dot{\theta}\tau/2$ (where $\dot{\theta} = L/mr_c^2$) must be a fractional multiple of π (that is, equal to π/n) where $n = 1, 2, 3, \ldots$ is an integer. One can show that the planets in the solar system have orbits that are stable as well as closed because the inverse distance squared force law obeys these conditions; see Problem 8.14.

EXAMPLE 8.12

For the case of a planet orbiting the sun, use the small oscillation period formula and show that it is possible to obtain Kepler's third law of motion.

Solution

According to the universal law of gravitation, $F(r_c) = -Gm_s m/r_c^2$, and $F'(r_c) = 2Gm_s m/r_c^3$, so that (8.9.5) becomes

$$\tau = \frac{2\pi}{\omega} = 2\pi \sqrt{\frac{m}{-2Gm_s m/r_c^3 + 3Gm_s m/r_c^3}} = 2\pi \sqrt{r_c^3/Gm_s}. \qquad (8.9.6)$$

If both sides are squared, the result is equivalent to Kepler's third law. The small oscillation formula (8.9.5) says that the radius of an orbit increases and decreases with a certain period. For small oscillations, we can get an approximate value of the period of oscillation about an equilibrium position. We did something similar for a molecule potential in Chapter 3. Similarly, planets are bound to the sun through a gravitational potential similar to that of Figure 8.7 in Section 8.6, and it is not surprising that such period exits.

■ Chapter 8 Problems

8.1 Following Equation (8.2.3), show that the curl of a central force vanishes.

8.2 Consider a body of mass m moving under the action of a central force in the form $f(r) = -ar^2$, where $a = 108 \text{ N/m}^2$. If at time $t = 0$, the body's initial position, tangential velocity and angle are 1 m, 6 m/s, and 0 rad, respectively, obtain plots of $r(t)$, $v(t)$, $\theta(t)$, and $r(\theta)$. Assume $m = 1$ kg.

8.3 Consider the body discussed in Problem 8.2 that moves under the action of a central force in the form $f(r) = -ar^2$, where $a = 108 \text{ N/m}$. Obtain the period of its orbit. As in Problem 8.2, assume that at time $t = 0$, the body's initial position, tangential velocity, and angle are 1 m, 6 m/s, and 0 rad, and that the body's mass $m = 1$ kg.

8.4 Under what conditions does Equation (8.4.5) give the period of a mass performing simple harmonic motion at the end of a spring of constant k, assuming Hooke's law applies? Show that in fact it does.

8.5 Show the solution for a body's orbit in the absence of a force of Example 8.4; i.e., $r = C \csc \theta$, is a solution to the orbit equations of motion $m\ddot{r} = F(r) + L^2/mr^3$ and $\dot{\theta} = L/mr^2$. (*Hint:* Solve for $\theta(t)$ from the last equation with the help of the first and substitute it back into the first to obtain $r(t)$.)

8.6 Use Equation (8.4.2) to show that the energy of a body of mass m orbiting the sun is given by Equation (8.6.17).

8.7 Show that when a body orbits the sun in a circular orbit, since the gravitational pull and the centrifugal force balance out, the body's orbital radius is given by Equation (8.6.18).

8.8 Obtain the moon's apogee distance as it orbits Earth. How long is the moon's elliptical orbit's semimajor axis?

8.9 Kepler third law, $\tau^2/a^3 = 1\,\mathrm{yr}^2/AU^3$, applies for the planets in the solar system. What is the corresponding law for the case of the moon orbiting Earth? Given that the moon orbits Earth about once every 27.32 days, what is the moon's semimajor axis distance?

8.10 Saturn's mass is 95.18 Earth masses. One of its moons, Titan, has an orbital sidereal period of 15.945 days. Estimate its distance from Saturn.

8.11 Escape velocity is defined to be the minimum velocity needed in order to escape a planet's gravitational pull, in the absence of any other gravitational body; in doing so, all of the initial energy is spent in the process. Obtain an expression for the escape velocity of Earth.

8.12 Orbital velocity is defined to be the velocity needed in order for a body to remain in orbit around a second body. Obtain an expression for the velocity needed by a satellite in order to remain in orbit near Earth's surface.

8.13 Pluto's average distance from the sun, $a = \dfrac{1}{2}(r_{min} + r_{max})$, is about $5913.5 \times 10^6\,\mathrm{km}$ and its orbital eccentricity is $e = 0.248$. Obtain Pluto's perihelion and aphelion distances and its orbital speeds at those positions. Make a polar plot of Pluto's orbit around the sun under these conditions.

8.14 Prove whether the universal law of gravitation obeys the conditions that lead to both stable as well as closed orbits.

■ Additional Problems

8.15 Using Equations (8.6.9 and 8.6.10), obtain the equations for r_{min} and r_{max} in terms of the semimajor axis and the eccentricity.

8.16 Consider a body of mass m moving under the action of a central force Yukawa-like potential of the form $V(r) = -ce^{-\alpha r}/r$, where $c = 30\ \text{Nm}^2$ and $\alpha = 0.01\ \text{m}^{-1}$.

    a. Obtain the differential equation of motion obeyed by $r(t)$.

    b. If at time $t = 0$, the body's initial position, tangential velocity, and angle are 1.5 m, 4 m/s, and 0 rad, respectively, obtain plots of $r(t)$, $v(t)$, $\theta(t)$, and $r(\theta)$ during the first nine seconds. Assume $m = 1$ kg.

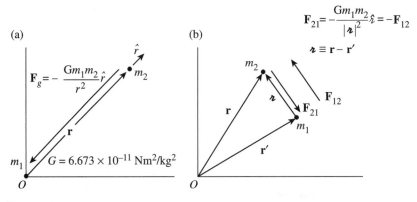

9 Gravitation

■ 9.1 Gravitational Force and Gravitational Field

According to Newton's universal law of gravitation, a body of mass m_1 located at the origin attracts another of mass m_2 with a force that varies as the inverse square of the distance r between them:

$$\mathbf{F}_g = -\frac{Gm_1m_2}{r^2}\hat{r}, \tag{9.1.1}$$

where $G = 6.673 \times 10^{-11}\,\dfrac{\text{Nm}^2}{\text{kg}^2}$ is the universal gravitational constant. As shown in Figure 9.1(a), the force points in the direction of $-\hat{r}$, that is, toward m_1. It is convenient to write a more general form of the force. This is done if the masses m_1 and m_2 are located at positions $\mathbf{r}'$ and $\mathbf{r}$, respectively, from the origin, as shown in Figure 9.1(b). We say that the gravitational force on mass m_2 due to mass m_1 is

$$\mathbf{F}_{21} = -\frac{Gm_1m_2}{\imath^2}\hat{\imath} = -\frac{Gm_1m_2}{\imath^3}\boldsymbol{\imath}, \tag{9.1.2}$$

(a)   (b)

$$\mathbf{F}_{21} = -\frac{Gm_1m_2}{|\boldsymbol{\imath}|^2}\hat{\imath} = -\mathbf{F}_{12}$$

$$\boldsymbol{\imath} \equiv \mathbf{r} - \mathbf{r}'$$

$$\mathbf{F}_g = -\frac{Gm_1m_2}{r^2}\hat{r}$$

$$G = 6.673 \times 10^{-11}\ \text{Nm}^2/\text{kg}^2$$

FIGURE 9.1 The universal law of gravitation (a) with one mass at the origin, and (b) with both masses away from the origin.

where $\boldsymbol{\imath} \equiv \mathbf{r} - \mathbf{r}'$, with $\hat{\imath} = \boldsymbol{\imath}/\imath$ and $\imath \equiv |\boldsymbol{\imath}|$, has been defined to take into account the displacement of mass m_1 from the origin. In the limit as $\mathbf{r}' \rightarrow 0$, $\boldsymbol{\imath} \rightarrow \mathbf{r}$, and (9.1.2) becomes identical to (9.1.1). The usefulness of (9.1.2) is that it shows that $\mathbf{F}_{21}$ points in the direction toward m_1, as seen in Figure 9.1(b). Of course, by Newton's third law of motion we also have that $\mathbf{F}_{21} = -\mathbf{F}_{12}$, where $\mathbf{F}_{12}$ is the force on mass m_1 due to mass m_2, which points toward m_2.

It is possible to extend (9.1.2) to a discrete system of particles with masses m_i for $i = 1, 2, \ldots$ exerting a gravitational force on a separate particle of mass m. This is illustrated in Figure 9.2.

The gravitational force at position $\mathbf{r}$ due to the discrete system of particles with total mass $M = \sum_i m_i$ is written as

$$\mathbf{F}(\mathbf{r}) = -Gm\sum_i \frac{m_i}{\imath_i^2}\hat{\imath}_i = -Gm\sum_i \frac{m_i}{\imath_i^3}\boldsymbol{\imath}_i, \tag{9.1.3}$$

where $\boldsymbol{\imath}_i \equiv \mathbf{r} - \mathbf{r}_i$. In obtaining (9.1.3), we have used the *superposition principle*; that is, the total force due to the cluster is the sum of all the individual forces due to each mass m_i that makes up the cluster. If the collection of particles producing the gravitational force is continuous whose total mass is $M = \int \rho(\mathbf{r}')dV'$, where $\rho(\mathbf{r}')$ is the volume density at position $\mathbf{r}'$, then in (9.1.3) one can make the replacement $m_i \rightarrow dm'$, and $\sum_i m_i \rightarrow \int dm'$, where dm' can take on the expressions $\rho(r')dV'$ or $\sigma(r')dA'$

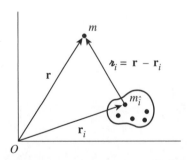

| FIGURE 9.2 Discrete system of particles acting on a single mass.

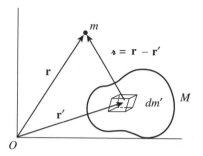

| FIGURE 9.3 Continuous system of particles acting on a single mass.

or $\lambda(r')\,dl'$, depending on whether we are dealing with three-, two-, or one-dimensional densities, respectively. (We will come back to this in Section 9.3.) Referring to Figure 9.3, and assuming a three-dimensional distribution, the total contribution to the force at $\mathbf{r}$ on m due to the extended mass M is

$$\mathbf{F}(\mathbf{r}) = -Gm\int\frac{\rho(\mathbf{r}')}{\imath^2}\hat{\imath}\,dV' = -Gm\int\frac{\rho(\mathbf{r}')}{\imath^3}\boldsymbol{\imath}\,dV', \qquad (9.1.4)$$

where $\boldsymbol{\imath} \equiv \mathbf{r} - \mathbf{r}'$.

If the mass M is of uniform or constant density, then we can make the replacement $\rho(\mathbf{r}') \rightarrow \rho$, and ρ can be taken out of the integral in (9.1.4). Finally, the gravitational field at position $\mathbf{r}$ is defined as

$$\mathbf{g}(\mathbf{r}) = \frac{\mathbf{F}(\mathbf{r})}{m}, \qquad (9.1.5)$$

which has units of acceleration and is obtained from Equations (9.1.3 and 9.1.4) for the discrete and continuous systems, respectively.

■ 9.2   Gravitational Potential Energy and Gravitational Potential

In the preceding section, the gravitational force as well as the gravitational field is a vector. In this section, we deal with gravitational quantities that are scalars. As discussed in Chapter 8, if the gravitational force is a central force, i.e., $\mathbf{F}(r) = F(r)\hat{r}$, the potential energy is given by the $V(r) = -\int \mathbf{F}(r) \cdot d\mathbf{r}$ and involves only the magnitude of the force or $V(r) = -\int F(r)\,dr$. Since $\imath_i \equiv \mathbf{r} - \mathbf{r}_i$ and $d\imath = d\mathbf{r}$ or $dr = d\imath$, for the discrete system of particles, from (9.1.3) we have

$$V(r) = Gm\sum_i \int \frac{m_i}{\imath_i^2} dr = Gm\sum_i \int \frac{m_i}{\imath_i^2} d\imath = -Gm\sum_i \frac{m_i}{\imath_i}. \qquad (9.2.1)$$

Similarly, for the continuous distribution, from (9.1.4)

$$V(r) = Gm\int \rho(\mathbf{r}')dV' \int \frac{1}{\imath^2} dr = -Gm\int \frac{\rho(\mathbf{r}')}{\imath} dV'. \qquad (9.2.2)$$

These are the scalar gravitational potential energies $V(r)$ of each respective configuration. In a similar way, it is possible to define the scalar gravitational potential

$$\phi(r) = -\int \mathbf{g}(r) \cdot d\mathbf{r}, \qquad (9.2.3)$$

where similar to (9.1.5) this definition shows us that

$$\phi(r) = V(r)/m. \qquad (9.2.4)$$

Furthermore, since the scalar gravitational potential yields the gravitational force through the expression, $\mathbf{F}(r) = -\nabla V(r)$, it is easy to see that the gravitational field is obtained from the gravitational scalar potential through

$$\mathbf{g}(r) = -\nabla\phi(r). \qquad (9.2.5)$$

For the special case of a central gravitational field, we can use spherical symmetry and write $\mathbf{g}(r) = g(r)\hat{r} = -\nabla\phi(r) = -\hat{r}\,\partial\phi(r)/\partial r$; thus,

$$g(r) = -\frac{\partial\phi(r)}{\partial r} = -\frac{d\phi(r)}{dr}, \qquad (9.2.6)$$

as in a one-dimensional case. Finally, integrals may involve different geometries and may use volume dV, areal dA, and linear dl elements. These are discussed in Appendix D.

■ 9.3 Examples on Gravitation

According to (9.1.5), if we obtain the gravitational field $\mathbf{g}(\mathbf{r})$, we can find the force $\mathbf{F}(\mathbf{r})$. The potential can also be obtained from (9.2.3). Therefore, most of the examples involve finding $\mathbf{g}(\mathbf{r})$ or $\mathbf{F}(\mathbf{r})$.

EXAMPLE 9.1

Figure 9.4 shows three masses, m_0, m_1, and m_2 at locations 0, x_1, and x_2, respectively. (a) Assuming points masses, obtain the gravitational forces exerted on m_1 and m_2 due to m_0. (b) Obtain the total gravitational force exerted on m_0 due to m_1 and m_2. (c) What is the gravitational field at the location of m_0 due to m_1 and m_2?

Solution

(a) This is a discrete system in one dimension. Equation (9.1.1) can be applied. The force on m_1 due to m_0 is $\mathbf{F}_{10} = -\hat{i}Gm_0m_1/x_1^2$, and that on m_2 due to m_0 is $\mathbf{F}_{20} = -\hat{i}Gm_0m_2/x_2^2$.

(b) The gravitational force exerted on m_0 due to m_1 is $\mathbf{F}_{01} = -\mathbf{F}_{10}$, and that of m_2 on m_0 is $\mathbf{F}_{02} = -\mathbf{F}_{20}$. The total force on m_0 is the sum of these two, or $\mathbf{F}_{total} = -\mathbf{F}_{10} - \mathbf{F}_{20} = m_0[Gm_1/x_1^2 + Gm_2/x_2^2]\hat{i}$.

(c) The gravitational field at the position of m_0 is
$\mathbf{g}_{total} = \mathbf{F}_{total}/m_0 = [Gm_1/x_1^2 + Gm_2/x_2^2]\hat{i}$.

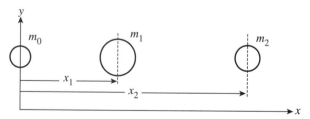

I FIGURE 9.4 Three particle system in one dimension.

EXAMPLE 9.2

Figure 9.5 shows three masses of the previous example, m_0, m_1, and m_2, located at $(0,0)$, (x_1, y_1), and (x_2, y_2). (a) Find the total gravitational force on m_0 due to the other two masses. (b) What is the main behavior of the force in the limit $m_1 = m_2$, $x_1 \sim x_2 = x$, $y_1 \sim y_2 = y$, with $y \gg x$? (c) What is the total gravitational potential energy of m_0? (d) What is the main behavior of the potential energy in the same limit as in (b)?

Solution

(a) We will use (9.1.3) with $\mathbf{r} = 0$, $\boldsymbol{\imath}_1 = -\mathbf{r}_1 = -(x_1\hat{i} + y_1\hat{j})$, and $\boldsymbol{\imath}_2 = -\mathbf{r}_2 = -(x_2\hat{i} + y_2\hat{j})$. The force on m_0 due to m_1 is $\mathbf{F}_{01} = -Gm_0m_1\boldsymbol{\imath}_1/\imath_1^3 = Gm_0m_1(x_1\hat{i} + y_1\hat{j})/(x_1^2 + y_1^2)^{3/2}$, and similarly that of m_2 on m_0, $\mathbf{F}_{02} = Gm_0m_2(x_2\hat{i} + y_2\hat{j})/(x_2^2 + y_2^2)^{3/2}$. The total force is the sum of these two

$$\mathbf{F}_{total} = Gm_0 \left\{ \left[\frac{m_1 x_1}{(x_1^2 + y_1^2)^{3/2}} + \frac{m_2 x_2}{(x_2^2 + y_2^2)^{3/2}} \right]\hat{i} + \left[\frac{m_1 y_1}{(x_1^2 + y_1^2)^{3/2}} + \frac{m_2 y_2}{(x_2^2 + y_2^2)^{3/2}} \right]\hat{j} \right\}.$$

$$(9.3.1)$$

(b) For very large values of $y = y_1 = y_2$ with $y \gg x$ ($x = x_1 = x_2$), we can ignore the $\hat{i}$ component since it decays as $1/y^3$. The larger $\hat{j}$ component behaves as $1/y^2$, so that for $m_1 = m_2$, $\mathbf{F}_{total} \sim \hat{j}2Gm_0m_1/y^2$.

(c) The gravitational potential energy is obtained from (9.2.1); that is,

$$V_{total} = V_{01} + V_{02} = -Gm_0 \left[\frac{m_1}{(x_1^2 + y_1^2)^{1/2}} + \frac{m_2}{(x_2^2 + y_2^2)^{1/2}} \right].$$

$$(9.3.2)$$

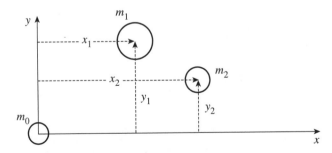

▌ FIGURE 9.5   Three-particle system in two dimensions.

(d) In the limit as in (b), we can write $V_{total} = -\int \mathbf{F}_{total} \cdot (\hat{j}\,dy)$ or $V_{total} \sim -2Gm_0m_1/y$, which can also be obtained from (9.3.2) in the $y = y_1 = y_2$, $y \gg x$ limit.

EXAMPLE 9.3

(a) Find the gravitational field due to a thin uniform rod of mass M and length L at a point $\mathbf{r}$ above its center and perpendicular to it.

(b) What is the behavior of the resulting expression in the separate limits of when $L \to 0$, $y \to \infty$, and $y \to 0$?

Solution

(a) Referring to Figure 9.6, we write the one-dimensional form of (9.1.4), with the use of (9.1.5)

$$\mathbf{g}(\mathbf{r}) = -G\int \frac{\lambda(\mathbf{r}')}{\imath^3}\boldsymbol{\imath}\,dl', \qquad (9.3.3)$$

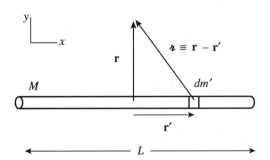

FIGURE 9.6   Example 9.3 for the field due to a one-dimensional continuous mass system.

where we have used $dm' = \lambda(r')dl'$ as described in Appendix D. Also note that for uniform density $\lambda(r') \to \lambda = M/L$. From the figure, we see that we can take $\mathbf{r} = y\hat{j}$, $\mathbf{r}' = x\hat{i}$, and so $\boldsymbol{\imath} = -x\hat{i} + y\hat{j}$ with $-L/2 < x < L/2$, and (9.3.3) becomes

$$\mathbf{g}(\mathbf{r}) = -G\lambda \int_{-L/2}^{L/2} \frac{(-x\hat{i} + y\hat{j})}{(x^2 + y^2)^{3/2}} dx = -G\lambda[-\hat{i}I_1 + \hat{j}yI_2], \qquad (9.3.4)$$

where

$$I_1 = \int_{-L/2}^{L/2} \frac{x\,dx}{(x^2 + y^2)^{3/2}}, \text{ and } I_2 = \int_{-L/2}^{L/2} \frac{dx}{(x^2 + y^2)^{3/2}}. \qquad (9.3.5)$$

The first integral can be done if we define $u = x^2 + y^2$ and $du = 2xdx$, so that $I_1 = \int_{-L/2}^{L/2} du/u^{3/2} = 1/\sqrt{u} = 1/\sqrt{x^2 + y^2}\Big|_{-L/2}^{L/2} = 0$. This makes sense

because of the symmetry in the problem. Each side of the rod has an opposing attractive field in the x direction, so the x components of the field cancel each other out. The second integral is of the form of the integral expression (B.10) with $m = 3/2$, so that we find

$$I_2 = \frac{x}{y^2(x^2 + y^2)^{1/2}}\Big|_{-L/2}^{L/2} = \frac{L}{y^2[(L/2)^2 + y^2]^{1/2}}. \qquad (9.3.6)$$

Going back to (9.3.4) we have that

$$\mathbf{g}(\mathbf{r}) = -\hat{j}\frac{G\lambda L}{y[(L/2)^2 + y^2]^{1/2}} = -\hat{j}\frac{GM}{y[(L/2)^2 + y^2]^{1/2}}. \qquad (9.3.7)$$

(b) In the limits of $L \to 0$ or $y \to \infty$, $\mathbf{g}(\mathbf{r}) \to -\hat{j}GM/y^2$; that is, the rod behaves like a point mass. In the limit of $y \to 0$, $\mathbf{g}(\mathbf{r}) \to -\infty$ because of the proximity to the rod. The gravitational force does become large for small enough distances, as can be seen from (9.1.1) as well.

EXAMPLE 9.4

(a) Find the gravitational field due to a uniform disk of mass M and radius r at a point z above its center and perpendicular to the disk's plane. (b) What is the behavior of the field in the limits of $z \to 0$ and $z \to \infty$ with $z \gg R$?

Solution

(a) Referring to Figure 9.7, we write the two-dimensional form of (9.1.4)

$$\mathbf{g}(\mathbf{r}) = -G \int \frac{\sigma(\mathbf{r}')}{\imath^3} \boldsymbol{\imath} \, dA', \tag{9.3.8}$$

where $dm' = \sigma(r')dA'$, as described in Appendix D, with $\sigma = M/\pi R^2$ is the constant areal density, and because of the symmetry, we use cylindrical coordinates.

Here let $\mathbf{r} = z\hat{k}$, $\mathbf{r}' = \rho\hat{\rho}$, so that $\boldsymbol{\imath} = \mathbf{r} - \mathbf{r}' = z\hat{k} - \rho\hat{\rho}$, and $dA' = \rho \, d\rho \, d\phi$. Equation (9.3.8) becomes

$$\mathbf{g}(\mathbf{r}) = -G\sigma \int \frac{\rho \, d\rho \, d\phi(z\hat{k} - \rho\hat{\rho})}{(z^2 + \rho^2)^{3/2}} = -G\sigma[z\hat{k} I_1 - \mathbf{I}_2], \tag{9.3.9}$$

where we have defined the integrals

$$I_1 = \int_0^R \frac{\rho \, d\rho}{(z^2 + \rho^2)^{3/2}} \int_0^{2\pi} d\phi = 2\pi \int_0^R \frac{\rho \, d\rho}{(z^2 + \rho^2)^{3/2}}, \text{ and } \mathbf{I}_2 = \int_0^R \frac{\rho^2 \, d\rho}{(z^2 + \rho^2)^{3/2}} \int_0^{2\pi} \hat{\rho} \, d\phi. \tag{9.3.10}$$

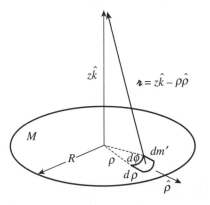

I FIGURE 9.7 Example 9.4 for the field due to a uniform disk.

In $\mathbf{I}_2$ since $\hat{\rho} = \boldsymbol{\rho}/\rho = (x\hat{i} + y\hat{j})/\rho = (\rho\cos\phi\hat{i} + \rho\sin\phi\hat{j})/\rho$ depends on the angle ϕ, the

integral $\int_0^{2\pi} \hat{\rho}d\phi = \int_0^{2\pi} (\cos\phi\hat{i} + \sin\phi\hat{j})d\phi = 0$, and therefore $\mathbf{I}_2 = 0$. In the integral I_1, we

let $u = z^2 + \rho^2$, $du = 2\rho d\rho$, and $I_1 = 2\pi[\int du/u^{3/2}]/2 = -2\pi/\sqrt{u}$, or

$$I_1 = -\frac{2\pi}{(z^2 + \rho^2)^{1/2}}\Big|_0^R = 2\pi\left[\frac{1}{z} - \frac{1}{(z^2 + R^2)^{1/2}}\right],$$

so that we finally get

$$\mathbf{g}(\mathbf{r}) = -2\pi G\sigma\hat{k}\left[1 - \frac{z}{(z^2 + R^2)^{1/2}}\right] = -\frac{2GM}{R^2}\left[1 - \frac{z}{(z^2 + R^2)^{1/2}}\right]\hat{k}. \tag{9.3.11}$$

The gravitational field, therefore, points toward the center of the disk. This formula can be made to work for a distance z below the disk, in which case replace $\hat{k} \rightarrow -\hat{k}$, but still $z > 0$ in (9.3.11).

(b) In the limit as $z \rightarrow 0$, it is easy to see that $\mathbf{g}(\mathbf{r}) \rightarrow -2GM\hat{k}/R^2$, which is a finite constant value. We do see that if r is small, the field gets large because the mass would be concentrated in a smaller area, and vice versa as r gets large. In the limit as $z \rightarrow \infty$ where $z \gg R$, we can expand the quantity $z/(z^2 + R^2)^{1/2} = 1/[1 + (R/z)^2]^{1/2} \sim 1 - (R/z)^2/2$, so that from (9.3.11), $\mathbf{g}(\mathbf{r}) \rightarrow -GM\hat{k}/z^2$. This shows that far away from the disk, it looks as though it were a point charge. Such result is expected and is useful in determining the correct field behavior at large distances away from the source.

EXAMPLE 9.5

Find the gravitational field due to a uniform sphere of mass M and radius r at a point $z \geq R$ from its center.

Solution
Consider Figure 9.8, where the sphere is thought to be composed of disks of mass dm whose radii change from $y = 0$ at $z' = \pm R$, to $y = R$ at $z' = 0$. We already know the gravitational field at a distance z due to a disk of radius y. It is given by (9.3.11), which we write as

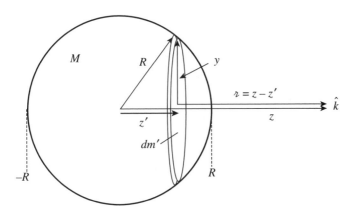

I FIGURE 9.8   Example 9.5 for the field due to a uniform sphere.

$$dg_{disk}(\mathbf{r}) = -\frac{2Gdm'}{y^2}\left[1 - \frac{\imath}{(\imath^2 + y^2)^{1/2}}\right]\hat{k}. \tag{9.3.12}$$

The mass element $dm' = \rho dV = \rho A dz' = \rho\pi y^2 dz'$, also $\imath = z - z'$, and $y^2 = R^2 - z'^2$, so that substituting these into (9.3.12) and simplifying, we get

$$dg_{disk}(\mathbf{r}) = -2G\rho\pi dz'\left[1 - \frac{z - z'}{(z^2 + R^2 - 2zz')^{1/2}}\right]\hat{k}, \tag{9.3.13}$$

and the total gravitational field at $z > R$ due to the sphere is

$$\mathbf{g} = \int dg_{disk}(\mathbf{r}) = -2G\rho\pi\hat{k}\int_{-R}^{R}\left[1 - \frac{z - z'}{(z^2 + R^2 - 2zz')^{1/2}}\right]dz' = -2G\rho\pi\hat{k}[I_1 - zI_2 + I_3],$$

$$\tag{9.3.14}$$

where we have defined the integrals

$$I_1 = \int_{-R}^{R}dz' = 2R,\; I_2 = \int_{-R}^{R}\frac{dz'}{(z^2 + R^2 - 2zz')^{1/2}},\; I_3 = \int_{-R}^{R}\frac{z'dz'}{(z^2 + R^2 - 2zz')^{1/2}}. \tag{9.3.15}$$

If in the second integral we let $u = z^2 + R^2 - 2zz'$, $du = -2z\,dz'$, then $I_2 = -u^{1/2}/z$, or

$$I_2 = -\frac{1}{z}(z^2 + R^2 - 2zz')^{1/2}\Big|_{-R}^{R} = -\frac{1}{z}[(z - R) - (z + R)] = \frac{2R}{z}. \tag{9.3.16a}$$

The last integral, I_3, is of the form $\int x\,dx/\sqrt{ax+b}$, with $a = -2z$ and $b = z^2 + R^2$, so that letting $u = x$, $dv = dx/(ax+b)^{1/2}$, and integrating by parts, $\int u\,dv = uv - \int v\,du$, we get $\int x\,dx/\sqrt{ax+b} = 2(ax - 2b)\sqrt{ax+b}/3a^2$, or

$$I_3 = 2\left[\frac{-2zz' - 2(z^2 + R^2)}{3(-2z)^2}\sqrt{-2zz' + z^2 + R^2}\right]_{-R}^{R}$$

$$= \frac{(-2zR - 2z^2 - 2R^2)(z - R) - (2zR - 2z^2 - 2R^2)(z + R)}{6z^2}$$

$$= \frac{(-4z^2R + 4Rz^2 + 4R^3)}{6z^2} = \frac{2R^3}{3z^2} \tag{9.3.16b}$$

Putting I_1, I_2, and I_3 from (9.3.15 and 9.3.16) into (9.3.14), and using $\rho = M/(4\pi R^3/3)$ for the sphere, we obtain

$$\mathbf{g}_{z \geq R} = -2G\pi\frac{3M}{4\pi R^3}\hat{k}\left[2R - 2R + \frac{2R^3}{3z^2}\right] = -\frac{GM}{z^2}\hat{k}. \tag{9.3.17}$$

This is an important result because it is independent of the sphere radius. It tells us that a uniform sphere of constant density and mass M produces a gravitational field outside of it that is identical to that of a point mass at a distance z from its center.

<hr/>

EXAMPLE 9.6

Find the gravitational field due to a uniform ring of mass M and radius r at a point z from its center and perpendicular to the ring's plane.

···

Solution

Consider Figure 9.9, the ring's element of mass $dm' = \lambda R\,d\phi$, located at $\mathbf{r}' = R\hat{\rho}$, which produces a field contribution at $\mathbf{r} = z\hat{k}$. We have $\boldsymbol{\imath} = z\hat{k} - R\hat{\rho}$ and $\imath^2 = z^2 + R^2$, since $\hat{k}\cdot\hat{\rho} = 0$ in cylindrical polar coordinates. The coordinates are also shown in the lower-right part of the figure. Here the uniform ring density is $\lambda = M/2\pi R$.

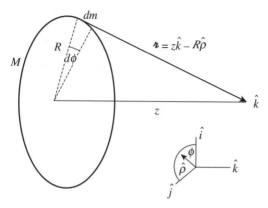

FIGURE 9.9   Example 9.6 for the field due to a uniform ring.

The gravitational field is

$$\mathbf{g} = -G\int\frac{\boldsymbol{\imath}\,dm'}{\imath^3} = -G\lambda R\int\limits_0^{2\pi}\frac{(z\hat{k} - R\hat{\rho})d\phi}{(z^2 + R^2)^{3/2}}. \tag{9.3.18}$$

The cylindrical coordinate unit vector $\hat{\rho} = \hat{i}\cos\theta + \hat{j}\sin\theta$; the second integral is therefore $\int\limits_0^{2\pi}\hat{\rho}\,d\theta = 0$, and the remaining integral over ϕ gives a contribution of 2π. Substituting for λ, the final result is

$$\mathbf{g} = -\frac{GM\,z\hat{k}}{(z^2 + R^2)^{3/2}}. \tag{9.3.19}$$

EXAMPLE 9.7

Consider a uniform spherical shell of mass M and radius R. (a) Find its gravitational field at a point $z \geq R$ from its center. (b) What is the gravitational field due to the spherical shell at $z < R$?

Solution

(a) The spherical shell shown in Figure 9.10 has element of mass $dm' = \sigma dA$, where the shell's surface uniform density is $\sigma = M/4\pi R^2$.

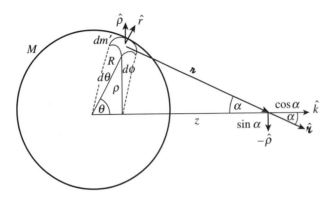

FIGURE 9.10 Example 9.7 for the field due to a uniform spherical shell.

In spherical coordinates, $dA = (\rho d\phi)(R d\theta)$, where $\rho = R\sin\theta$. From the figure, we notice that $\hat{\imath} = \hat{k}\cos\alpha - \hat{\rho}\sin\alpha$, but also since $\mathbf{r}' = R\hat{r}$ and $\mathbf{r} = z\hat{k}$, then $\mathbf{\imath} = \mathbf{r} - \mathbf{r}' = z\hat{k} - R\mathbf{r}$ or $\imath^2 = z^2 + R^2 - 2zR\hat{r} \cdot \hat{k}$, where $\mathbf{r} \cdot \hat{k} = \cos\theta$, so that we write the gravitational field as

$$\mathbf{g}_{z \geq R} = -G \int \frac{\hat{\imath}}{\imath^2} dm' = -G\sigma R^2 \int_0^\pi d\theta \int_0^{2\pi} d\phi \left[\frac{(\hat{k}\cos\alpha - \hat{\rho}\sin\alpha)\sin\theta}{z^2 + R^2 - 2zR\cos\theta} \right]. \quad (9.3.20)$$

Here, since $\hat{\rho}$ is the only quantity that depends on ϕ, we can write it in terms of Cartesian coordinates as $\hat{\rho} = \hat{\imath}\cos\phi + \hat{j}\sin\phi$ and it is seen that $\int_0^{2\pi} (\hat{\imath}\cos\phi + \hat{j}\sin\phi) d\phi = 0$. This shows that there is no contribution to the field in the $\hat{\rho}$ direction. In this way, (9.3.20) becomes

$$\mathbf{g}_{z \geq R} = -G\sigma R^2 \hat{k} \left[\int_0^{2\pi} d\phi \right] \int_0^\pi \frac{\cos\alpha\sin\theta}{z^2 + R^2 - 2zR\cos\theta} d\theta. \quad (9.3.21)$$

From this we can write $\mathbf{r}' = R\hat{r} = z\hat{k} - \imath \Rightarrow R^2 = z^2 + \imath^2 - 2z\imath\cos\alpha$, since $\mathbf{r} \cdot \imath = \cos\alpha$. Solving this for $\cos\alpha$, we get

$$\cos\alpha = \frac{R^2 - z^2 - \imath^2}{-2z\imath}. \tag{9.3.22a}$$

Furthermore, we can change integration variables as follows, since $\imath^2 = z^2 + R^2 - 2zR\cos\theta$ or

$$2\imath d\imath = 2zR\sin\theta d\theta \Rightarrow \sin\theta d\theta = \imath d\imath/zR, \tag{9.3.22b}$$

where the integration ranges change from $0 \leq \theta \leq \pi$ to $z - R \leq \imath \leq z + R$. Substituting (9.3.22) into (9.3.21), we get

$$\mathbf{g}_{z \geq R} = -\frac{2\pi G\sigma R^2 \hat{k}}{2z^2 R} \int_{z-R}^{z+R} \left(\frac{\imath^2 + z^2 - R^2}{\imath} \right) \frac{\imath d\imath}{\imath^2} = -\frac{GM\hat{k}}{4Rz^2} I, \tag{9.3.23}$$

where the integral is

$$I = \int_{z-R}^{z+R} \left(1 + \frac{z^2 - R^2}{\imath^2} \right) d\imath = \left[\imath - (z^2 - R^2)\frac{1}{\imath} \right]_{z-R}^{z+R}$$

$$= 2R - (z^2 - R^2)\left(\frac{1}{z+R} - \frac{1}{z-R} \right) = 4R, \tag{9.3.24}$$

and (9.3.23) yields

$$\mathbf{g}_{z \geq R} = -\frac{GM\hat{k}}{z^2}. \tag{9.3.25}$$

This result is significant because it says that the gravitational field outside a spherical shell is the same as that of a point mass. This is similar to the field outside a sphere of uniform density, Equation (9.3.17).

(b) In order to obtain the field inside the spherical shell, except for the fact that now $z < R$, the process is identical in every respect until we get to the limits in (9.3.23). Since $\imath^2 = z^2 + R^2 - 2zR\cos\theta$ with $0 \leq \theta \leq \pi$, as we make the change in variable from θ to $\imath$, we now have that $R - z \le \imath \le R + z$. The field is given by (9.3.23) with

$$I \to \int_{R-z}^{R+z} \left(1 + \frac{z^2 - R^2}{\imath^2}\right) d\imath = \left[\imath - (z^2 - R^2)\frac{1}{\imath}\right]_{R-z}^{R+z}$$

$$= 2z - (z^2 - R^2)\left(\frac{1}{R + z} - \frac{1}{R - z}\right) = 0.$$

(9.3.26)

Thus $\mathbf{g}_{z<R} = 0$ everywhere inside the spherical shell.

EXAMPLE 9.8

Find the gravitational field due to a uniform sphere of mass M and radius r at (a) a point $z < R$ from its center, and (b) show that when $z = R$, the formula simplifies to that of Equation (9.3.17). Summarize the results.

Solution

(a) The result of Example 9.7 is very useful to answer this problem. A sphere can be thought of many shells of varying radii in the range $0 < r < R$. However, the field at $z < R$ will only have contributions from shells that have radii in the range $0 < r < z$. Those shells in the range $z < r < R$ will not contribute to the total field due to their null contribution within, as found in the previous example. This is illustrated in Figure 9.11, where only the mass within the dashed spherical shell contributes to the total field at $z < R$.

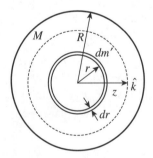

| FIGURE 9.11   Example 9.8 for the field inside and outside of a uniform sphere.

If we let the shell mass be $dm' = \rho dV = \rho A dr = \rho 4\pi r^2 dr$, where $\rho = M/(4\pi R^3/3)$, then any shell contributes amount $dg_{r \leq z} = -\hat{k}Gdm'/z^2$ outside its radius $r \leq z$ for $z < R$. The uniform sphere field can, therefore, be obtained as

$$\mathbf{g}_{z<R} = -\frac{\hat{k}G}{z^2}\int_0^z dm' = -\frac{\hat{k}G}{z^2}\rho 4\pi \int_0^z r^2 dr = -\frac{\hat{k}G}{z^2}\rho 4\pi \left[\frac{z^3}{3}\right] = -\frac{GM}{R^3}z\hat{k}. \qquad (9.3.27)$$

(b) At $z = R$, this result agrees with that of (9.3.17) on the surface of the sphere. Thus, the gravitational field due to a solid uniform density sphere of mass M and radius r can be summarized as follows:

$$\mathbf{g}_{sphere} = \begin{cases} -\dfrac{GM}{z^2}\hat{k} & z \geq R \\ -\dfrac{GM}{R^3}z\hat{k} & 0 < z \leq R. \end{cases} \qquad (9.3.28)$$

EXAMPLE 9.9

The gravitational potential from a uniform spherical shell, of mass M and radius r, can be found using Equation (9.2.3). (a) Find the potential at point $r \geq R$ from its center. (b) Find the gravitational potential at $r < R$. Assume the potential reference point is at infinity.

Solution
From Example 9.7, the spherical shell gravitational field can be summarized as follows:

$$\mathbf{g}_{spherical\ shell} = \begin{cases} -\dfrac{GM}{r^2}\mathbf{r} & r \geq R \\ 0 & 0 < z \leq R, \end{cases} \qquad (9.3.29)$$

so that for outside the shell we have, using $d\mathbf{r} = \hat{r}dr$,

$$(a)\ \phi(R \leq r \leq \infty) = -\int \mathbf{g}(r)\cdot d\mathbf{r} = GM\int_\infty^r \frac{1}{r^2}dr = -\frac{GM}{r}, \qquad (9.3.30)$$

which shows that the potential outside the spherical shell is that of a point mass.

For inside we have

(b) $\phi(0 \le r \le R) = -\int \mathbf{g}(r) \cdot d\mathbf{r} = -\int_{\infty}^{R} \mathbf{g}_{out}(r) \cdot d\mathbf{r} - \int_{R}^{r} \mathbf{g}_{in}(r) \cdot d\mathbf{r}$

$$= GM\left[\int_{\infty}^{R} \frac{1}{r^2}dr + \int_{R}^{r} 0 \cdot dr \right] = -\frac{GM}{R}, \tag{9.3.31}$$

which indicates that while there is no gravitational field, the gravitational potential is constant throughout the inside of the spherical shell. Notice also from (9.3.30) that at $r = R$, the potential is continuous.

EXAMPLE 9.10

(a) What is the form of the gravitational field of Earth, and what is a good approximation for its value near the surface? (b) Repeat for the gravitational potential.

Solution

(a) The gravitational field of Earth can be taken to be that of a solid sphere; that is, $\mathbf{g} = g_E \hat{r}$, with $g_E = -Gm_E/r^2$. Using Earth's radius R_E, this can be written as

$$g_E = -\frac{Gm_E}{(R_E + h)^2} = -\frac{Gm_E}{R_E^2(1 + h/R_E)^2} = -\frac{Gm_E}{R_E^2}\left(1 - 2\frac{h}{R_E} + \cdots\right), \tag{9.3.32}$$

where h is a height from the surface of Earth and we have used a Taylor expansion. On the surface of Earth, $h \simeq 0$, so that

$$g_E \simeq -\frac{Gm_E}{R_E^2} = -\frac{(6.67 \times 10^{-11}\text{Nm}^2/\text{kg}^2)(5.98 \times 10^{24}\text{ kg})}{(6.37 \times 10^6\text{ m})^2} \simeq -9.8\frac{\text{m}}{\text{s}^2}.$$

(b) This is a central force problem, so that, from (9.2.6),

$$\phi_E(r) = -\int g_E dr = -\frac{Gm_E}{r} = -\frac{Gm_E}{(R_E + h)} = -\frac{Gm_E}{R_E(1 + h/R_E)}$$

$$= -\frac{Gm_E}{R_E}(1 - h/R_E + (h/R_E)^2 + \cdots). \tag{9.3.33}$$

which to first order in h becomes

$$\phi_E(r) \simeq \phi_{0E} + g_E h,$$ (9.3.34)

where $\phi_{0E} \equiv -Gm_E/R_E$ is Earth's reference potential from where the measurement is made.

■ 9.4   Gauss' Law of Gravitation

Recall from Chapter 5 the definition of a field flux, which for the case of the gravitational field we write as

$$\Phi = \int \mathbf{g} \cdot \hat{n} \, dA,$$ (9.4.1)

where A is the area crossed by the field and $\hat{n}$ is its direction, as shown in Figure 9.12.

We also define the gravitational field flux associated with a closed surface as

$$\Phi_g = \oint \mathbf{g} \cdot \hat{n} \, dA.$$ (9.4.2)

In this expression, the dot product guarantees that only components that are parallel to the surface direction, as shown in Figure 9.12, contribute to the integral.

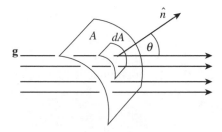

| FIGURE 9.12   Gravitational flux through a surface area.

EXAMPLE 9.11

Obtain the gravitational field flux due to a mass m inside a spherical surface of radius r.

Solution

The gravitational field due to the point mass is given by $\mathbf{g} = -Gm\hat{r}/r^2$, and the spherical surface element is $\hat{n}\,dA = \mathbf{r}(r\,d\theta)(r\sin\theta\,d\phi)$, so that the total field flux is

$$\Phi_g = \oint \mathbf{g}\cdot\hat{n}\,dA = -Gm\oint \frac{\hat{r}\cdot\hat{r}}{r^2}r^2\sin\theta\,d\theta\,d\phi = -Gm\int_0^\pi \sin\theta\,d\theta \int_0^{2\pi} d\phi = -4\pi Gm, \quad (9.4.3)$$

which shows that the flux through the closed surface is actually a measure of the mass inside and which produces the field.

From this example we see that in a highly symmetric situation we could write $\mathbf{g} = g\hat{r}$ as is the case of spherical symmetry. Since $\mathbf{g}$ is symmetric, it has the same value at a fixed radius r from the origin, and then performing (9.4.2) gives

$$\Phi_g = \oint \mathbf{g}\cdot\hat{n}\,dA = g\oint \hat{r}\cdot\hat{n}\,dA = gr^2\oint \sin\theta\,d\theta\,d\phi = 4\pi gr^2. \quad (9.4.4)$$

Because this result must be identical to (9.4.3), we can equate them to get

$$4\pi gr^2 = -4\pi Gm \quad\Rightarrow\quad g = -\frac{Gm}{r^2}, \quad \therefore \quad \mathbf{g} = -\frac{Gm}{r^2}\hat{r}; \quad (9.4.5)$$

i.e., we get back the field due to the point mass. This simple exercise actually points to a totally different way of obtaining the gravitational field in highly symmetric cases. Thus, in essence, Equation (9.4.3) is Gauss' law as applied to a point mass. Gauss' law is more generally stated as follows:

$$\oint \mathbf{g}\cdot\hat{n}\,dA = -4\pi Gm_{enc}. \quad (9.4.6)$$

It expresses the fact that the gravitational flux through any surface enclosing a mass is proportional to the negative of the enclosed mass, m_{enc}. The constant of proportionality is $4\pi G$. The law is, once again, most useful in situations of high symmetry.

EXAMPLE 9.12

A hollow spherical shell of inner radius a and outer radius b with total mass M has a nonuniform density $\rho = k/r^2$ in the region $a \leq r \leq b$.

(a) Obtain the value of the constant k. Find the gravitational field in the regions

(b) $r < a$,

(c) $a < r < b$, and

(d) $r > b$.

(e) Plot the field as a function of r with suitable parameter values.

Solution

Gauss' law provides us with a simple but powerful method to tackle this problem. Figure 9.13(a) shows three Gaussian surfaces I, II, and III, shown by the dashed lines, to be used in the corresponding regions $r < a$, $a < r < b$, and $r > b$, respectively. The mass M is concentrated in the shaded region II. In all cases we let $\mathbf{g} = g\hat{r}$ due to spherical symmetry.

(a) We first obtain the constant k by integrating the density over the volume

$$M = \int \rho dV = k \int_a^b \frac{r^2 \sin\theta\, d\theta\, d\phi\, dr}{r^2} = 4\pi k(b - a) \quad \Rightarrow \quad k = M/[4\pi(b - a)]. \quad (9.4.7)$$

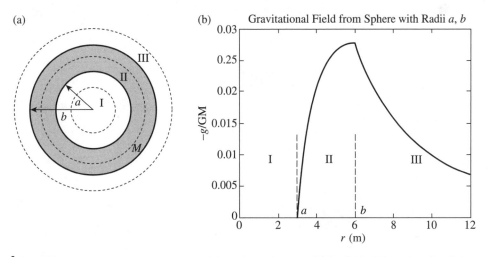

(a)

(b) Gravitational Field from Sphere with Radii a, b

FIGURE 9.13 Example 9.12 for the field of (a) a hollow sphere, and (b) the field within each region of the sphere.

(b) In region I, the right-hand side of (9.4.6) is zero because there is no mass enclosed; thus, $g_{r<a} = 0$.

(c) In region II, $\oint \mathbf{g} \cdot \hat{n} \, dA = g4\pi r^2$, and $m_{enc} = k \displaystyle\int_a^r \dfrac{r^2 \sin\theta \, d\theta \, d\phi \, dr}{r^2} = 4\pi k(r - a)$, so that by (9.4.6) we equate these two results to get $g_{a<r<b} = -[4\pi Gk(r - a)]/r^2$.

(d) In region III, once again $\oint \mathbf{g} \cdot \hat{n} \, dA = g4\pi r^2$, and $m_{enc} = k \displaystyle\int_a^b \dfrac{r^2 \sin\theta \, d\theta \, d\phi \, dr}{r^2}$ $= 4\pi k(b - a)$, and equating these two results gives $g_{r>b} = -[4\pi Gk(b - a)]/r^2$. In summary, with the use of (9.4.7) we have

$$\mathbf{g} = \begin{cases} 0 & r < a \\[2mm] -\dfrac{GM(r - a)}{(b - a)r^2}\hat{r} & a \leq r \leq b. \\[2mm] -\dfrac{GM}{r^2}\hat{r} & r \geq b \end{cases} \qquad (9.4.8)$$

Notice that in region III, the sphere's field behaves like that from a point mass.

(e) A plot of the magnitude $-g/GM$ versus r is shown in Figure 9.13(b), where each region is labeled and for which the following MATLAB code gauss_sphere.m was used.

SCRIPT

```
%gauss_sphere.m - plots the gravitational field for a sphere of mass M and
%inner radius a and outer radius b.
clear;
a=3; b=6; N=250; %a, b in meters, and points to plot
rmin=0; rmax=2*b; dr=(rmax-rmin)/N; %ranges
for i=1:N
    r(i)=rmin+(i-1)*dr; g(i)=0;
    if r(i)>=a & r(i)<b g(i)=(r(i)-a)/(b-a)/r(i)^2;
    else if r(i)>=b g(i)=1/r(i)^2;
        end
    end
end
```

```
[c,j]=max(g);%get maximum of g and its index
plot(r,g,'k',a,0,'r.',b,0,'r.')
text(a*(1+0.03),0.0008,'a','FontSize',14)
text(b*(1+0.03),0.0008,'b','FontSize',14)
line([a,a],[0,c/2],'Color','r','LineStyle','-')
line([b,b],[0,c/2],'Color','r','LineStyle','-')
text(a/2,c/3,'I','FontSize',14)
text((a+b)/2,c/3,'II','FontSize',14)
text(b+(rmax-b)/2,c/3,'III','FontSize',14)
xlabel('r(m)','FontSize',14);ylabel('-g/GM','FontSize',14);
title('Gravitational Field from Sphere with radii a,b','FontSize',14)
```

■ 9.5   Binary Mass System Simulation

In this section we consider the dynamical interaction between two masses, m_1 and m_2. The masses are free to move under the gravitational force law. We let $\mathbf{r}_i$ designate the coordinate for mass i, as shown in Figure 9.14.

Referring to the figure, the force on m_2 due to m_1 is

$$\mathbf{f}_{21} = -\frac{Gm_1m_2\mathbf{r}_{21}}{r_{21}^3}, \tag{9.5.1}$$

and the force on m_1 due to m_2 is

$$\mathbf{f}_{12} = -\frac{Gm_1m_2\mathbf{r}_{12}}{r_{12}^3}. \tag{9.5.2}$$

We can, therefore, write Newton's second law for each mass, respectively, as follows:

$$m_1\frac{d^2\mathbf{r}_1}{dt^2} = -\frac{Gm_1m_2\mathbf{r}_{12}}{r_{12}^3} \quad \text{and} \quad m_2\frac{d^2\mathbf{r}_2}{dt^2} = -\frac{Gm_1m_2\mathbf{r}_{21}}{r_{21}^3}. \tag{9.5.3}$$

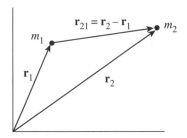

▎ FIGURE 9.14

We would like to simulate the motion of this system and will use two methods to do it. The first method uses the center of mass–relative coordinate concept to simplify the problem. The second method is the straight brute-force numerical method of solving the coupled differential equations computationally. Both of these methods produce equivalent results and are instructively useful, but the first method is less numerically intensive.

A. Center of Mass—Relative Coordinate Method

In this method it is convenient to define the center of mass and relative coordinates by

$$\mathbf{r}_{cm} \equiv \frac{m_1\mathbf{r}_1 + m_2\mathbf{r}_2}{m} \text{ and } \mathbf{r} \equiv \mathbf{r}_{21} = -\mathbf{r}_{12} = \mathbf{r}_2 - \mathbf{r}_1, \tag{9.5.4}$$

respectively, where the total mass is $m \equiv m_1 + m_2$. In matrix form this is

$$\begin{pmatrix} \mathbf{r}_{cm} \\ \mathbf{r} \end{pmatrix} = \begin{pmatrix} m_1/m & m_2/m \\ -1 & 1 \end{pmatrix} \begin{pmatrix} \mathbf{r}_1 \\ \mathbf{r}_2 \end{pmatrix}. \tag{9.5.5}$$

This matrix can be inverted (see Appendix B) to express $\mathbf{r}_1$ and $\mathbf{r}_2$ in terms of $\mathbf{r}_{cm}$ and $\mathbf{r}$ as

$$\mathbf{r}_1 = \mathbf{r}_{cm} - \frac{m_2}{m}\mathbf{r} \quad \text{and} \quad \mathbf{r}_2 = \mathbf{r}_{cm} + \frac{m_1}{m}\mathbf{r}, \tag{9.5.6}$$

as shown in Figure 9.15.

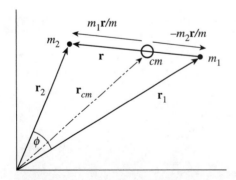

| FIGURE 9.15   Center of a mass of a binary mass system.

Next, multiplying the first of (9.5.3) by m_2 and subtracting it from the second multiplied by m_1, we have

$$\mu\frac{d^2\mathbf{r}}{dt^2} = -\frac{Gm_1m_2\mathbf{r}}{r^3}, \tag{9.5.7}$$

where we have defined the reduced mass μ as

$$\mu = \frac{m_1m_2}{(m_1 + m_2)} \quad \text{or} \quad \frac{1}{\mu} = \frac{1}{m_1} + \frac{1}{m_2}. \tag{9.5.8}$$

We have managed to convert two equations of motion, one for each mass' coordinates, to one equation for the relative coordinate $\mathbf{r}$. Equation (9.5.7) corresponds to the motion of the reduced mass under the action of a central force $\mathbf{F}(r) = F(r)\hat{r}$, where $F(r) = -G\mu(m_1 + m_2)/r^2$, so that we can write

$$\mu\frac{d^2\mathbf{r}}{dt^2} = -\frac{G\mu(m_1 + m_2)\mathbf{r}}{r^3}, \quad \text{or} \quad \frac{d^2\mathbf{r}}{dt^2} = -\frac{G(m_1 + m_2)\hat{r}}{r^2}. \tag{9.5.9}$$

We studied the central force problem in Chapter 8. So from (8.6.8) we can write the corresponding expressions for the solutions to (9.5.7 and 9.5.9); that is,

$$r = -\frac{L_\mu^2}{\mu K_\mu\left(1 - \dfrac{u_0 L_\mu^2}{\mu K_\mu}\cos\theta\right)}, \tag{9.5.10}$$

where the angular momentum associated with the reduced mass is $L_\mu = \mu v r$, and the central force constant $K_\mu = -Gm_1m_2 = -G\mu(m_1 + m_2)$, as indicated in (9.5.7). The quantity in (9.5.10),

$$-\frac{L_\mu^2}{\mu K_\mu} = \frac{\mu^2 v^2 r^2}{\mu Gm_1m_2} = \frac{\mu v^2 r^2}{Gm_1m_2} = \frac{v^2 r^2}{G(m_1 + m_2)}, \tag{9.5.11}$$

provides some simplification, and using (8.6.8) we get

$$r = \frac{v^2 r^2}{G(m_1 + m_2)\left(1 + \dfrac{u_0 v^2 r^2}{G(m_1 + m_2)}\cos\theta\right)} = r_{min}\left(\frac{1 + e}{1 + e\cos\theta}\right) \tag{9.5.12}$$

for the solution of the relative coordinate variable. Here, $r_{min} = v^2 r^2 / [G(m_1 + m_2)(1 + e)]$, and $e = u_0 v^2 r^2 / [G(m_1 + m_2)]$ similar to what was done in Chapter 8. This, in essence, represents the analytical solution to the binary mass motion problem. Here, $r(\theta)$ is the position versus angle of the reduced mass in a central field potential $V(r) = -G\mu(m_1 + m_2)/r$. The orbit is an ellipse with eccentricity e and, as before, r_{min} is the closest approach distance. The farthest approach distance is r_{max}, which is given by the previous expression (8.6.10). Furthermore, the ellipse will have a semimajor axis

$$a = \frac{r_{min} + r_{max}}{2}. \tag{9.5.13}$$

Once the relative coordinate is known, the coordinates of each individual mass can be obtained from (9.5.6). Each of the masses m_1 and m_2 will trace its own characteristic elliptical orbit with respective r_{1min}, r_{1max} and r_{2min}, r_{2max}, as well as respective semimajor axes given by

$$a_1 = \frac{r_{1min} + r_{1max}}{2} \quad \text{and} \quad a_2 = \frac{r_{2min} + r_{2max}}{2}. \tag{9.5.14}$$

From Figure 9.15, using the law of cosines, we see that $r = \sqrt{r_1^2 + r_2^2 - r_1 r_2 \cos\phi}$. This is useful because if, for example, the center of mass coordinate is placed at the origin, and because the masses are always on opposite sides of the center of mass, we let $\phi = \pi$, then we see that the magnitude $r = r_1 + r_2$. This can also be seen from (9.5.6) for the magnitudes, if $\mathbf{r}_{cm}$ is set to zero. Thus, we can see that a relation between the semimajor axes follows as

$$a = a_1 + a_2. \tag{9.5.15}$$

An example of this is shown in Figure 9.16 for the major axes of the reduced mass orbit (the largest orbit), the smaller binary mass orbit (the middle orbit), and the larger binary mass orbit (the smallest orbit). The scale used is shown in the upper-left corner of the figure. Also, notice that with the information available in the figure, one can find the eccentricity of each ellipse using Equation (8.6.10), which relates the eccentricity of the ellipse and the minimum and maximum distances from the focus point, to get

$$e = \frac{r_{max} - r_{min}}{r_{max} + r_{min}} = \frac{2r_{max} - 2a}{2a}, \tag{9.5.16}$$

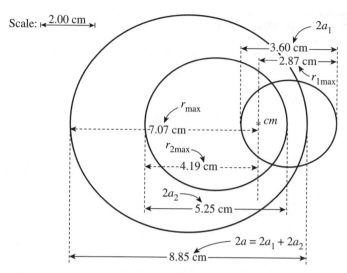

Scale: $\mapsto$ 2.00 cm

FIGURE 9.16   A binary system with each mass (middle ellipse = low mass, small ellipse = large mass) orbiting their common center of mass with their respective semimajor axes. The reduced mass's orbit is also shown (large orbit).

where we have used the previously defined average for the semimajor axis, $a = (r_{max} + r_{min})/2$ from Chapter 8, Section 6.

Using the smaller ellipse values we get $e = [2(2.87) - 3.6]/3.6 \sim 0.6$. If we repeat the calculation for the other two ellipses, we will find that all the orbits have the same eccentricity. This is an inherent property of a binary system.

Notice that if we take $m_1 \gg m_2$, then m_2 can be ignored in (9.5.12). Furthermore, in this case, from (9.5.9), since $\mu \to m_2$ the equation of motion becomes $m_2 d^2\mathbf{r}/dt^2 \to -Gm_2m_1\hat{r}/r^2$, and from (9.5.6) we can set $\mathbf{r}_1 \to \mathbf{r}_{cm} \equiv 0$; therefore, $\mathbf{r}_2 \to \mathbf{r}$, and similarly from (9.5.15) $a \to a_2$. Thus the only motion that would matter is that of the less massive (m_2) about the more massive (m_1). This limiting case is similar to planetary motion about the sun, where the sun's mass is much greater than a planet's mass and remains essentially at rest while the planet orbits it. Similarly, we know that for $m_1 \gg m_2$ from (8.7.5), Kepler's third law is $\tau^2/a^3 = 4\pi^2/Gm_1$. However, when the masses become comparable the correct form of Kepler's third law for the period of revolution replaces a single mass with the sum of the masses; i.e.,

$$\tau^2 = \frac{4\pi^2 a^3}{Gm} = \frac{4\pi^2(a_1 + a_2)^3}{G(m_1 + m_2)}. \tag{9.5.17}$$

So each mass (m_1, m_2) takes the same amount of time to go around the center of mass, including the reduced mass (μ) itself as well.

EXAMPLE 9.13

A binary system is characterized by an eccentricity of $e = 0.6$ with masses $m_1 = 3m_s$ and $m_2 = 2m_s$, where m_s is a solar mass. (a) If the center of mass is assumed to be at the origin and the relative coordinate has a minimum radius of $2.5\,AU$, simulate the orbital motion of the reduced mass as well as that of each of the masses m_1 and m_2. (b) What is the orbital period of this system?

Solution

(a) From (9.5.12) the reduced mass' orbit is given by $r = 2.5[(1 + 0.6)/(1 + 0.6\cos\theta)]R$, where R is an astronomical unit. With this r, the orbits of m_1 and m_2 are given by (9.5.6) as $r_1 = -(2r/5)R$ and $r_2 = (3r/5)R$. The simulation is done using the MATLAB script binary1.m, as shown in Figure 9.17.

   The figure shows the orbits of each of the masses as detailed by the legend. The reduced mass orbit is the largest. The period calculated as in part (b) is also shown. The actual animated motion can more readily be appreciated by running the script binary1.m that follows.

(b) The orbital period is, from (9.5.17), $\tau = \tau_0\sqrt{((a_1 + a_2)^3)/(m_1 + m_2)}$ or $\tau = \sqrt{a^3/5}\,\tau_0$, where $\tau_0 = 2\pi\sqrt{R^3/(Gm_s)} = 1\,\text{yr}$. Since $a = (r_{min} + r_{max})/2$ and $r_{max} = r_{min}(1 + e)/(1 - e)$ or $r_{max} = 2.5(1 + 0.6)/(1 - 0.6) = 10\,AU$ then $a = 6.25\,AU$ and $\tau \sim 6.988\,\text{yr}$.

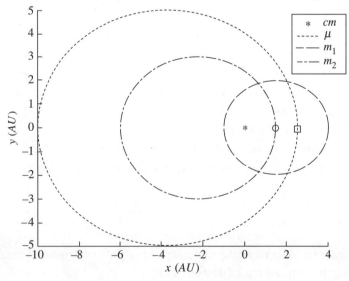

Binary System, Center of Mass–Relative Coordinate Method: tau = 6.988 tau_0, $m_1 = 3m_s$, $m_2 = 2m_s$

| FIGURE 9.17   Binary system simulation of Example 9.13 using analytic formulas.

SCRIPT

```
%binary1.m - binary star system given the eccentricity - using center of mass-
%relative coordinate concept
clear;
m=5;m1=3;m2=m-m1;  %initial total, and individual masses (in Ms)
rcm=0; rmin=2.5;   %center of mass, rmin for relative coordinate in AU
e=0.6;             %given eccentricity (e=0 => circle)
%e=0.0;
th=[0:0.05:2*pi];
r=rmin*(1+e)./(1+e*cos(th));   %use orbit formula for reduced mass motion
rmax=rmin*(1+e)/(1-e);%use formula for rmax
%Note: reduced mass, m1 and m2, all have the same period - pick one
a=(rmin+rmax)/2;  %semimajor axis of the full reduced mass orbit
tau=sqrt(a^3/m);  %tau in units of tau_0 (tau_0=1 year)
str=cat(2,'Binary System, CM-Rel. Coord. Method: tau = ',num2str(tau,4),' tau_0,',...
' m_1 = ',num2str(m1,2),' m_s, m_2 = ',num2str(m2,2),' m_s');
r1=rcm-m2*r/m; r2=rcm+m1*r/m;%m1, m2 positions
[nr,nc]=size(r);            %get rows and columns of r
%polar plots are similar to reg plots of x, y with x=r.*cos(th); y=r.*sin(th), etc;
  for i=1:nc
     clf;
     hold on
     plot(0,rcm,'*')
     polar(th,r,'r:')        %show full paths
     polar(th,r1,'k-')
     polar(th,r2,'b-.')
     polar(th(i),r(i),'rs')  %place animation symbols
     polar(th(i),r2(i),'bo')
     polar(th(i),r1(i),'k.')
     polar(th(i),r2(i),'bo')
     pause(0.025)
  end
xlabel('x (AU)','FontSize',14),ylabel('y (AU)','FontSize',14)
title(str,'FontSize',11)
h=legend('cm','\mu','m_1','m_2'); set(h,'FontSize',12)
```

B. Numerical Method

We first rewrite (9.5.3) as follows:

$$\frac{d^2(x_1\hat{i} + y_1\hat{j})}{dt^2} = \frac{Gm_2((x_2 - x_1)\hat{i} + (y_2 - y_1)\hat{j})}{[(x_2 - x_1)^2 + (y_2 - y_1)^2]^{3/2}} \tag{9.5.18a}$$

for m_1, and

$$\frac{d^2(x_2\hat{i} + y_2\hat{j})}{dt^2} = -\frac{Gm_1((x_2 - x_1)\hat{i} + (y_2 - y_1)\hat{j})}{[(x_2 - x_1)^2 + (y_2 - y_1)^2]^{3/2}} \tag{9.5.18b}$$

for m_2, where we have used $\mathbf{r}_i = x_i\hat{i} + y_i\hat{j}$, $\mathbf{r}_{ij} = \mathbf{r}_i - \mathbf{r}_j$, $\mathbf{r}_{ji} = -\mathbf{r}_{ij}$, and $r_{ij} = r_{ji} \equiv \sqrt{(x_i - x_j)^2 + (y_i - y_j)^2}$, for i, j taking on the possible values 1 and 2 with $i \neq j$. Equations (9.5.18) consist of coupled differential equations for the x, y coordinates of each mass. However, the numerical solution method employs the derivative of the coordinate as another differential equation for the velocity, as we saw in Chapter 8. Thus Equations (9.5.18) become

$$\dot{v}_{1x} = \frac{Gm_2}{r^3}(x_2 - x_1), \ \dot{v}_{1y} = \frac{Gm_2}{r^3}(y_2 - y_1); \ \dot{v}_{2x} = -\frac{Gm_1}{r^3}(x_2 - x_1), \ \dot{v}_{2y} = -\frac{Gm_1}{r^3}(y_2 - y_1),$$

(9.5.19a)

for the derivative of the velocities, where we have defined $r \equiv \sqrt{(x_2 - x_1)^2 + (y_2 - y_1)^2}$, with the auxiliary equations

$$\dot{x}_1 = v_{1x}, \ \dot{y}_1 = v_{1y}, \ \dot{x}_2 = v_{2x}, \ \dot{y}_2 = v_{2y},$$

(9.5.19b)

for the derivatives of the coordinates. So there are four differential equations for each mass, for a total of eight coupled differential equations. We will use astronomical distance (R) units for positions and solar mass (m_s) units for the masses. In order for this to work, the unit of time has to be large, like years. For example, looking at the first of (9.5.19a), the acceleration of the first particle in the x direction is

$$\frac{d}{d\bar{t}}\bar{v}_{1x} = \frac{4\pi^2 \bar{m}_2(\bar{x}_2 - \bar{x}_1)}{[(\bar{x}_2 - \bar{x}_1)^2 + (\bar{y}_2 - \bar{y}_1)^2]^{3/2}},$$

(9.5.20)

where we have used $x_i = R\bar{x}_i$, $y_i = R\bar{y}_i$, $v_{1x} = R\bar{v}_{1x}/\tau_0$, $m_2 = m_s\bar{m}_2$, and $t = \tau_0\bar{t}$. We also have that $\tau_0^2 Gm_s/R^3 \equiv 4\pi^2$, which makes it possible to define the same time units used by Kepler; i.e., $\tau_0 \equiv 2\pi\sqrt{R^3/Gm_s} \approx 3.14 \times 10^7$ s or 1 year. In a similar way we can simply transform each of the relations to dimensionless units.

EXAMPLE 9.14

Consider the binary system of Example 9.13. (a) Reproduce the simulation of the orbital motion of each of the masses m_1 and m_2. (b) Obtain plots of the position and speed of each mass versus time and deduce an orbital period. (c) Does the orbital period of this system agree with that of Example 9.13? Be sure to work in units of $R(AU)$ for distance, τ_0 (yr) for time, and $R/\tau_0(AU/yr)$ for speed.

Solution

(a) In order to reproduce the motion, initial values are needed for the positions and the speeds in the x and y directions for each mass; that is, we need $x_{10}, y_{10}, v_{1x0}, v_{1y0}$ and $x_{20}, y_{20}, v_{2x0}, v_{2y0}$. From the information available and Figure 9.17, we have the initial relative coordinate $(x_0, y_0) = \mathbf{r}_0 \equiv (\bar{r}_0, 0)R$ where $\bar{r}_0 = 2.5$, and since $\mathbf{r}_{cm} = 0$, then from the relative coordinate equation (9.5.6) we get $\mathbf{r}_{10} = (x_{10}, y_{10}) = -m_2\mathbf{r}_0/m = (-2.5(2/5), 0)R = (-1, 0)R$. In addition, $\mathbf{r}_{cm} = 0 \Rightarrow \mathbf{r}_{20} = -m_1\mathbf{r}_{10}/m_2 = (1.5, 0)R = (x_{20}, y_{20})$. For the speeds, notice that from (8.8.2) we have a circular orbit if the initial velocity is

$$\dot{r}_c = \sqrt{\frac{Gm_s\bar{m}}{r_0}} = \sqrt{\frac{Gm_s R^2(\bar{m}_1 + \bar{m}_2)}{R^3}\frac{}{\bar{r}_0}} = \frac{2\pi R}{\tau_0}\sqrt{\frac{(\bar{m}_1 + \bar{m}_2)}{\bar{r}_0}}$$

$$= \bar{r}_c\frac{R}{\tau_0} \Rightarrow \bar{r}_c = 2\pi\sqrt{\frac{(\bar{m}_1 + \bar{m}_2)}{\bar{r}_0}} \tag{9.5.21}$$

where we have used $\tau_0 = 2\pi\sqrt{R^3/Gm_s}$ as before and the masses have been written in terms of the solar mass ($m = \bar{m}m_s$). The quantity $\bar{r}_c$ is the minimum initial velocity for the relative coordinate in units of $R/\tau_0 = AU/yr$ that's needed for circular motion. The deviation from a circle is due to the eccentricity, which can be incorporated into the initial speed through (8.8.4) in terms of the circular velocity; that is,

$$\bar{r}_0 = \bar{r}_c\sqrt{1 - e}. \tag{9.5.22}$$

This is actually the y component of velocity for the relative coordinate inherent in Example 9.13. The x component of the velocity is zero as the reduced mass starts at zero angle in that example. Now, the initial values of the m_1 and m_2 velocities can be obtained; that is, similar to (9.5.6), we have

$$\dot{\mathbf{r}}_1 = \dot{\mathbf{r}}_{cm} - \frac{m_2}{m}\dot{\mathbf{r}}, \tag{9.5.23}$$

where we have used $\dot{\mathbf{r}}_{cm} \equiv 0$. So the initial values of the speed components for m_1 are

$$\bar{v}_{1x0} = 0 \quad \text{and} \quad \bar{v}_{1y0} = -\frac{\bar{m}_2\bar{r}_0}{\bar{m}} = -\frac{\bar{m}_2}{\bar{m}}2\pi\sqrt{\frac{(\bar{m}_1 + \bar{m}_2)}{\bar{r}_0}}\sqrt{1 - e}. \tag{9.5.24}$$

Furthermore, these can also be used to obtain the initial values of the speed components for m_2; that is, from (9.5.4) we have

$$\dot{\mathbf{r}}_{cm} \equiv \frac{m_1\dot{\mathbf{r}}_1 + m_2\dot{\mathbf{r}}_2}{m} = 0 \Rightarrow \dot{\mathbf{r}}_2 = -\frac{m_1}{m_2}\dot{\mathbf{r}}_1, \Rightarrow \overline{v}_{2x0} = 0 \text{ and } \overline{v}_{2y0} = -\frac{m_1}{m_2}\overline{v}_{1y0}. \quad (9.5.25)$$

All this is incorporated by the MATLAB script `binary2.m`, including the period as in Example 9.12, as shown in Figure 9.18. Once again, to see the animation the script needs to be run.

(b) The script is also able to obtain a plot of the positions and the speeds of the masses m_1, m_2 versus time. These are shown in Figure 9.19.

From the figure we can see that the plot is for a full period of about $7\tau_0$. The black dotted line corresponds to m_1 with an overall smaller distance to travel on its orbit than m_2 and therefore smaller speed. In a period the positions (speeds) go from their minima (maxima) to their maxima (minima), as is expected in noncircular orbits.

(c) The period does agree with the analytic value obtained in Example 9.12.

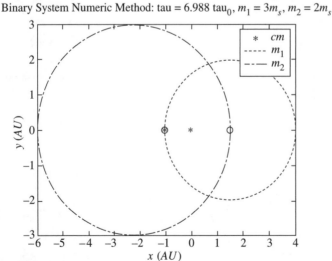

Binary System Numeric Method: tau = 6.988 tau$_0$, $m_1 = 3m_s$, $m_2 = 2m_s$

FIGURE 9.18 Binary system simulation of Example 9.14 using a numeric approach.

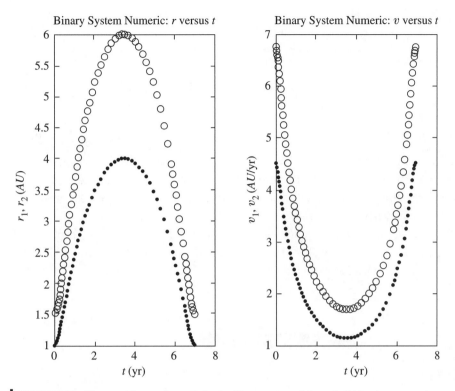

FIGURE 9.19   Mass positions and speeds for the binary system of Example 9.14 using a numeric approach.

The script `binary2.m` follows. It simulates the binary system with an eight-dimensional array $w(1, 2, \ldots, 8) = \overline{x}_1, \overline{v}_{1x}, \overline{y}_1, \overline{v}_{1y}, \overline{x}_2, \overline{v}_{2x}, \overline{y}_2, \overline{v}_{2y}$ to denote the variables. Thus, the MATLAB script solves the equations:

$$\dot{w}(2) = \frac{4\pi^2 \overline{m}_2 (w(5) - w(1))}{w_r^3}, \ \dot{w}(4) = \frac{4\pi^2 \overline{m}_2 (w(7) - w(3))}{w_r^3}, \qquad (9.5.26a)$$

corresponding to $\overline{v}_{1x}$, and $\overline{v}_{1y}$ in (9.5.19a). We also have

$$\dot{w}(6) = -\frac{4\pi^2 \overline{m}_1 (w(5) - w(1))}{w_r^3}, \ \dot{w}(8) = -\frac{4\pi^2 \overline{m}_1 (w(7) - w(3))}{w_r^3}, \qquad (9.5.26b)$$

for $\bar{v}_{2x}$ and $\bar{v}_{2y}$ in (9.5.19a). Here we have defined

$$w_r \equiv \sqrt{(w(5) - w(1))^2 + (w(7) - w(3))^2}$$

corresponding to the dimensionless form of r in (9.5.19a). The last equations of (9.5.19b) become

$$\dot{w}(1) = w(2), \dot{w}(3) = w(4), \dot{w}(5) = w(6), \dot{w}(7) = w(8). \qquad (9.5.26c)$$

The w array used by the script is really a two-dimensional array. One dimension is for tracking the preceding definitions, and the other dimension is for time. The numerical approach used by the script is the Runge–Kutta ode45 scheme. After the call to ode45 is made, one can plot $w(3)$ vs. $w(1)$ to get m_1's orbit and $w(7)$ vs. $w(5)$ for m_2's orbit, and so on, as shown by the script. The script uses the initial values for the positions and the speeds as explained in part (a) of this example, as follows. The derivatives are done by the following auxiliary m-file binary2_der.m, which is called by the main script.

SCRIPT

```
%binary2.m - program to simulate the binary mass system in a fully numerical way
clear;
rcm=0.0;              %center of mass, time unit
c=4*pi^2;            %constant
m=5;m1=3;m2=m-m1;    %initial total, and individual masses (in Ms)
tmax=6.988;          %maximum time in units of tau_0 (tau_0=1 yr)
%tmax=1.77;
x10=-1;y10=0;x20=-m1*x10/m2;y20=0;
x00=x20-x10;         %relative coordinate circular orbit parameter
e=0.6;               %given eccentricity (e=0 => circle)
%e=0.0;
d=sqrt(1+e); %e > 0 =>deviation from circular orbit
v1x0=0;v1y0=d*(-m2*2*pi*sqrt(m/x00)/m);v2x0=0;v2y0=-m1*v1y0/m2;
ic1=[x10,v1x0,y10,v1y0,x20,v2x0,y20,v2y0]; %initial conditions
%Use MATLAB's Runge-Kutta (4,5) formula (uncomment/comment as needed)
opt=odeset('AbsTol',1.e-7,'RelTol',1.e-4);%user set Tolerances
[t,w]=ode45('binary2_der',[0.0,tmax],ic1,opt,c,m1,m2);%with set tolerance
%[t,w]=ode45('binary2_der',[0.0,tmax],ic1,[],c,m1,m2);%with default tolerance
%calculate period obtained in units of tau_0
r1=sqrt(w(:,1).^2+w(:,3).^2); r2=sqrt(w(:,5).^2+w(:,7).^2);%get radii
rmin1=min(r1); rmax1=max(r1); rmin2=min(r2); rmax2=max(r2);%get min, max r's
a1=(rmin1+rmax1)/2; a2=(rmin2+rmax2)/2; a=a1+a2;%get the a's
tau=sqrt(a^3/m); %get the period
str=cat(2,'Binary Syst. Numeric Method: tau = ',num2str(tau,4),' tau_0, ',...
```

```
' m_1 = ',num2str(m1,2),' m_s, m_2 = ',num2str(m2,2),' m_s');
[nr,nc]=size(w);        %get rows and columns of r
figure
for i=1:nr
    clf;
    plot(rcm,rcm,'*')% center of mass
    hold on
    plot(w(:,1),w(:,3),'k:',w(:,5),w(:,7),'b-.')
    plot(w(i,1),w(i,3),'k.','MarkerSize',20)
    plot(w(i,5),w(i,7),'bo','MarkerSize',8)
    axis([-6.0 4.0 -3.0 3.0]);
    pause(0.0125)
end
xlabel('x (AU)','FontSize',14),ylabel('y (AU)','FontSize',14)
title(str,'FontSize',12)
h=legend('cm','m_1','m_2'); set(h,'FontSize',12)
figure
subplot(1,2,1)%plot radii vs time
plot(t,r1,'k.',t,r2,'bo')
xlabel('t (yr)','FontSize',14),ylabel('r_1,r_2 (AU)','FontSize',14)
title('Binary Syst. Numeric: r versus t ','FontSize',12)
subplot(1,2,2)%plot speeds vs time
v1=sqrt(w(:,2).^2+w(:,4).^2); v2=sqrt(w(:,6).^2+w(:,8).^2);%get speeds
plot(t,v1,'k.',t,v2,'bo')
xlabel('t (yr)','FontSize',14),ylabel('v_1,v_2 (AU/yr)','FontSize',14)
title('Binary Syst. Numeric: v versus t ','FontSize',12)
```

FUNCTION

```
%binary2_der.m: returns the derivatives for binary mass system
function derivs = binary1_der( t, w, flag,c,m1,m2)
% c=4*pi^2, m1=mass of 1st body, m2=mass of 2nd body
% Entries in the vector of dependent variables are:
% w(1,2,...8)=x1,v1x,y1,v1y,x2,v2x,y2,v2y
wr=sqrt((w(5)-w(1)).^2+(w(7)-w(3)).^2);
derivs = [w(2); c*m2*(w(5)-w(1))./wr.^3; w(4); c*m2*(w(7)-w(3))./wr.^3;...
          w(6); -c*m1*(w(5)-w(1))./wr.^3;w(8); -c*m1*(w(7)-w(3))./wr.^3];
```

■ Chapter 9 Problems

9.1 In Example 9.1, what is the total gravitational potential of m_0 in the presence of the masses m_1 and m_2?

9.2 Consider three masses, m_0, m_1, and m_2 located at $(x, 0)$, $(0, y)$, and $(0, -y)$, respectively.

a. Find the total gravitational force on m_0 due to the other two masses.

b. Suppose that $m_1 = m_2$ and that $y \gg x$, if m_0 were allowed to move, show that the motion would be oscillatory about the origin. What would the frequency of oscillation be?

9.3 a. Find the gravitational potential due to the uniform density rod of mass M and length L at a point $\mathbf{r}$ above its center and perpendicular to it, as shown in Example 9.3.

b. What is the behavior of the potential limits when $L \to 0$, $y \to \infty$, and $y \to 0$?

9.4 a. Let $(a, 0)$ be the coordinates of mass m_1 and let $(0, b)$ be the coordinates of mass m_2. Find the gravitational field at a general position $\mathbf{r} = (x, y)$ due to these masses.

b. What does the resulting expression give in the event that $x = y = c$, $m_1 = m_2$, and $a = b = c$?

9.5 a. Find the gravitational potential due to a uniform disk of mass M and radius r at a point z above its center and perpendicular to it.

b. What is the behavior of the potential in the limits of $z \to 0$ and $z \to \infty$ with $z \gg R$?

9.6 If the ring's gravitational field of Example 9.6 is considered to be a small contribution due to a disk, after identifying the element of mass, the contribution can be integrated over the radius and should yield the disk result of Example 9.4. Obtain the expected result.

9.7 The gravitational potential from a uniform solid sphere, of mass M and radius r, can be found using Equation (9.2.3).

a. Find the potential at point $r \geq R$ from its center.

b. Find the gravitational potential at $r < R$. Assume the potential reference point is at infinity.

9.8 Use Gauss' law to obtain (a) the gravitational field at a point $z \geq R$ from the center of a uniform spherical shell of mass M and radius r, and (b) repeat for $z < R$. Compare your results with those obtained in Example 9.7.

9.9 Use Gauss' law to obtain (a) the gravitational field at a point $z \geq R$ from the center of a uniform sphere of mass M and radius r, and (b) repeat for $z < R$. Compare your results with those obtained in Example 9.8.

9.10 Use the information provided in Figure 9.16 to obtain the eccentricity of each orbit. What is your observation about binary mass systems?

9.11 A certain binary star system is observed to have velocities $v_1 = 1.257\,AU/yr$ and $v_2 = 5.027AU/yr$ for the respective masses. Its measured period is $\tau = 5\,yr$. Assuming the orbits are circular, find the mass of each star in units of solar masses.

9.12 In the numerical part of Section 9.5, if one of the masses is very large, it is possible to simplify the numerical set of equations needed to solve because the more massive body's coordinates are essentially constant.

   a. Modify scripts `binary2.m` and the related `binary2_der.m` in order to simulate the motion of a small mass body about another more massive one. Assume the massive body is equivalent to one solar mass, and take the initial position of the small body as $(x_0, y_0) = (2.0, 0.0)$ and eccentricity e.

   b. Obtain a plot of the position and speed versus time for $e = 0$ and explain your results. Can you obtain the orbital period from your plots?

   c. What is the orbital period?

   d. What changes in parts (b) and (c) are needed if the eccentricity is increased?

9.13 A binary system's period is known to be 50 yr with an associated semimajor axis of 20 AU. If one of the masses is known to be 2.2 m_s, what is the value of the other mass, and what are their respective distances from their common center of mass?

■ Additional Problems

9.14 Express Equation (9.5.17) in terms of astronomical units.

9.15 The period associated with a binary system is known to be 50 years and the average distance between the masses is 20 AU. Given that one of the masses has a value of 2.2 m_s, find the value of the second mass and their average distances from their known common center of mass.

10 Rutherford Scattering

■ 10.1 Introduction

Ernest Rutherford (1871–1937) was born near Nelson, New Zealand, and was a British physicist. He studied at Christchurch University, moved to Cambridge, UK (1895), and in 1898 became professor of physics at McGill University, Canada. In 1907 he became professor at the University of Manchester in Manchester, England, and in 1911 he developed the modern concept of the atom. In 1912 he showed that the positive charge of an atom was concentrated in the central core of an atom or nucleus. Rutherford discovered the atomic nucleus by experimenting with α-particles. The α-particles were originally discovered by Henri Becquerel in 1896. They are a natural form of radiation produced by radioactive uranium. They consist of doubly ionized helium atoms and therefore have a positive charge of $2\,q_e$, where $q_e \equiv 1.602 \times 10^{-19}$ C is the positive value of the electronic charge. Alpha particles have a mass of $m_\alpha = 4\,u$, where $1\,u = 1.66 \times 10^{-27}$ kg. At Rutherford's urging H. Geiger and E. Marsden carried out experiments (*Phil. Mag.* Vol. 25, 605 [1913]) where α-particles were directed through thin foils of gold and silver. It was found that the charged particles were occasionally deflected by very large angles. Because the particles have a large mass compared to electrons, collisions with electrons can't produce large scattering angles; thus, Rutherford concluded that the projectile α-particles were undergoing collisions with a positively charged nuclear target. Rutherford deduced that a large repulsive electric force was possible at short distances if the nucleus is concentrated at a point on the order of about 1 fm in radius (a femtometer, 1 fm $= 1 \times 10^{-15}$ m) and of positive charge Zq_e where Z is the number of protons.

In 1919 Rutherford became professor at Cambridge University and director of the Cavendish Laboratory in England. He received the Nobel Prize for Chemistry in 1908, was knighted in 1914, and was made a peer in 1931. (A peer is a member of any of the five degrees of the nobility in Great Britain and Ireland: duke, marquis, earl, viscount, and baron.)

■ 10.2 Coulomb's Law and a Charge's Motion Under a Repulsive Electric Force

According to Coulomb's Law, a particle of charge q_1, located at the origin, interacts with another of charge q_2 according to a force that varies as the inverse of the square of the distance r between them

$$\mathbf{F}_e = \frac{kq_1q_2}{r^2}\hat{r}, \tag{10.2.1}$$

where $k = 1/4\pi\varepsilon_0 = 8.988 \times 10^9 \, \text{Nm}^2/\text{C}^2$ is the electric force constant where the permittivity of free space $\varepsilon_0 \equiv 8.854 \times 10^{-12} \, \text{C}^2/\text{Nm}^2$ has been used. As shown in Figure 10.1, the force is positive and repulsive and points in the direction of $\hat{r}$ if the charges have the same sign.

However, if q_1 and q_2 have opposite charge signs, the force is attractive and points in the direction of $-\hat{r}$. For attractive forces, the problem is similar to that of planetary motion and everything studied in Chapter 9 applies if the replacement $Gm_1m_2 \to kq_1q_2$ is made. In the case of Rutherford scattering, the force is repulsive and the solution to the equation of motion is different. If the charge q_1 is considered to be at rest due to being very massive compared to the mass of q_2, then we can write Newton's second law for q_2 alone as

$$m\mathbf{a} = \frac{K}{r^2}\hat{r}, \tag{10.2.2}$$

where we have let $K = kq_1q_2$ and $m_2 = m$ and $q_1, q_2 > 0$. This is a central force problem for which $K > 0$. The solution to this problem is provided by the results of Chapter 8, where from (8.6.4) and (8.6.8,9) since $r \equiv 1/u$

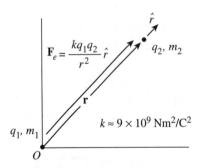

FIGURE 10.1   The Coulomb force between two charges with one located at the origin.

$$\frac{d^2u}{d\theta^2} + u = -\frac{m}{L^2u^2}Ku^2 = -\frac{mK}{L^2} \Rightarrow r = r_{min}\left(\frac{1+e}{1+e\cos\theta}\right),\qquad (10.2.3)$$

with the closest approach distance and eccentricity given by

$$r_{min} \equiv -\frac{L^2}{mK}\frac{1}{(1+e)}\ \text{and}\ e \equiv -\frac{L^2}{mKr_0}.\qquad (10.2.4)$$

Notice that if K is negative, then e and r_{min} are always positive, and since from (8.6.16) the eccentricity is also

$$e = \pm\sqrt{1 + \frac{2EL^2}{mK^2}},\qquad (10.2.5)$$

where $E = mv^2/2 - |K|/r$. This energy has a range of $-mK^2/2L^2 < E < \infty$, as discussed in Chapter 8. The bound orbit energy regime where $E < 0$ is responsible for elliptical orbits ($0 < e < 1$). When $E = 0$, the orbit becomes parabolic and unbound ($|e| = 1$). In the region where $E > 0$, the orbit is also unbound and hyperbolic ($|e| > 1$). However, if K is positive as is the case in Rutherford scattering, then from (10.2.4) $e < 0$ and from (10.2.5), since $E = mv^2/2 + |K|/r > 0$, then $|e| > 1$, and therefore

$$r_{min} \equiv -\frac{L^2}{mK}\frac{1}{(1+e)} = \frac{L^2}{m|K|}\frac{1}{(|e|-1)},\qquad (10.2.6)$$

which is always positive. Given the above conditions, it is useful to look at possible orbits for $r(\theta)$ from (10.2.3) for various eccentricity values to see how the α particle orbit is extracted. This is shown in Figure 10.2.

From the figure we see that any hyperbola ($|e| > 1$) can describe a scattering orbit because the curves are open. For example, looking at the red line case of $e = -1.2$, one can imagine a projectile particle coming in from the bottom-right asymptote, being repelled by a second target particle located at the origin, and finally the original particle leaving from the top right-asymptote. Similarly, one can repeat the argument for the mirror image hyperbola, but for an attractive force; that is, the projectile could be coming in from the bottom left, be attracted by a target particle near the origin, and leave off the top left, when the particle carries high energy; otherwise it will be trapped in an elliptical orbit. The curves in Figure 10.2 were created by the MATLAB script conic1.m that follows, which provides an output with colored lines.

Plot: $r = r\min*(1 + e)/(1 + e* \cos(\theta))$ vs. θ, varying e

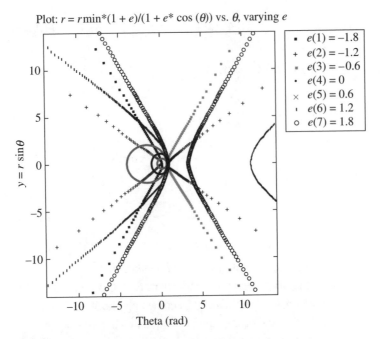

■	$e(1) = -1.8$
+	$e(2) = -1.2$
▪	$e(3) = -0.6$
·	$e(4) = 0$
×	$e(5) = 0.6$
ι	$e(6) = 1.2$
○	$e(7) = 1.8$

| FIGURE 10.2   Plot of Equation (10.2.3) for various eccentricities.

SCRIPT

```
%conic1.m - script for possible conic section curves with rmim and
% eccentricity e.
clear;
L=14;                    %for viewing
th=[0:0.001:2*pi];       %angle range
rmin=1; N=7;             %any rmin, and curves to draw
emin=-1.8;emax=1.8;      %range of e
de=(emax-emin)/(N-1);    %step size
e=[emin:de:emax];
c=get(gca,'ColorOrder');%get available colors to use
stp=cat(2,'o*vsdph');    %plot symbols (more are .ox+*sdv^><ph )
cz=size(c);
for i=1:1:N
r=rmin*(1+e(i))./(1+e(i)*cos(th));
x=r.*cos(th);y=r.*sin(th);m=length(x);ms=5;%count points to plot
j=mod(i,cz(1))+1;        %only cz(1) colors available so 1 < j < 7
plot(x(1:ms:m),y(1:ms:m),stp(j),'MarkerSize',2.5,'Color',c(j,:))%in color order
axis ([-L L -L L])
hold on
```

```
str=cat(2,'e(',num2str(i),')=',num2str(e(i)));%curve label
sz=size(str); sc(i,1:sz(2))=str;%store labels in sc array according to length
end
str2=cat(2,'plot: r = rmin*(1+e)/(1+e*cos(\theta)) vs \theta, varying e');
title(str2,'FontSize',14)
h=legend(sc,-1);          % place all labels as legend
set(h,'FontSize',12)
xlabel('x=rcos(\theta))','FontSize',14)
ylabel('y=rsin(\theta) (Arbitrary Units)','FontSize',14)
```

As previously discussed, the proper orbit of interest, however, has a negative eccentricity. Also important in this discussion is the angle between the asymptotes, i.e., the angle difference between the incoming and outgoing directions. This is referred to as the *scattering angle* Θ. The scattering angle can be obtained by finding the asymptotic points, that is, the angles at which the denominator of (10.2.3) becomes zero:

$$1 + e\cos\theta = 0 \quad \Rightarrow \quad \cos\theta = -\frac{1}{e}. \tag{10.2.7}$$

Since the $\cos\theta$ is a harmonic function, there are various angles at which this condition can be satisfied. A way to solve the problem is to plot the orbit of the projectile to get an idea what region of space it comes in from and similarly what region of space it goes out of. A plot of the $\cos\theta$ versus θ can be superimposed on a plot of the line $-1/e$ to get an estimate of where the angle intersections are. These angles can then be guesses to a numerical approach to get the more accurate numerical roots of the equation. In fact, this is precisely what has been done in Figure 10.3 for the case of $e = -1.2$.

In Figure 10.3(a), the incoming asymptote angle lies in the range $[-\pi/2, 0]$ and the outgoing asymptote lies in the range $[0, \pi/2]$. After the following MATLAB script conic2.m, used to do the plots, is run, one can get the incoming and outgoing angles as measured from the $+x$-axis by inquiring for the variables thmin and thmax (simply type their name within MATLAB), which are $\theta_{min} = -0.5857$ rad and $\theta_{max} = 0.5857$ rad, respectively. Because the $\cos\theta$ is symmetric, we let $\alpha \equiv |\theta_{min}| = |\theta_{max}|$; that is, using (10.2.7) with $e = -|e|$, the asymptotes of the projectile path are more explicitly written as

$$\cos\alpha = -\frac{1}{e} = \frac{1}{|e|}. \tag{10.2.8}$$

(a)

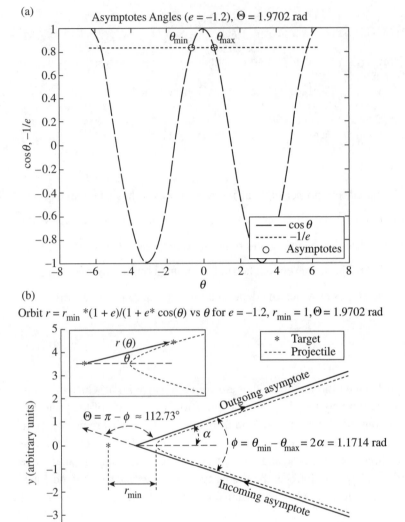

(b)

FIGURE 10.3   (a) Plot of the cosine of the angle from Equation (10.2.7), and (b) orbit of a projectile with the incoming and outgoing asymptotes shown.

From Figure 10.3(b) we also have $\phi \equiv \theta_{max} - \theta_{min} \equiv 2\alpha = 1.1714$ rad, so that the scattering angle is $\Theta \equiv \pi - 2\alpha = 1.9702$ rad $\approx 112.73°$. The figure also includes some extra information that has been added after the script is run, i.e., the inset that shows how $r(\theta)$ describes the projectile's path from the target's location. The angle θ is measured with respect to the $+x$-axis. It is convenient to rotate the orbital path counterclockwise by angle α in order to have incidence parallel to the x-axis and measure the impact parameter from it. This will be discussed in the next section. Finally, also added to Figure 10.3(b) is r_{min}, i.e., the distance of closest approach, which is represented by the shortest straight line distance from the center of the target to a point on the projectile particle's curved path.

SCRIPT

```
%conic2.m - script for conic section projectile hyperbolic orbit
%with negative eccentricity ye.
clear; warning off;
ye=-1.2; %eccentricity
rmin=1;
L=5;%for window viewing purpose
% find theta limits for the assymptotes of r for this ye
% these occur at the zeros of the denominator of the r(theta) equation
% opt=optimset('TolX',1.0e-10); %optional convergence criteria
% thmin=fzero(inline('ye*cos(x)+1'),[-pi/2,0],opt,ye)+1.e-3;
% thmax=fzero(inline('ye*cos(x)+1'),[0,pi/2],opt,ye)-1.e-3;
thmin=fzero(inline('ye*cos(x)+1'),[-pi/2,0],[],ye)+1.e-5;
thmax=fzero(inline('ye*cos(x)+1'),[0,pi/2],[],ye)-1.e-5;
THETA=pi-(thmax-thmin);%scattering angle
%--------show asymptote angles----------
figure%plots the cos(x) versus -1/ye to show assymptote angles
x=[-2*pi:.1:2*pi];
plot(x,cos(x),'b--')
hold on
line([-2*pi,2*pi],[-1/ye,-1/ye],'Color','red')
plot(thmin,-1/ye,'ko',thmax,-1/ye,'ko')
text(thmin*(2+.3),(-1/ye)*(1+0.1),'\theta_{min}')
text(thmax*(1+0.1),(-1/ye)*(1+0.1),'\theta_{max}')
xlabel('\theta','FontSize',14),ylabel('cos\theta, -1/e','FontSize',14)
str=cat(2,' Asymptotes Angles (e = ',num2str(ye),'), \Theta=',num2str(THETA),' rad');
title(str,'FontSize',12)
h=legend('cos\theta','-1/e','Asymptotes',4); set(h,'FontSize',12)
%---------------------------------------
th=[thmin:0.001:thmax]; %range of orbit plot
r=rmin*(1+ye)./(1+ye*cos(th));
[nr,nc]=size(r);              %get rows and columns of r
```

```
%polar plots are similar to reg plots of x, y with x=r.*cos(th); y=r.*sin(th), etc;
figure
for i=1:25:nc
    clf;
    hold on
    plot(0,0,'*')
    polar(th,r,'r:')        %show full path
    polar(th(i),r(i),'r.') %place animation symbols
    axis ([-L/L L -L L])
    pause(0.025)
 end
xlabel('x (Arbitrary Units)','FontSize',14)
ylabel('y (Arbitrary Units)','FontSize',14)
str=cat(2,'Orbit: r = r_{min}*(1+e)/(1+e*cos(\theta) vs \theta for e = ',...
    num2str(ye),', r_{min}=',num2str(rmin),', \Theta=',num2str(THETA),' rad');
title(str,'FontSize',12)
h=legend('Target','Projectile'); set(h,'FontSize',12)
```

■ 10.3 Simulation of an α-Particle Incident on a Heavy Target

In this section we consider an α-particle with a kinetic energy $E_k = 5 \times 10^6$ eV and an impact parameter of $b = 20$ fm, that's incident on a gold foil target, as shown in Figure 10.4.

From the figure, it can be seen that the particle's energy (E_k) as well as the angular momentum ($L = mvb$) play an important role in the scattering. We would like to (a) simulate the particle's trajectory, (b) find the scattering angle Θ, and (c) obtain the orbit's eccentricity as well as the distance of closest approach. We will look at this problem in two ways. The first is through the analytic solution discussed in the previous section, i.e., Equation (10.2.3). We will simulate the motion, find the scattering angle, the eccentricity, and the distance of closest approach. In the second way, a straight numerical simulation will be performed to provide a complementary view for

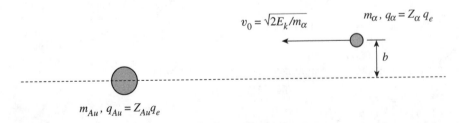

I FIGURE 10.4 Alpha particle with impact parameter directed at a target.

a better understanding. To make the problem amenable to numerical computation, we would like to use a set of convenient units for time, distance, and mass. To this end, if we first look at the equation of motion (10.2.2) and let the target and the projectile charge be $q_1 = Z_t q_e$, $q_2 = Z_p q_e$, respectively, we get

$$m_p \frac{d}{dt}(v_x \hat{i} + v_y \hat{j}) = \frac{kq_e^2 Z_t Z_p}{(x^2 + y^2)^{3/2}}(x\hat{i} + y\hat{j}). \tag{10.3.1}$$

where m_p is a general projectile mass (which later will be taken as the α-particle mass). Letting $(x, y) = (\bar{x}, \bar{y})a_b$ with $a_b = 1$ fm the unit of distance, $t = \bar{t}\tau$ with τ a unit of time, $(v_x, v_y) = (\bar{v}_x, \bar{v}_y)v_b$ with $v_b \equiv a_b/\tau$ the speed unit, and $m_p = \bar{m}m_\alpha$ in units of the α-particle mass $m_\alpha = 4\,u$, Equation (10.2.1) becomes

$$\frac{d}{d\bar{t}}(\bar{v}_x \hat{i} + \bar{v}_y \hat{j}) = \left[\frac{kq_e^2 \tau^2}{m_\alpha a_b^3} \right] \frac{\overline{K}(\bar{x}\hat{i} + \bar{y}\hat{j})}{\bar{m}(\bar{x}^2 + \bar{y}^2)^{3/2}}, \tag{10.3.2}$$

where $\overline{K} \equiv Z_t Z_p$ (later $Z_p \to Z_\alpha$.) Noticing that the quantity in brackets on the right has no dimensions, we are free to let its value be unity. Doing so yields our sought unit of time

$$\frac{kq_e^2 \tau^2}{m_\alpha a_b^3} \equiv 1 \Rightarrow \tau = \sqrt{\frac{m_\alpha a_b^3}{kq_e^2}} = \sqrt{\frac{4(1.66 \times 10^{-27}\,\text{kg})(10^{-15}\,\text{m})^3}{\left(9 \times 10^9\, \dfrac{\text{Nm}^2}{\text{C}^2} \right)(1.602 \times 10^{-19}\,\text{C})^2}} = 1.695 \times 10^{-22}s. \tag{10.3.3}$$

This allows us to find the value of our speed unit as

$$v_b = a_b/\tau = 1 \times 10^{-15}\,\text{m}/1.695 \times 10^{-22}\,\text{s} = 5.898 \times 10^6\,\text{m/s} = 0.01965c, \tag{10.3.4}$$

where the speed of light is $c = 3 \times 10^8$ m/s. Equation (10.3.2) is actually ready to be incorporated in a suitable numerical solution to the differential equation, which will be done later. We first look at the analytic solution simulation.

A. Analytic Solution Simulation

We use the previously defined units. From (10.2.4 and 10.2.5), while the eccentricity is already a dimensionless quantity, it does use quantities whose dimensions we need to consider. We can write for our problem

$$e = -sign(K)\sqrt{1 + \frac{2EL^2}{mK^2}} = -sign(K)\sqrt{1 + \frac{2\left(\dfrac{1}{2}m_p v_0^2\right)(m_p v_0 b)^2}{m_p K^2}}, \tag{10.3.5}$$

which if we use the dimensionless units discussed, we get

$$e = -sign(K)\sqrt{1 + \left[\frac{m_\alpha a_b^3}{kq_e^2\tau^2}\right]^2 \frac{\overline{m}^2\overline{v}_0^4\overline{b}^2}{\overline{K}^2}}, \tag{10.3.6}$$

with the quantity within the square brackets equal to unity, as before. Because the impact parameter for the problem has been given, we can find its dimensionless value, $b = \overline{b}a_b$, that is, $\overline{b} = 20$. The initial speed can be obtained from the incident energy of the projectile. If the energy is given in electron volts (eV) and the mass in eV/c^2, we have

$$v_0 = \overline{v}_0 v_b = \sqrt{\frac{2E(\text{eV})}{m_p\left(\frac{\text{eV}}{c^2}\right)}} = c\sqrt{\frac{2E(\text{eV})}{\overline{m}m_\alpha c^2(\text{eV})}} \Rightarrow \overline{v}_0 = \frac{c}{v_b}\sqrt{\frac{2E(\text{eV})}{\overline{m}E_\alpha(\text{eV})}}, \tag{10.3.7}$$

where the rest mass energy of the α-particle is well known, $E_\alpha \equiv m_\alpha c^2 = 3730 \times 10^6$ eV. The last quantity to be found is $r_{\min} = \overline{r}_{\min}a_b$. From (10.2.4) and the units defined above, we have

$$r_{\min} = -\frac{L^2}{mK}\frac{1}{(1 + e)} = -\frac{m_p v_0^2 b^2}{K(1 + e)} = -\frac{\overline{m}\overline{v}_0^2\overline{b}^2}{\overline{K}(1 + e)}a_b \Rightarrow \overline{r}_{\min} = -\frac{\overline{m}\overline{v}_0^2\overline{b}^2}{\overline{K}(1 + e)}. \tag{10.3.8}$$

Finally, the quantity $\overline{K} = Z_{Au}Z_\alpha$ where $Z_{Au} = 79, Z_\alpha = 4$ and because the projectile is the α-particle, $\overline{m} = 1$. The MATLAB script conic3.m executes Equation (10.2.3) for $r(\theta)$ with the preceding analytic formulas and performs the simulation. The results are shown on Figure 10.5.

The eccentricity works out to $e = -1.33$, the distance of closest approach is $r_{\min} = 53.1\,a_b = 53.1$ fm, and the scattering angle is $\Theta = 97.4°$. The curve has been shifted counterclockwise by $|\theta_{\min}| = 0.721$ rad $\sim 41.3°$ to align the incident particle asymptote with the x-axis as previously mentioned. The following script conic3.m, which contains all the details discussed, animates the particle motion when run.

SCRIPT

```
%conic3.m - script uses the conic section curve formula with rmim and
%eccentricity ye. Case of Rutherford scattering, given the initial
%projectile energy and impact parameter.
clear; warning off;
rcm=0.0;                    %target position
za=2; zt=79;                %za=projectile, zt=target charges (2->alpha, 79->gold)
```

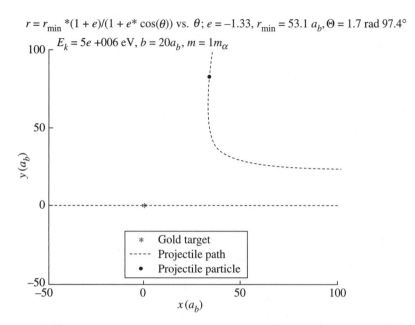

$r = r_{min} *(1+e)/(1+e* \cos(\theta))$ vs. θ; $e = -1.33$, $r_{min} = 53.1\ a_b$, $\Theta = 1.7$ rad $97.4°$

$E_k = 5e +006$ eV, $b = 20a_b$, $m = 1m_\alpha$

FIGURE 10.5   Plot of alpha particle projectile path onto a gold target. Equations (10.2.3) and (10.3.8) have been used.

```
m=1;                      %projectile mass in units of alpha particle mass
vb=0.01965;               %velocity units (a_b/tau_b) (in inits of c=light speed)
ma=3730e6;                %alpha particle mass energy in eV
Ene=5e6;                  %initial projectile energy in eV
v0=sqrt(2*Ene/m/ma)/vb;   %initial speed in units of vb
b=20;                     %impact parameter
K=za*zt;                  %dimensionless force constant
ye=-sign(K)*sqrt(1+m^2*v0^4*b^2/K^2);% eccentricity
rmin=-m*v0^2*b^2/K/(1+ye);
L=100;%for window viewing purpose
% find theta limits for the asymptotes of r for this ye
% these occur at the zeros of the denominator of the r(theta) equation
thmin=fzero(inline('ye*cos(x)+1'),[-pi/2,0],[],ye)+1.e-5;
thmax=fzero(inline('ye*cos(x)+1'),[0,pi/2],[],ye)-1.e-5;
THETA=pi-(thmax-thmin);%scattering angle
%----------------
%show the roots of the denominator 1+ye*cos(th)
figure
x=[-2*pi:.1:2*pi];
plot(x,cos(x))
hold on
```

```
line([-2*pi,2*pi],[-1/ye,-1/ye],'Color','red')
plot(thmin,-1/ye,'ko',thmax,-1/ye,'ko')
text(thmin*(2+.3),(-1/ye)*(1+0.1),'\theta_{min}')
text(thmax*(1+0.1),(-1/ye)*(1+0.1),'\theta_{max}')
xlabel('\theta','FontSize',13),ylabel('cos\theta, -1/e','FontSize',13)
str=cat(2,' Asymptotes Angles (e = ',num2str(ye,3),'), \Theta=',num2str(THETA,3),' rad');
title(str,'FontSize',13)
h=legend('cos\theta','-1/e','Asymptotes',0);
set(h,'FontSize',13)
%----------------
th=[thmin:0.025:thmax];
r=rmin*(1+ye)./(1+ye*cos(th));
ths=th-thmin;% rotate orbital path counterclockwise by 'thmin'
            % to align asymptote with + x-axis
[nr,nc]=size(r);  %get rows and columns of r
figure
  for i=1:1:nc-10  %play all minus 10 to freeze particle
     clf;
     hold on
     plot(rcm,rcm,'*')
     polar(ths,r,'k:')          %show full path
     polar(ths(i),r(i),'b.')  %place animation symbols
     axis ([-L/2 L -L/2 L])
     line([-L/2,L],[0,0],'Color','red','LineStyle',':')
     pause(0.025)
  end
xlabel('x (a_b)','FontSize',13), ylabel('y (a_b)','FontSize',13)
str=cat(2,'r = r_{min}*(1+e)/(1+e*cos(\theta)) vs \theta; e = ',...
    num2str(ye,3),', r_{min}= ',num2str(rmin,3),' a_b, \Theta=',...
    num2str(THETA,3),' rad = ',num2str(THETA*360/2/pi,3),'^o');
str2=cat(2,'E_k= ',num2str(Ene,3),' eV,', b= ',num2str(b,3),' a_b',...
    ', m= ',num2str(m,3),' m_\alpha');
text(-L/2*(1-0.1),L*(1-0.1),str2)
title(str,'FontSize',11)
h=legend('Gold Target','Projectile Path','Projectile Particle',0);
set(h,'FontSize',13)
```

B. Numerical Solution Simulation

The numerical way of simulating the dynamics of the scattering begins by converting (10.3.2) into two coupled differential equations for the velocity components in the same units used above, to get

$$\frac{d\,\bar{v}_x}{d\,\bar{t}} = \frac{\bar{K}\bar{x}}{m(\bar{x}^2 + \bar{y}^2)^{3/2}} \text{ and } \frac{d\,\bar{v}_y}{d\,\bar{t}} = \frac{\bar{K}\bar{y}}{m(\bar{x}^2 + \bar{y}^2)^{3/2}} \qquad (10.3.9)$$

which are coupled to the two differential equations for the position

$$\frac{d\overline{x}}{d\overline{t}} = \overline{v}_x, \text{ and } \frac{d\overline{y}}{d\overline{t}} = \overline{v}_y, \tag{10.3.10}$$

with suitable initial condition values for $(\overline{x}_0, \overline{v}_{x0}, \overline{y}_0, \overline{v}_{y0})$. Since the projectile particle comes from far away, we take $\overline{x}_0 = \overline{x}_{far}$ (a large number). The impact parameter enables us to write $\overline{y}_0 = \overline{b}$ and because it is assumed the particle comes in parallel to the x-axis, we take $\overline{v}_{y0} = 0$. Finally, we use energy conservation for the initial velocity in the x-direction. The initial energy is

$$E_i = \frac{1}{2}m_p v_{xi}^2 + \frac{K}{x_i} = \frac{1}{2}m_p v_{x0}^2 + \frac{K}{x_0}. \tag{10.3.11}$$

Since $x_i = \infty$, we can solve for the initial speeds as

$$v_{x0} = \sqrt{v_i^2 - \frac{K}{2m_p x_i}} \quad \Rightarrow \quad \overline{v}_{x0} = \pm\sqrt{\overline{v}_i^2 - \frac{K}{2\overline{m}\,\overline{x}_{far}}}, \tag{10.3.12}$$

where we choose the negative root if the particle travels toward the negative direction and we let $\overline{v}_i = \overline{v}_0$ as in (10.3.7). The rest of the quantities have been detailed previously. In order to obtain the minimum radius, we take the minimum distance calculated from the target, that is,

$$\overline{r}_{min} = \min(\sqrt{\overline{x}(\overline{t})^2 + \overline{y}(\overline{t})^2}), \tag{10.3.13}$$

in units of a_b. There are two asymptotes. The first is the line $\overline{y} = \overline{b}$, which is parallel to the x-axis. The second is trickier to find. After a very long time has passed, the projectile will have left the target's influence, so that we take the maximum array values calculated for $\overline{x}$ and $\overline{y}$, say $\overline{x}_N, \overline{y}_N$, respectively, and the slope of the asymptote at this point is

$$s_A \approx \frac{\overline{y}_N - \overline{y}_{N-1}}{\overline{x}_N - \overline{x}_{N-1}}, \tag{10.3.14}$$

so that the outgoing asymptote line is

$$\overline{y} - \overline{y}_N = s_A(\overline{x} - \overline{x}_N). \tag{10.3.15}$$

The angle between the incoming and outgoing asymptotes is the scattering angle. Because the incoming projectile is taken parallel to the x-axis, the scattering angle is related to the outgoing asymptote's tilt, measured from the $+x$-axis, subtracted from π (see Figure 10.3); that is,

$$\Theta = \pi - \arctan \, (outgoing\ asymptote) \, = \pi - \arctan \, (s_A). \qquad (10.3.16)$$

Finally, the eccentricity is calculated as in (10.3.6). The results are shown in Figure 10.6.

The eccentricity works out to $e = -1.33$ in agreement to the analytical method. The distance of closest approach is $r_{min} = 52.8 \, a_b = 52.8$ fm, which is less than the analytic method by 0.3, and which corresponds to a difference of $100[2(53.1 - 52.8)/(53.1 + 52.8)] = 0.57\%$. The scattering angle obtained is $\Theta = 97.2°$ and is less than the analytic method by 0.2 or about 0.2% difference. The curve shown is, however, straight out of the numerical work with the initial condition details explained earlier. Overall, the agreement between the two methods is very good, and doing the two approaches is very instructive. The MATLAB script used to obtain Figure 10.6 is ruther.m and follows. The script simulates the projectile particle's path by solving the differential Equations (10.3.9 and 10.3.10) with initial con-

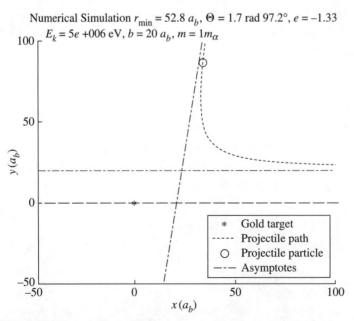

I FIGURE 10.6 Numerical simulation of a projectile alpha particle onto a gold target.

ditions discussed above. A four-dimensional array $w(1, 2, \ldots, 4) = \bar{x}, \bar{v}_x, \bar{y}, \bar{v}_y$ to denote the variables. Thus, the MATLAB script solves the equations:

$$\dot{w}(2) = \frac{\overline{K}w(1)}{\overline{m}(w(1)^2 + w(3)^2)^{3/2}} \text{ and } \dot{w}(4) = \frac{\overline{K}w(3)}{\overline{m}(w(1)^2 + w(3)^2)^{3/2}} \quad (10.3.17)$$

which are coupled to the two differential equations for the position

$$\dot{w}(1) = w(2) \text{ and } \dot{w}(3) = w(4). \quad (10.3.18)$$

The w array used by the script is really a two-dimensional array. One dimension is for tracking the preceding definitions, and the other dimension is for time. The numerical approach used by the script is the Runge–Kutta ode45 scheme. After the call to ode45 is made, one can plot $w(3)$ vs. $w(1)$ to the projectile's path as shown by the script. The derivatives are done by the auxiliary m-file ruther_der.m, which follows, and which is called by the main script. The simulation uses a large value for $\bar{x}_0$ and for the maximum time in order to get accurate proper asymptotes and thereby an accurate scattering angle. The script, when it is run by the user, does animate the projectile motion.

SCRIPT

```
%ruther.m - program to simulate the Rutherford Scattering alpha particle trajectory
%numerically, given the initial projectile energy and impact parameter.
clear;
rcm=0.0;                %target position
za=2; zt=79;            %za=projectile, zt=target charges (2->alpha, 79->gold)
K=za*zt;                %dimensionless force constant
m=1;                    %projectile mass in units of alpha particle mass
vb=0.01965;             %velocity units (in inits of c=light speed)
ma=3730e6;              %alpha particle mass energy in eV
Ene=5e6;                %initial projectile energy in eV
v0=sqrt(2*Ene/m/ma)/vb; %initial speed in units of vb
b=20;                   %impact parameter
L=100;                  %for window viewing purpose
x10=60*L;               %initial x position - far away (a_b=1 Fermi units)
y10=b;                  %initial y position is impact parameter
v1x0=-sign(x10)*sqrt(v0^2-K/x10/m/2); %initial x speed based on energy conservation
                        %toward target, opposite sign to x10
v1y0=0.0;               %zero y-speed - projectile coming in horizontally
tmax=2500;              %max calculation time in units of tau_b (tau_b~1.7e-22 sec)
ic1=[x10,v1x0,y10,v1y0]; %initial conditions
opt=odeset('AbsTol',1.e-9,'RelTol',1.e-6);%user set Tolerances
```

```
[t,w]=ode45('ruther_der',[0.0,tmax],ic1,opt,K,m);%with set tolerance
% str=cat(2,'y versus x - binary system tau=',num2str(tau,4),' tau_0');
[nr,nc]=size(w);%get rows and columns of r
%for i=1:nr
 for i=round(7*nr/16):round(13*nr/16) %not all points animated
    clf;
    hold on
    plot(rcm,rcm,'*')
    plot(w(:,1),w(:,3),'k:') %plot y versus x
    axis ([-L/2 L -L/2 L])
    plot(w(i,1),w(i,3),'ro','MarkerSize',10)
    pause(0.0125)
 end
xlabel('x (a_b)','FontSize',13), ylabel('y (a_b)','FontSize',13)
str=cat(2,'E_k= ',num2str(Ene,3),' eV',', b= ',num2str(b,3),' a_b',...
    ', m= ',num2str(m,3),' m_\alpha');
text(-L/2*(1-0.1),L*(1-0.1),str,'FontSize',10)
%== asymptotes next ==
x=w(:,1);y=w(:,3);                  %let x, y take their names
lx=length(x);                       %x array length
sa=(y(lx)-y(lx-1))/(x(lx)-x(lx-1)); %outgoing asymptote slope (use last point)
xa=-L/2:L;                          %variable to plot outgoing asymptote
ya=y(lx)+sa*(xa-x(lx));             %asymptote line
plot (xa,ya,'b-.')                  %outgoing asymptote
line([-L/2,L],[b,b],'Color','b','LineStyle','-.') %incoming asymptote
if(sa>=0) THETA=pi-atan(sa); end    %first quadrant case
if(sa <0) THETA=pi-(atan(abs(1/sa))+pi/2); end %2nd quadrant case
r=sqrt(x.^2+y.^2); rmin=min(r);     %get r(t) and rmin
%=====================
ye=-sign(K)*sqrt(1+m^2*v1x0^4*b^2/K^2);% eccentricity
str2=cat(2,'Numerical Simulation, r_{min}= ',num2str(rmin,3),' a_b, \Theta=',...
    num2str(THETA,3),' rad = ',num2str(THETA*360/2/pi,3),'^o',', e= ',num2str(ye,3));
title(str2,'FontSize',11)
line([-L/2,L],[0,0],'Color','red','LineStyle','-')
h=legend('Gold Target','Projectile Path','Projectile Particle',' Asymptotes',4);
set(h,'FontSize',13)
```

FUNCTION

```
%ruther_der.m: returns the derivatives for Rutherford scattering trajectory
function derivs = ruther_der( t, w, flag,K,m)
% za=projectile charge, zt=charge of target, m=projectile mass
% Entries in the vector of dependent variables are:
% w(1,2,...8)=x1,v1x,y1,v1y,x2,v2x,y2,v2y
wr=sqrt(w(1).^2 + w(3).^2);
derivs = [w(2); K*w(1)./wr.^3/m; w(4);K*w(3)./wr.^3/m];
```

■ 10.4   Rutherford Scattering and Scattering Cross-Section

The scattering cross-section is a measure of the size of the target area seen by the projectile particle. Referring to Figure 10.7, the angle of interest is the scattering angle

$$\Theta = \pi - 2\alpha , \tag{10.4.1}$$

because it has full information about the interaction between the projectile and the target. This angle can be measured experimentally as well.

Both angles Θ and α have been discussed before in Section 10.2 in connection to Figure 10.3. Basically, Θ is the angle between the incoming and outgoing asymptotes associated with the projectile particle path and $\alpha = (\pi - \Theta)/2$. This relation along with (10.2.8) allows us to write

$$\tan\left(\frac{\Theta}{2}\right) = \tan\left(\frac{\pi}{2} - \alpha\right) = \frac{\sin\left(\dfrac{\pi}{2} - \alpha\right)}{\cos\left(\dfrac{\pi}{2} - \alpha\right)} = \frac{\cos\alpha}{\sin\alpha} = \frac{1/e}{\sqrt{1 - (1/e)^2}} = \frac{1}{\sqrt{e^2 - 1}}, \tag{10.4.2}$$

where we have used the identity $\sin\alpha = \sqrt{1 - \cos^2\alpha}$. From (10.2.5) we see that

$$|e| = \sqrt{1 + \frac{2EL^2}{m_pK^2}} \Rightarrow e^2 = 1 + \frac{2EL^2}{m_pK^2}, \tag{10.4.3}$$

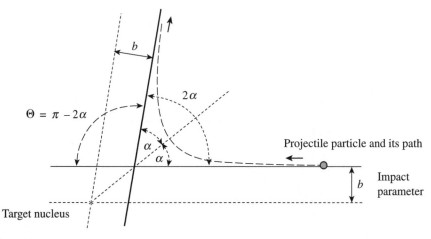

I FIGURE 10.7 Alpha particle asymptotes and the scattering angle.

where m_p is the projectile mass. Substituting this into (10.4.2), we find

$$\tan(\Theta/2) = \sqrt{\frac{m_p K^2}{2EL^2}}. \tag{10.4.4}$$

When the projectile particle is far away the energy is all kinetic, its initial speed is v_0, and so $E = m_p v_0^2/2$ and its angular momentum is $L = m_p v_0 b$, which when substituted into (10.4.4) get

$$\tan(\Theta/2) = \frac{K}{m_p b v_0^2} \quad \Rightarrow \quad b = \frac{K}{m_p v_0^2 \tan(\Theta/2)}, \tag{10.4.5}$$

so that the impact parameter is expressed in terms of the experimentally measured scattering angle. However, usually there are many projectile particles. Some come in with impact parameters $b + db$ and are scattered through angles $\Theta - d\Theta$ as shown in Figure 10.8.

The number of particles scattered through angles Θ and $\Theta + d\Theta$ is written as

$$\frac{dN}{d\Theta} = Nn \frac{d\sigma}{d\Theta}, \tag{10.4.6}$$

where N is the number of projectile particles, n is the number of nuclei scatterers, and $d\sigma/d\Theta$ is the differential scattering cross-section. From Figure 10.8, we can write the scattering cross-sectional area as

$$d\sigma = 2\pi b |db|, \tag{10.4.7}$$

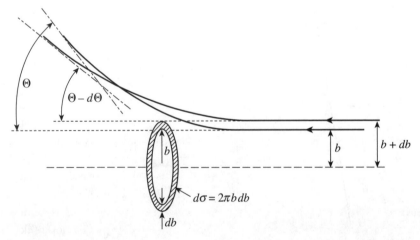

$$d\sigma = 2\pi b \, db$$

I FIGURE 10.8   A change in impact parameter has an associated change in the scattering angle.

where the absolute value is used since db can be negative, as will be seen next. From (10.4.5)

$$db = -\frac{K[d\tan(\Theta/2)/d\Theta]d\Theta}{m_p v_0^2 \tan^2(\Theta/2)} = -\frac{K[\sec^2(\Theta/2)]d\Theta/2}{m_p v_0^2 \tan^2(\Theta/2)} = -\frac{Kd\Theta}{2m_p v_0^2 \sin^2(\Theta/2)},$$

$$(10.4.8)$$

with the minus sign indicating that as b increases ($db > 0$), Θ decreases ($d\Theta < 0$). This is so because the electric force decreases as the inverse squared of the distance between the charges. The projectile feels less repulsion for a larger impact parameter and consequently it experiences less scattering. Equation (10.4.7) becomes

$$d\sigma = \frac{\pi K^2 d\Theta}{m_p^2 v_0^4 \tan(\Theta/2)\sin^2(\Theta/2)},$$

$$(10.4.9)$$

or with the trigonometric identity $2\sin A\cos A = \sin 2A$, the scattering cross-section simplifies to

$$\sigma(\Theta) \equiv \frac{d\sigma}{d\Theta} = 2\pi\left(\frac{K}{2m_p v_0^2}\right)^2 \frac{\sin\Theta}{\sin^4(\Theta/2)},$$

$$(10.4.10)$$

where we have used the notation where the differential scattering cross-section $d\sigma/d\Theta$ is sometimes known as $\sigma(\Theta)$. Finally, from (10.4.6) the number of projectile particles scattered through angles Θ and $\Theta + d\Theta$ is

$$\frac{dN}{d\Theta} = 2\pi Nn\left(\frac{K}{2m_p v_0^2}\right)^2 \frac{\sin\Theta}{\sin^4(\Theta/2)}.$$

$$(10.4.11)$$

This formula predicts how scattering occurs for a repulsive inverse squared distance force law. Ernest Rutherford and coworkers first confirmed this formula in 1912 while investigating the inner structure of the atom. The experiment consisted of aiming α-particles at a gold foil. Sir J. J. Thomson had proposed a model of the atom in which the positive charge was distributed evenly throughout the atom. Thus the atom remained uniformly neutral. That model was known as the plum pudding model of the atom. Rutherford's experiments showed that a fraction of the α-particles actually experienced large angle scattering, unlike what would be expected from the Thomson's uniformly distributed charge model. In fact, some α-particles rebounded back, making a 180° turn! This led Rutherford to propose the nuclear model. Here the positive charge is at core of the atom while the electrons orbit it. This is not unlike what happens in the solar system. The force that binds the electrons, however, is an electric force rather than gravitational. Certainly the experiments carried out by Rutherford and coworkers confirmed the nuclear model. The simulations performed in previous sections of this chapter illustrate the path taken by the α-particles as they are repelled by the large nuclear positive charge of gold. The positive charge is itself held together by nonelectrical nuclear forces in a very small space of a radius on the order of 1 fm in size, which also involves the neutron.

EXAMPLE 10.1

Alpha particles of energy $E_k = 5 \times 10^6$ eV are incident on a gold foil. Obtain a plot of the scattering cross-section versus scattering angle.

Solution

From (10.4.10) in our dimensionless units used in previous sections, we have

$$\sigma(\Theta) = 2\pi \left(\frac{\overline{K} k q_e^2}{2 \overline{m} m_\alpha v_b^2 \overline{v}_0^2} \right)^2 \frac{\sin\Theta}{\sin^4(\Theta/2)} = 2\pi \left(\frac{\overline{K}}{2\overline{m}\,\overline{v}_0^2} \right)^2 \left(a_b \frac{k q_e^2 \tau^2}{m_\alpha a_b^3} \right)^2 \frac{\sin\Theta}{\sin^4(\Theta/2)} = \overline{\sigma}(\Theta) a_b^2$$

(10.4.12)

where once again, $k q_e^2 \tau^2 / m_\alpha a_b^3 \equiv 1$. We see that $\sigma(\Theta)$ has units of area as expected, and the dimensionless scattering cross-section is

$$\overline{\sigma}(\Theta) \equiv 2\pi \left(\frac{\overline{K}}{2\,\overline{m}\,\overline{v}_0^2} \right)^2 \frac{\sin\Theta}{\sin^4(\Theta/2)},$$

(10.4.13)

where $\overline{K} \equiv Z_t Z_\alpha$ with $(Z_t, Z_\alpha) = (79, 4)$, $\overline{m} = 1$, and $\overline{v}_0$ as in (10.3.7). The plot of $\sigma(\Theta)$ versus Θ is produced in Figure 10.9. The scattering cross-section decreases as the scattering angle increases. This means that large angle scattering implies a small area, as in the case of nuclear size target.

The MATLAB script ruther_cross1.m simply executes Equation (10.4.12) and produces the desired plot. A listing of the code follows.

SCRIPT

```
%ruther_cross1.m - program to do the scattering cross-section versus scattering
%angle in Rutherford Scattering
clear; warning off;
m=1;                      %projectile mass in units of alpha particle mass
vb=0.01965;               %velocity units (a_b/tau_b) (in inits of c=light speed)
ma=3730e6;                %alpha particle mass energy in eV
```

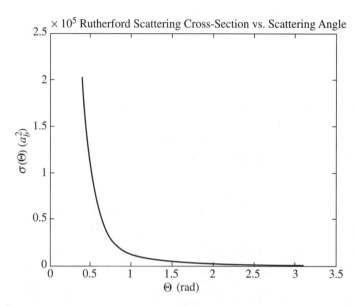

I FIGURE 10.9   Evaluation of Equation (10.4.13) for the scattering cross-section.

```
Ene=5e6;                    %initial projectile energy in eV
v0=sqrt(2*Ene/m/ma)/vb;     %initial speed in units of vb
za=2;zt=79;                 %za=projectile, zt=target charges (2->alpha, 79->gold)
K=za*zt;                    %dimensionless force constant
thsc=.4:.1:pi;
%Scatt. Cross-section
sigma=K^2*2*pi*sin(thsc)./sin(thsc/2).^4/4/m^2/v0^4;
plot(thsc,sigma,'k')
xlabel('\Theta (Radians)','FontSize',14)
ylabel('\sigma(\Theta) (a_b^2)','FontSize',14)
title('Rutherford Scattering Cross-section versus Scattering Angle')
```

EXAMPLE 10.2

An experiment is to be carried out using alpha particles of energy $E_k = 5 \times 10^6$ eV. The particles are to be directed at various nuclear targets, beginning with aluminum and all the way through gold. A fixed impact parameter of $b = 20$ fm is to be used. Obtain plots showing the scattering angle produced versus the atomic number, the scattering cross-section versus the atomic number, and the scattering cross-section versus the calculated scattering angle. Explain your observations.

···

Solution

A MATLAB script `ruther_cross2.m` (listed on the next page) was written to carry out the plots. Dimensionless units are used as before. The impact parameter is expressed as $b = 20a_b$. The scattering cross-section is found as in (10.4.13) of the previous example, and the rest of the parameters are the same. The scattering angle Θ for each target is found through (10.4.1) with the help of (10.2.8). The results are shown in Figure 10.10.

The upper-left plot of the figure shows that the higher the atomic number, the higher the scattering angle Θ. The upper-right figure shows that because higher Z nuclei produce higher Θ, then the $\overline{\sigma}(\Theta)$ has to decrease with Z. Finally, the lower-left plot is consistent with Figure 10.9 and is consistent with the upper plots of the figure.

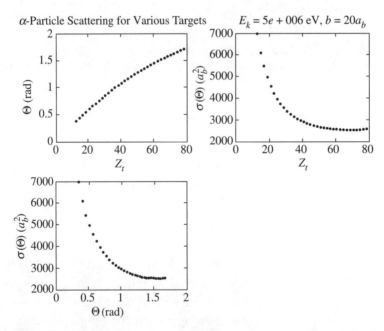

FIGURE 10.10   Example 10.2 for (a) the scattering angle, (b) the scattering cross-section versus atomic number, and (c) the scattering cross-section versus scattering angle.

SCRIPT

```
%ruther_cross2.m - program to do the scattering cross-section versus atomic
%number in Rutherford Scattering
clear; warning off;
m=1;                    %projectile mass in units of alpha particle mass
vb=0.01965;             %velocity units (a_b/tau_b) (in inits of c=light speed)
ma=3730e6;              %alpha particle mass energy in eV
Ene=5e6;                %initial projectile energy in eV
v0=sqrt(2*Ene/m/ma)/vb; %initial speed in units of vb
b=20;                   %impact parameter
za=2;                   %za=projectile(2->alpha)
ztmin=13; ztmax=79;     %from Aluminum to Gold
j=0;                    %initial counter
for iz=ztmin:2:ztmax
    j=j+1;
    zt(j)=iz;               %target array
    K=za*zt(j);             %dimensionless force constant
    ye=-sign(K)*sqrt(1+m^2*v0^4*b^2/K^2);% eccentricity
% can find min.max angle limits for the assymptotes of r for this ye
% these occur at the zeros of the denominator of the r(theta) equation
    %thmin=fzero(inline('ye*cos(x)+1'),[-pi/2,0],[],ye)+1.e-5;
    %thmax=fzero(inline('ye*cos(x)+1'),[0,pi/2],[],ye)-1.e-5;
    %th(j)=thmax-thmin;   %angle from +x axis
    %thsc(j)=pi-th(j);    %total scattering angle produced by target species
%best is to find alpha angle assymptote - faster way, done only once
    alpha=fzero(inline('ye*cos(x)+1'),[0,pi/2],[],ye)-1.e-5;
    thsc(j)=pi-2*alpha;  %total scattering angle produced by target species
%Scatt. Cross-section
    sigma(j)=K^2*2*pi*sin(thsc(j))/sin(thsc(j)/2)^4/4/m^2/v0^4;
end
str=cat(2,'E_k= ',num2str(Ene,3),' eV',', b= ',num2str(b,3),' a_b');
subplot(2,2,1)
plot(zt(:),thsc(:),'k.','MarkerSize',5)
xlabel('Z_t','FontSize',14),ylabel('\Theta (Radians)','FontSize',14)
title('\alpha Particle Scattering for various Targets','FontSize',12)
subplot(2,2,2)
plot(zt(:),sigma(:),'b.','MarkerSize',5)
xlabel('Z_t','FontSize',14),ylabel('\sigma(\Theta) (a_b^2)','FontSize',14)
title(str,'FontSize',12)
subplot(2,2,3)
plot(thsc(:),sigma(:),'r.','MarkerSize',5)
xlabel('\Theta (Radians)','FontSize',14)
ylabel('\sigma(\Theta) (a_b^2)','FontSize',14)
```

EXAMPLE 10.3

In 1913, H. Geiger and E. Marsden published results on the number of scintillations obtained versus scattering angle for the case of silver and gold foil targets. (*Phil. Mag.* Vol. 25, 605 [1913]) Their data are shown in Figure 10.11.

Obtain a comparison between Rutherford's theory and the experiment.

Solution

Equation (10.4.11) gives the number of projectile particles scattered through angles Θ and $\Theta + d\Theta$, given that there are N projectile particles, and n nuclei scatterers. However, the number of particles that strike the area ds of the detector at angle Θ is

$$N(\Theta) \equiv \frac{dN}{ds} = \frac{(dN/d\Theta)d\Theta}{ds}, \tag{10.4.14}$$

where $ds = (2\pi r \sin\Theta)(r d\Theta)$, as shown in Figure 10.12, and where r is the distance to the detector.

Table II.
Variation of Scattering with Angle. (Collected results)

I. Angle of deflexion, ϕ.	II. $\dfrac{1}{\sin^4\phi/2*}$	III. Silver. Number of scintil-lations, N.	IV. $\dfrac{N}{\sin^4\phi/2*}$	V. Gold. Number of scintil-lations, N.	VI. $\dfrac{N}{\sin^4\phi/2*}$
150	1·15	22·2	19·3	33·1	28·8
135	1·38	27·4	19·8	43·0	31·2
120	1·79	33·0	18·4	51·9	29·0
105	2·53	47·3	18·7	69·5	27·5
75	7·25	136	18·8	211	29·1
60	16·0	320	20·0	477	29·8
45	46·6	989	21·2	1435	30·8
37·5......	93·7	1760	18·8	3300	35·3
30	223	5260	23·6	7800	35·0
22·5......	690	20300	29·4	27300	39·6
15	3445	105400	30·6	132000	38·4
30	223	5·3	0·024	3·1	0·014
22·5......	690	16·6	0·024	8·4	0·012
15	3445	93·0	0·027	48·2	0·014
10	17330	508	0·028	200	0·0115
7·5	54650	1710	0·031	607	0·011
5	276300	. . .	. . .	3320	0·012

FIGURE 10.11   Table II from H. Geiger and E. Marsden, *Phil. Mag.* Vol. 25, 605 (1913), reprinted with permission from Taylor & Francis, http://www.tandf.co.uk. The angle ϕ refers to our scattering angle Θ.

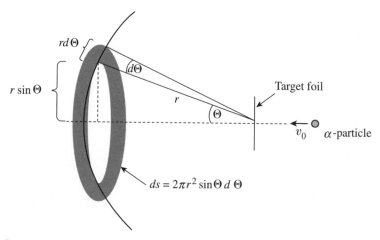

FIGURE 10.12   Particles scattered between Θ and $\Theta + d\Theta$ strike area $ds = (2\pi r \sin \Theta) 9 r d\Theta$.

Substituting (10.4.11) into this, get the number of particles scattered per unit area striking the screen at Θ, and measured in the experiment as

$$N(\Theta) = \frac{Nn}{r^2} \left(\frac{K}{2m_p v_0^2} \right)^2 \frac{1}{\sin^4(\Theta/2)}. \qquad (10.4.15)$$

This quantity is actually dimensionless and can be calculated as follows:

$$N(\Theta) = \frac{Nn}{\overline{r}^2 a_b^2} \left(\frac{\overline{K}}{2\overline{m}\,\overline{v}_0^2} \right)^2 \left(a_b \frac{kq_e^2 \tau^2}{m_a a_b^3} \right)^2 \frac{1}{\sin^4(\Theta/2)} = \frac{Nn}{\overline{r}^2} \left(\frac{\overline{K}}{2\overline{m}\,\overline{v}_0^2} \right)^2 \frac{1}{\sin^4(\Theta/2)} \qquad (10.4.16)$$

using dimensionless units as before. However, because the actual value of the initial speed, the detector distance, the number of α-particles, and the number of nuclei are not provided in the experimental data, we assume an energy of about 1×10^6 eV and multiply the whole theoretical expression by a constant whose value is determined in such a way as to minimize the discrepancy between the theoretical curve and the experimental one. In Figure 10.13, this has been done in such a way that we actually plot

$$N(\Theta) = c_r \left(\frac{\overline{K}}{2\,\overline{m}\,\overline{v}_0^2} \right)^2 \frac{1}{\sin^4(\Theta/2)}, \qquad (10.4.17)$$

where $c_r \equiv cons(Nn/\overline{r}^2)$ with "cons" a constant. For c_r we use the values $(0.019, 0.01)$ for silver and gold, respectively. The rest of the constants are similar to the previous example. The comparison between theory and experiment is shown in Figure 10.13 with the vertical axis on a logarithmic scale.

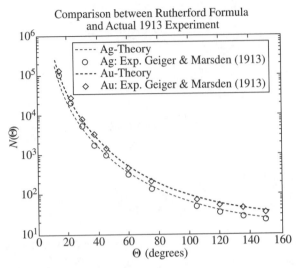

Comparison between Rutherford Formula
and Actual 1913 Experiment

----- Ag-Theory
  ○   Ag: Exp. Geiger & Marsden (1913)
----- Au-Theory
  ◇   Au: Exp. Geiger & Marsden (1913)

$N(\Theta)$

Θ (degrees)

| FIGURE 10.13   Comparison between Rutherford theory and the actual experimental data of Figure 10.11.

As can be seen, the comparison between theory and experiment is very good. Here it has been assumed that the gold and silver target particles do not recoil. This assumption is based on the much larger mass of the target than the projectile particles; however, later in Chapter 11, the recoil aspect of the target will be considered. The following MATLAB script gm_ruther.m has been used to perform the comparison.

SCRIPT

```
%gm_ruther.m - program to do a comparison between the Geiger and Marsden 1913
%experimental data and the Rutherford Scattering formula for the number of particles
clear; warning off;
m=1;                  %projectile mass in units of alpha particle mass
za=2;                 %za=projectile
zt=[47 79];           %zt=target charges (47-> silver, 79->gold)
K=za*zt;              %dimensionless force constant
vb=0.01965;           %velocity units (a_b/tau_b) (in inits of c=light speed)
ma=3730e6;            %alpha particle mass energy in eV
Ene=1e6;              %assume this energy
v0=sqrt(2*Ene/m/ma)/vb; %initial speed in units of vb
factor=K.^2/4/m^2/v0^4; %regular cross-section factor
cr=[0.019 0.01];      %fit coef for comparison with experiment
%================== Experimental Data =========================
%Actual Data from H. Geiger & E. Marsden, Phil. Mag. Vol.25, 605 (1913)
%for angle in degrees and number of scintillations obtained for Ag and Au
dth=[15 22.5 30 37.5 45 60 75 105 120 135 150]; % Exp. angle in Degrees from paper
rth=dth*2*pi/360;     %convert angle to radians
```

```
rsig(1,:)=[105400 20300 5260 1760  989 320 136 47.3 33.0 27.4 22.2];%Silver
rsig(2,:)=[132000 27300 7800 3300 1435 477 211 69.5 51.9 43.0 33.1];%Gold
%=====================================================================
N=50;thmin=min(rth-0.05);thmax=max(rth+0.05);ths=(thmax-thmin)/(N-1);
th=thmin:ths:thmax;      %uses experimental angle range in radians, N points
thd=th*360/2/pi;         %angle in degrees, use for plotting
for i=1:2                %do formula on silver (i=1) and gold (i=2)
   for j=1:N
      Nth(i,j)=cr(i)*factor(i)/sin(th(j)/2)^4;
   end
end
semilogy(thd(:),Nth(1,:),'r:',dth(:),rsig(1,:),'bo') %Ag
hold on
semilogy(thd(:),Nth(2,:),'k:',dth(:),rsig(2,:),'md') %Au
xlabel('\Theta (degrees)','FontSize',14)
ylabel('N(\Theta)','FontSize',14)
title('Comparison between Rutherford Formula and Actual 1913 Experiment')
h=legend('Ag-Theory','Ag: Exp. Geiger & Marsden (1913)','Au-Theory',
   'Au: Exp. Geiger & Marsden (1913)',1);
set(h,'FontSize',13)
```

EXAMPLE 10.4

An α-particle carrying an energy of 5 MeV incident on a gold foil is seen to be scattered by 90°. (a) Find the α-particle's impact parameter, and (b) its distance of closest approach to the gold nucleus.

....

Solution

(a) From (10.4.5), we have for $\Theta = \pi/2$, and with $E_k = m_p v_0^2/2$, we have

$$b = \frac{K}{m_p v_0^2 \tan(\Theta/2)} \rightarrow \frac{k q_e^2 Z_{Au} Z_\alpha}{2 E_k \tan(\pi/4)}$$

$$= \frac{9 \times 10^9 \frac{\text{Nm}^2}{\text{C}^2}(1.602 \times 10^{-19} \text{ C})^2(79)(2)}{2(5 \times 10^6 \text{ eV})\left(\frac{1.602 \times 10^{-19} \text{ J}}{\text{eV}}\right)\tan(\pi/4)} = 22.78 \times 10^{-15}\text{m}.$$

(10.4.18)

(b) From (10.3.8) and (10.4.5) we can write

$$r_{\min} = -\frac{m_p v_0^2 b^2}{K(1+e)} = -\frac{2 E_k \tan(\Theta/2) b^2}{K(1+e)\tan(\Theta/2)} = -\frac{b^2}{b\tan(\Theta/2)(1+e)} = -\frac{b}{\tan(\Theta/2)(1+e)}.$$

(10.4.19)

From Section 10.2 we have $\cos \alpha = -1/e = 1/e$ and $\alpha = (\pi - \Theta)/2 = (\pi - \pi/2)/2 = \pi/4$, which implies $e = -1/\cos(\alpha) = -\sqrt{2}$. Thus, (10.4.19) gives

$$r_{min} = -\frac{b}{\tan(45°)(1+e)} = \frac{b}{|1 - \sqrt{2}|} = \frac{22.78 \times 10^{-15} \text{ m}}{|1 - \sqrt{2}|} = 5.5 \times 10^{-15} \text{ m}. \quad (10.4.20)$$

■ Chapter 10 Problems

10.1 What is the value of the electric force between an α-particle and an A_u nucleus when their separation distance is 1 fm?

10.2 If an α-particle with an initial energy of 1 MeV is directed toward a gold nucleus with zero impact parameter value, how close could the particle get to the gold?

10.3 Use the analytic solution method to (a) simulate a 10 MeV α-particle's trajectory that's incident on a silver target with an impact parameter of 10 fm, (b) find the scattering angle, and (c) obtain the orbit's eccentricity as well as the distance of closest approach.

10.4 Use the numerical solution method to (a) simulate a 10 MeV α-particle's trajectory that's incident on a silver target with an impact parameter of 10 fm, (b) find the scattering angle, and (c) obtain the orbit's eccentricity as well as the distance of closest approach.

10.5 Alpha particles of energy $E_k = 5 \times 10^6$ eV are incident on a silver foil. Obtain a plot of the scattering cross-section versus scattering angle and compare it to that of a gold foil target.

10.6 An α-particle carrying an energy of 10 MeV incident on a silver foil is seen to be scattered by 60°. Find (a) the α-particle's impact parameter, and (b) its distance of closest approach to the silver nucleus.

10.7 Figure 10.7 indicates that there is a mathematical relationship between the impact parameter and the distance of closest approach. Obtain this relationship and show that it is consistent with Equation (10.4.5). (*Hint:* Use the understanding of hyperbolas, as well Appendix C.)

11 | Systems of Particles

■ 11.1 Introduction

Thus far we have mainly dealt with the dynamics of systems in terms of a single particle—for example, a mass vibrating at the end of a spring, or a planet going around the sun, or even an α-particle interacting with a nucleus. While we have had some examples of systems of particles, like the case of interacting spring–mass systems in Chapter 4, and gravitation, including binary systems, in Chapter 9, we need to have a more general understanding. In this chapter we discuss more specifically the underlying principles involved in the dynamics associated with systems of more than one particle. In particular, we work with N particles that may be acted upon by external, as well as internal, forces. We will again revisit the concepts of center of mass, momentum, and energy. We will also consider variable mass rocket propulsion and particle collisions (elastic and inelastic).

■ 11.2 Center of Mass and Center of Gravity

A. Discrete System

We start by considering a system of N particles, as shown in Figure 11.1(a).

Let the masses and positions of the particles be labeled $m_1, m_2, m_3, m_4, \cdots, m_{N-1}, m_N$, and $\mathbf{r}_1, \mathbf{r}_2, \mathbf{r}_3, \mathbf{r}_4, \cdots, \mathbf{r}_{N-1}, \mathbf{r}_N$, respectively. The masses are not necessarily considered to be at rest, so that their velocities are $\mathbf{v}_1, \mathbf{v}_2, \mathbf{v}_3, \mathbf{v}_4, \cdots, \mathbf{v}_{N-1}, \mathbf{v}_N$, and associated accelerations are $\mathbf{a}_1, \mathbf{a}_2, \mathbf{a}_3, \mathbf{a}_4, \cdots, \mathbf{a}_{N-1}, \mathbf{a}_N$. The center of mass (cm) position, velocity, and acceleration of this N particle system are defined as

$$\mathbf{R}_{cm} \equiv \frac{1}{M} \sum_i m_i \mathbf{r}_i, \quad \mathbf{v}_{cm} = \dot{\mathbf{R}}_{cm} \equiv \frac{1}{M} \sum_i m_i \mathbf{v}_i, \quad \mathbf{a}_{cm} = \dot{\mathbf{v}}_{cm} = \ddot{\mathbf{R}}_{cm} \equiv \frac{1}{M} \sum_i m_i \mathbf{a}_i,$$

$$(11.2.1)$$

where $M \equiv \sum_i m_i$, $\mathbf{v}_i = \dot{\mathbf{r}}_i$, and $\mathbf{a}_i = \dot{\mathbf{v}}_i = \ddot{\mathbf{r}}$ have been used. Furthermore, in these

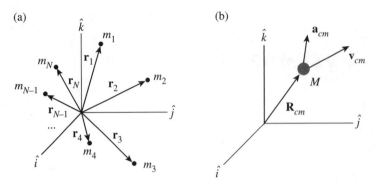

| FIGURE 11.1

expressions it is understood that each vector variable can be further expressed in terms of its respective x, y, and z components. The quantities in (11.2.1) correspond to the representation of the N-particle system in terms of the center of mass quantities; that is, the whole particle system is treated as if it were a single mass M with center of mass coordinates, as shown in Figure 11.1(b).

Equation (11.2.1) can be used to obtain the center of gravity (cg) analogues

$$\mathbf{R}_{cg} \equiv \frac{\sum_i m_i g_i \mathbf{r}_i}{\sum_i m_i g_i}, \quad \mathbf{v}_{cg} \equiv \frac{\sum_i m_i g_i \mathbf{v}_i}{\sum_i m_i g_i}, \quad \mathbf{a}_{cg} \equiv \frac{\sum_i m_i g_i \mathbf{a}_i}{\sum_i m_i g_i}, \tag{11.2.2}$$

where the value of the gravitational acceleration at the position of m_i is $g_i = g(r_i)$. In the case when the gravitational field is uniform throughout the system of particles' spatial extent, Equation (11.2.2) becomes identical to (11.2.1) because $g(r_i) \rightarrow g$ factors out.

Continuous System

The above discrete system formulas can be applied to an extended, not necessarily rigid, body with a total mass M if we consider it to be composed of infinitesimal mass elements. One such element of mass dm located at position $\mathbf{r}$ from the origin is shown in Figure 11.2(a).

In this case, the center of mass quantities of interest are given by

$$\mathbf{R}_{cm} \equiv \frac{\int \mathbf{r}\,dm}{M}, \quad \mathbf{v}_{cm} \equiv \frac{\int \mathbf{v}\,dm}{M}, \quad \mathbf{a}_{cm} \equiv \frac{\int \mathbf{a}\,dm}{M}, \tag{11.2.3}$$

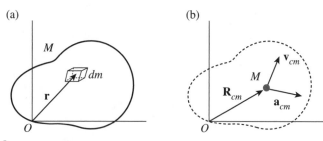

FIGURE 11.2

where we now have used $M = \int dm$ as shown in Figure 11.2(b). The center of gravity analogues are

$$\mathbf{R}_{cg} \equiv \frac{\int \mathbf{r}g(r)dm}{\int g(r)dm}, \quad \mathbf{v}_{cg} \equiv \frac{\int \mathbf{v}g(r)dm}{\int g(r)dm}, \quad \mathbf{a}_{cg} \equiv \frac{\int \mathbf{a}g(r)dm}{\int g(r)dm}, \quad (11.2.4)$$

which, once again, in the case of a constant gravitational field, become equivalent to (11.2.3).

EXAMPLE 11.1

The positions of three particles of masses $m_1 = 1$ kg, $m_2 = 2$ kg, and $m_3 = 3$ kg are given as a function of time by $\mathbf{r}_1 = \left(3\ \mathrm{m} + 2\frac{\mathrm{m}}{\mathrm{s}^2}t^2\right)\hat{i} + 4\ \mathrm{m}\hat{j}$, $\mathbf{r}_2 = (-2\ \mathrm{m} + 1\ \mathrm{m}\cdot \mathrm{s}/t)\hat{i} + 2\frac{\mathrm{m}}{\mathrm{s}}t\hat{j}$, and $\mathbf{r}_3 = 1\ \mathrm{m}\hat{i} - 3\frac{\mathrm{m}}{\mathrm{s}^2}t^2\hat{j}$, respectively. (a) Obtain the center of mass position, velocity, and acceleration associated with the three-particle system as a function of time. (b) Evaluate the expressions at time $t = 1$ s.

Solution

The total mass of the three particles is $M = 6$ kg. From (11.2.1), the center of mass position is

$$\mathbf{R}_{cm} \equiv \frac{1\left[\left(3\mathrm{m} + 2\frac{\mathrm{m}}{\mathrm{s}^2}t^2\right)\hat{i} + 4\ \mathrm{m}\hat{j}\right] + 2\left[\left(-2\mathrm{m} + \frac{1\ \mathrm{m}\cdot \mathrm{s}}{t}\right)\hat{i} + 2\frac{\mathrm{m}}{\mathrm{s}}t\hat{j}\right] + 3\left[1\ \mathrm{m}\hat{i} - 3\frac{\mathrm{m}}{\mathrm{s}^2}t^2\hat{j}\right]}{6}$$

$$= \frac{1}{3}\left\{\left(1\ \mathrm{m} + 1\frac{\mathrm{m}}{\mathrm{s}^2}t^2 + 1\ \mathrm{m}\cdot \mathrm{s}\frac{1}{t}\right)\hat{i} + \left(2\ \mathrm{m} + 2\frac{\mathrm{m}}{\mathrm{s}}t - \frac{9}{2}\frac{\mathrm{m}}{\mathrm{s}^2}t^2\right)\hat{j}\right\}$$

The velocities are $\ddot{\mathbf{r}}_1 = \dfrac{4\ \text{m}}{\text{s}^2}\, t\hat{i},\ \ddot{\mathbf{r}}_2 = -\dfrac{1\ \text{m}\cdot\text{s}}{t^2}\hat{i} + \dfrac{2\ \text{m}}{\text{s}}\hat{j},$ and $\ddot{\mathbf{r}}_3 = -\dfrac{6\ \text{m}}{\text{s}^2}t\hat{j}.$ The accelerations

are $\ddot{\mathbf{r}}_1 = 4\hat{i}\,\dfrac{\text{m}}{\text{s}^2},\ \ddot{\mathbf{r}}_2 = \dfrac{2\ \text{m}\cdot\text{s}}{t^3}\hat{i},$ and $\ddot{\mathbf{r}}_3 = -\dfrac{6\ \text{m}}{\text{s}^2}\hat{j}.$ Thus, the center of mass velocity is

$$\mathbf{v}_{cm} = \frac{\dfrac{4\ \text{m}}{\text{s}^2}t\hat{i} + 2\left[-\dfrac{1\ \text{m}\cdot\text{s}}{t^2}\hat{i} + \dfrac{2\ \text{m}}{\text{s}}\hat{j}\right] + 3\left[-\dfrac{6\ \text{m}}{\text{s}^2}t\hat{j}\right]}{6}$$

$$= \frac{1}{3}\left(\left[2\dfrac{\text{m}}{\text{s}^2}t - \dfrac{1\ \text{m}\cdot\text{s}}{t^2}\right]\hat{i} + \left[2\dfrac{\text{m}}{\text{s}} - 9\dfrac{\text{m}}{\text{s}^2}t\right]\hat{j}\right),$$

and the center of mass acceleration is

$$\mathbf{a}_{cm} = \frac{4\dfrac{\text{m}}{\text{s}^2}\hat{i} + 2\dfrac{2\ \text{m}\cdot\text{s}}{t^3}\hat{i} + 3\left[-6\dfrac{\text{m}}{\text{s}^2}\hat{j}\right]}{6} = \frac{1}{3}\left(\left[2\dfrac{\text{m}}{\text{s}^2} + \dfrac{2\ \text{m}\cdot\text{s}}{t^3}\right]\hat{i} - 9\dfrac{\text{m}}{\text{s}^2}\hat{j}\right)$$

Notice that these results are consistent with the alternative expressions $\mathbf{v}_{cm} = \dot{\mathbf{R}}_{cm}$, and $\mathbf{a}_{cm} = \dot{\mathbf{v}}_{cm}$ of (11.2.1).

(b) We can evaluate the above expressions at $t = 1$ s to obtain

$\mathbf{r}_1 = [5\hat{i} + 4\hat{j}]$ m, $\mathbf{r}_2 = [-\hat{i} + 2\hat{j}]$ m, $\mathbf{r}_3 = [\hat{i} - 3\hat{j}]$ m for the positions; $\dot{\mathbf{r}}_1 = 4\hat{i}\,\dfrac{\text{m}}{\text{s}}$,

$\dot{\mathbf{r}}_2 = [-\hat{i} + 2\hat{j}]\dfrac{\text{m}}{\text{s}},\ \dot{\mathbf{r}}_3 = -6\hat{j}\,\dfrac{\text{m}}{\text{s}}$ for the velocities; and $\ddot{\mathbf{r}}_1 = 4\hat{i}\,\dfrac{\text{m}}{\text{s}^2},\ \ddot{\mathbf{r}}_2 = 2\hat{i}\,\dfrac{\text{m}}{\text{s}^2},\ \ddot{\mathbf{r}}_3 = -6\hat{j}\,\dfrac{\text{m}}{\text{s}^2}$,

for the accelerations. Also at $t = 1$ s, $\mathbf{R}_{cm} = \left(\hat{i} - \dfrac{1}{6}\hat{j}\right)$ m, $\mathbf{v}_{cm} = \dfrac{1}{3}(\hat{i} - 7\hat{j})\dfrac{\text{m}}{\text{s}}$, and

$\mathbf{a}_{cm} = \left(\dfrac{4}{3}\hat{i} - 3\hat{j}\right)\dfrac{\text{m}}{\text{s}^2}.$

All of these results can be verified with the symbolic capability of MATLAB as described in the tutorial in Appendix A. The script particl.m shown next is followed by the actual output (file particl.txt), where each matrix row represents a vector. The results are identical to the preceding ones, analytically and numerically.

```
%particl.m - given the mass and initial time-dependent positions of 3 particles,
%this script calculates the velocities, accelerations and their center of mass
%values. Type "help symbolic" within the command line
%if need help on symbolic functions
clear;
format compact;              %Suppress extra line-feeds on outputs
syms t;                      %declare t as a symbol
t = sym('t', 'real');            %let t, and the masses be real
m=[1, 2, 3]                      %mass vector
M=sum(m)                     %total mass
r=[3+2*t^2, 4, 0;-2+1/t, 2*t, 0;1, -3*t^2, 0]%given position vectors
v=diff(r, t, 1)                  %velocity vector
a=diff(r, t, 2)                  %acceleration vector
rcm=m*r/M                    %center of mass coordinate
vcm=m*v/M                    %center of mass velocity
acm=m*a/M                    %center of mass acceleration
%=================== evaluate =====================
t=1;                         %set the time at which to evaluate expressions
r=eval(r)
v=eval(v)
a=eval(a)
rcm=eval(rcm)
vcm=eval(vcm)
acm=eval(acm)
```

```
particl.txt
>> particl
m =
   1    2    3
M =
   6
r =
[ 3+2*t^2,     4,     0]
[ -2+1/t,    2*t,     0]
[     1,   -3*t^2,    0]
v =
[  4*t,     0,     0]
[ -1/t^2,    2,     0]
[   0,   -6*t,     0]
a =
[   4,     0,     0]
[ 2/t^3,    0,     0]
[   0,    -6,     0]
```

```
rcm =
[ 1/3+1/3*t^2+1/3/t, 2/3+2/3*t-3/2*t^2,          0]
vcm =
[ 2/3*t-1/3/t^2,     2/3-3*t,       0]
acm =
[ 2/3+2/3/t^3,          -3,       0]
r =
    5    4    0
   -1    2    0
    1   -3    0
v =
    4    0    0
   -1    2    0
    0   -6    0
a =
    4    0    0
    2    0    0
    0   -6    0
rcm =
   1.0000   -0.1667        0
vcm =
   0.3333   -2.3333        0
acm =
   1.3333   -3.0000        0
>>
```

EXAMPLE 11.2

Obtain the center of mass of a uniform density circular disk quadrant of mass M and radius R.

Solution

Write the center of mass as $\mathbf{R}_{cm} \equiv \dfrac{1}{M}\displaystyle\int_0^R \mathbf{r}\,dm = \dfrac{1}{M}(x_{cm}\hat{i} + y_{cm}\hat{j})$, where $x_{cm} = \dfrac{1}{M}\displaystyle\int_0^R x\,dm$, and $y_{cm} = \dfrac{1}{M}\displaystyle\int_0^R y\,dm$. Following Figure 11.3(a), for the first integral, $x = R\cos\theta$, $dx = -R\sin\theta\,d\theta$, $y = R\sin\theta$ and $dm = \sigma y\,dx$, so that

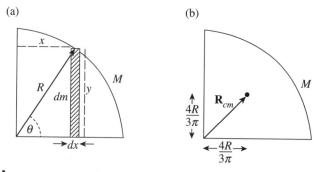

(a)   (b)

FIGURE 11.3 Quadrant of a uniform circular disk whose center of mass is calculated in Example 11.2.

$$x_{cm} = -\frac{1}{M}\sigma R^3 \int_{\pi/2}^{0} \sin^2\theta \cos\theta \, d\theta = -\frac{1}{M}\sigma R^3 \left[\frac{\sin^3\theta}{3}\right]_{\pi/2}^{0} = \frac{\sigma R^3}{3M}. \text{ Using}$$

$\sigma = \dfrac{M}{(\pi R^2/4)}$, we finally have $x_{cm} = \dfrac{4\,R}{3\,\pi}$.

By symmetry or by a similar process, the second integral yields $y_{cm} = \dfrac{4\,R}{3\,\pi}$.

Thus, the center of mass of the quadrant is given by $\mathbf{R}_{cm} = \dfrac{4\,R}{3\,\pi}(\hat{i} + \hat{j})$, as shown

in Figure 11.3(b).

■ 11.3   Multiparticle Systems

In this section we study the dynamics of a system of particles. We describe the system's forces, linear momentum, torques, angular momentum, and energy (kinetic, potential, and total). Whenever possible, terms in the description of the system are simplified to make use of the simpler center of mass variables.

A. External and Internal Forces

Consider, for example, the system of three particles of Figure 11.4. Each of the particles is acted upon by an external force $\mathbf{F}_1$, $\mathbf{F}_2$, and $\mathbf{F}_3$, respectively, as shown. In addition, the particles are assumed to be interacting and, therefore, exert internal forces on each other. The internal forces on particle 1 due to particles 2 and 3 are $\mathbf{F}_{12}$ and $\mathbf{F}_{13}$, respectively. Similarly, the internal forces on particle 2 due to particles 1 and 3 are $\mathbf{F}_{21}$

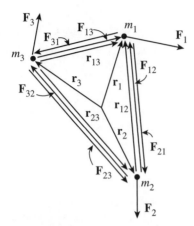

FIGURE 11.4   System of particles acted on by external and internal forces.

and $\mathbf{F}_{23}$; those on particle 3 due to particles 1 and 2 are $\mathbf{F}_{31}$ and $\mathbf{F}_{32}$. The quantities $\mathbf{r}_{12}$, $\mathbf{r}_{13}$, $\mathbf{r}_{23}$ are the relative coordinates between particle pairs $(1, 2)$, $(1, 3)$, and $(2, 3)$, respectively.

Each of the particles is subject to the net force $(\mathbf{F}^T)$, which is the sum of the external and internal forces, or

$$\mathbf{F}_1^T = \mathbf{F}_1 + \mathbf{F}_{12} + \mathbf{F}_{13} = m_1\ddot{\mathbf{r}}_1, \tag{11.3.1a}$$

for particle 1, and similarly for particles 2 and 3

$$\mathbf{F}_2^T = \mathbf{F}_2 + \mathbf{F}_{21} + \mathbf{F}_{23} = m_2\ddot{\mathbf{r}}_2, \quad \mathbf{F}_3^T = \mathbf{F}_3 + \mathbf{F}_{31} + \mathbf{F}_{32} = m_3\ddot{\mathbf{r}}_3. \tag{11.3.1b}$$

Furthermore, each of the internal forces, such as $\mathbf{F}_{12}$, is a function of the particle coordinates $\mathbf{r}_1$ and $\mathbf{r}_2$; similarly $\mathbf{F}_{13}$ is a function of the particle coordinates $\mathbf{r}_1$ and $\mathbf{r}_3$, and so on. Thus Equation (11.3.1) represents a multicoupled system of differential equations for the particle coordinates $\mathbf{r}_1$, $\mathbf{r}_2$, $\mathbf{r}_3$. Equation (11.3.1) can be extended to a system of N particles. Thus, the total or net force on particle i due to its external force as well as the internal forces on it from the other $N - 1$ particles is

$$\mathbf{F}_i^T = \mathbf{F}_i + \sum_{k \neq i}^{N} \mathbf{F}_{ik} = m_i\ddot{\mathbf{r}}_i, \tag{11.3.2}$$

where $\mathbf{F}_i$ is the external force on particle i, and $\mathbf{F}_{ik}$ is the internal force on particle i due to the particle k. As can be seen, the N particle case is even more difficult to solve.

It is at this point that the center of mass coordinate defined in (11.2.1) plays a significant role. Going back to the three-particle system, if we consider the sum of each of their total forces (11.3.1), we get

$$m_1\ddot{\mathbf{r}}_1 + m_2\ddot{\mathbf{r}}_2 + m_3\ddot{\mathbf{r}}_3 = \mathbf{F}_1 + \mathbf{F}_2 + \mathbf{F}_3 + \mathbf{F}_{12} + \mathbf{F}_{13} + \mathbf{F}_{21} + \mathbf{F}_{23} + \mathbf{F}_{31} + \mathbf{F}_{32}; \qquad (11.3.3a)$$

however, as illustrated in Figure 11.4, by Newton's third law, the internal forces obey the relations $\mathbf{F}_{12} = -\mathbf{F}_{21}$, $\mathbf{F}_{13} = -\mathbf{F}_{31}$, and $\mathbf{F}_{23} = -\mathbf{F}_{32}$, so that the preceding equation simplifies to

$$m_1\ddot{\mathbf{r}}_1 + m_2\ddot{\mathbf{r}}_2 + m_3\ddot{\mathbf{r}}_3 = \mathbf{F}_1 + \mathbf{F}_2 + \mathbf{F}_3. \qquad (11.3.3b)$$

For the N particle system, we have $\mathbf{F}_{ij} = -\mathbf{F}_{ji}$, and the sum of all the total forces (11.3.2) of each individual particle gives

$$\sum_i \mathbf{F}_i = \sum_i m_i\ddot{\mathbf{r}}_i \quad \Rightarrow \quad M\ddot{\mathbf{R}}_{cm} = \mathbf{F}, \qquad (11.3.4a)$$

where the sum of all the internal forces

$$\sum_i \sum_{k \neq i}^{N} \mathbf{F}_{ik} = 0 \qquad (11.3.4b)$$

has been used, and $\mathbf{F} = \sum_i \mathbf{F}_i$ is the total net external force acting on the system. Equation (11.3.4) introduces the notion that the dynamics of any system of particles can be studied without regard to the internal forces if we look at the center of mass motion of that system.

B. Linear Momentum

Based on this understanding, we next look at the momentum and energy of a system of N particles. In Chapter 1, Newton's laws of motion were reviewed. In particular, the second law can be restated as follows:

$$\mathbf{F} = \frac{d\mathbf{p}}{dt} = \frac{d}{dt}(m\mathbf{v}) = m\frac{d\mathbf{v}}{dt} + \mathbf{v}\frac{dm}{dt}, \qquad (11.3.5)$$

where the second term contributes in a variable mass problem. For a system of N particles, from (11.3.2), the total net force on particle i due to its external force and the internal forces on it from the $N-1$ particles is written as

$$\frac{d\mathbf{p}_i}{dt} = \mathbf{F}_i^T = \mathbf{F}_i + \sum_{k \neq i}^{N} \mathbf{F}_{ik}, \tag{11.3.6}$$

where $\mathbf{p}_i$ is the momentum of the ith article. Summing this over all the particles and using (11.3.4), we find the total rate of change of the momentum of the system as

$$\frac{d\mathbf{P}}{dt} = \sum_i \frac{d\mathbf{p}_i}{dt} = \mathbf{F} = M\ddot{\mathbf{R}}_{cm}, \tag{11.3.7}$$

which with (11.3.4) says that the total system momentum is given by

$$\mathbf{P} = \sum_i \mathbf{p}_i = M\mathbf{v}_{cm}, \tag{11.3.8}$$

which is the center of mass momentum. From (11.3.7), in the absence of an external force, we can see that

$$\frac{d\mathbf{P}}{dt} = 0 \Rightarrow \mathbf{P} = \text{constant.} \tag{11.3.9a}$$

Since the total particle system mass is constant, the constancy of the total momentum applies to both, the total momentum of the system as well as to the center of mass velocity of the system of particles

$$\sum_i \mathbf{p}_i = \text{constant,} \quad \mathbf{v}_{cm} = \text{constant.} \tag{11.3.9b}$$

A deeper understanding of linear momentum associated with N particles can be obtained if we express the position of the ith particle in terms of the vectors $\mathbf{r}'_i$ and $\mathbf{R}_{cm}$ shown in Figure 11.5.

That is, $\mathbf{r}'_i$ is the relative position of the ith particle with respect to the center of mass location. We write

$$\mathbf{r}_i = \mathbf{R}_{cm} + \mathbf{r}'_i, \tag{11.3.10}$$

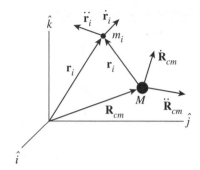

I FIGURE 11.5   Position of a particle of mass m_i with respect to the center of mass M.

followed by the first and second time derivatives and a multiplication by m_i to get

$$m_i \dot{\mathbf{r}}_i = m_i \dot{\mathbf{R}}_{cm} + m_i \dot{\mathbf{r}}' = \mathbf{p}_i, \text{ and } m_i \ddot{\mathbf{r}}_i = m_i \ddot{\mathbf{R}}_{cm} + m_i \ddot{\mathbf{r}}'_i = \mathbf{F}_i^T, \qquad (11.3.11)$$

where we have identified the first with the total particle's momentum and the second with the particle's net force (11.3.6). Summing these expressions over all the N particles and comparing with Equations (11.3.7 and 11.3.8), we must have that

$$\mathbf{P} = M\dot{\mathbf{R}}_{cm} + \mathbf{P}' = M\dot{\mathbf{R}}_{cm}, \text{ and } \mathbf{F} = M\ddot{\mathbf{R}}_{cm} + \frac{d\mathbf{P}'}{dt} = M\ddot{\mathbf{R}}_{cm} = \frac{d\mathbf{P}}{dt}, \qquad (11.3.12)$$

where $\mathbf{P}' \equiv \sum_i m_i \mathbf{v}'_i$ is the total relative momentum and $\mathbf{v}'_i = \dot{\mathbf{r}}'_i$ is the relative velocity with respect to the center of mass motion. This result has two implications. The first is that

$$\mathbf{P}' = \sum_i m_i \dot{\mathbf{r}}'_i = \sum_i \mathbf{p}'_i = \frac{d}{dt} \sum_i m_i \mathbf{r}'_i = 0; \qquad (11.3.13)$$

that is, the total internal momentum of the system is zero. The second is obtained by recalling the result from (11.3.4), which with (11.3.11) implies that

$$\frac{d\mathbf{P}'}{dt} = \frac{d}{dt} \sum_i \mathbf{p}'_i = \sum_i \sum_{k \neq i}^N \mathbf{F}_{ik} = 0. \qquad (11.3.14)$$

This means that we can identify the time rate of change of the relative momentum of particle i, with the sum of all the internal forces on it due to the other $N - 1$ particles

$$\frac{d\mathbf{p}'_i}{dt} = \sum_{k \neq i}^{N} \mathbf{F}_{ik} = m_i \ddot{\mathbf{r}}'_i. \tag{11.3.15}$$

Furthermore, comparing the second of (11.3.11) with (11.3.6), we must also have that

$$\mathbf{F}_i = m_i \ddot{\mathbf{R}}_{cm}, \tag{11.3.16}$$

which means that the effect of an external force on particle i is equivalent to the center of mass acceleration of the system of particles multiplied by the particle's mass. Finally, it is useful to note that (11.3.13) does imply that the relative position of the internal center of mass is a constant; that is,

$$\sum_i m_i \mathbf{r}'_i = \text{constant}. \tag{11.3.17}$$

In fact, the value of the constant can be found since from (11.3.10), $\mathbf{r}'_i = \mathbf{r}_i - \mathbf{R}_{cm}$, so that

$$\sum_i m_i \mathbf{r}'_i = \sum_i m_i (\mathbf{r}_i - \mathbf{R}_{cm}) = \sum_i m_i \mathbf{r}_i - \sum_i m_i \mathbf{R}_{cm} = M\mathbf{R}_{cm} - M\mathbf{R}_{cm} = 0,$$

$$\tag{11.3.18}$$

as should be because as far as the relative coordinate is concerned, the system of particles' center of mass is the relative origin. In conclusion, for a system of particles, its relative center of mass is zero, and so are its first and second derivatives as seen by Equations (11.3.13 and 11.3.14).

C. Angular Momentum and Torque

Angular momentum was discussed in Chapter 7 in connection to the rotating Earth, and also in Chapter 8 in connection with its relationship to a central force. In particular, referring to Figure 11.5, the angular momentum of particle i with respect to the origin is

$$\mathbf{L}_i = m_i \mathbf{r}_i \times \dot{\mathbf{r}}_i = \mathbf{r}_i \times \mathbf{p}_i. \tag{11.3.19}$$

The total angular momentum of a system of N particles is obtained by performing the sum of the individual momenta

$$\mathbf{L} = \sum_i m_i \mathbf{r}_i \times \dot{\mathbf{r}}_i. \tag{11.3.20a}$$

If we use (11.3.10), this can be rewritten as

$$\mathbf{L} = \sum_i m_i (\mathbf{R}_{cm} + \mathbf{r}'_i) \times (\ _{cm} + \ '_i)$$

$$= \mathbf{R}_{cm} \times \ _{cm}\sum_i m_i + \mathbf{R}_{cm} \times \left(\sum_i m_i \ '_i\right) + \left(\sum_i m_i \mathbf{r}'_i\right) \times \ _{cm} + \sum_i m_i \mathbf{r}'_i \times \ '_i; \tag{11.3.20b}$$

however, according to (11.3.13 and 11.3.18), the second and third terms in parentheses vanish, and we find that

$$\mathbf{L} = \mathbf{R}_{cm} \times \mathbf{P} + \sum_i \mathbf{r}'_i \times \mathbf{p}'_i. \tag{11.3.20c}$$

This means that the total angular momentum of a system of particles is equal to the sum of the center of mass angular momentum about the origin and the total internal angular momentum about the center of mass. A torque can be obtained from the angular momentum as

$$\boldsymbol{\tau} = \frac{d\mathbf{L}}{dt} = \dot{\mathbf{R}}_{cm} \times \mathbf{P} + \mathbf{R}_{cm} \times \dot{\mathbf{P}} + \sum_i \dot{\mathbf{r}}'_i \times \mathbf{p}'_i + \sum_i \mathbf{r}'_i \times \dot{\mathbf{p}}'_i = \mathbf{R}_{cm} \times \dot{\mathbf{P}} + \sum_i \boldsymbol{\tau}'_i, \tag{11.3.21}$$

where we have defined the internal torques, $\boldsymbol{\tau}'_i \equiv \mathbf{r}'_i \times \dot{\mathbf{p}}'_i$ about the center of mass. Also, the second term on the right of the first equal sign vanishes because $\dot{\mathbf{R}}_{cm} \times \mathbf{P} = M\dot{\mathbf{R}}_{cm} \times \dot{\mathbf{R}}_{cm} = 0$, and similarly for the third term. With the use of (11.3.15), the sum of the internal torques can be rewritten as

$$\sum_i \boldsymbol{\tau}'_i = \sum_i \mathbf{r}'_i \times \dot{\mathbf{p}}'_i = \sum_i \sum_{k \neq i} \mathbf{r}'_i \times \mathbf{F}_{ik}. \tag{11.3.22}$$

It is worth looking at this term in more detail. For example, for the system of three particles shown in Figure 11.6, this term becomes

$$\sum_{i}^{3} \sum_{k \neq i}^{3} \mathbf{r}'_i \times \mathbf{F}_{ik} = \mathbf{r}'_1 \times (\mathbf{F}_{12} + \mathbf{F}_{13}) + \mathbf{r}'_2 \times (\mathbf{F}_{21} + \mathbf{F}_{23}) + \mathbf{r}'_3 \times (\mathbf{F}_{31} + \mathbf{F}_{32})$$

$$= (\mathbf{r}'_1 - \mathbf{r}'_2) \times \mathbf{F}_{12} + (\mathbf{r}'_1 - \mathbf{r}'_3) \times \mathbf{F}_{13} + (\mathbf{r}'_2 - \mathbf{r}'_3) \times \mathbf{F}_{23}$$

$$= \sum_{i}^{3} \sum_{k > i}^{3} (\mathbf{r}'_i - \mathbf{r}'_k) \times \mathbf{F}_{ik} = \sum_{i}^{3} \sum_{k > i}^{3} \mathbf{r}'_{ik} \times \mathbf{F}_{ik}, \qquad (11.3.23)$$

where we have used $\mathbf{r}'_{ik} = \mathbf{r}'_i - \mathbf{r}'_k$ and $\mathbf{F}_{ij} = -\mathbf{F}_{ji}$ as in Equations (11.3.3 and 11.3.4) and consistent with Figure 11.6.

Also, Figure 11.6 suggests that the internal forces are either parallel or antiparallel to the relative vector difference between the particles; i.e., $\mathbf{F}_{ij} = s_{ij}|\mathbf{F}_{ij}|\hat{r}'_{ij}$, where s_{ij} is $+1$ (if parallel) or -1 (if antiparallel). Certainly this is the case for central internal forces. In this case, extending the above results to a system of N particles, the sum of the internal torques of (11.3.21) is

$$\sum_{i} \boldsymbol{\tau}'_i = \sum_{i} \sum_{k \neq i} \mathbf{r}'_i \times \mathbf{F}_{ik} = \sum_{i} \sum_{k > i} \mathbf{r}'_{ik} \times \mathbf{F}_{ik} = \sum_{i} \sum_{k > i} s_{ik}|\mathbf{F}_{ik}|\, \mathbf{r}'_{ik} \times \hat{r}'_{ik} = 0,$$

$$(11.3.24)$$

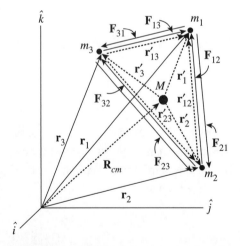

FIGURE 11.6   System of three particles showing the various coordinates and internal forces.

so that the system's net torque, if the internal forces are central and the external forces do not affect the internal angular momentum (see the following comments), with the use of (11.3.7), is

$$\boldsymbol{\tau} = \frac{d\mathbf{L}}{dt} = \mathbf{R}_{cm} \times \dot{\mathbf{P}} = \mathbf{R}_{cm} \times \mathbf{F}, \tag{11.3.25a}$$

and is the same as the total external torque applied on the system without any contribution from the internal torques. However, when no information is available regarding internal forces, a more general expression of the total torque on the system is, using (11.3.20a), $\boldsymbol{\tau} = \dfrac{d}{dt} \sum_i m_i \mathbf{r}_i \times \dot{\mathbf{r}}_i = \sum_i m_i (\dot{\mathbf{r}}_i \times \dot{\mathbf{r}}_i + \mathbf{r}_i \times \ddot{\mathbf{r}}_i).$ But since $\dot{\mathbf{r}}_i \times \dot{\mathbf{r}}_i = 0$, this becomes

$$\boldsymbol{\tau} = \sum_i \mathbf{r}_i \times \dot{\mathbf{p}}_i, \tag{11.3.25b}$$

which by (11.3.6) does include the effect of internal torques, if any are present. Notice that if we use (11.3.6) into (11.3.25b), we get $\boldsymbol{\tau} = \sum_i \mathbf{r}_i \times \left(\mathbf{F}_i + \sum_{k \neq i}^N \mathbf{F}_{ik} \right)$ or $\boldsymbol{\tau} = \sum_i \mathbf{r}_i \times \mathbf{F}_i + \sum_i \sum_{k \neq i}^N \mathbf{r}_i \times \mathbf{F}_{ik}.$ If one assumes that $\mathbf{F}_{ij} = -\mathbf{F}_{ji}$, the last term in this torque expression turns out to be identical to the result in (11.3.23 and 11.3.24) and vanishes because, as can be seen from Figure 11.6, $\mathbf{r}'_{ik} = \mathbf{r}'_i - \mathbf{r}'_k = \mathbf{r}_i - \mathbf{r}_k = \mathbf{r}_{ik}.$ Thus, we can also write

$$\boldsymbol{\tau} = \sum_i \mathbf{r}_i \times \mathbf{F}_i, \tag{11.3.25c}$$

which involves only the coordinate of each particle and the particle's associated external force. However, even this formula differs from (11.3.25a). That is, $\boldsymbol{\tau} = \sum_i \mathbf{r}_i \times \mathbf{F}_i = \sum_i (\mathbf{R}_{cm} + \mathbf{r}') \times \mathbf{F}_i,$ or $\boldsymbol{\tau} = \mathbf{R}_{cm} \times \sum_i \mathbf{F}_i + \sum_i \mathbf{r}' \times \mathbf{F}_i = \mathbf{R}_{cm} \times \mathbf{F} + \sum_i \mathbf{r}' \times \mathbf{F}_i,$ which contains contributions due to internal torques associated with the application of external forces, as evidenced by the last term. In this expression, the external forces affect the internal angular momentum and give rise to an internal torque. Only in the absence of such internal torques is (11.3.25a) obtainable from (11.3.25c). The most general torque expression is, therefore, (11.3.25b).

D. Kinetic Energy

Continuing with the preceding system of N particles, we next consider the work done on the system in moving it from an initial position $\boldsymbol{\alpha}$ to a final position $\boldsymbol{\beta}$, where all the particles' positions are specified at both locations; that is, $\boldsymbol{\alpha} \equiv \{\mathbf{r}_{1\alpha}, \mathbf{r}_{2\alpha}, \cdots, \mathbf{r}_{N\alpha}\}$ and $\boldsymbol{\beta} \equiv \{\mathbf{r}_{1\beta}, \mathbf{r}_{2\beta}, \cdots, \mathbf{r}_{N\beta}\}$ with associated velocities $\{\mathbf{v}_{1\alpha}, \mathbf{v}_{2\alpha}, \cdots, \mathbf{v}_{N\alpha}\}$ and $\{\mathbf{v}_{1\beta}, \mathbf{v}_{2\beta}, \cdots, \mathbf{v}_{N\beta}\}$, respectively. The total work done in doing so is written as a sum over all the individual works done on each particle

$$W = \sum_i \int_{\mathbf{r}_{i\alpha}}^{\mathbf{r}_{i\beta}} \mathbf{F}_i^T \cdot d\mathbf{r}_i = \sum_i \int_{\mathbf{r}_{i\alpha}}^{\mathbf{r}_{i\beta}} \frac{d\mathbf{p}_i}{dt} \cdot d\mathbf{r}_i = \sum_i m_i \int_{\mathbf{r}_{i\alpha}}^{\mathbf{r}_{i\beta}} \frac{d\mathbf{v}_i}{dt} \cdot d\mathbf{r}_i, \tag{11.3.26a}$$

where we have used the definitions (11.3.2 and 11.3.6). This can be rearranged to get

$$W = \sum_i \int_{\mathbf{r}_{i\alpha}}^{\mathbf{r}_{i\beta}} \frac{d\mathbf{v}_i}{dt} \cdot \frac{d\mathbf{r}_i}{dt} dt = \sum_i m_i \int_{\mathbf{v}_{i\beta}}^{\mathbf{v}_{i\beta}} \mathbf{v}_i \cdot d\mathbf{v}_i = \sum_i m_i \left. \frac{\mathbf{v}_i^2}{2} \right|_{\mathbf{v}_{i\beta}}^{\mathbf{v}_{i\beta}} = \sum_i t_i \Big|_\alpha^\beta = \Delta T, \tag{11.3.26b}$$

where we have defined the kinetic energy of particle i as $t_i \equiv m_i \mathbf{v}_i^2 / 2 = \mathbf{p}_i^2 / 2 m_i$. This means that the work done on the system is equal to the change in kinetic energy of the system. The total kinetic energy of the system is given by

$$T = \sum_i t_i = \sum_i \frac{p_i^2}{2 m_i} = \frac{1}{2} \sum_i m_i \dot{r}_i^2. \tag{11.3.27}$$

At this point it is convenient to use (11.3.10) to write $\mathbf{v}_i = \dot{\mathbf{R}}_{cm} + \dot{\mathbf{r}}'$, so that

$$\begin{aligned} T &= \frac{1}{2} \sum_i m_i \left(\dot{\mathbf{R}}_{cm} + \dot{\mathbf{r}} \right) \cdot \left(\dot{\mathbf{R}}_{cm} + \dot{\mathbf{r}}' \right) \\ &= \frac{1}{2} \left\{ \sum_i m_i \dot{\mathbf{R}}_{cm}^2 + 2 \left(\sum_i m_i \dot{\mathbf{r}} \right) \cdot \dot{\mathbf{R}}_{cm} + \sum_i m_i \dot{\mathbf{r}}'^2 \right\}; \end{aligned} \tag{11.3.28}$$

however, by (11.3.18) the middle term vanishes and we have

$$T = \sum_i \frac{p_i^2}{2\,m_i} = \frac{1}{2}Mv_{cm}^2 + \frac{1}{2}\sum_i m_i \dot{r}^2. \tag{11.3.29}$$

This result means that the kinetic energy of the system of N particles is equal to the sum of the kinetic energy of the whole system moving with the center of mass velocity and the internal kinetic energy of the particles moving with respect to the center of mass.

E. Potential Energy

Following the work of Chapter 6 on the potential energy function, and according to the total force on a particle in a system of particles (11.3.2), assuming the forces are conservative, we can define the change in potential energy of each particle as

$$V_i^T = -\int_{\mathbf{r}_{i\alpha}}^{\mathbf{r}_{i\beta}} \mathbf{F}_i^T \cdot d\mathbf{r}_i = -\left[\int_{\mathbf{r}_{i\alpha}}^{\mathbf{r}_{i\beta}} \mathbf{F}_i \cdot d\mathbf{r}_i + \sum_{k \neq i} \int_{\mathbf{r}_{i\alpha}}^{\mathbf{r}_{i\beta}} \mathbf{F}_{ik} \cdot d\mathbf{r}_i \right]. \tag{11.3.30a}$$

and the total potential energy change of the system as

$$\sum_i V_i^T = -\sum_i \int_{\mathbf{r}_{i\alpha}}^{\mathbf{r}_{i\beta}} \mathbf{F}_i^T \cdot d\mathbf{r}_i = -\left[\sum_i \int_{\mathbf{r}_{i\alpha}}^{\mathbf{r}_{i\beta}} \mathbf{F}_i \cdot d\mathbf{r}_i + \sum_i \sum_{k \neq i} \int_{\mathbf{r}_{i\alpha}}^{\mathbf{r}_{i\beta}} \mathbf{F}_{ik} \cdot d\mathbf{r}_i \right], \tag{11.3.30b}$$

where we have followed the notation for the limits introduced in the kinetic energy subsection. Each of the forces, external and internal, has an associated respective potential

$$\mathbf{F}_i = -\nabla_i V_i, \quad \mathbf{F}_{ij} = -\nabla_i V_{ij} = -\mathbf{F}_{ji} = \nabla_j V_{ji} = \nabla_j V_{ij}, \tag{11.3.31}$$

where we have used $V_{ij} = V_{ji}$ because it depends only on the coordinate magnitude. We keep in mind that the external force, $\mathbf{F}_i$, on particle i depends on the particle coordinate $\mathbf{r}_i$, but that the internal force, $\mathbf{F}_{ij}$, between particles i and j depends on coordinates $\mathbf{r}_i$ and $\mathbf{r}_j$. Therefore, a change in the external potential is given by

$$dV_i = \nabla_i V_i \cdot d\mathbf{r} = -\mathbf{F}_i \cdot d\mathbf{r}, \tag{11.3.32a}$$

and following the second of (11.3.31) a change in the internal potential is

$$dV_{ij} = \boldsymbol{\nabla}_i V_{ij} \cdot d\mathbf{r}_i + \boldsymbol{\nabla}_j V_{ij} \cdot d\mathbf{r}_j = -\mathbf{F}_{ij} \cdot (d\mathbf{r}_i - d\mathbf{r}_j) = -\mathbf{F}_{ij} \cdot d\mathbf{r}_{ij}, \quad (11.3.32b)$$

where we have used $d\mathbf{r}_{ij} = d\mathbf{r}_i - d\mathbf{r}_j$ for the change in position vectors of particles i and j. The development that led to (11.3.23) enables us to see that we can similarly write the second term inside brackets of (11.3.30b) in the form

$$\sum_i \sum_{k \neq i} \int_{\mathbf{r}_{i\alpha}}^{\mathbf{r}_{i\beta}} \mathbf{F}_{ik} \cdot d\mathbf{r}_i = \sum_i \sum_{k>i} \int_{\mathbf{r}_{i\alpha}}^{\mathbf{r}_{i\beta}} (d\mathbf{r}_i - d\mathbf{r}_k) \cdot \mathbf{F}_{ik} = \sum_i \sum_{k>i} \int_{\mathbf{r}_{i\alpha}}^{\mathbf{r}_{i\beta}} d\mathbf{r}_{ik} \cdot \mathbf{F}_{ik}. \qquad (11.3.33)$$

Using this into (11.3.30b) followed by a substitution of the corresponding expressions of (11.3.32) into, we get the total change in potential energy

$$\sum_i V_i^T = -\sum_i \int_{\mathbf{r}_{i\alpha}}^{\mathbf{r}_{i\beta}} \mathbf{F}_i^T \cdot d\mathbf{r}_i = \sum_i \int_{\mathbf{r}_{i\alpha}}^{\mathbf{r}_{i\beta}} dV_i + \sum_i \sum_{k>1} \int_{\mathbf{r}_{i\alpha}}^{\mathbf{r}_{i\beta}} dV_{ik} = \sum_i \left[V_i + \sum_{k>1} V_{ik} \right]_{\mathbf{r}_{i\alpha}}^{\mathbf{r}_{i\beta}}$$

$$= \sum_i \left(u_{i\beta} - u_{i\alpha} \right) = \sum_i \Delta u_i = \left[\sum_i V_i + \sum_i \sum_{k>1} V_{ik} \right]_\alpha^\beta = U_\beta - U_\alpha = \Delta U, \qquad (11.3.34a)$$

where we have defined the potential energy of particle i at a given position as

$$u_i = V_i + \sum_{k>i} V_{ik}. \qquad (11.3.34b)$$

This means that the potential energy of the ith particle equals the sum of its potential energy associated with the external force and the potential energy due to the internal interactions between it and the rest of the particles. In the preceding expression, we have also defined the total potential energy of the N particle system as

$$U \equiv \sum_i u_i = \sum_i V_i + \sum_i \sum_{k>i} V_{ik}. \qquad (11.3.34c)$$

F. Total Energy

The results of the previous two subsections can be used to obtain the total energy change associated with a system of N particles. In particular, notice that from (11.3.26 and 11.3.29) and (11.3.34)

$$\Delta T + \Delta U = 0, \qquad (11.3.35)$$

which indicates that, in the absence of any dissipative forces, for the system of particles, the total energy E, i.e., the sum of kinetic and potential energies, Equations (11.3.29 and 11.3.34c), respectively,

$$E = T + U = \sum_i \frac{p_i^2}{2\,m_i} + \sum_i u_i$$

$$= \frac{1}{2}Mv_{cm}^2 + \frac{1}{2}\sum_i m_i r'^2 + \sum_i V_i + \sum_i \sum_{k>i} V_{ik} = \text{constant}, \qquad (11.3.36)$$

is conserved.

EXAMPLE 11.3

For the system of three particles with $m_1 = 1$ kg, $m_2 = 2$ kg, and $m_3 = 3$ kg of Example 11.1, find the system's (a) linear momentum, (b) angular momentum, (c) kinetic energy, (d) net force, and (e) net torque at time $t = 1$ s.

Solution

From Example 11.1, at $t = 1$ s, the positions, velocities, accelerations, as well as their center of mass values, are known.

(a) The individual linear momenta are $\mathbf{p}_1 = m_1\dot{\mathbf{r}}_1 = (1)(4\,\hat{\imath}) = 4\,\hat{\imath}\,$kg m/s, $\mathbf{p}_2 = m_2\dot{\mathbf{r}}_2 = (2)(-\hat{\imath} + 2\hat{\jmath}) = [-2\hat{\imath} + 4\hat{\jmath}]$ kg m/s, and $\mathbf{p}_3 = m_3\dot{\mathbf{r}}_3 = (3)(-6\hat{\jmath}) = -18\hat{\jmath}$ kg m/s. The total system momentum is from (11.3.8), $\mathbf{P} = \mathbf{p}_1 + \mathbf{p}_2 + \mathbf{p}_3 = (2\hat{\imath} - 14\hat{\jmath})$ kg m/s, which agrees with $\mathbf{P} = M\mathbf{v}_{cm} = (6)\frac{1}{3}(\hat{\imath} - 7\hat{\jmath})$ kg m/s.

(b) From (11.3.20), $\mathbf{L} = \sum_i m_i\mathbf{r}_i \times \dot{\mathbf{r}}_i = \mathbf{r}_1 \times \mathbf{p}_1 + \mathbf{r}_2 \times \mathbf{p}_2 + \mathbf{r}_3 \times \mathbf{p}_3$, or

$\mathbf{L} = (5\hat{\imath} + 4\hat{\jmath}) \times 4\hat{\imath} + (-\hat{\imath} + 2\hat{\jmath}) \times (-2\hat{\imath} + 4\hat{\jmath}) + (\hat{\imath} - 3\hat{\jmath}) \times (-18\hat{\jmath})$ or
$\mathbf{L} = -16\,\hat{k} + (-4\,\hat{k} + 4\,\hat{k}) - 18\,\hat{k} = -34\,\hat{k}\,$kg m^2/s.

(c) The kinetic energies are obtained from (11.3.29),

$$T = \sum_i \frac{p_i^2}{2\,m_i} = \frac{1}{2}(m_1 v_1^2 + m_2 v_2^2 + m_2 v_2^2) \text{ or}$$

$$T = \frac{1}{2}[(1)(4\hat{i} \cdot 4\hat{i}) + (2)(-\hat{i} + 2\hat{j}) \cdot (-\hat{i} + 2\hat{j}) + (3)(-6\hat{j}) \cdot (-6\hat{j})] =$$
$$\frac{1}{2}(16 + 10 + 108) = 67 \text{ J}.$$

(d) Each of the forces in (11.3.2) is $\mathbf{F}_1^T = m_1 \mathbf{a}_1 = (1)(4\hat{i}) = 4\hat{i}$ N, $\mathbf{F}_2^T = m_2 \mathbf{a}_2 = (2)(2\hat{i}) = 4\hat{i}$ N, and $\mathbf{F}_3^T = m_3 \mathbf{a}_3 = (3)(-6\hat{j}) = -18\hat{j}$ N.

The sum of these is $\mathbf{F}^T = \mathbf{F}_1^T + \mathbf{F}_2^T + \mathbf{F}_3^T = 4\hat{i} + 4\hat{i} - 18\hat{j} = (8\hat{i} - 18\hat{j})$ N, which agrees with (11.3.4) $\mathbf{F} = M\ddot{\mathbf{R}}_{cm} = (6 \text{ kg})\left(\frac{4}{3}\hat{i} - 3\hat{j} \frac{\text{m}}{\text{s}^2}\right)$, as expected.

(e) We note that each of the forces on the particles may have an internal force contribution that is not obvious from the previous result, since in the sum of all forces the contribution due to the internal forces cancel. Also, the total torque may be affected by unknown internal torques. Thus, we use the most general (11.3.25b) to obtain the total torque on the system $\boldsymbol{\tau} = \mathbf{r}_1 \times \mathbf{F}_1^T + \mathbf{r}_2 \times \mathbf{F}_2^T + \mathbf{r}_3 \times \mathbf{F}_3^T =$
$(5\hat{i} + 4\hat{j}) \times (4\hat{i}) + (-\hat{i} + 2\hat{j}) \times (4\hat{i}) + (\hat{i} - 3\hat{j}) \times (-18\hat{j})$ or

$\boldsymbol{\tau} = -16\hat{k} - 8\hat{k} - 18\hat{k} = -42\,\hat{k}$, which could be shown to equal $\dot{\mathbf{L}}$ if the time dependence of $\mathbf{L}$ had been calculated above. Finally, we don't have enough information to perform the calculation based on either (11.3.25a) or (11.3.25c) because the forces involved in these equations are purely external and would have to be explicitly given or be available.

All of the above results can be reproduced using the MATLAB script partic2.m shown below followed by the actual output (file partic2.txt), where each matrix row represents a vector corresponding to the components of each particle's calculated quantity. For the energies, the matrix row columns correspond to the individual particle energies.

SCRIPT

```
%partic2.m - given the mass and initial positions, velocities, and
%accelerations of 3 particles, this script calculates their linear
%and angular momenta, energies, forces, and torques
%help symbolic        %type this if need help on symbolic functions
clear;
format compact;       %Suppress extra line-feeds on outputs
```

```
u=[1, 1, 1];          %vector used for component sum
m=[1, 2, 3];          %mass vector
M=sum(m);          %total mass
r =[5, 4, 0;-1, 2, 0;1, -3, 0]; %position vectors
v =[4, 0, 0;-1, 2, 0;0, -6, 0]; %velocity vectors
a =[4, 0, 0;2, 0, 0;0, -6, 0];  %acceleration vectors
rcm =[1.0, -0.1667, 0];   %center of mass position vector
vcm =[0.3333, -2.3333, 0];   %center of mass velocity vector
acm =[1.3333, -3.0000, 0];   %center of mass acceleration vector
for i=1:3
  p(i, :)=m(i)*v(i, :);          %momentum calculation
  l(i, :)=cross(r(i, :), p(i, :));   %angular momentum calculation
  e(i)=dot(p(i, :), p(i, :))/2/m(i); %kinetic energies
  f(i, :)=m(i)*a(i, :);          %forces
  tau(i, :)=cross(r(i, :), f(i, :));   %torques
end
p                   %obtained momenta
P=u*p                   %net momentum 1st way
P=M*vcm                   %P center of mass way
l                   %obtained angular momenta
L=u*l                   %net angular momentum
e                   %energies
E=sum(e(:))                   %total kinetic energy
f                   %obtained forces
F=u*f                   %net force 1st way
F=M*acm                   %F center of mass way
tau                   %obtained torques
Tau=u*tau                   %Net torque
```

OUTPUT

```
partic2.txt
>> partic2
p =
   4    0    0
  -2    4    0
   0  -18    0
P =
   2  -14    0
P =
  1.9998 -13.9998    0
l =
   0    0  -16
   0    0    0
   0    0  -18
```

```
L =
    0    0  -34
e =
    8    5   54
E =
   67
f =
    4    0    0
    4    0    0
    0  -18    0
F =
    8  -18    0
F =
    7.9998 -18.0000       0
tau =
    0    0  -16
    0    0   -8
    0    0  -18
Tau =
    0    0  -42
>>
```

EXAMPLE 11.4

A projectile is fired at an angle θ with an initial velocity of v_0. At the highest point, the projectile explodes into two fragments of masses m_1 and m_2, with an extra energy E_0. Each fragment continues in the original horizontal direction. By how much distance are the fragments separated when they land?

....

Solution

The separation is given by $d = R_1 - R_2 = (v_{1x} - v_{2x})t_{top}$, where v_{1x} and v_{2x} are the sought fragment velocities in the x-direction of m_1 and m_2, respectively, and t_{top} is the time for the masses to fall back to the ground from a height H. Since the masses have no y-velocity components, we have that $H = gt_{top}^2/2$, where H is obtained by applying energy conservation to the original projectile ($m = m_1 + m_2$), $mgH = mv_y^2/2$ with $v_y = v_0\sin\theta$. Thus, $t_{top} = v_0\sin\theta/g$. The fragments have only x-velocity components, and momentum conservation gives

$$m_1 v_{1x} + m_2 v_{2x} = mv_{cm} = mv_{0x} \quad \Rightarrow \quad v_{2x} = \frac{mv_{0x} - m_1 v_{1x}}{m_2},$$

Where $v_{cm} = v_{0x}$ is the center of mass (projectile) velocity at the top with $v_{0x} = v_0 \cos\theta$. Energy conservation leads to

$$\frac{1}{2}mv_{0x}^2 + E_0 = \frac{1}{2}m_1 v_{1x}^2 + \frac{1}{2}m_2 v_{2x}^2.$$

Substituting for v_{2x}, solving for v_{1x} in the resulting quadratic, and choosing the positive root, get

$$v_{1x} = v_{0x} + \sqrt{\frac{2\,m_2 E_0}{mm_1}},$$

which when substituted back into the expression for v_{2x}, find

$$v_{2x} = v_{0x} - \sqrt{\frac{2\,m_1 E_0}{mm_2}}.$$

Finally, obtain $d = (v_{1x} - v_{2x})t_{top} = \dfrac{v_0}{g}\sin\theta \sqrt{\dfrac{2\,E_0}{m}}\left(\sqrt{\dfrac{m_2}{m_1}} - \sqrt{\dfrac{m_1}{m_2}}\right)$.

■ 11.4   Variable Mass Rocket Motion

Rocket propulsion is based on the principle that the rate of change of momentum is equivalent to an external net force, and if the external net force is zero, the change in momentum is also zero and so momentum must be conserved. Consider Figure 11.7(a), where a rocket of mass m is shown to be under the action of an external force **F**; it is ejecting gas with a speed **u** relative to the rocket, and it has a time-dependent speed **v**.

Let t be the time just before the rocket fires, and let the initial momentum of the rocket be

$$\mathbf{p}_i = m\mathbf{v}. \tag{11.4.1}$$

As soon as the rocket begins to burn and eject gas at speed **u** relative to it, its mass decreases as a function of time. Let the mass and velocity of the rocket at time $t + dt$ be $m + dm$ and $\mathbf{v} + d\mathbf{v}$, respectively, where $dm = -|dm|$ is a small decrease in rocket mass due to its loss of fuel in the short time dt. The gas mass exhausted is $-dm$ and has velocity $\mathbf{v} + \mathbf{u}$ relative to the ground, as shown in Figure 11.7(b). The total final momentum of the rocket–exhausted-gas system is

$$\mathbf{p}_f = (m + dm)(\mathbf{v} + d\mathbf{v}) - dm(\mathbf{v} + \mathbf{u}) \approx m\mathbf{v} + m\,d\mathbf{v} - \mathbf{u}\,dm, \tag{11.4.2}$$

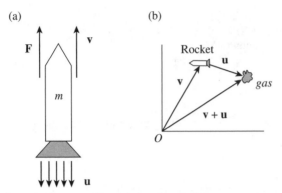

(a)   (b)

FIGURE 11.7   Rocket of mass m is under the action of an external force $\mathbf{F}$ ejecting gas with a speed $\mathbf{u}$ relative to the rocket, and it has a time dependent speed $\mathbf{v}$.

where the smaller term $dm\,d\mathbf{v}$ has been neglected. The net force on the system is by Newton's second law

$$\mathbf{F} = \lim_{\Delta t \to 0} \frac{\Delta \mathbf{p}}{\Delta t} = \lim_{\Delta t \to 0} \frac{\mathbf{p}_f - \mathbf{p}_i}{\Delta t} = \frac{(m\mathbf{v} + m\,d\mathbf{v} - \mathbf{u}\,dm) - m\mathbf{v}}{dt} = m\frac{d\mathbf{v}}{dt} - \mathbf{u}\frac{dm}{dt},$$
(11.4.3a)

or

$$m\frac{d\mathbf{v}}{dt} = \mathbf{u}\frac{dm}{dt} + \mathbf{F},$$
(11.4.3b)

which is known as the *rocket equation*.

EXAMPLE 11.5

Assuming a constant gas exhaust speed, obtain the velocity as a function of time of a rocket in free space if its initial speed is v_0.

Solution

In free space, there are no external forces acting on the rocket and $\mathbf{F} = 0$. From (11.4.3), using $\mathbf{v} = v\hat{i}$ and $\mathbf{u} = -u\hat{i}$, where u is constant, get

$$m\frac{dv}{dt} = \left(-\frac{dm}{dt}\right)u,$$
(11.4.4)

where the quantity on the right is known as the *thrust* and acts to propel the rocket forward since $dm/dt = -|dm/dt|$ is a negative quantity. The preceding differential equation can be solved to yield

$$\int_{v_0}^{v} dv = -u \int_{m_0}^{m} \frac{dm}{m} \quad \Rightarrow \quad v = v_0 - u\ln\left(\frac{m}{m_0}\right) = v_0 + u\ln\left(\frac{m_0}{m}\right). \quad (11.4.5)$$

The quantity $m_0/m > 1$ is known as the *fuel-to-payload ratio*, where m is the mass of the rocket's structure, including the engine, and m_0 is the initial fuel plus m. Thus, one needs a large amount of fuel in order to have a large fuel-to-payload ratio and, therefore, achieve large speeds. One can also increase u by choosing a high-quality fuel.

The preceding is an example of a single-stage rocket. A two-stage rocket is one where two fuel containers are part of the total initial mass (m_0). Starting at v_0, after the first fuel container is emptied and the rocket achieves speed $v_1 = v_0 + u\ln(m_0/m_{f1})$, where m_{f1} is the final mass at the end of the first stage, the first fuel container is subsequently discarded. The remaining rocket mass (m_a) is propelled as the fuel in the second container begins to burn. The final velocity at the end of the second burnout, $v_2 = v_1 + u\ln(m_a/m_{f2})$, where m_{f2} is the final mass at the end of the second stage, can be made higher than the velocity achieved in a single-stage rocket with the same amount of total fuel. This process can be extended to multistage rockets so that fuel containers can be jettisoned after their fuel is burned out, and the total remaining mass is much less than that of a single-stage rocket.

EXAMPLE 11.6

Obtain the equation for the velocity as a function of time of a rocket fired vertically upward near the surface of Earth. What is the condition for liftoff?

Solution

On Earth's surface we can set $\mathbf{F} = -mg\hat{k}$ (11.4.3), using $\mathbf{v} = v\hat{k}$ and $\mathbf{u} = -u\hat{k}$, where u is assumed constant, and get in one dimension

$$m\frac{dv}{dt} = \left(-\frac{dm}{dt}\right)u - mg. \quad (11.4.6)$$

Next, multiplying both sides of (11.4.6) by dt, dividing by m, and integrating get

$$v = v_0 - u\ln\left(\frac{m}{m_0}\right) - gt = v_0 + u\ln\left(\frac{m_0}{m}\right) - gt. \tag{11.4.7}$$

The liftoff condition occurs when in (11.4.6) we set $dv/dt = 0$; that is, if at $t = 0$, we let the fuel burn rate be the constant $\alpha \equiv -dm/dt = |dm/dt|$, then the equation tells us that in order to achieve liftoff we must have

$$\alpha u \geq mg, \tag{11.4.8}$$

which means that the thrust must, at least, be equal to the rocket's weight.

11.5 Rocket Takeoff from the Surface of a Massive Body-Motion Simulation

In this section we employ (11.4.3b) in order to simulate the motion of a rocket take-off from a body of mass M. Given a certain amount of fuel as a percentage of the total initial rocket mass, the rocket is launched and the fuel burns out at a later time. The rocket can subsequently escape, stay in orbit, or perform projectile motion and ultimately free fall back down. The external force becomes Chapter 9's gravitational force of attraction between the massive body and the rocket, $\mathbf{F} = -(GMm/r^2)\hat{r}$, where $G = 6.673 \times 10^{-11}\ \mathrm{Nm^2/kg^2}$. The rocket equation becomes

$$m\frac{d\mathbf{v}}{dt} = \mathbf{u}\frac{dm}{dt} - \frac{GMm}{r^2}\hat{r}. \tag{11.5.1}$$

We next write the rocket's velocity as $\mathbf{v} = v_x\hat{i} + v_y\hat{j}$, and the burned propellant velocity as $\mathbf{u} = -u_x\hat{i} - u_y\hat{j}$, because the gas is exhausted out the back end of the rocket and propels the rocket forward. We also write the gravitational force's direction as $\hat{r} = \mathbf{r}/r$, with the vector $\mathbf{r} = x\hat{i} + y\hat{j}$, and its magnitude as $r = (x^2 + y^2)^{1/2}$. We will assume that the burnt fuel mass ejection rate is constant; that is, $\alpha = -dm/dt$, which is positive, as alluded to in the previous section, but is only present during the time that fuel burns. After fuel burnout, α is no longer present. Incorporating these definitions, the rocket equation (11.5.1), therefore, becomes three time-dependent coupled differential equations for $x(t)$, $y(t)$, and $m(t)$

$$\frac{dv_x}{dt} = \frac{\alpha\Theta_t u_x}{m(t)} - \frac{GMx}{(x^2 + y^2)^{3/2}}, \frac{dv_y}{dt} = \frac{\alpha\Theta_t u_y}{m(t)} - \frac{GMy}{(x^2 + y^2)^{3/2}}, \frac{dm}{dt} = -\alpha\Theta_t, \tag{11.5.2}$$

with the usual understanding that $v_{x,y} = d(x,y)/dt$, and where we have included a step function with the following properties

$$\Theta_t \equiv \Theta(t - t_{f\max}) = \begin{cases} 1 & 0 < t < t_{f\max} \\ 0 & t_{f\max} \le t < \infty \end{cases}, \tag{11.5.3}$$

in order to cut off the mass burn rate as soon as the propellant is consumed. The rocket's thrust is active up to the time $t_{f\max}$, because this is the fuel burnout time. We can find this time as follows:

$$-\frac{dm}{dt} = \alpha \Theta_t \quad \Rightarrow \quad -\int_{m_i}^{m_f} dm = \alpha \int_0^{t_{f\max}} dt \quad \Rightarrow \quad t_{f\max} = \frac{m_i - m_f}{\alpha} = \frac{m_p}{\alpha},$$

$$\tag{11.5.4}$$

where the initial rocket mass, payload, and fuel is m_i; the final rocket mass after fuel burnout is m_f, and the total propellant mass is m_p. If the rocket's engine thrust value is provided, and the speed, u, at which the engine is capable of ejecting the gas is known, the mass burn rate can be found from

$$thrust = \alpha u \quad \Rightarrow \quad \alpha = \frac{thrust}{u} = \frac{thrust}{\sqrt{u_x^2 + u_y^2}}. \tag{11.5.5}$$

In order to carry out the simulation, it is convenient to use appropriate units, in a similar way to our previous simulations of Chapters 9 and 10. We use the following units: mass, m_0; length, R; time, τ; speed $v_0 = R/\tau$; and force, $F_0 = m_0 R/\tau^2$. Thus, for example, in the first of (11.5.2) we can write $v_x = v_0 \overline{v}_x$, $t = \overline{t}\tau$, $x = \overline{x}\tau$, $y = \overline{y}R$, $u_x = v_0 \overline{u}\cos\theta$, $\alpha = \overline{\alpha}m_0/\tau$, $m = \overline{m}m_0$, and $M = \overline{M}m_0$, with all the barred quantities being dimensionless, to get from (11.5.2)

$$\frac{R\,d\overline{v}_x}{\tau^2\,dt} = \frac{m_0 R}{m_0 \tau^2}\frac{\overline{\alpha}\Theta_t \overline{u}\cos\theta}{\overline{M}(t)} - \frac{Rm_0 G\overline{M}\,\overline{x}}{R^3(\overline{x}^2 + \overline{y}^2)^{3/2}} \Rightarrow \frac{d\overline{v}_x}{dt} = \frac{\overline{\alpha}\Theta_t \overline{u}\cos\theta}{\overline{M}(t)} - \left(\frac{Gm_0\tau^2}{R^3}\right)\frac{\overline{M}\,\overline{x}}{(\overline{x}^2 + \overline{y}^2)^{3/2}}.$$

$$\tag{11.5.6}$$

This expression actually gives the value of our time units, that is

$$\frac{Gm_0\tau^2}{R^3} \equiv 1 \quad \Rightarrow \quad \tau = \sqrt{\frac{R^3}{Gm_0}}, \tag{11.5.7}$$

so that (11.5.6) can be simplified. A similar process produces the rest of (11.5.2) in dimensionless units, and all together they are

$$\frac{d\bar{v}_x}{dt} = \frac{\bar{\alpha}\Theta_t\bar{u}\cos\theta}{\bar{m}(t)} - \frac{\overline{M}\,\bar{x}}{(\bar{x}^2 + \bar{y}^2)^{3/2}}, \quad \frac{d\bar{v}_y}{dt} = \frac{\bar{\alpha}\Theta_t\bar{u}\sin\theta}{\bar{m}(t)} - \frac{\overline{M}\,\bar{y}}{(\bar{x}^2 + \bar{y}^2)^{3/2}}, \quad \frac{d\bar{m}}{dt} = -\bar{\alpha}\Theta_t.$$

(11.5.8)

In these equations, θ represents the angle at which the gas is ejected, and corresponds to the launch angle in the present simulation. The fuel consumption rate is from (11.5.5) where $thrust = \overline{thrust}\,F_0$, to get

$$\alpha = \bar{\alpha}\frac{m_0}{\tau} = \frac{\overline{thrust}\quad F_0}{\sqrt{\bar{u}_x^2 + \bar{u}_y^2}\,v_0} \quad \Rightarrow \quad \bar{\alpha} = \frac{\overline{thrust}}{\sqrt{\bar{u}_x^2 + \bar{u}_y^2}}.$$

(11.5.9a)

The fuel burnout time from (11.5.4) is similarly written in these units as

$$\bar{t}_{f\max} = \frac{\bar{m}_i - \bar{m}_f}{\bar{\alpha}} = \frac{\bar{m}_p}{\bar{\alpha}}.$$

(11.5.9b)

We are now ready to perform the simulation. We will assume that the massive body is Earth, $M = 5.98 \times 10^{24}$ kg, $R = 6.37 \times 10^6$ m, which will be used as the units of mass and distance. The unit of time from (11.5.7) is

$$\tau = \sqrt{\frac{(6.37 \times 10^6 \text{ m})^3}{6.673 \times 10^{-11} \text{ Nm}^2/\text{kg}^2 \cdot 5.98 \times 10^{24} \text{ kg}}} \approx 805 \text{ s}.$$

(11.5.10)

We use the following MATLAB script rocket.m to perform the simulation. The script begins with a total initial rocket mass (fuel and payload) of $m_i = 2.8 \times 10^6$ kg. The propellant fuel mass is a high fractional part of this mass, as seen within the script. Varying this amount affects the burnout time (11.5.9b). The thrust used is a few times larger than the initial rocket weight at the surface of the Earth in order to achieve liftoff as in (11.4.7). The script makes use of arrays to store the variables: $w[1, 2, 3, 4, 5] = \{\bar{x}, \bar{v}_x, \bar{y}, \bar{v}_y, \overline{M}\}$. The three differential equations (11.5.8) become five differential equations

$$\frac{dw[1]}{dt} = w[2], \quad \frac{dw[2]}{dt} = \frac{\bar{\alpha}\Theta_t\bar{u}\cos\theta}{w[5]} - \frac{\overline{M}\,w[1]}{(w[1]^2 + w[3]^2)^{3/2}}$$

$$\frac{dw[3]}{dt} = w[4], \quad \frac{dw[4]}{dt} = \frac{\bar{\alpha}\Theta_t\bar{u}\cos\theta}{w[5]} - \frac{\overline{M}\,w[3]}{(w[1]^2 + w[3]^2)^{3/2}}$$

(11.5.11)

$$\frac{dw[5]}{dt} = -\bar{\alpha}\Theta_t;$$

the extra two differential equations have to be added because $d(x, y)/dt = v_{x, y}$ and the numerical solution is carried out using the Runge–Kutta method. As in previous simulations of this sort, the derivatives are declared in their own separate m-file, here named rocket_der.m, which also follows. For the simulation to work well, we need a numerical step function that will perform the prescription from (11.5.3). To this end, we use the following function

$$\Theta(t - t_{f\max}) \approx \frac{1}{1 + e^{100(t - t_{f\max})}}, \tag{11.5.12}$$

whose denominator is very close to unity when $t < t_{f\max}$, but is very large when $t > t_{f\max}$, so that it mimics the desired step function needed for the calculation. The script stepf.m, also included in the following listing , does this. If needed, the script can create the inverse step as well. Figure 11.8 contains the results of the simulation. The real-time motion of the rocket is best seen while running the script from within MATLAB.

The initial values used for the coordinates are as follows: $[x, y, v_x, v_y, m] = [0, 1, 0, 0, m_i]$; that is, the rocket is at rest on its launch pad at the surface of Earth (the circle in Figure 11.8a) at the North Pole with the previously mentioned initial mass that includes the fuel mass and the payload. The rocket's trajectory is shown in Figure 11.8(a) where the launching angle used is 55° and the thrust is given on Figure 11.8(a) with a value of 4.13×10^7 N, which is 1.5 times its rest weight, as seen in the script rocket.m. The exhaust velocity was chosen as the fraction $u = 0.35 v_0 = 0.35 R/\tau \approx 2770$ m/s. The range of the projectile is calculated by the absolute magnitude of the difference in the latitude angles between takeoff and landing, $\theta = |\tan^{-1}(y_{take-off}/x_{take-off}) - \tan^{-1}(y_{landing}/x_{landing})|$, so that the arc length gives $Range = R\theta$. For the conditions of Figure 11.8(a), the range is 8.52×10^6 m. The upper section of Figure 11.8(b) shows the magnitude of the rocket's velocity, $v = \sqrt{v_x^2 + v_y^2}$, as well as the v_x and v_y velocity components as a function of time. The lower section of the figure shows the magnitude of the position, $r = \sqrt{x^2 + y^2}$ (from the center of Earth), as well as the x and y components as a function of time. The rocket's thrust is active for a short time up to $t_{f\max} = (0.224)(805 \text{ s}) \approx 180$ s until the fuel runs out. It then begins to slow down, as shown by the speed curve bending in the upper section of Figure 11.8(b), as it continues to climb under the action of Earth's gravity until at about 2.4τ when it reaches its maximum distance (lower section of the figure) of about $1.7R$ from the center of Earth, where the velocity magnitude becomes small again. At the maximum height, v_y becomes zero. It regains speed once more, and v_y becomes negative, as the rocket free-falls to the ground. The trajectory plot stops as soon as the rocket reaches Earth's radius (ground). The rocket's round trip time is shown to be 4.4τ, or a little more than 3500 s. The details of the motion are affected by the rocket's fuel mass, the launching

(a)   Rocket Simulation: $\tau = 805$ s, $m_i = 2.8E + 006$ kg, $m_p = 2.69E + 006$ kg

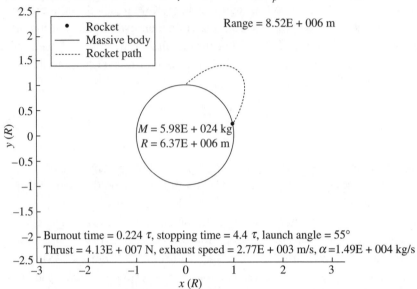

(b)

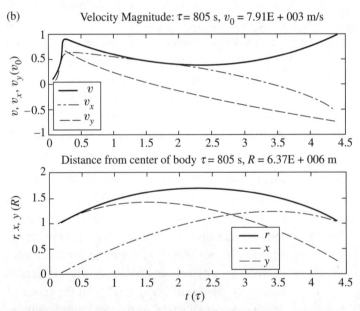

FIGURE 11.8   (a) Snapshot of the real-time simulation of a rocket takeoff from Earth, and (b) position and velocity components versus time.

angle, the thrust, and the engine exhaust speed. The script allows the user to vary any of the parameters as wished. The initial position as speed of the rocket can also be changed so as to simulate a different set of circumstances other than takeoff.

SCRIPT

```
%rocket.m
%program to solve the variable mass rocket equation in the
%presence of a gravitational force from a massive body
clear;
G=6.67e-11;              %universal gravitational constant (Nm^2/kg^2)
R=6.37e6;               %unit of distance (m) - massive body radius
m0=5.98e24;             %unit of mass (kg) - could be the massive body
M=5.98e24/m0;           %massive body mass in units of m0
tau=sqrt(R^3/(G*m0));   %unit of time in seconds
v0=R/tau; F0=m0*R/tau^2; %unit of speed and unit of force
mi=2.8e6/m0;            %initial payload+fuel mass in units of m0
ff=0.96; mp=mi*ff;      %fuel fraction, and fuel mass
Thrust=1.5*(G*mi*M/R^2)*m0^2; %Let the Thrust be # times initial mass weight
Thrust=Thrust/F0;       %Thrust in units of force
u=0.35;                 %gas exhaust velocity in units of v0
an=55; th=an*2*pi/360;      %angle of burn determines launch angle
ux=u*cos(th); uy=u*sin(th); %exhaust velocity components in units of v0
alpha=(Thrust/u);       %alpha in units of m0/tau
mf=mi-mp;               %final mass is payload mass (after fuel burnout)
tfmax=(mi-mf)/alpha;        %fuel burnout time in units of tau
tmax=50*tfmax;          %simulation run time
x0=0;y0=1;vx0=0;vy0=0;      %initial positions, speeds
ic1=[x0;vx0;y0;vy0;mi];     %initial conditions: position, velocity, rocket mass
%Use MATLAB's Runge-Kutta (4, 5) formula (uncomment/comment as needed)
%opt=odeset('AbsTol', 1.e-8, 'RelTol', 1.e-5);%user set Tolerances
%[t, w]=ode45('rocket_der', [0.0, tmax], ic1, opt, alpha, ux, uy, M, tfmax);%with set
tolerance
[t, w]=ode45('rocket_der', [0.0, tmax], ic1, [], alpha, ux, uy, M, tfmax);%default
tolerance
L=2.5*sqrt(x0^2+y0^2);      %window size
h=[0:0.025:2*pi];
x=cos(h);y=sin(h);          %massive body perimeter
%plot(x, y, 'g'); hold on    %massive body plot if needed here
%plot(w(:, 1), w(:, 3))      %use this to plot all the x, y points calculated
n=length(t);            %size of the time array
for i=1:n               %Loop to pick the points that lie above ground
 if sqrt(w(i, 1)^2+w(i, 3)^2) >= 0.99 %ground is 1.0 R, include points slightly below
   nn=i;
   t1(i)=t(i);
   x1(i)=w(i, 1); y1(i)=w(i, 3);
   vx1(i)=w(i, 2); vy1(i)=w(i, 4);
 else
```

```
   break;
 end
end
%==============================%plot the magnitude of the velocity vs time
v1=sqrt(vx1.^2+vy1.^2);
r1=sqrt(x1.^2+y1.^2);
subplot(2, 1, 1)
plot(t1, v1, 'k'), hold on
plot(t1, vx1, 'b-.'), plot(t1, vy1, 'r-')
ylabel('v, v_x, v_y (v_0)', 'FontSize', 14)
str=cat(2, 'Velocity Magnitude:', ' \tau=', num2str(tau, 3), ...
    ' s, v_0=', num2str(v0, 3), ' m/s');
title(str, 'FontSize', 12)
h=legend('v', 'v_x', 'v_y', 3); set(h, 'FontSize', 12)
subplot(2, 1, 2)
plot(t1, r1, 'k'), hold on
plot(t1, x1, 'b-.'), plot(t1, y1, 'r-')
xlabel('t (\tau)', 'FontSize', 14);ylabel('r, x, y (R)', 'FontSize', 14)
str=cat(2, 'Distance from center of body:', ' \tau=', num2str(tau, 3), ...
    ' s, R=', num2str(R, 3), ' m');
title(str, 'FontSize', 12)
h=legend('r', 'x', 'y', 0); set(h, 'FontSize', 12)
%==============================%simulate the points that lie above ground
figure
for i=1:nn
 clf;
 hold on
 plot(x1(i), y1(i), 'k.')    %rocket position
 plot(x, y, 'b');          %draw the massive body
 axis ([-L L -L L])        %windows size
 axis equal          %square window
 pause(0.0125)
end
plot(x1, y1, 'r:')         %trace the rocket path
xlabel('x (R)', 'FontSize', 14);ylabel('y (R)', 'FontSize', 14)
h=legend('Rocket', 'Massive Body', ' Rocket path', 2); set(h, 'FontSize', 14)
str=cat(2, 'Rocket Simulation:', ' \tau=', num2str(tau, 3), ...
    ' s, m_i=', num2str(mi*m0, 3), ' kg, m_p=', num2str(mp*m0, 3), ' kg');
title(str, 'FontSize', 12)
str2=cat(2, 'burnout time=', num2str(tfmax, 3), ' \tau, stopping time=', ...
    num2str(t1(nn), 3), ' \tau', ', launch angle=', num2str(an, 3), '^o');
str3=cat(2, 'Thrust=', num2str(Thrust*F0, 3), ' N, exhaust speed=', ...
    num2str(u*v0, 3), ' m/s, \alpha=', num2str(alpha*m0/tau, 3), ' kg/s');
str4=cat(2, 'M=', num2str(M*m0, 3), ' kg');
str5=cat(2, ' R=', num2str(R, 3), ' m');
text(-L*(1+0.2), -L*(1-0.2), str2, 'FontSize', 10)
text(-L*(1+0.2), -L*(1-0.1), str3, 'FontSize', 10)
text(-0.75, 0, str4, 'FontSize', 10)
```

```
text(-0.75, -0.25, str5, 'FontSize', 10)
warning off;
Range=R*abs(atan(y1(nn)/x1(nn))-atan(y0/x0));
str6=cat(2, ' Range=', num2str(Range, 3), ' m');
text(L*(1-0.5), L*(1-0.1), str6, 'FontSize', 10)
```

FUNCTION

```
% rocket_der.m: returns the derivatives for the rocket equations
function derivs = rocket_der( t, w, flag, alpha, ux, uy, M, tfmax)
% w(1):x, w(2):vx, w(3):y, w(4):vy, w(5):m
% fuel runs out after tfmax, so use a step function to stop burn rate alpha.
wr=sqrt(w(1).^2+w(3).^2);
tmp=alpha*stepf(t, tfmax, 1);
derivs = [w(2);ux*tmp./w(5)-M*w(1)./wr.^3;w(4);...
        uy*tmp./w(5)-M*w(3)./wr.^3;-tmp;];
```

SCRIPT

```
%stepf.m
function step=stepf(x, s, ss)
% This is a step function stepf(x, s, ss)
% x is the input array, s the value where the step occurs
% if ss=1 the step occurs at 's' from 1 to 0 for all x
% if ss=2 the step occurs at 's' from 0 to 1 for all x
% if ss is neither 1 nor 2 the function returns the value of zero
if ss==1
step=1./(1+exp(100*(x-s)));
elseif ss==2
    step=1-1./(1+exp(100*(x-s)));
  else
    step=0.0;
end
```

■ 11.6   Center of Mass Frame Revisited

The idea of using the center of mass–relative coordinate concept as an alternative way to consider problems involving interacting systems has been discussed before. In Chapter 5, Section 13, the center of mass concept was introduced in connection to the motion of a system of masses interacting through a spring force in the absence of any walls. Similarly, in Chapter 9, Section 5, a binary mass system made use of the center of mass and relative coordinates in order to simulate the motion of the pair of masses interacting through a gravitational force. Here we revisit the concept for a two-particle system interacting through a general internal force $\mathbf{f}$ and under general external force $\mathbf{F}$. Recalling Chapter 9, Equation (9.5.4), the center of mass ($\mathbf{R}_{cm}$) and relative ($\mathbf{r}$) coordinates for two particles of masses m_1 and m_2 with $m = m_1 + m_2$ and $\mu = 1/(m_1^{-1} + m_2^{-1}) = m_1 m_2/m$, as shown in Figure 11.9, are given by

$$\mathbf{R}_{cm} \equiv \frac{m_1\mathbf{r}_1 + m_2\mathbf{r}_2}{m} \text{ and } \mathbf{r} \equiv \mathbf{r}_{21} = -\mathbf{r}_{12} = \mathbf{r}_2 - \mathbf{r}_1, \tag{11.6.1}$$

where the mass positions are related to these coordinates by

$$\mathbf{r}_1 = \mathbf{R}_{cm} - \frac{\mu}{m_1}\mathbf{r} \quad \text{and} \quad \mathbf{r}_2 = \mathbf{R}_{cm} + \frac{\mu}{m_2}\mathbf{r}. \tag{11.6.2}$$

We next apply Newton's second law to (11.6.2) to get

$$m_1\ddot{\mathbf{r}}_1 = m_1\ddot{\mathbf{R}}_{cm} - \mu\ddot{\mathbf{r}} \quad \text{and} \quad m_2\ddot{\mathbf{r}}_2 = m_2\ddot{\mathbf{R}}_{cm} + \mu\ddot{\mathbf{r}}. \tag{11.6.3}$$

On the right of each expression, the first term describes the effect of the external force on each particle, and the second term describes the effect of the internal force between particles 1 and 2. If we let $\mathbf{F}_i \equiv m_i\ddot{R}_{cm}$ be the external force on the ith particle, and $\mathbf{F}_{21} \equiv \mu\ddot{\mathbf{r}}$ be the internal force on particle 2 due to particle 1, with the internal force on particle 1 due to particle 2 given by $\mathbf{F}_{12} = -\mathbf{F}_{21}$, then the total force on the ith particle is

$$m_1\ddot{\mathbf{r}}_1 = \mathbf{F}_1^T = \mathbf{F}_1 + \mathbf{F}_{12}, \text{ and } m_2\ddot{\mathbf{r}}_2 = \mathbf{F}_2^T = \mathbf{F}_2 + \mathbf{F}_{21}, \tag{11.6.4}$$

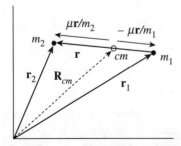

I FIGURE 11.9   Center of mass and relative coordinates associated with a two-particle system.

which is similar to (11.3.6) but for two particles; additionally, in the present case we know what $\mathbf{F}_{ij}$ is. If we add the two forces we get

$$\mathbf{F} = m_1\mathbf{r}_1 + m_2\mathbf{r}_2 = \mathbf{F}_1 + \mathbf{F}_{12} + \mathbf{F}_2 + \mathbf{F}_{21} = \mathbf{F}_1 + \mathbf{F}_{12} + \mathbf{F}_2 - \mathbf{F}_{12} = \mathbf{F}_1 + \mathbf{F}_2 = m\ddot{\mathbf{R}}_{cm},$$
$$(11.6.5)$$

consistent with (11.3.7), as expected; that is, the center of mass motion is determined by the external force alone. If we now multiply the first of (11.6.4) by m_2 and subtract that from the result of multiplying the second of (11.6.4) by m_1, we get

$$m_1 m_2(\ddot{\mathbf{r}}_2 - \ddot{\mathbf{r}}_1) = m_1 m_2\ddot{\mathbf{R}}_{cm} - m_2 m_1\ddot{\mathbf{R}}_{cm} + m_1\mathbf{F}_{21} - m_2\mathbf{F}_{12} = (m_1 + m_2)\mathbf{F}_{21}, \qquad (11.6.6a)$$

where we have used $\mathbf{F}_i = m_i\ddot{R}_{cm}$ and $\mathbf{F}_{12} = -\mathbf{F}_{21}$. This result can also be written as

$$\mathbf{F}_{21} \equiv \mathbf{f} = \mu\ddot{\mathbf{r}}; \qquad (11.6.6b)$$

that is, the relative motion between two bodies is like a one-body problem whose mass is that of the reduced mass. Here, $\mathbf{f}$ is the internal force between the two bodies. In summary, the motions of two interacting particles, m_1 and m_2 with coordinates $\mathbf{r}_1$ and $\mathbf{r}_2$, respectively, in the presence of an external force, can be thought of as two separate particles, m and μ, with coordinates $\mathbf{R}_{cm}$ and $\mathbf{r}$, respectively, one moving in the presence of the external force, $\mathbf{F}$, and the other moving in the presence of the internal force, $\mathbf{f}$, respectively.

EXAMPLE 11.7

Chapter 5, Section 15 introduced a one-dimensional model of a two-atom molecule. In that model, the atoms interact through a hypothetical force of the form, $F(x) = 3A/x^4 - 2B/x^3$. Here, x is the separation (oriented perpendicularly to the force due to gravity) between the two atoms as a function of time, $A = u_0 a_0^3$, $B = u_0 a_0^2$, with u_0 a unit of molecular energy, and a_0 a unit of molecular distance. Describe the motion of the molecule as it free falls from a certain height on Earth's surface, and implement the preceding model in order to approximate the motion of each atom as a function of time. Assume that each atom has equal mass, as opposed to previous assumptions where one atom was assumed to be very massive compared to the other.

Solution

The molecule's center of mass obeys Equation (11.6.5), where the external force is $\mathbf{F} = -mg\hat{j} = m\ddot{\mathbf{R}}$. This external force will only affect the displacement of the molecule in the y direction, and in that case the center of mass coordinate is described by the kinematics equations of motion for free fall. Thus we write $\mathbf{R}_{cm} = y_{cm}\hat{j}$, where $y_{cm} = y_0 + v_{0y}t - gt^2/2$. Following (11.6.2), we let x be the relative displacement between the atoms, and because they have the same mass $m_1 = m = m_2$, $\mu = m/2$, we write the coordinates of each atom as $\mathbf{r}_1 = y_{cm}\hat{j} - x\hat{i}/2$, and $\mathbf{r}_2 = y_{cm}\hat{j} + x\hat{i}/2$. The equation of motion for the reduced mass (11.6.6) becomes

$$f = 3A/x^4 - 2B/x^3 = \mu\ddot{x}. \tag{11.6.7}$$

This equation can be conveniently solved approximately using the method of successive approximations of Chapter 4, Sections 14 and 15. In fact, we already have an approximate solution for (11.6.7). It is given by Equation (4.15.14) as

$$x = x_b + A_1\cos(\omega_0 t) + \frac{a_2 A_1^2}{6\omega_0^2}(3 - \cos(2\omega_0 t)), \tag{11.6.8}$$

where the bond length is $x_b = 3A/2B = 3a_0/2$,

$$A_1 = (3\omega_0^2/2\,a_2)[-1 \pm \sqrt{1 + (4\,a_2 x_i/3\omega_0^2)}],$$

with x_i some initial position away from equilibrium, $\omega_0 = \sqrt{k/\mu}$ the vibration frequency about the bond length, where $k \equiv 32\,u_0/81\,a_0^2$, $a_2 = c/2\mu$, and $c \equiv 512\,u_0/243\,a_0^3$. Notice that we have used μ instead of a single atom's mass here because both of the atoms move about a common center of mass, as opposed to one of the atoms moving about a very massive one. A plot of the motion can be carried out if we employ units similar to those previously used in Chapter 4. That is, we express distance as $x = \bar{x}a_0$, time as $t = \bar{t}\tau_0$, where $\tau_0 = 2\pi/\omega_0 = 2\pi\sqrt{81\mu a_0^2/32u_0}$. The preceding solution simplifies as in Equation (4.15.17),

$$\bar{x} = \bar{x}_b + \bar{A}_1\cos(2\pi\bar{t}) + \frac{4\,\bar{A}_1^2}{9}(3 - \cos(4\pi\bar{t}))$$

with $\bar{A}_1 = \frac{9}{16}\left(-1 + \sqrt{1 + \frac{32}{9}\bar{x}_i}\right)$. Similarly for the center of mass variable $y_{cm} = \bar{y}_{cm}a_0 = \bar{y}_0 a_0 + \bar{v}_{0y}\bar{t}(a_0\tau_0/\tau_0) - \bar{g}\bar{t}^2(a_0\tau_0^2/\tau_0^2)/2$, and if for convenience one picks magnitude values of $a_0 = 1$ and a magnitude value for

$u_0 = (4\pi^2)81\,\mu a_0^2/32$, then the magnitude of $\tau_0 = 1$ as well, and the dimensionless value of the acceleration due to gravity remains unchanged, $\bar{g} = g(\tau_0^2/a_0) = 9.8$. The rest of the parameters are shown in the MATLAB script molec_mu.m that follows, whose results are plotted in Figure 11.10. The figure shows the x coordinates of the atoms in the molecule as they vibrate about the bond length and the molecule free-falls. The molecule accelerates on the way down so that the molecule atoms' path appears stretched.

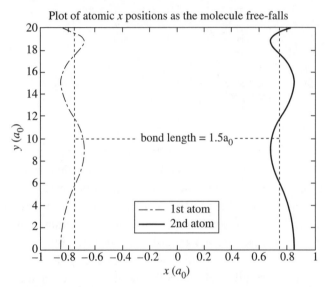

FIGURE 11.10 Example 11.7 for a freely falling vibrating molecule oriented perpendicularly to the force of gravity.

SCRIPT

```
%molec_mu.m
%program to plot the coordinates of the atoms of a molecule. The molecular
%potential model use is a simple 1/x^4-1/x^3 type. The method of successive
%approximations has been used for the relative coordinate. The center of
%mass is under the action due to gravity.
clear;
xcm=0.0;        %x center of mass position
v0y=0.0;y0=20.0;g=9.8;%initial y-velocity, and y-position, g=gravity
tmax=v0y/g+sqrt((v0y/g)^2+2*y0/g);%time to reach ground is used for tmax
xb=3/2;         %equilibrium relative coordinate (molecule bond length)
```

```
xi=0.2;      %initial position measured from equilibrium
t=[0:0.01:tmax];%time array
A1=9*(-1+sqrt(1+32*xi/9))/16;        %A1, at t=0 x-xb=xi
x=xb+A1*cos(2*pi*t)+4*A1^2*(3-cos(4*pi*t))/9; %relative coordinate
ycm=y0+v0y*t-0.5*g*t.^2;             %y-center of mass position
x1=xcm-0.5*x; x2=xcm+0.5*x;          %coordinates of the atoms
plot (x1, ycm, 'r-.')                %1st atom x, y positions
hold on
plot (x2, ycm, 'k')                  %2nd atom x, y positions
xlabel('x (a_0)', 'FontSize', 13), ylabel('y (a_0)', 'FontSize', 13)
title('Plot of atomic x positions as the molecule free falls', 'FontSize', 14)
line([-xb/2, -xb/2], [0, y0], 'color', 'blue', 'LineStyle', ':') %atom1 ave x
line([xb/2, xb/2], [0, y0], 'color', 'blue', 'LineStyle', ':')  %atom2 ave x
line([-xb/2, -xb/2*(1-0.6)], [y0/2, y0/2], 'LineStyle', ':')  %bond length
line([xb/2, xb/2*(1-0.6)], [y0/2, y0/2], 'LineStyle', ':')   %bond length
str=cat(2, ' bond length=', num2str(xb, 3), 'a_0');
text(-xb/2*(1-0.6), y0/2, str, 'FontSize', 14);
h=legend('1st atom', '2nd atom', 0); set(h, 'FontSize', 14)
```

■ 11.7 Collisions

In this section, although we do discuss impulse, we mainly deal with collisions of two particles in the absence of any external forces, in which case momentum conservation is possible. It is worth noting that it is possible to use momentum conservation, even in the presence of external forces, if the collision is brief; that is, the momentum transfer in the collision is much greater than the external impulse from the sum of the external forces. Collisions can be elastic or inelastic. In the event of elastic collisions, both momentum and energy are conserved. While inelastic collisions do conserve momentum, kinetic energy is not conserved due to friction as well as heating losses that affect the collision or result from the collision processes. Next we consider the collision of particles moving in two dimensions and look at the one-dimensional special case of head-on collisions. We also look at the collision process from the laboratory and center of mass frames of reference.

A. Impulse

Newton's second law states that

$$\mathbf{F} = \frac{d\mathbf{P}}{dt}.$$

(11.7.1)

The impulse over an infinitesimal time is defined by multiplying this force by dt. Furthermore, performing an average of the impulse, we get

$$\frac{1}{(t_f - t_i)} \int_{t_i}^{t_f} \mathbf{F} dt = \mathbf{F}_{ave} = \frac{1}{(t_f - t_i)_i} \int_{\mathbf{p}_i}^{\mathbf{p}_f} d\mathbf{P} = \frac{\Delta \mathbf{P}}{\Delta t}. \tag{11.7.2}$$

This equation expresses the fact that applying an average force on a body, over a time interval, results in the change in momentum of that body. It also expresses the fact that if the average force is zero, the momentum of the body remains constant or is conserved.

EXAMPLE 11.8

A 5 kg mass falls from a height of 3 m to the ground. If the magnitude of the force exerted by the floor as a function of time is described by Figure 11.11, assuming there are no other forces present and that the collision is elastic, what is the value of F_0?

Solution

The velocity of the mass just before it strikes the floor is obtained from energy conservation, $v_i = \sqrt{2 gh}$ or $\mathbf{v}_i = (-\sqrt{gh}\hat{j})$ m/s. If there are no other forces present, the rebound produces a final velocity of equal magnitude and opposite direction; that is, $\mathbf{v}_f = (\sqrt{gh}\hat{j})$ m/s. Letting t_f be the final time of 10×10^{-3} s, the impulse area in Figure 11.11 is

$$\int_0^{t_f} \mathbf{F} dt = 2(\mathbf{F}_0 t_f/2) = m\Delta \mathbf{v} = 2 m\sqrt{gh}\hat{j}$$

or with $\mathbf{F}_0 = F_0\hat{j}$, we can solve to get

$$F_0 = 2(5 \text{ kg})(\sqrt{9.8 \cdot 3}(\text{m/s})^2/(10 \times 10^{-3} \text{ s}) = 5422.18 \text{ N}.$$

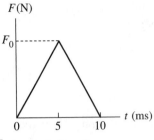

I FIGURE 11.11

B. Elastic Collisions in Two Dimensions—Laboratory Frame of Reference

Consider the collision between two masses, m_1, is taken to be initially moving with velocity $\mathbf{v}_{1i}$ along the $+x$-axis, and m_2 initially at rest. After the collision, m_1 moves with velocity $\mathbf{v}_{1f}$ at an angle θ_1 from the $+x$-axis, and m_2 moves with velocity $\mathbf{v}_{2f}$ at an angle θ_2 from the $+x$-axis (Figure 11.12). This kind of collision analysis is known as the laboratory frame of reference, which remains at rest throughout the collision process.

This picture treats the two-dimensional collision process as simply as possible. It actually is capable of including the case when m_2 is initially moving. The way it does so is that collisions are viewed from the reference frame moving along the initial velocity of m_2. In that frame, m_2 is initially at rest (see Problem 11.17), as we have assumed. According to the momentum conservation concept, the total momentum of the system remains constant before ($\sum \mathbf{p}_i$) and after ($\sum \mathbf{p}_f$); that is, $\sum \mathbf{p}_i = \sum \mathbf{p}_f$, where the sum is over the particles, or in component form

$$m_1 v_{1i} = m_1 v_{1f}\cos\theta_1 + m_2 v_{2f}\cos\theta_2, \qquad (11.7.3a)$$

Laboratory Frame of Reference

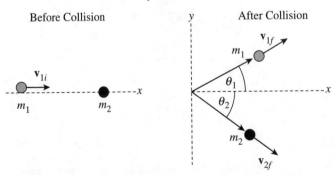

I FIGURE 11.12 Two-particle glancing angle collision in the laboratory frame of reference.

and

$$0 = m_1 v_{1f} \sin\theta_1 - m_2 v_{2f} \sin\theta_2. \tag{11.7.3b}$$

This corresponds to two equations with four unknowns, i.e., v_{1f}, v_{2f}, θ_1, and θ_2. Thus we need more information. This comes from the energy conservation equation for elastic collisions,

$$\frac{1}{2}m_1 v_{1i}^2 = \frac{1}{2}m_1 v_{1f}^2 + \frac{1}{2}m_2 v_{2f}^2, \tag{11.7.4}$$

which provides a way to obtain a third unknown. This means that an experiment would have to be conducted in order to measure the fourth unknown. We begin by obtaining solutions to the final velocities v_{1f} and v_{2f} in terms of one of the angles, say, θ_1. First we rearrange (11.7.3) as

$$v_{1i} - v_{1f}\cos\theta_1 = \frac{m_2}{m_1}v_{2f}\cos\theta_2 \quad \text{and} \quad m_1 v_{1f}\sin\theta_1 = m_2 v_{2f}\sin\theta_2, \tag{11.7.5}$$

which if we square each of these and add them together, we can eliminate the θ_2 variable, to get

$$v_{1i}^2 - 2v_{1i}v_{1f}\cos\theta_1 + v_{1f}^2 = \left(\frac{m_2}{m_1}\right)^2 v_{2f}^2. \tag{11.7.6}$$

From Equation (11.7.4) we can solve for v_{2f} in terms of v_{1f} to get

$$v_{2f}^2 = \frac{m_1}{m_2}(v_{1i}^2 - v_{1f}^2). \tag{11.7.7}$$

This can be substituted back into (11.7.6) to obtain a quadratic equation for v_{1f}

$$v_{1f}^2\left(1 + \frac{m_2}{m_1}\right) - 2v_{1i}v_{1f}\cos\theta_1 + v_{1i}^2\left(1 - \frac{m_2}{m_1}\right) = 0, \tag{11.7.8}$$

whose solution is

$$v_{1f} = \frac{m_1 v_{1i}}{m_1 + m_2}\cos\theta_1 \pm \sqrt{\left(\frac{m_1 v_{1i}}{m_1 + m_2}\cos\theta_1\right)^2 - \left(\frac{m_1 - m_2}{m_1 + m_2}\right)v_{1i}^2},$$

and can be rearranged as

$$v_{1f} = \frac{m_1 v_{1i}}{m_1 + m_2}\left[\cos\theta_1 \pm \sqrt{\cos^2\theta_1 - \frac{m_1^2 - m_2^2}{m_1^2}}\right] = \frac{m_1 v_{1i}}{m_1 + m_2}\left[\cos\theta_1 \pm \sqrt{\left(\frac{m_2}{m_1}\right)^2 - \sin^2\theta_1}\right],$$

(11.7.9)

where the '+' ('−') sign is to be taken for glancing angle (head-on) collisions. This in turn can be substituted into (11.7.7) to yield v_{1f}. Thus, (11.7.7 and 11.7.9) give v_{1f}, and v_{2f} as a function of θ_1. Finally, from (11.7.3) we can get two expressions for the remaining unknown, θ_2 as

$$\sin\theta_2 = \frac{m_1 v_{1f}}{m_2 v_{2f}}\sin\theta_1 \quad \text{and} \quad \tan\theta_2 = \frac{v_{1f}\sin\theta_1}{v_{1i} - v_{1f}\cos\theta_1}.$$

(11.7.10)

We should notice that in (11.7.9), the square root quantity must be real. This means that in two dimensions, we must have that

$$\cos^2\theta_1 \geq \frac{m_1^2 - m_2^2}{m_1^2} \quad \Rightarrow \quad \cos^2\theta_{1max} = 1 - \left(\frac{m_2}{m_1}\right)^2,$$

(11.7.11)

with the equality occurring when $\cos^2\theta$ has its minimum value; that is, θ_1 reaches its maximum value θ_{1max}, since $\cos\theta_1$ decreases as θ_1 increases, and so $\theta_1 \leq \theta_{1max}$. A plot of θ_{1max} versus the m_2/m_1 ratio is given in Figure 11.13 for when $m_2 < m_1$. The simple

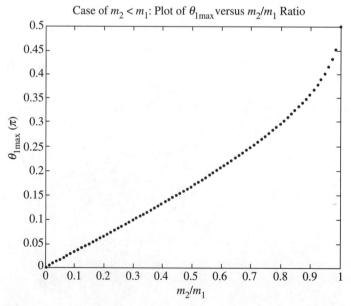

Case of $m_2 < m_1$: Plot of θ_{1max} versus m_2/m_1 Ratio

| FIGURE 11.13   A plot of θ_{1max} if Equation (11.7.11) versus the m_2/m_1 ratio.

code used in `theta_max.m` also follows. Finally, when $m_1 < m_2$, the RHS of the inequality in (11.7.11) will always be satisfied because $\cos^2\theta_1 - [1 - (m_2/m_1)^2] > 0$, and so $0 \le \theta_1 \le \pi$ in this case.

SCRIPT

```
%theta_max.m - plots the maximum scattering angle theta_1 versus the
%m2/m1 ratio
clear;
x=[0.0:0.0125:1];
y=acos(sqrt(1-x.^2))/pi;
plot(x, y, 'b.')
xlabel('m_2/m_1', 'FontSize', 13)
ylabel('\theta_{1max} (\pi)', 'FontSize', 13)
str=cat(2, 'case of m_2 < m_1: Plot of \theta_{1max} versus m_2/m_1 ratio');
title(str, 'FontSize', 13)
```

EXAMPLE 11.9

Obtain the final velocity expressions, $\mathbf{v}_{1f}$ and $\mathbf{v}_{2f}$, of two particles of masses m_1 and m_2 after undergoing a head-on elastic collision (a) if m_1 is initially moving with velocity $\mathbf{v}_{1i}$ and m_2 is initially at rest, and (b) assuming both particles are initially moving with velocities $\mathbf{v}_{1i}$ and $\mathbf{v}_{2i}$, respectively. (c) Using the result from (b), what can we say about the two particles' relative velocities before and after the collision?

Solution

(a) Because this is a one-dimensional problem, the momentum and energy conservation equations are given by (11.7.3 and 11.7.4) with $\theta_1 = \theta_2 = 0$. We can, therefore, take the limit of the negative root of (11.7.9) as $\theta_1 \to 0$ to get

$$v_{1f} \to \frac{m_1 v_{1i}}{m_1 + m_2}\left[1 - \left(\frac{m_2}{m_1}\right)\right] = \left(\frac{m_1 - m_2}{m_1 + m_2}\right)v_{1i}, \tag{11.7.12}$$

for the first particle. Substituting this expressions into (11.7.7), get

$$v_{2f}^2 = \frac{m_1}{m_2}(v_{1i}^2 - v_{1f}^2) = \frac{m_1}{m_2}\left[v_{1i}^2 - \left(\frac{m_1 - m_2}{m_1 + m_2}v_{1i}\right)^2\right]$$

$$= \frac{m_1}{m_2(m_1 + m_2)^2}v_{1i}^2[(m_1 + m_2)^2 - (m_1 - m_2)^2],$$

which simplifies to

$$v_{2f} = \frac{2\,m_1 v_{1i}}{m_1 + m_2}.$$ (11.7.13)

(b) As discussed in connection with Figure 11.12 as well as the more general Problem (11.17), the results (11.7.12 and 11.7.13) can be thought of having been obtained in a frame of reference in which m_2 is at rest. This means that, if in those expressions, we make the replacements $v_{1i} \rightarrow v_{1i} - v_{2i}$, $v_{1f} \rightarrow v_{1f} - v_{2i}$, and $v_{2f} \rightarrow v_{2f} - v_{2i}$, then (11.7.12 and 11.7.13) become

$$v_{1f} - v_{2i} = \left(\frac{m_1 - m_2}{m_1 + m_2}\right)(v_{1i} - v_{2i}) \text{ and } v_{2f} - v_{2i} = \frac{2\,m_1}{m_1 + m_2}(v_{1i} - v_{2i}),$$

which can be used to get simplified expressions for v_{1f} and v_{2f}, respectively,

$$v_{1f} = \left(\frac{m_1 - m_2}{m_1 + m_2}\right)v_{1i} + \frac{2\,m_2}{m_1 + m_2}v_{2i},$$

and

$$v_{2f} = \frac{2\,m_1}{m_1 + m_2}v_{1i} + \left(\frac{m_2 - m_1}{m_1 + m_2}\right)v_{2i}.$$ (11.7.14a)

(c) Regarding the relative velocities between the particles before and after, we can subtract the above two expressions to get

$$v_{2f} - v_{1f} = v_{1i}\left(\frac{2\,m_1}{m_1 + m_2} - \frac{m_1 - m_2}{m_1 + m_2}\right) + v_{2i}\left(\frac{m_2 - m_1}{m_1 + m_2} - \frac{2\,m_2}{m_1 + m_2}\right)$$

$$= v_{1i} - v_{2i};$$

that is, the particles' final and initial relative velocities are equal in magnitude and opposite in direction

$$v_{rf} = -v_{ri},$$ (11.7.14b)

where $v_{r(i,\,f)} \equiv v_{1(i,\,f)} - v_{2(i,\,f)}$ is the initial (v_{ri}) and final (v_{rf}) relative velocity between the particles.

EXAMPLE 11.10

(a) Obtain the final velocity expressions, $\mathbf{v}_{1f}$ and $\mathbf{v}_{2f}$, of two particles of equal mass after undergoing a glancing angle elastic collision if one of the particles is initially moving with velocity $\mathbf{v}_{1i}$ and the second particle is initially at rest.

(b) What is the relationship between each particle's scattering angle?

Solution

(a) Using the positive root of expression (11.7.9) with $m_1 = m_2$, we see that

$$v_{1f} \to v_{1i}\cos\theta_1. \tag{11.7.15}$$

Substituting this result into (11.7.7), we get for the final speed of the second particle

$$v_{2f}^2 \to \frac{m_1}{m_2}v_{1i}^2(1 - \cos^2\theta_1) \quad \Rightarrow \quad v_{2f} = v_{1i}\sin\theta_1. \tag{11.7.16}$$

(b) Using the first of (11.7.10), we have that

$$\sin\theta_2 \to \frac{v_{1i}\cos\theta_1}{v_{1i}\sin\theta_1}\sin\theta_1 \quad \Rightarrow \quad \sin\theta_2 = \cos\theta_1 = \sin\left(\frac{\pi}{2} - \theta_1\right), \tag{11.7.17a}$$

or $\theta_2 = \dfrac{\pi}{2} - \theta_1$, so that

$$\theta_1 + \theta_2 = \frac{\pi}{2}; \tag{11.7.17b}$$

that is, when the masses are equal, their scattering angles are such that the particles leave at 90° from each other.

EXAMPLE 11.11

(a) Here we refer to the $+x$ direction as east and the $+y$ as north. A 0.5 kg particle moving at 2 m/s east collides elastically with a second 0.75 kg particle initially at rest. If after the collision, the first particle is seen leaving at an angle of 35° measured north of east, what are the final velocities of each particle and the exit angle of the second particle?

(b) A 0.5 kg particle coming from the northwest, initially moving at 1.5 m/s toward the origin at an angle of $\theta_{10} = 40°$ above the $-x$-axis, collides with a second 1.5 kg particle coming from the southwest, initially moving at 0.75 m/s toward the origin at an angle of $\theta_{20} = 20°$ below the $-x$-axis. If after the collision, the first particle is seen leaving at an angle of 30° north of east, what are the final velocities of each particle and the exit angle of the second particle?

(c) Draw a vector diagram showing the collision process in (b).

(d) Perform a check to see that both energy and momentum are indeed conserved.

Solution

(a) From (11.7.9) $v_{1f} = \dfrac{0.5}{1.25}2\left(\cos 35° + \sqrt{\left(\dfrac{0.75}{0.5}\right)^2 - \sin^2 35°}\right) = 1.764$ m/s,

and using this into (11.7.7), get $v_{2f} = \sqrt{\left(\dfrac{0.5}{0.75}\right)(2^2 - 1.764^2)} = 0.769$ m/s.

Finally, from the first of (11.7.10), we get
$\theta_2 = \sin^{-1}\left\{\left(\dfrac{0.5}{0.75}\right)\left(\dfrac{1.764}{0.769}\right)\sin 35°\right\} = 61.259°.$

(b) The vector diagram of the collision process has been carried out with the MATLAB script ecoll_2d.m, which uses the auxiliary function ecoll_2dfun.m; both are listed below. The files work with the nonlinear equation solver mmfsolve.m (see the text's website.) The results are shown in Figure 11.14. The energy momentum output check is in the file ecoll_2d.txt and also follows here, indicating energy and momentum conservation. As can be seen, the final velocities are $v_{1f} = 1.55$ m/s, $v_{2f} = 0.718$ m/s, and the final exit angle for the second particle is $\theta_2 = 26.7°$. The energy and momentum before and after work out nicely, as expected, giving credence to the numerical results output shown in the figure.

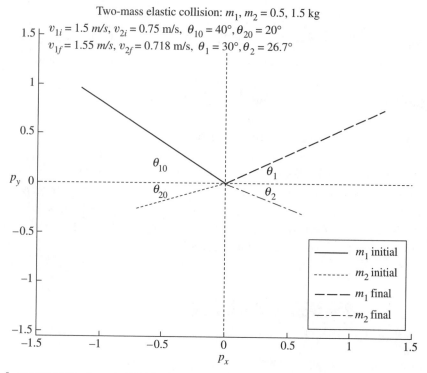

Two-mass elastic collision: $m_1, m_2 = 0.5, 1.5$ kg
$v_{1i} = 1.5$ m/s, $v_{2i} = 0.75$ m/s, $\theta_{10} = 40°, \theta_{20} = 20°$
$v_{1f} = 1.55$ m/s, $v_{2f} = 0.718$ m/s, $\theta_1 = 30°, \theta_2 = 26.7°$

FIGURE 11.14   Example 11.11 for the elastic collision of two particles entering and leaving at the angles shown in the laboratory frame.

The scripts solve the complete version of the system of Equations (11.7.3 and 11.7.4); that is,

$$m_1 v_{1ix} + m_2 v_{2ix} = m_1 v_{1f}\cos\theta_1 + m_2 v_{2f}\cos\theta_2,$$

$$-m_1 v_{1iy} + m_2 v_{2iy} = m_1 v_{1f}\sin\theta_1 - m_2 v_{2f}\sin\theta_2, \qquad (11.7.18)$$

$$\frac{1}{2}m_1 v_{1i}^2 + \frac{1}{2}m_2 v_{2i}^2 = \frac{1}{2}m_1 v_{if}^2 + \frac{1}{2}m_2 v_{2f}^2,$$

through a nonlinear equation solver (see comments within the script mmfsolve.m). The process needs good starting guesses for the unknowns, v_{1f}, v_{2f}, and θ_2. Different sets of parameters need different starting guesses; for lack of better guesses, the main script ecoll_2d.m presently uses $v_{1f} \approx v_{1i}/2$,

$$v_{2f} = \sqrt{\frac{m_1}{m_2}(v_{1i}^2 - v_{1f}^2)} \approx v_{1i}\sqrt{\frac{3m_1}{4m_2}},$$ and $\theta_2 \approx \frac{\pi}{2} - \theta_1$. The advantage of the scripts is that they can be modified, if needed, in order to solve for unknowns other than v_{1f}, v_{2f}, and θ_2 by redefining the vector function X within ecoll_2dfun.m, which is presently geared toward the unknowns: $X(1 \cdots 3) = \{v_{1f}, v_{2f}, \theta_2\}$. Should a different set of unknowns be desired, for example, masses or initial velocity, the script can be modified accordingly, with the main calling script's guesses for this vector modified as needed along with the proper input values for the known quantities. There is a caveat associated with the numerical work, and that is to keep a close eye on the energy and momentum conservation output check. If the system does not converge appropriately, it will be evident in the lack of conservation. The present system of Equations (11.7.18) can be solved analytically as well (see Problem 11.19).

OUTPUT

```
file ecoll_2d.txt - energy, momentum check results
>> ecoll_2d
Momentum, Energy Check: m1= 0.5000, m2= 1.5000
v1i= 1.5000, v2i= 0.7500, th10=40.0000, th20=20.0000
v1f= 1.5463, v2f= 0.7180, th1=30.0000, th2=26.7001
Ei= 0.9844, Ef= 0.9844, Pxi= 1.6317 Pxf= 1.6317 Pyi=-0.0973 Pyf=-0.0973
```

SCRIPT

```
%ecoll_2d.m - two-dimensional collision script to find the final velocities
%of two particles and their scattering angles.
%needed input is the initial velocities, their masses, and one of the
%scattered angles. This version was made from 'ecoll_2d0.m', but it's based
%on the use of the mmfsolve.m if the optimization toolbox in not available
clear;
global v1i v2i m1 m2 tp10 tp20 tp1
%th10=0.0;th20=0.0;th1=35;m1=0.5; m2=0.75; v1i=2; v2i=0.0;%example of m2 at rest
%th10=20;th20=10;th1=15;m1=0.5; m2=0.75; v1i=2; v2i=1.5;%masses, initial angles, speeds
th10=40;th20=20;th1=30;m1=0.5; m2=1.5; v1i=1.5; v2i=0.75;%masses, initial angles, speeds
tp10=th10*pi/180;tp20=th20*pi/180;tp1=th1*pi/180;   %convert angles to radians
v1fg=v1i/2;v2fg=v1i*sqrt(0.75*m1/m2);tp2g=pi/2-tp1; %guesses - change as needed
%opts =mmfsolve('Display', 'off');         %Turn off Display
```

```
opts =mmfsolve('FunTol', 1e-7, 'MaxIter', 100);        %alternate options
X0=[v1fg;v2fg;tp2g;];
X=mmfsolve(@ecoll_2dfun, X0, opts);
v1f=X(1);v2f=X(2);tp2=X(3);th2=tp2*360/2/pi;
%------------- plotting ------------------
line([-v1i*cos(tp10), 0], [v1i*sin(tp10), 0], 'Color', 'k')        %initial p m1
line([-v2i*cos(tp20), 0], [-v2i*sin(tp20), 0], 'Color', 'm', 'LineStyle', ':')%initial p m2
line([0, v1f*cos(tp1)], [0, v1f*sin(tp1)], 'Color', 'b', 'LineStyle', '-')  %scattered m1
line([0, v2f*cos(tp2)], [0, -v2f*sin(tp2)], 'Color', 'r', 'LineStyle', '-.') %scattered m2
h=legend('m_1 initial', 'm_2 initial', 'm_1 final', 'm_2 final', 4);
set(h, 'FontSize', 11)
va=max(v1i, v2i);vb=max(v1f, v2f);                 %pick higher values for axis
line([-va, vb], [0, 0], 'LineStyle', ':')          %shows x axis
line([0, 0], [-va, va], 'LineStyle', ':')          %shows y axis
axis([-va va -vb vb])
xlabel('p_x', 'FontSize', 14)
ylabel('p_y', 'FontSize', 14)
str0=cat(2, 'Two mass elastic collision: m1, m2 = ', num2str(m1, 3), ...
      ', ', num2str(m2, 3), ' kg');
str1=cat(2, 'v_{1i}=', num2str(v1i, 3), 'm/s, v_{2i}=', num2str(v2i, 3), ...
    'm/s, \theta_{10}=', num2str(th10, 3), '^o, \theta_{20}=', ...
    num2str(th20, 3), '^o');
str2=cat(2, 'v_{1f}=', num2str(v1f, 3), 'm/s, v_{2f}=', num2str(v2f, 3), ...
    'm/s, \theta_1=', num2str(th1, 3), '^o, \theta_2=', num2str(th2, 3), '^o');
title(str0, 'FontSize', 14);
text(-va*(1-.05), vb*(1-0.05), str1, 'FontSize', [11]);
text(-va*(1-.05), vb*(1-0.175), str2, 'FontSize', [11]);
text(-v1i*cos(tp10)*(1-0.5), v1i*sin(tp10)*(1-.85), '\theta_{10}', 'FontSize', [9])
if v2i~= 0,
text(-v2i*cos(tp20)*(1-0.5), -v2i*sin(tp20)*(1-0.85), '\theta_{20}', 'FontSize', [9])
end
text(v1f*cos(tp1)*(1-0.75), v1f*sin(tp1)*(1-0.92), '\theta_1', 'FontSize', [9])
text(v2f*cos(tp2)*(1-0.75), -v2f*sin(tp2)*(1-0.92), '\theta_2', 'FontSize', [9])
%------------- test results, Momentum, Energy ----------------
Ei=(m1*v1i^2+m2*v2i^2)/2;
Ef=(m1*v1f^2+m2*v2f^2)/2;
Pxi=m1*v1i*cos(tp10)+m2*v2i*cos(tp20);
Pxf=m1*v1f*cos(tp1)+m2*v2f*cos(tp2);
Pyi=-m1*v1i*sin(tp10)+m2*v2i*sin(tp20);
Pyf=m1*v1f*sin(tp1)-m2*v2f*sin(tp2);
fprintf('Momentum, Energy Check: m1=%7.4f, m2=%7.4f\n', m1, m2)
fprintf('v1i=%7.4f, v2i=%7.4f, th10=%7.4f, th20=%7.4f\n', v1i, v2i, th10, th20)
fprintf('v1f=%7.4f, v2f=%7.4f, th1=%7.4f, th2=%7.4f\n', v1f, v2f, th1, th2)
fprintf('Ei=%7.4f, Ef=%7.4f, Pxi=%7.4f Pxf=%7.4f Pyi=%7.4f Pyf=%7.4f\n', ...
    Ei, Ef, Pxi, Pxf, Pyi, Pyf);
```

FUNCTION

```
%ecoll_2dfun.m finds the zeros of the momentum, energy equations
%in the 2d elastic collision in order to find the unknowns.
%This version was made from 'ecoll_2dfun0.m' but is based
%on the use of the mmfsolve.m if the optimization toolbox in not available
function froot =ecoll_2dfun(X)%use this line with a "global" statement
global v1i v2i m1 m2 tp10 tp20 tp1
%X(1)=v1f, X(2)=v2f, X(3)=tp2
%energy equation: m1*v1i^2+m2*v2i^2-m1*v1f^2-m2*v2f^2=0
%x-momentum: m1*v1i*cos(tp10)+m2*v2i*cos(tp20)-m1*v1f*cos(tp1)-
m2*v2f*cos(tp2)=0
%y-momentum: -m1*v1i*sin(tp10)+m2*v2i*sin(tp20)-
m1*v1f*sin(tp1)+m2*v2f*sin(tp2)=0
froot = [m1.*v1i.^2+m2.*v2i.^2-m1.*X(1).^2-m2.*X(2).^2;...
  m1.*v1i.*cos(tp10)+m2.*v2i.*cos(tp20)-m1.*X(1).*cos(tp1)-
m2.*X(2).*cos(X(3));...
  -m1.*v1i.*sin(tp10)+m2.*v2i.*sin(tp20)-
m1.*X(1).*sin(tp1)+m2.*X(2).*sin(X(3));];
```

C. Center-of-Mass Frame of Reference

In the preceding subsection, we have studied collisions from the point of view of the laboratory reference (Lab) frame that is at rest throughout the collision process. Collisions take on a particularly useful symmetry and interesting form when viewed from the center-of-mass (CM) frame of reference. Recall from Equation (11.3.9) that the center-of-mass velocity of a system of particles, in the absence of external forces, is similar to the conservation of linear momentum; that is, it remains constant before and after collision. Referring to the preceding laboratory frame of reference subsection as well as to Figure 11.12, the CM velocity of the system of two colliding particles is given by

$$\mathbf{v}_{cm} = \frac{m_1\mathbf{v}_{1i} + m_2\mathbf{v}_{2i}}{m} = \frac{m_1\mathbf{v}_{1f} + m_2\mathbf{v}_{2f}}{m}, \tag{11.7.19}$$

where $m = m_1 + m_2$. Furthermore, following (11.6.1 and 11.6.2), we can write the time derivatives of the particles' positions or

$$\mathbf{v}_{1i} = \mathbf{v}_{cm} - \frac{\mu}{m_1}\mathbf{v}_i = \mathbf{v}_{cm} + \mathbf{v}'_{1i}, \text{ and } \mathbf{v}_{2i} = \mathbf{v}_{cm} + \frac{\mu}{m_2}\mathbf{v}_i = \mathbf{v}_{cm} + \mathbf{v}'_{2i}, \tag{11.7.20a}$$

for the initial velocities, and

$$\mathbf{v}_{1f} = \mathbf{v}_{cm} - \frac{\mu}{m_1}\mathbf{v}_f = \mathbf{v}_{cm} + \mathbf{v}'_{1f}, \text{ and } \mathbf{v}_{2f} = \mathbf{v}_{cm} + \frac{\mu}{m_2}\mathbf{v}_f = \mathbf{v}_{cm} + \mathbf{v}'_{2f}, \quad (11.7.20b)$$

for the final velocities. Here

$$\mathbf{v}_i \equiv \mathbf{v}_{2i} - \mathbf{v}_{1i}, \quad \mathbf{v}_f \equiv \mathbf{v}_{2f} - \mathbf{v}_{1f} \qquad (11.7.20c)$$

are the initial and final relative velocities between the two particles, μ is the reduced mass, and the primed velocities are the particle velocities with respect to the CM. Comparing the left side and right side of each of the Equations (11.7.20a and 11.7.20b), we see that

$$\mathbf{v}'_{1i} = -\frac{\mu}{m_1}\mathbf{v}_i, \quad \mathbf{v}'_{2i} = \frac{\mu}{m_2}\mathbf{v}_i, \quad \text{and} \quad \mathbf{v}'_{1f} = -\frac{\mu}{m_1}\mathbf{v}_f, \quad \mathbf{v}'_{2f} = \frac{\mu}{m_2}\mathbf{v}_f. \qquad (11.7.20d)$$

These equations express the relationship between the velocities in the CM and those in the Lab frame. The final particle velocities, as seen from the CM reference frame, are opposite in direction. This is illustrated in Figure 11.15(a) along with the particle Lab velocities $\mathbf{v}_{1f}$ and $\mathbf{v}_{2f}$. Also shown in the figure are the CM velocities, $\mathbf{v}'_{1f}$, $\mathbf{v}'_{2f}$

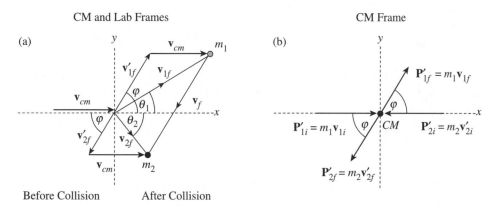

FIGURE 11.15   (a) Relation between the center of mass frame and the laboratory frame. The center of mass velocity is constant. (b) The final particle momenta, as seen from the CM reference frame, are opposite in direction.

and the Lab relative velocity $\mathbf{v}_f$. The angle φ is the scattering angle made after the collision, as seen in the CM frame of reference. Notice from (11.7.20a and 11.7.20b) that

$$m_1\mathbf{v}'_{1i} + m_2\mathbf{v}'_{2i} = 0 = m_1\mathbf{v}'_{1f} + m_2\mathbf{v}'_{2f}, \tag{11.7.21}$$

which indicates that the total momentum in the CM reference frame is zero. This is consistent with the result obtained before as expressed in Equation (11.3.13); the particles' relative momenta are equal and opposite, before and after collision. Also, since the center of mass velocity is constant, from (11.3.19) the total kinetic energies of the particles before and after are

$$T_i = \frac{1}{2}m_1v_{1i}^2 + \frac{1}{2}m_2v_{2i}^2 \qquad\qquad T_f = \frac{1}{2}m_1v_{1f}^2 + \frac{1}{2}m_2v_{1f}^2$$

$$= \frac{1}{2}mv_{cm}^2 + \frac{1}{2}m_1v'^2_{1i} + \frac{1}{2}m_2v'^2_{2i}, \qquad = \frac{1}{2}mv_{cm}^2 + \frac{1}{2}m_1v'^2_{1f} + \frac{1}{2}m_2v'^2_{2f},$$

$$\tag{11.7.22a}$$

which, for elastic collisions, are equal. Thus, we find that the CM kinetic energies are also equal

$$T'_i = \frac{1}{2}m_1v'^2_{1i} + \frac{1}{2}m_2v'^2_{2i} = T'_f = \frac{1}{2}m_1v'^2_{1f} + \frac{1}{2}m_2v'^2_{2f}. \tag{11.7.22b}$$

If in this expression we substitute in for the velocity values of (11.7.20d), we get $\mu^2v_i^2/m_1 + \mu^2v_i^2/m_2 = \mu^2v_f^2/m_1 + \mu^2v_f^2/m_2$, or,

$$v_i = v_f. \tag{11.7.23}$$

This says that the magnitude of the initial and final relative particles' velocities are equal. This is not surprising in view of a similar result in one dimension, as shown in Equation (11.7.15). If we let $\mathbf{v}_{1i}$ lie along the x direction, and take $\mathbf{v}_{2i} \equiv 0$, as done in the previous subsection, then the relative initial velocity and the initial CM frame velocities become

$$\mathbf{v}_i = \mathbf{v}_{2i} - \mathbf{v}_{1i} = -\mathbf{v}_{1i}, \ \mathbf{v}'_{1i} = -\frac{\mu}{m_1}\mathbf{v}_i = \frac{\mu}{m_1}\mathbf{v}_{1i}, \ \mathbf{v}'_{2i} = \frac{\mu}{m_2}\mathbf{v}_i = -\frac{\mu}{m_2}\mathbf{v}_{1i}; \tag{11.7.24}$$

therefore, the initial momentum of each particle lies in the x direction, while the final momentum of each particle lies at the angle φ from the x-axis, as illustrated in Figure 11.15(b).

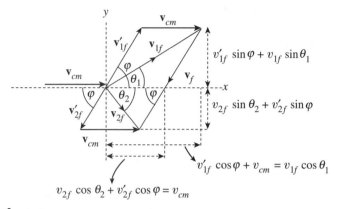

$v'_{1f} \sin\varphi + v_{1f} \sin\theta_1$

$v_{2f} \sin\theta_2 + v'_{2f} \sin\varphi$

$v'_{1f} \cos\varphi + v_{cm} = v_{1f} \cos\theta_1$

$v_{2f} \cos\theta_2 + v'_{2f} \cos\varphi = v_{cm}$

I FIGURE 11.16   Relationship of the CM frame scattering angle φ to θ_1 and θ_2 of the Lab frame.

It is of interest to relate the CM frame scattering angle φ to θ_1 and θ_2 of the Lab frame. To do this, we refer to Figure 11.16, from which we get that

$$\tan\theta_1 = \frac{v'_{1f}\sin\varphi}{v'_{1f}\cos\varphi + v_{cm}} = \frac{\dfrac{\mu}{m_1}v_f\sin\varphi}{\dfrac{\mu}{m_1}v_f\cos\varphi + v_{cm}}, \qquad (11.7.25a)$$

since only velocity magnitudes are involved. Further, using (11.7.19 and 11.7.24), with $\mathbf{v}_{2i} \equiv 0$, then the speeds $v_{cm} = m_1 v_{1i}/m$ and $v_i = v_{1i}$, so that

$$\tan\theta_1 = \frac{\dfrac{\mu}{m_1}v_f\sin\varphi}{\dfrac{\mu}{m_1}v_f\cos\varphi + \dfrac{m_1 v_{1i}}{m}} = \frac{\sin\varphi}{\cos\varphi + \dfrac{m_1 v_{1i}}{m_2 v_f}}. \qquad (11.7.25b)$$

Using (11.7.23) and (11.7.24) we see that for the final relative speed we can also write $v_f = v_i = v_{1i}$, and the above expression simplifies to

$$\tan\theta_1 = \frac{\dfrac{\mu}{m_1}v_f\sin\varphi}{\dfrac{\mu}{m_1}v_f\cos\varphi + \dfrac{m_1 v_{1i}}{m}} = \frac{\sin\varphi}{\cos\varphi + \dfrac{m_1}{m_2}}. \qquad (11.7.25c)$$

Similarly, from Figure 11.16 we have that

$$v_{2f}\sin\theta_2 = v'_{2f}\sin\varphi, \text{ and } v_{2f}\cos\theta_2 = v_{cm} - v'_{2f}\cos\varphi, \tag{11.7.26a}$$

and taking their ratio leads to

$$\tan\theta_2 = \frac{v'_{2f}\sin\varphi}{v_{cm} - v'_{2f}\cos\varphi} = \frac{\dfrac{\mu}{m_2}v_f\sin\varphi}{\dfrac{m_1 v_{1i}}{m} - \dfrac{\mu}{m_2}v_f\cos\varphi} = \frac{\sin\varphi}{\dfrac{v_{1i}}{v_f} - \cos\varphi} = \frac{\sin\varphi}{1 - \cos\varphi}. \tag{11.7.26b}$$

Using $1 - \cos\varphi = 2\cos^2\dfrac{\varphi}{2}$, and $\sin\varphi = 2\sin\dfrac{\varphi}{2}\cos\dfrac{\varphi}{2}$, we can simplify this to get

$$\tan\theta_2 = \frac{\cos\dfrac{\varphi}{2}}{\sin\dfrac{\varphi}{2}} = \frac{\sin\left(\dfrac{\pi}{2} - \dfrac{\varphi}{2}\right)}{\cos\left(\dfrac{\pi}{2} - \dfrac{\varphi}{2}\right)} = \tan\left(\dfrac{\pi}{2} - \dfrac{\varphi}{2}\right) \quad \Rightarrow \quad \theta_2 = \dfrac{\pi}{2} - \dfrac{\varphi}{2}. \tag{11.7.26c}$$

Notice from (11.7.25c) that in the event that the particle masses $m_1 = m_2$, then

$$\tan\theta_1 = \frac{\sin\varphi}{\cos\varphi + \dfrac{m_1}{m_2}} \rightarrow \frac{\sin\varphi}{\cos\varphi + 1} = \frac{\sin\varphi}{2\cos^2(\varphi/2)}$$
$$\tag{11.7.27}$$
$$= \frac{2\sin(\varphi/2)\cos(\varphi/2)}{2\cos^2(\varphi/2)} = \tan(\varphi/2) \quad \Rightarrow \quad \theta_1 \rightarrow \varphi/2,$$

which when combined with (11.7.26c) gives $\theta_1 + \theta_2 = \pi/2$, in agreement with our previous result (11.7.17b) of Example 11.10 when $m_1 = m_2$. Finally, Equation (11.7.25c) gives θ_1, if we know φ; it is possible to find an expression for φ if we know θ_1. Squaring (11.7.25c) and rearranging, we get

$$\left(\cos^2\varphi + 2\frac{m_1}{m_2}\cos\varphi + \left(\frac{m_1}{m_2}\right)^2\right)\tan^2\theta_1 = \sin^2\varphi = 1 - \cos^2\varphi.$$

Simplifying this and noticing that $1 + \tan^2\theta_1 = \sec^2\theta_1$, as well as $\tan^2\theta_1/\sec^2\theta_1 = \sin^2\theta_1$, this becomes a quadratic equation for $\cos\varphi$

$$0 = \cos^2\varphi + 2\frac{m_1}{m_2}\sin^2\theta_1\cos\varphi - \frac{(1 - (m_1/m_2)^2\tan^2\theta_1)}{\sec^2\theta_1}$$

$$= \cos^2\varphi + 2\frac{m_1}{m_2}\sin^2\theta_1\cos\varphi - (\cos^2\theta_1 - (m_1/m_2)^2\sin^2\theta_1), \qquad (11.7.28a)$$

with solution

$$\cos\varphi = -\frac{m_1}{m_2}\sin^2\theta_1 \pm \sqrt{\left(\frac{m_1}{m_2}\sin^2\theta_1\right)^2 + \cos^2\theta_1 - (m_1/m_2)^2\sin^2\theta_1}$$

$$= -\frac{m_1}{m_2}\sin^2\theta_1 \pm \cos\theta_1\sqrt{1 - (m_1/m_2)^2\sin^2\theta_1}. \qquad (11.7.28b)$$

If we choose the positive root, this expression gives, when $m_1 = m_2$

$$\cos\varphi = -\sin^2\theta_1 + \cos^2\theta_1 = 1 - 2\sin^2\theta_1 = \cos(2\theta_1) \quad \Rightarrow \quad \varphi = 2\theta_1,$$
$$(11.7.28c)$$

in agreement with (11.7.27).

EXAMPLE 11.12

From the information given in Example 11.11(a), use the preceding CM frame concepts to obtain the final velocities of each particle as well as the exit angle of the second particle.

Solution

In Example 11.11(a), the particle masses are $m_1 = 0.5$ kg, $m_2 = 0.75$ kg, $v_{1i} = 2\hat{i}$ m/s, $v_{2i} = 0$, and the exit angle of the first particle is 35 degrees north of rast. Here, $m = m_1 + m_2 = (5/4)$ kg, $m_1/m_2 = 2/3$, $\mu/m_1 = m_2/m = (3/5)$, $\mu/m_2 = m_1/m = (2/5)$. From (11.7.19) we have that the center of mass velocity is $v_{cm} = m_1 v_{1i}/m = (2/5)(2\hat{i}) = 0.8\hat{i}$ m/s and from (11.7.20), the initial relative velocity is $v_i = v_{2i} - v_{1i} = -v_{1i} = -2\hat{i}$ m/s. From (11.7.23), we can find the magnitude $v_f = v_i = 2$ m/s, and from Figure 11.16, we can see that we must have

$\mathbf{v}_f = -v_f\cos\varphi\hat{i} - v_f\sin\varphi\hat{j} = (-2\cos\varphi\hat{i} - 2\sin\varphi\hat{j})$ m/s. The CM frame final angle φ can be obtained from (11.7.28), as

$$\varphi = \cos^{-1}\left(-\frac{2}{3}\sin^2 35° + \cos 35°\sqrt{1 - (2/3)^2\sin^2 35°}\right) = 57.48°. \tag{11.7.29a}$$

This gives the final relative velocity as

$$\mathbf{v}_f = (-2\cos 57.48°\hat{i} - 2\sin 57.48°\hat{j}) \text{ m/s} = -1.08\,\hat{i} - 1.69\,\hat{j}. \tag{11.7.29b}$$

Equation (11.7.20) allows us to write

$$\mathbf{v}'_{1f} = -\mu\mathbf{v}_f/m_1 = -(3/5)(-1.08\,\hat{i} - 1.69\,\hat{j}) = (0.65\,\hat{i} + 1.01\,\hat{j}) \text{ m/s}, \tag{11.7.29c}$$

and

$$\mathbf{v}'_{2f} = \mu\mathbf{v}_f/m_2 = (2/5)(-1.08\,\hat{i} - 1.69\,\hat{j}) = (-0.43\,\hat{i} - 0.68\,\hat{j}) \text{ m/s}, \tag{11.7.29d}$$

along with

$$\mathbf{v}_{1f} = \mathbf{v}_{cm} + \mathbf{v}'_{1f} = (0.8\,\hat{i} + 0.65)\,\hat{i} + 1.01\,\hat{j} \Rightarrow v_{1f} = \sqrt{1.45^2 + 1.01^2} = 1.77 \text{ m/s}, \tag{11.7.30a}$$

and

$$\mathbf{v}_{2f} = \mathbf{v}_{cm} + \mathbf{v}'_{2f} = (0.8\,\hat{i} - 0.43)\,\hat{i} - 0.68\,\hat{j} \Rightarrow v_{2f} = \sqrt{0.37^2 + 0.68^2} = 0.77 \text{ m/s}. \tag{11.7.30b}$$

Finally, the exit angle of the second particle is from (11.7.26b)

$$\theta_2 = \tan^{-1}\left(\frac{\sin 57.48°}{1 - \cos 57.48°}\right) = 61.26°. \tag{11.7.30c}$$

These results agree with those of Example 11.11(a), as expected. The slight differences are due to round-off errors.

EXAMPLE 11.13

For the system of particles in Example 11.11(b), use the CM frame concepts to obtain the final velocities of each particle as well as the exit angle of the second particle.

Solution

For this part, to refer to Figure 11.17 and realize that the relation (11.7.25a) now applies for angles $\theta_{1c} = \theta_1 - \theta_c$ and φ, where θ_c is the angle $\mathbf{v}_{cm}$ makes with the x-axis as shown, so that we have

$$\tan\theta_{1c} = \frac{\dfrac{\mu}{m_1}v_f\sin\varphi}{\dfrac{\mu}{m_1}v_f\cos\varphi + v_{cm}} = \frac{\sin\varphi}{\cos\varphi + \dfrac{m_1 v_{cm}}{\mu v_f}}. \qquad (11.7.31)$$

Since the Lab frame angle θ_1 is given, to use the CM frame formulation, we need φ in terms of θ_{1c}. This is obtained from (11.7.31) in a process identical to the derivation of (11.7.28b), and it suffices to make the replacement $m_1/m_2 \to m_1 v_{cm}/\mu v_f$ in that equation to get

$$\cos\varphi = -\frac{m_1 v_{cm}}{\mu v_f}\sin^2\theta_{1c} \pm \cos\theta_{1c}\sqrt{1 - (m_1 v_{cm}/\mu v_f)^2\sin^2\theta_{1c}}. \qquad (11.7.32)$$

According to the information given, $m_1 = 0.5$ kg coming from the northwest, initially moving at $v_{1i} = 1.5$ m/s toward the origin at an angle of $\theta_{10} = 40°$ above the $-x$-axis, $m_2 = 1.5$ kg particle coming from the southwest, and moving at $v_{2i} = 0.75$ m/s toward the origin at an angle of $\theta_{20} = 20°$ below the $-x$-axis and, after the collision, the first particle is seen leaving at an angle of 30° north of east. The initial velocity components are then

$$v_{1ix} = 1.5\cos 40° = 1.149 \text{ m/s}, \qquad v_{2ix} = 0.75\cos 20° = 0.705 \text{ m/s},$$
$$v_{1iy} = -1.5\sin 40° = -0.964 \text{ m/s}, \qquad v_{2iy} = (0.75\sin 20°) = 0.257 \text{ m/s}.$$

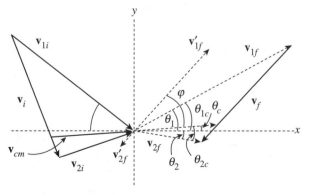

FIGURE 11.17   Example 11.13 for a system of two particles. The CM frame is used to obtain the final velocities and the exit angle of the second particle.

Thus, with $m = m_1 + m_2 = 2$ kg, the center of mass velocity components can be found:

$$v_{cmx} = (m_1 v_{1ix} + m_2 v_{2ix})/m = ((0.5)(1.149) + (1.5)(0.705))/2$$

$$= 0.816 \text{ m/s}$$

$$v_{cmy} = (m_1 v_{1iy} + m_2 v_{2iy})/m = ((0.5)(-0.964) + (1.5)(0.257))/2$$

$$= -0.048 \text{ m/s}$$

$$\Rightarrow \quad v_{cm} = \sqrt{0.816^2 + 0.048^2} = 0.817 \text{ m/s}, \quad \theta_c = \tan^{-1}(-0.048/0.816) = -3.366°.$$

The negative sign of θ_c means that the angle is below the $+x$-axis. With this, because after the collision, the Lab frame exit angle given is 30 degrees north of east, then $\theta_{1c} = 30 - (-3.366) = 33.366°$. From (11.7.23), the relative speed $v_f = v_i = \sqrt{(v_{2ix} - v_{1ix})^2 + (v_{2iy} - v_{1iy})^2} = \sqrt{(0.705 - 1.149)^2 + (0.257 - (-0.964))^2} = 1.299$ m/s, which with $\mu = (0.5)(1.5)/2 = 0.375$ kg, and using (11.7.32), we get

$$\cos\varphi = -\frac{(0.5)(0.817)}{(0.375)(1.299)}\sin^2 33.366 + \cos 33.366\sqrt{1 - ((0.5)(0.817)/[(0.375)(1.299)])^2 \sin^2 33.366}$$

$$= 0.483 \quad \Rightarrow \quad \varphi = \cos^{-1}(0.483) = 61.118°.$$

The final velocities can be obtained from (11.7.20b), where $\mathbf{v}'_{1f} = -\mu\mathbf{v}_f/m_1$, and $\mathbf{v}'_{2f} = \mu\mathbf{v}_f/m_2$, and where $\mathbf{v}_f$ makes the angle $\varphi + \theta_c = 61.118 + (-3.366) = 57.782°$ with respect to the x-axis, as seen in Figure 11.17. We have $\mu/m_1 = 0.375/0.5 = 0.75$, $\mu/m_2 = 0.375/0.5 = 0.25$, and according to the figure, $v_{fx} = -(1.299)\cos 57.782 = -0.693$ m/s and $v_{fy} = -(1.299)\sin 57.782 = -1.099$ m/s for the x- and y-components of the final relative velocity, respectively. Furthermore, from (11.7.20)

$$v_{1fx} = v_{cmx} + v'_{1fx} = 0.816 - (0.75)(-0.693) = 1.336 \text{ m/s}$$

$$\Rightarrow \quad v_{1f} = \sqrt{1.336^2 + 0.776^2} = 1.545 \text{ m/s}$$

$$v_{1fy} = v_{cmy} + v'_{1fy} = -0.048 - (0.75)(-1.099) = 0.776 \text{ m/s}$$

$$v_{2fx} = v_{cmx} + v'_{2fx} = 0.816 + (0.25)(-0.693) = 0.643 \text{ m/s}$$

$$\Rightarrow \quad v_{2f} = \sqrt{0.643^2 + 0.323^2} = 0.720 \text{ m/s},$$

$$v_{2fy} = v_{cmy} + v'_{2fy} = -0.048 + (0.25)(-1.099) = -0.323 \text{ m/s}$$

for the final particle speeds. The final exit angle of the second particle can be seen from Figure 11.17 to be given by $\theta_2 = \theta_{2c} - \theta_c$, where θ_{2c} can be obtained from the first of (11.7.26b), using

$$v'_{2f} = \mu v_f / m_2 = (0.25)\sqrt{0.693^2 + 1.099^2} = 0.325 \text{ m/s, and}$$

$$\theta_{2c} = \tan^{-1}\left(\frac{v'_{2f}\sin\varphi}{v_{cm} - v'_{2f}\cos\varphi}\right) = \tan^{-1}\left(\frac{(0.325)\sin 61.118°}{(0.817) - (0.325)\cos 61.118°}\right) = 23.354°.$$

Finally, $\theta_2 = 23.354° - (-3.416°) = 26.77°$. These values are in agreement with the values obtained in Example 11.11(b), and the slight differences in the decimals are due to round-off errors (see Problem 11.20).

D. Kinetic Energies in the CM Frame of Reference

We now wish to obtain relations that express the particles' initial and final CM frame and final Lab frame energies in terms of the initial Lab frame energy. We also need expressions of the energies in terms of the CM frame exit angle φ and the Lab frame angles θ_1 and θ_2. For convenience we again take $\mathbf{v}_{2i} \equiv 0$. The initial Lab frame energy is

$$T_i = T_{1i} = \frac{1}{2}m_1 v_{1i}^2 \tag{11.7.33}$$

since from (11.7.24), for the magnitudes $v'_{1i} = \mu v_{1i}/m_1$ and $v'_{2i} = \mu v_{1i}/m_2$, the initial CM frame energy (11.7.22b) becomes

$$T'_i = \frac{1}{2}m_1\left(\frac{\mu}{m_1}\right)^2 v_{1i}^2 + \frac{1}{2}m_2\left(\frac{\mu}{m_2}\right)^2 v_{1i}^2 = \frac{1}{2}\mu v_{1i}^2 = \frac{m_2}{m}T_i. \tag{11.7.34}$$

This shows that $T' \leq T_i$, as should be, since $T_i = mv_{cm}^2/2 + T'_i$. The final CM particles' energies are

$$T'_{1f} = \frac{1}{2}m_1 v'^2_{1f} = \frac{1}{2}m_1\left(\frac{\mu}{m_1}\right)^2 v_{1i}^2 = \left(\frac{m_2}{m_1 + m_2}\right)^2 T_i \tag{11.7.35a}$$

and

$$T'_{2f} = \frac{1}{2}m_2 v'^2_{2f} = \frac{1}{2}m_2\left(\frac{\mu}{m_2}\right)^2 v_{1i}^2 = \frac{m_1 m_2}{(m_1 + m_2)^2}T_i, \tag{11.7.35b}$$

where $v_f = v_i = v_{1i}$ from (11.7.23) has been used. To obtain the final Lab frame particle energies, in terms of T_i we begin with

$$\frac{T_{1f}}{T_i} = \frac{m_1 v_{1f}^2/2}{m_1 v_{1i}^2/2} = \left(\frac{v_{1f}}{v_{1i}}\right)^2. \tag{11.7.36}$$

Recall from (11.7.20b) that $\mathbf{v}_{1f} = \mathbf{v}_{cm} - \mu \mathbf{v}_f/m_1$, and according to Figure 11.16, we can write

$$v_{1f}^2 = v_{cm}^2 + \left(\frac{\mu}{m_1}\right)^2 v_f^2 - 2 v_{cm}\frac{\mu}{m_1}v_f\cos(\pi - \varphi) = v_{cm}^2 + \left(\frac{\mu}{m_1}\right)^2 v_f^2 + 2 v_{cm}\frac{\mu}{m_1}v_f\cos\varphi$$

$$= v_{1i}^2\left[\left(\frac{m_1}{m}\right)^2 + \left(\frac{m_2}{m}\right)^2 + 2\frac{m_1 m_2}{m^2}\cos\varphi\right] = v_{1i}^2\left[\frac{m^2 - 2 m_1 m_2}{m^2} + 2\frac{m_1 m_2}{m^2}\cos\varphi\right] \tag{11.7.37}$$

$$\Rightarrow \quad \frac{T_{1f}}{T_i} = \frac{v_{1f}^2}{v_{1i}^2} = 1 - 2\frac{m_1 m_2}{m^2}(1 - \cos\varphi),$$

for the final energy of the first particle, where, as before, we have also used $v_{cm} = m_1 v_{1i}/m$, $v_i = v_{1i}$, and $v_f = v_i$, with $v_{2i} \equiv 0$. A similar relation can be obtained for the final energy of the second particle. We use the energy conservation idea, $T_i = T_{1i} = T_{1f} + T_{2f}$, which gives

$$\frac{T_{2f}}{T_i} = 1 - \frac{T_{1f}}{T_i}, \tag{11.7.38a}$$

or

$$\frac{T_{2f}}{T_i} = 2\frac{m_1 m_2}{m^2}(1 - \cos\varphi). \tag{11.7.38b}$$

We can express these energy ratios in terms of the Lab angle θ_2 if we use (11.7.26c), so that $\cos\varphi = \cos(\pi - 2\theta_2) = 1 - 2\cos^2\theta_2$. Thus (11.7.37 and 11.7.38b) become

$$\frac{T_{1f}}{T_i} = 1 - \frac{4 m_1 m_2}{m^2}\cos^2\theta_2 \quad \text{and} \quad \frac{T_{2f}}{T_i} = \frac{4 m_1 m_2}{m^2}\cos^2\theta_2. \tag{11.7.39}$$

Finally, the energy ratio T_{1f}/T_i can be expressed in terms of the Lab angle θ_1, by the use of Expression (11.7.9), to get

$$\frac{T_{1f}}{T_i} = \left(\frac{m_1}{m_1 + m_2}\right)^2 \left[\cos\theta_1 \pm \sqrt{\left(\frac{m_2}{m_1}\right)^2 - \sin^2\theta_1}\right]^2, \tag{11.7.40}$$

and T_{2f}/T_i is given in terms of θ_1 according to (11.7.38a). The sign chosen inside the square brackets is according to the discussion in Subsection (11.7B).

EXAMPLE 11.14

For the system of particles in Example 11.11(b), use the CM frame energy concept to obtain the final velocities of each particle.

....

Solution

Here the center of mass speed that includes both particles is needed, and we will need to use the first line of expression (11.7.37). Using most of the needed quantities as calculated in Example 11.13, we have

$$v_{1f} = \sqrt{v_{cm}^2 + \left(\frac{\mu}{m_1}\right)^2 v_f^2 + 2\,v_{cm}\frac{\mu}{m_1}v_f\cos\varphi}$$

$$= \sqrt{(0.817)^2 + (0.75)^2(1.299)^2 + 2(0.817)(0.75)(1.299)\cos 61.118°} = 1.545 \text{ m/s},$$

for the final speed of the first particle. Using (11.7.38a), the final kinetic energy is

$$T_{2f} = T_i - T_{1f} = \frac{1}{2}[m_1(v_{1i}^2 - v_{1f}^2) + m_2 v_{2i}^2]$$

$$= \frac{1}{2}[(0.5)((1.5)^2 - (1.545)^2) + (1.5)(0.75)^2] = 0.388 \text{ J};$$

therefore, the final velocity for the second particles is $v_{2f} =$

$\sqrt{\dfrac{2}{m_2}T_{2f}} = \sqrt{\dfrac{2}{1.5}(0.388)} = 0.719$ m/s, which is in close agreement to the previous result,

and slight differences are due to round-off error (see Problem 11.20).

E. Inelastic Collisions in Two Dimensions

Inelastic collisions occur when particles collide in such a way that frictional losses and heat production in the process cause energy not to be conserved. An example of such collision is a two-vehicle crash in which both cars incur damage and energy goes into twisting metal, as well as sound and heat creation during the collision. In such cases, although kinetic energy is not conserved, in the absence of external forces momentum conservation still takes place. As noted before, it is also possible to use momentum conservation, even in the presence of external forces, as long as the impulse from them is negligible compared to the momentum transfer in the collision. Next, again referring to east as the $+x$ direction and north as the $+y$ direction, consider the situation regarding two particles of respective masses m_1 and m_2, in which one is traveling toward the origin along the x-axis and collides with another particle, also traveling toward to origin, at angle θ south of west, as shown in Figure 11.18. After the collision, the particles stick together and the combined mass $m = m_1 + m_2$ exits at an angle φ. If the initial particles' velocities are known, we can use momentum conservation to obtain the final speed of the composite particle as well as the exit angle.

We can write the x and y momentum conservation equations

$$m_1 v_{1i} + m_2 v_{2i} \cos\theta = m v_f \cos\varphi, \qquad m_2 v_{2i} \sin\theta = m v_f \sin\varphi. \qquad (11.7.41)$$

Taking the ratio of the second to the first, we can find the relation for the exit angle as

$$\tan\varphi = \frac{m_2 v_{2i} \sin\theta}{m_1 v_{1i} + m_2 v_{2i} \cos\theta}. \qquad (11.7.42)$$

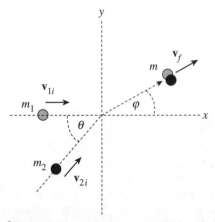

I FIGURE 11.18   Inelastic collision of two particles that stick together after the collision.

Adding the square of the expressions in (11.7.41), we get

$$v_f^2 m^2 = m_2^2 v_{2i}^2 \sin^2\theta + (m_1 v_{1i} + m_2 v_{2i} \cos\theta)^2$$

$$= m_2^2 v_{2i}^2 + m_1^2 v_{1i}^2 + 2\, m_1 m_2 v_{1i} v_{2i} \cos\theta \qquad (11.7.43)$$

$$\Rightarrow \quad v_f = \frac{\sqrt{m_1^2 v_{1i}^2 + m_2^2 v_{2i}^2 + 2\, m_1 m_2 v_{1i} v_{2i} \cos\theta}}{m},$$

which is the final velocity of the composite mass. This result has built in the situation in which the collision takes place in one dimension, in which case $\theta \to 0$, $\varphi \to 0$, and

$$v_f \to \frac{m_1 v_{1i} + m_2 v_{2i}}{m}. \qquad (11.7.44)$$

EXAMPLE 11.15

Two identical masses undergo an inelastic collision in space, as shown in Figure 11.18, after which they are seen to leave as a combined mass. If they are traveling at the same initial speed, (a) what is the combined mass exit angle? (b) Obtain an expression for the fraction of the total energy loss after the collision versus the initial collision angle. Explain your results.

....

Solution

(a) From (11.7.42) we take $m_1 = m_2 = m/2$, and $v_{1i} = v_{2i} = v_i$ to get $\tan\varphi = \sin\theta/(1 + \cos\theta) = \tan(\theta/2)$ or $\varphi = \theta/2$.

(b) From (11.7.43) the final speed simplifies to

$$v_f = \frac{(m/2)v_i\sqrt{2(1 + \cos\theta)}}{m} = v_i \cos\frac{\theta}{2},$$ and the final kinetic energy is

$T_f = \frac{1}{2}mv_f^2 = \frac{1}{2}mv_i^2 \cos^2\frac{\theta}{2}$. The initial kinetic energy is

$T_i = \frac{1}{2}m_1 v_{1i}^2 + \frac{1}{2}m_2 v_{2i}^2 \to \frac{1}{2}mv_i^2$, and the fraction of energy loss, or

$$\frac{T_i - T_f}{T_i} = 1 - \cos^2\frac{\theta}{2}.$$ This means that the maximum energy loss occurs

when $\theta = \pi$ because the particles travel in opposite directions, in which case the final velocity of the combined mass is zero. The minimum energy loss occurs during a head-on collision when $\theta = 0°$. In this case, since the particles travel at the same speed and in the same direction, there is actually no contact and therefore no energy loss; their initial and final speeds are identical.

■ 11.8   Rutherford Scattering Revisited

In the study of Rutherford scattering in Chapter 10, α-particles collide with heavy nuclei found in gold and silver foil targets. In the study of the scattering process, we treated the target nuclei as fixed centers due to their much larger mass than the projectile α-particle. In fact, when we wrote the equation of motion of the α-particle in Equation (10.2.2), we wrote it as

$$m_\alpha \mathbf{a} = \frac{kq_1q_2}{r^2}\hat{r}, \qquad (11.8.1)$$

where $k = 1/4\pi\varepsilon_0 = 8.988 \times 10^9 \, \mathrm{Nm^2/C^2}$ is the electric force constant where the permittivity of free space $\varepsilon_0 \equiv 8.854 \times 10^{-12} \, \mathrm{C^2/Nm^2}$ and $\mathbf{r}$ is the projectile position with respect to the nucleus at rest. This assumes that the target nuclei have infinite mass and therefore do not recoil. This assumption led to the scattering cross-section of Equation (10.4.10)

$$\sigma(\Theta) \equiv \frac{d\sigma}{d\Theta} = 2\,\pi\left(\frac{K}{2\,m_\alpha v_0^2}\right)^2 \frac{\sin\Theta}{\sin^4(\Theta/2)}. \qquad (11.8.2)$$

In fact, the target nuclei have a finite mass, so that the nuclei do in fact experience motion as they interact with the α-particle. Thus, the situation is that of a binary problem in which two bodies interact through the Coulomb force. Following the procedure of Section 11.6, we can view this problem as a reduced mass problem where rather than solving an equation for each particle, we solve a single equation for a particle with reduced mass μ interacting through an internal force $\mathbf{f}$ as described in Equation (11.6.6)

$$\mathbf{f} = \mu\ddot{\mathbf{r}}, \qquad (11.8.3)$$

where $\mathbf{r} = \mathbf{r}_2 - \mathbf{r}_1$ is the relative coordinate of the particles' positions as a function of time, $\mu = m_1 m_2/m$ is the reduced mass, $m = m_1 + m_2$ is the total mass, and $\mathbf{f} = \dfrac{kq_1q_2}{|\mathbf{r}|^2}\hat{r}$ is the internal force between the two particles. The motion takes place about a common center of mass, so that if we follow the procedure of Chapter 10 to obtain (11.8.2) as well as employing the CM frame of reference of Subsection (11.7C), we see that to include the nuclear finite mass motion it suffices to make the replacement $m_\alpha \rightarrow \mu$, and $\Theta \rightarrow \varphi$, where φ and Θ are the CM frame and Lab frame scattering angles, respectively, to write for the scattering cross-section

$$\sigma(\Theta) \rightarrow \sigma(\varphi) \equiv \frac{d\sigma}{d\varphi} = 2\pi\left(\frac{K}{2\mu v_0^2}\right)^2 \frac{\sin\varphi}{\sin^4(\varphi/2)}, \qquad (11.8.4)$$

with $0 < \varphi \leq \pi$.

EXAMPLE 11.16

Projectile gold particles of energy $E_k = 5 \times 10^6$ eV are incident on a gold foil. Obtain a plot of the scattering cross-section versus scattering angle. Compare the cross-section with the case when the target nucleus is considered to be at rest.

Solution
From (11.8.4) , since target and projectile particles have the same mass, then $m_1 = m_2 = M$, $\mu = M/2$, and for the case of equal mass, from (11.7.27) $\varphi = 2\Theta$, so that the scattering cross-section becomes

$$d\sigma = 2\pi\left(\frac{K}{Mv_0^2}\right)^2 \frac{\sin\varphi}{\sin^4(\varphi/2)}d\varphi \quad \rightarrow \quad 2\pi\left(\frac{K}{Mv_0^2}\right)^2 \frac{\sin 2\Theta}{\sin^4(\Theta)}2\,d\Theta,$$

or

$$\sigma(\Theta)_{m_1=m_2} \equiv \frac{d\sigma}{d\Theta}\Big|_{m_1=m_2} = 4\pi\left(\frac{K}{Mv_0^2}\right)^2 \frac{\sin 2\Theta}{\sin^4(\Theta)} = 8\pi\left(\frac{K}{Mv_0^2}\right)^2 \frac{\sin\Theta\cos\Theta}{\sin^4(\Theta)}, \qquad (11.8.5)$$

now with $0 < \Theta \leq \pi/2$.

In the dimensionless units used in Chapter 10, we have

$$\sigma(\Theta)_{m_1 = m_2} = 8\,\pi\left(\frac{\overline{K}kq_e^2}{\overline{M}m_\alpha v_b^2 \overline{v}_0^2}\right)^2 \frac{\sin\Theta\cos\Theta}{\sin^4(\Theta)} = 8\,\pi\left(\frac{\overline{K}}{\overline{M}\overline{v}_0^2}\right)^2\left(a_b\frac{kq_e^2\tau^2}{m_\alpha a_b^3}\right)^2 \frac{\sin\Theta\cos\Theta}{\sin^4(\Theta)}$$

$$= \overline{\sigma}(\Theta)_{m_1 = m_2}a_b^2$$

where as before, we take $kq_e^2\tau^2/m_\alpha a_b^3 \equiv 1$ and we have

$$\overline{\sigma}(\Theta)_{m_1 = m_2} = 8\,\pi\left(\frac{\overline{K}}{\overline{M}\overline{v}_0^2}\right)^2 \frac{\sin\Theta\cos\Theta}{\sin^4(\Theta)}, \tag{11.8.6}$$

where $\overline{K} \equiv Z_1 Z_2$ with $(Z_1, Z_2) = (79, 79)$, $\overline{m} = 1$. We next obtain $\overline{v}_0$ as in (10.3.7); that is, $\overline{v}_0 = \dfrac{c}{v_b}\sqrt{\dfrac{2E(\text{eV})}{\overline{M}E_p(\text{eV})}}$, where v_b needs to be calculated. As in Equation (10.3.3), our unit of time is now

$$\frac{kq_e^2\tau^2}{m_\alpha a_b^3} \equiv 1 \Rightarrow \tau = \sqrt{\frac{m_\alpha a_b^3}{kq_e^2}} = \sqrt{\frac{79(1.66\times10^{-27}\,\text{kg})(10^{-15}\,\text{m})^3}{\left(9\times10^9\dfrac{\text{Nm}^2}{\text{C}^2}\right)(1.602\times10^{-19}\,\text{C})^2}} = 7.535\times10^{-22}\,\text{s},$$

and $v_b = a_b/\tau = 1\times10^{-15}\,\text{m}/7.535\times10^{-22}\,\text{s} = 1.327\times10^6\,\text{m/s} = 0.00442\,c$. Also, for the gold projectile we take as $E_p \equiv m_p c^2 \approx (197\text{ nucleons})(939\text{ MeV/nucleon}) = 184{,}983\times10^6\,\text{eV}$.

Equation (11.8.6) is to be compared to the case when the target nucleus is assumed stationary, for which, from (10.4.13), it is

$$\overline{\sigma}(\Theta) \equiv 2\,\pi\left(\frac{\overline{K}}{2\,\overline{M}\overline{v}_0^2}\right)^2 \frac{\sin\Theta}{\sin^4(\Theta/2)}. \tag{11.8.7}$$

Figure 11.19 contains the comparison sought on a semilog scale. The scattering cross-sections are close to each other for small angles, but at large angles the difference between them is more pronounced.

The MATLAB script `ruthercm_cross.m` was used to perform the comparison. The following script incorporates these formulas.

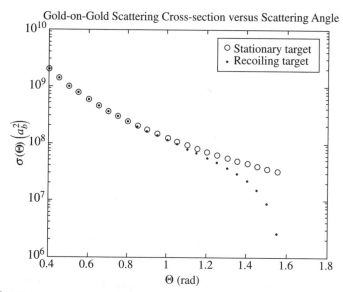

FIGURE 11.19   Semilog scale comparison of the Rutherford scattering cross-section for the cases with and without the inclusion of a target recoil.

SCRIPT

```
%ruthercm_cross.m - program to do the scattering cross-section versus scattering
%angle in Rutherford Scattering - modified version of Ch10's ruther_cross1.m
%here we include CM motion effect and compare to the regular case. The projectile
%particle here is gold, and the target particle is also gold.
clear; warning off;
m=1;                      %projectile mass in units of gold particle mass
vb=0.0044;                %velocity units (a_b/tau_b) (in inits of c=light speed)
ma=197*939e6;             %Gold particle mass energy in eV
Ene=5e6;                  %initial projectile energy in eV
v0=sqrt(2*Ene/m/ma)/vb;   %initial speed in units of vb
za=79;zt=79;              %za=projectile, zt=target charges (79->Gold)
K=za*zt;                  %dimensionless force constant
thsc=.4:.05:pi/2;
%Scatt. Cross-section - stationary target
sigma=K^2*2*pi*sin(thsc)./sin(thsc/2).^4/4/m^2/v0^4;
%Scatt. Cross-section - recoiling target
sigma_cm=K^2*8*pi*sin(thsc).*cos(thsc)./sin(thsc).^4/m^2/v0^4;
%plot(thsc, sigma, 'ko', thsc, sigma_cm, 'b.')
semilogy(thsc, sigma, 'ko', thsc, sigma_cm, 'b.')
xlabel('\Theta (Radians)', 'FontSize', 14)
ylabel('\sigma(\Theta) (a_b^2)', 'FontSize', 14)
str=cat(2, 'Gold on Gold Scattering Cross-section versus Scattering Angle');
title(str, 'FontSize', 12)
h=legend('Stationary Target', 'Recoiling Target'); set(h, 'FontSize', 14)
```

■ Chapter 11 Problems

11.1 In Example 11.1, show that the results are consistent with the definitions $\mathbf{v}_{cm} = \dot{\mathbf{R}}_{cm}$ and $\mathbf{a}_{cm} = \dot{\mathbf{v}}_{cm}$.

11.2 In Example 11.2, for the center of mass of a uniform disk quadrant of mass M and radius R, prove that the integral $y_{cm} = \dfrac{1}{M}\displaystyle\int_0^R y\,dm$ yields the result $4R/3\pi$.

11.3 Obtain the center of mass of a uniform density semi-circular disk of mass M and radius R.

11.4 Find the center of mass of a uniform density rectangular sheet of width a and height b.

11.5 The positions of three particles of masses $m_1 = (3/4)$ kg, $m_2 = (1/2)$ kg, and $m_3 = (1/4)$ kg are given as a function of time by $\mathbf{r}_1 = [(3 + t^2)\hat{i} + \hat{j}]$ m, $\mathbf{r}_2 = [(-2 - t)\hat{i} + t\hat{j} + (1/t)\hat{k}]$ m, and $\mathbf{r}_3 = [\hat{i} - t^2\hat{j} + (1/2)\,\hat{k}]$ m.

  a. Obtain the center of mass position, velocity, and acceleration associated with the three-particle system as a function of time.

  b. Evaluate the expressions at time $t = 2$ s.

11.6 For the system of three particles of Problem 11.5, find as a function of time the system's

  a. linear momentum

  b. angular momentum

  c. kinetic energy

  d. net force

  e. net torque

  f. evaluate the results at time $t = 2$ s.

11.7 In Examples 11.1 and 11.3, the time-dependent positions of three particles, $\mathbf{r}_1 = [(3 + 2t^2)\hat{i} + 4\hat{j}]$ m, $\mathbf{r}_2 = [(-2 + 1/t)\hat{i} + 2t\hat{j}]$ m, and $\mathbf{r}_3 = [\hat{i} - 3t^2\hat{j}]$ m, of masses $m_1 = 1$ kg, $m_2 = 2$ kg, and $m_3 = 3$ kg were employed. According to Equations (11.3.20a and 11.3.20c), their total angular momentum can be obtained from the general expression (i)

$\mathbf{L} = \sum_i m_i \mathbf{r}_i \times \dot{\mathbf{r}}_i$ as well as from (ii) $\mathbf{L} = \mathbf{R}_{cm} \times \mathbf{P} + \sum_i \mathbf{r}'_i \times \mathbf{p}'_i$. This second expression involves the relative coordinates, $\mathbf{r}_i = \mathbf{R}_{cm} + \mathbf{r}'_i$. The torques can be obtained from Equation (11.3.25b); that is, (iii) $\boldsymbol{\tau} = \sum_i \mathbf{r}_i \times \dot{\mathbf{p}}_i$ as well as from the general form (iv) $\boldsymbol{\tau} = d\mathbf{L}/dt$. As a function of time, perform the following:

a. Obtain the total angular momentum of the three particles using (i) and the total torque using (iii)

b. Obtain the total angular momentum using expression (ii)

c. Using the angular momentum from part (a), obtain the total torque using (iv)

d. Using the angular momentum from part (b), obtain the total torque using (iv).

e. Evaluate your expressions at time $t = 1$ s for $m_1 = 1$ kg, $m_2 = 2$ kg, and $m_3 = 3$ kg, and comment on your results.

11.8 The allies fire a cannon ball at the enemy located at position R. It is guaranteed to hit. The cannon ball, however, breaks up (at its highest point) into two equal masses. One falls at $r_1 = R/2$ from the cannon. Where does the other mass land?

11.9 A projectile is fired at an angle of 45 degrees with an initial energy E_0. At the highest point, the projectile explodes into two fragments (masses $m_1 + m_2 = m$), with an extra amount of energy E_0. One fragment ($m_1 = m/4$) travels straight down. Find the magnitude and direction of each fragment's velocity.

11.10 A 2.8×10^6 kg single-stage rocket is fired vertically upward near the surface of the Earth. If 2.1×10^6 kg of its initial mass is fuel and the rocket experiences a thrust of 37×10^6 N, with a constant gas exhaust speed of 2600 m/s, find the rocket's final speed after fuel burnout. Here assume that the acceleration due to gravity is g and that the fuel burn rate is constant.

11.11 A rocket experiences a single force that happens to be resistive, $\mathbf{F} = -b\mathbf{v}$, while in flight. Using Equation (11.4.3b) show that if the rocket starts with an initial speed v_0 and ejects mass at a constant rate of $\alpha = -dm/dt$, then the speed is given by $v = \alpha u/b + (m/m_0)^{b/\alpha}(v_0 - \alpha u/b)$.

11.12 A two-stage rocket is designed such that its initial mass is m_0, the mass of the first-stage payload is m_a, the mass of the first-stage fuel container is m_b, the mass of the second-stage payload is m_c, and the mass of the second-stage fuel container is m_d. If the rocket is in free space and $m_1 = m_a + m_b$, and $m_2 = m_c + m_d$, find the speed of the rocket at the end of the second stage in terms of m_0, m_a, m_1, m_2, and its initial speed v_0. Compare this result with that of a single-stage rocket with initial and final masses of m_0 and m_1, respectively, and identify how it's possible for the two stage-rocket to achieve larger final speed.

11.13 A 2.8×10^6 kg rocket can produce a thrust of 1.5 its weight on the moon. If the rocket's fuel mass makes up for 96% of its total mass and its exhaust speed is $0.3R_{Moon}/\tau$, where $\tau = \sqrt{R_{Moon}^3/Gm_{Moon}}$, find the rocket's fuel burnout time, assuming it is consumed at a constant rate.

11.14 a. Modify Equation (11.5.1) for a rocket taking off near the surface of the Earth, where, due to atmospheric effects, there is an additional force due to friction. Also express the resulting equation in dimensionless units. Assume the frictional force is of the form $\mathbf{F} = -b\rho(r)\mathbf{v}$, where $b = 1 \times 10^5$ m^3/s, $\mathbf{v}$ is the rocket's velocity vector, and $\rho(r)$ is the atmospheric air density versus height from the surface. For simplicity, use

$$\rho(r) = \rho_0 \exp\left(-\frac{m_N g}{KT}|r - R|\right),$$ where m_N is the mass of a nitrogen

molecule; g is the acceleration due to gravity on the surface of Earth; K is Boltzmann's constant; T a temperature of 300 Kelvin; R is Earth's radius; and $\rho_0 = 1.23$ kg/m^3, the density of air at sea level.

b. With the definitions of part (a) simulate the rocket's takeoff motion and obtain the analogue graphs of Section 11.5. Comment on your results regarding the projectile velocity curves, the projectile path, and its range compared to when friction is ignored.

11.15 Estimate the vibrational frequency (in Hertz) of a two-atom molecule interacting through a hypothetical internal force of the form, $F(x) = 3A/x^4 - 2B/x^3$, where $A = u_0 a_0^3$, $B = u_0 a_0^2$, with u_0 a unit of energy, and a_0 a unit of distance. For convenience, work with a hydrogen molecule. How many vibrations will such a molecule have made by the time it free falls for a distance of 1 m, if released from rest? How far will light have traveled during the time in which this molecule makes one vibration?

11.16 Starting from rest on a horizontal surface, a constant force is applied on a toy wagon of mass m. The wagon is let go after a time t. It then enters an incline, on which it is able to climb a height h. Assuming there is no friction, what is the power developed by the applied force as a function of time?

11.17 In the case of a two-dimensional elastic collision, show that the full momentum-energy conservation expressions $m_1\mathbf{v}_{1i} + m_2\mathbf{v}_{2i} = m_1\mathbf{v}_{1f} + m_2\mathbf{v}_{2f}$ and

$$\frac{1}{2}m_1v_{1i}^2 + \frac{1}{2}m_2v_{2i}^2 = \frac{1}{2}m_1v_{1f}^2 + \frac{1}{2}m_2v_{2f}^2,$$ where each of the particles m_1 and

m_2 are moving with initial and final velocities $\mathbf{v}_{1i}$, $\mathbf{v}_{2i}$, $\mathbf{v}_{1f}$, $\mathbf{v}_{2f}$, respectively, are equivalent to a system for which the velocities are viewed from the reference frame in which m_2 is initially at rest; that is,

$$m_1\mathbf{v}'_{1i} = m_1\mathbf{v}'_{1f} + m_2\mathbf{v}'_{2f}, \qquad \frac{1}{2}m_1v'^2_{1i} = \frac{1}{2}m_1v'^2_{if} + \frac{1}{2}m_2v'^2_{2f}, \qquad \text{where}$$

$\mathbf{v}'_{1i} \equiv \mathbf{v}_{1i} - \mathbf{v}_{2i}$, $\mathbf{v}'_{1f} \equiv \mathbf{v}_{1f} - \mathbf{v}_{2i}$, and $\mathbf{v}'_{2f} \equiv \mathbf{v}_{2f} - \mathbf{v}_{2i}$.

11.18 Derive Expressions (11.7.10).

11.19 Obtain analytic expressions for the final velocities v_{1f}, v_{2f}, and final exit angle θ_2 of the second particle in the system of Equations (11.7.18) for two particles of masses m_1 and m_2 with initial velocities v_{1i}, and v_{2i}, respectively, that collide elastically. Use the obtained expressions to confirm the numerical results of Example 11.11(b) given the problem's parameters: $m_1 = 0.5$ kg $v_{1i} = 1.5$ m/s, approaching the origin from the northwest at $\theta_{10} = 40$ degrees above the $-x$-axis; $m_2 = 1.5$ kg, $v_{2i} = 0.75$ m/s, approaching the origin from the southwest at an angle of $\theta_{20} = 20$ degrees below the $-x$-axis, and the first particle's exit angle is $\theta_1 = 30$ degrees north of east. Here let north and east be the positive y- and x-axes, respectively.

11.20 Perform the calculations of Examples 11.13 and 11.14 by means of a computer program or a MATLAB script, and show the results are identical to those of Example 11.11(b). Draw a vector diagram showing the collision process, and perform a check to see that energy is indeed conserved.

11.21 A particle of mass u_0 traveling along the x-axis undergoes an undetermined interaction that affects its energy and causes it to split into two fragments whose masses are $u_0/3$ and $2u_0/3$. After the split, the smaller fragment is seen leaving at an angle of 20 degrees north of east with a speed of 2×10^6 m/s while the large fragment is moving a quarter as fast. What is the exit angle of the larger fragment and what was the speed of the incoming particle? Compare the final and the initial energies, and discuss your answers.

■ Additional Problem

11.22 In a head-on elastic collision, a particle of mass m_1 moving with an initial velocity of v_{1i} collides with a second particle of mass m_2 that is initially at rest. Obtain an expression for the ratio $R = m_2/m_1$ for which the fraction of the total energy of m_2 is a maximum. Give a plot of this energy fraction and point out where the maximum is located. This is an example of what is referred to as *impedance matching*. (*Hint:* Consider the ratio between the final energy of m_2 to the total initial energy of the system.)

12 | Motion of Rigid Bodies

12.1 Introduction

Chapter 11 dealt with the underlying principles involved in the dynamics associated with systems of more than one particle. In particular, we worked with N particles that may be acted upon by external as well as internal forces. A rigid body can be thought of as an extended or continuous system of particles close to each other and whose relative distances are constrained in fixed positions. In reality, all bodies are composed of atoms that are in constant motion relative to each other; however, as an approximation, the atomic motion is ignored when considering the motion of the body as a collection of particles. Thus, when the body is acted upon by an external force, the body responds as a whole, but it is assumed that the body is nondeformable, that is, rigid. The rigid body can have both rotational and translational motion; as such, we will again revisit the concepts of center of mass, momentum, and energy, as well as discuss the concept of moment of inertia and how it is related to a rigid body's motion.

12.2 Center of Mass

For a discrete system of particles, discussed in the previous chapter, we wrote the center of mass as

$$\mathbf{R}_{cm} \equiv \frac{1}{M}\sum_i m_i \mathbf{r}_i, \tag{12.2.1}$$

where $M \equiv \sum_i m_i$. For a rigid body, which is thought of as a continuous system of particles with a total mass M, we consider summing over infinitesimal mass elements dm located at positions $\mathbf{r}$ from the origin, as shown in Figure 12.1(a).

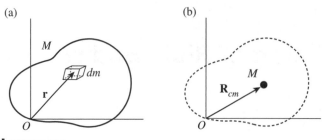

(a)   (b)

M M

dm $\mathbf{R}_{cm}$

$\mathbf{r}$

O O

| FIGURE 12.1

In that case, in the above formula we make the replacement $\sum_i m_i \mathbf{r}_i \rightarrow \int \mathbf{r}\,dm$, and the center of mass position, shown in Figure 12.1(b), can be obtained from

$$\mathbf{R}_{cm} \equiv \frac{\int \mathbf{r}\,dm}{M} = x_{cm}\hat{i} + y_{cm}\hat{j} + z_{xm}\hat{k}, \qquad (12.2.2)$$

where we have used $M = \int dm$. Since the element of mass $dm = \rho\,dV$ where ρ is the rigid body's density, we see that its center of mass coordinates are given by

$$x_{cm} = \frac{\int x\rho\,dV}{M}, \quad y_{cm} = \frac{\int y\rho\,dV}{M}, \quad z_{cm} = \frac{\int z\rho\,dV}{M}, \qquad (12.2.3)$$

with the body's total mass given by $M = \int \rho\,dV$. Depending on the geometry, the volume mass element $\rho\,dV$ may be replaced by the areal mass element $\sigma\,dA$ or the linear mass element $\lambda\,dl$, as discussed in Appendix D.

EXAMPLE 12.1

Prove that the center of mass of a thin, uniform density rod of length L is located at its halfway point.

....

Solution
As shown in Figure 12.2, because we are dealing with a thin rod, we ignore its cross-sectional area and rather than a volume density, we think of a linear density $\lambda = M/L$ with the rod oriented along the y axis. Ignoring the x and z coordinates, we have $y_{cm} = \frac{1}{M}\int y\rho\,dV \rightarrow \frac{1}{M}\int_0^L y\lambda\,dy = \frac{\lambda}{M}\left(\frac{y^2}{2}\right)\Big|_0^L = \frac{\lambda L^2}{M\,2} = \frac{L}{2}.$

❙ FIGURE 12.2

■ 12.3 Angular Momentum, Moment of Inertia, Inertia Tensor, and Torque

In this section, the equations associated with a rigid body's angular momentum are studied. The equations involve the inertia tensor, which is the rotational motion analogue quantity of the mass in linear motion. Under special conditions, the inertia tensor can be replaced by a simple moment of inertia of a rigid body; otherwise, the inertia tensor involves moments of inertia that are matrix elements of the inertia tensor. In this section, the relationship between a rigid body's angular momentum and torque is also discussed.

A. Angular Momentum and Moment of Inertia

In Figure 12.3, an extended rigid body is shown to be rotating about the origin with angular velocity $\boldsymbol{\omega}$ about an axis. The linear momentum of particle i at position $\mathbf{r}_i$ within the rigid body is $\mathbf{p}_i = m_i \mathbf{v}_i$. The angular momentum of the rigid body is equal to the total angular momentum of N such particles, which according to (11.3.19 and 11.3.20), is given by

$$\mathbf{L} = \sum_i \mathbf{r}_i \times \mathbf{p}_i, \tag{12.3.1}$$

where from this point on, the sum is taken over the N particles. From (7.6.11), for a rotating body with coordinates at the origin so that with $\mathbf{v}' = \dot{\mathbf{r}}' = 0 = \mathbf{v}_0$, and $\mathbf{r}' \to \mathbf{r}_i$, the velocity of the ith particle is given by

$$\mathbf{v}_i = \boldsymbol{\omega} \times \mathbf{r}_i, \tag{12.3.2}$$

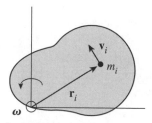

❙ FIGURE 12.3   Extended rigid body rotating about the origin with angular velocity $\boldsymbol{\omega}$ about an axis.

since $\boldsymbol{\omega}$ is the same for all particles. Thus (12.3.1) becomes

$$\mathbf{L} = \sum_i m_i \mathbf{r}_i \times (\boldsymbol{\omega} \times \mathbf{r}_i). \tag{12.3.3a}$$

Using the triple cross product from the inside back cover, we can write $\mathbf{r}_i \times (\boldsymbol{\omega} \times \mathbf{r}_i) = (\mathbf{r}_i \cdot \mathbf{r}_i)\boldsymbol{\omega} - (\mathbf{r}_i \cdot \boldsymbol{\omega})\mathbf{r}_i$, and the rigid body's angular momentum is finally given by

$$\mathbf{L} = \sum_i m_i[r_i^2 \boldsymbol{\omega} - (\mathbf{r}_i \cdot \boldsymbol{\omega})\mathbf{r}_i]. \tag{12.3.3b}$$

This is an important result, because it indicates that the body's angular momentum in general is not parallel to the angular velocity. In fact, individual particles' contributions to $\mathbf{L}$ have components along $\boldsymbol{\omega}$ as well as $\mathbf{r}_i$. The contributions to $\mathbf{L}$ other than those in the direction of $\boldsymbol{\omega}$ are due to particles whose position vectors do not lie perpendicular to $\boldsymbol{\omega}$. Although more will be said below, under certain conditions, it is possible for the angular momentum to be co-linear with $\boldsymbol{\omega}$. This can be seen if in (12.3.3b) we write the particle position vector in component form, one along $\boldsymbol{\omega}$ (say, $\boldsymbol{\omega} = \omega\hat{\parallel}$) and another perpendicular to it; that is, $\mathbf{r}_i = r_{\parallel,i}\hat{\parallel} + r_{\perp,i}\hat{\perp}$, then $\mathbf{r}_i \cdot \boldsymbol{\omega} = r_{\parallel,i}\omega$. Therefore, in the absence of parallel position components in the particle distribution ($r_{\parallel,i} = 0$, $\mathbf{r}_i \to r_{\perp,i}\hat{\perp}$), as in, for example, a lamina, plate, or disk, or even a series of these so as to compose a three-dimensional rigid body, then

$$\mathbf{L} \to I\boldsymbol{\omega}, \quad I = \sum_i m_i r_{\perp,i}^2, \tag{12.3.3c}$$

which is a specialized form of the angular momentum and the moment of inertia, I, as we discuss next.

B. Moment of Inertia Tensor

If (12.3.3b) is expanded in component form, it becomes

$$\mathbf{L} = \sum_i m_i[(x_i^2 + y_i^2 + z_i^2)(\omega_x\hat{i} + \omega_y\hat{j} + \omega_z\hat{k}) - (x_i\omega_x + y_i\omega_y + z_i\omega_z)(x_i\hat{i} + y_i\hat{j} + z_i\hat{k})], \tag{12.3.4}$$

which shows the components of the angular momentum. For the x-component we have

$$L_x = \sum_i m_i[(x_i^2 + y_i^2 + z_i^2)\omega_x - (x_i\omega_x + y_i\omega_y + z_i\omega_z)x_i], \tag{12.3.5}$$

and similarly for the y, z components. If we make use of the notation $(x_i, y_i, z_i) = (x_{i1}, x_{i2}, x_{i3})$, $(\omega_x, \omega_y, \omega_z) = (\omega_1, \omega_2, \omega_3)$,   and $(L_x, L_y, L_z) = (L_1, L_2, L_3)$, then we can generalize (12.3.5) for each of the three components of angular momentum as

$$L_\nu = \sum_i m_i \left[\left(\sum_\mu^3 x_{i\mu}^2 \right) \omega_\nu - \sum_\mu x_{i\mu} \omega_\mu x_{i\nu} \right], \tag{12.3.6a}$$

where the Greek letter indices μ, ν run over the coordinates, and the index i runs over the N particles. Since $\omega_\nu = \omega_\mu \delta_{\nu\mu}$, (12.3.6a) can be rewritten as

$$L_\nu = \sum_\mu \sum_i m_i \left[\delta_{\nu\mu} \sum_\gamma^3 x_{i\gamma}^2 - x_{i\nu} x_{i\mu} \right] \omega_\mu = \sum_\mu I_{\nu\mu} \omega_\mu, \tag{12.3.6b}$$

where we have defined the $\nu\mu^{\text{th}}$ element of the moment of inertia tensor $\{I\}$ as

$$I_{\nu\mu} \equiv \sum_i m_i \left[\delta_{\nu\mu} \sum_\gamma^3 x_{i\gamma}^2 - x_{i\nu} x_{i\mu} \right]; \tag{12.3.7a}$$

that is, for a discrete system of particles, the inertia tensor is

$$\{I\} = \begin{pmatrix} I_{xx} = \sum_i m_i(y_i^2 + z_i^2) & I_{xy} = -\sum_i m_i x_i y_i & I_{xz} = -\sum_i m_i x_i z_i \\ I_{yx} = -\sum_i m_i y_i x_i & I_{yy} = \sum_i m_i(x_i^2 + z_i^2) & I_{yz} = -\sum_i m_i y_i z_i \\ I_{zx} = -\sum_i m_i z_i x_i & I_{zy} = -\sum_i m_i z_i y_i & I_{zz} = \sum_i m_i(x_i^2 + y_i^2) \end{pmatrix} \tag{12.3.7b}$$

from which, for a continuous rigid body system, we have

$$\{I\} = \begin{pmatrix} \int (y^2 + z^2)dm & -\int xy\,dm & -\int xz\,dm \\ -\int yx\,dm & \int (x^2 + z^2)dm & -\int yz\,dm \\ -\int zx\,dm & -\int zy\,dm & \int (x^2 + y^2)dm \end{pmatrix}, \tag{12.3.7c}$$

where, depending on the rigid body geometry, the mass element is $dm = \rho dV, \sigma dA, \lambda dl$, for a volume, area, or line distribution, respectively. The diagonal elements of the inertia tensor, that is, I_{xx}, I_{yy}, I_{zz}, are called *moments of inertia* about the x-, y-, and z-axes, respectively, and the negative off-diagonal elements are called *products of inertia*. The inertia tensor is symmetric, as can be seen from (12.3.7), so that $I_{\nu\mu} = I_{\mu\nu}$. With the preceding inertia tensor, the total angular momentum components of (12.3.4) can be nicely written as

$$\mathbf{L} = \{I\}\boldsymbol{\omega} \quad \text{or} \quad \begin{pmatrix} L_x \\ L_y \\ L_z \end{pmatrix} = \begin{pmatrix} I_{xx} & I_{xy} & I_{xz} \\ I_{yx} & I_{yy} & I_{yz} \\ I_{zx} & I_{zy} & I_{zz} \end{pmatrix} \begin{pmatrix} \omega_x \\ \omega_y \\ \omega_z \end{pmatrix}. \tag{12.3.8}$$

Before dwelling on this result, it is convenient to consider the motion about a fixed axis, i.e., motion about an axis that is constrained at both ends so as not to be free to change direction.

C. Rotation About a Fixed Axis

The preceding results (12.3.3 and 12.3.8) underline the fact that in general, the angular momentum of a system is not necessarily co-linear with the angular velocity. Here, we take a simple example in which we delve into the consequences of not having the angular momentum co-linear with the angular velocity. We consider the rotational motion of a point mass at the end of a massless, rigid rod, as shown in Figure 12.4(a). The mass is constrained to rotate with a constant angular velocity $\boldsymbol{\omega}$ about an axis that makes an angle θ with respect to the rod. We wish to calculate the angular momentum associated with this system, as well as the resulting torque. Writing the position of the point mass in cylindrical coordinates shown in Figure 12.4(b) (see also Section 7.3) as

$$\mathbf{r} = R\hat{\rho} + z\hat{k}, \tag{12.3.9}$$

and with $\boldsymbol{\omega} = \omega_0\hat{k}$, the mass's linear velocity is

$$\mathbf{v} = \boldsymbol{\omega} \times \mathbf{r} = \omega_0\hat{k} \times (R\hat{\rho} + z\hat{k}) = R\omega_0\hat{\varphi}. \tag{12.3.10}$$

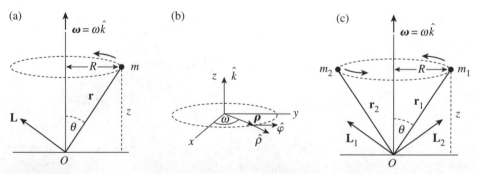

FIGURE 12.4 Rotational motion of a point mass at the end of a massless, rigid rod.

From (12.3.1), the angular momentum is

$$\mathbf{L} = m(R\hat{\rho} + z\hat{k}) \times (R\omega_0\hat{\varphi}) = mR\omega_0(R\hat{k} - z\hat{\rho})$$
$$= I\omega_0(\hat{k} - (z/R)\hat{\rho}) = I\omega_0(\hat{k} - \cot\theta\,\hat{\rho}), \tag{12.3.11}$$

where $R = r\sin\theta$, $z = r\cos\theta$ have been used, and $I = mR^2$ is the mass' moment of inertia about the z-axis. The angular momentum has components along the z and the ρ axes and is perpendicular to both $\mathbf{r}$ and $\mathbf{v}$ or $\mathbf{p}$. This result for the angular momentum can, of course, be obtained using (12.3.8), which we do next. The moment of inertia matrix elements are obtained for the one-particle system using (12.3.7b). Noting the Cartesian coordinates of the mass at the instant shown in Figure 12.4 as $(0, y, z)$, we have $I_{xx} = m(y^2 + z^2)$, $I_{xy} = -m(xy) = 0$, $I_{xz} = -m(xz) = 0$, $I_{yx} = I_{xy} = 0$, $I_{yy} = m(x^2 + z^2) = mz^2$, $I_{yz} = -myz$, $I_{zx} = I_{xz} = 0$, $I_{zy} = I_{yz} = -myz$, and $I_{zz} = m(x^2 + y^2) = my^2$. The angular momentum becomes

$$\begin{pmatrix} L_x \\ L_y \\ L_z \end{pmatrix} = \begin{pmatrix} m(y^2 + z^2) & 0 & 0 \\ 0 & mz^2 & -myz \\ 0 & -myz & my^2 \end{pmatrix} \begin{pmatrix} 0 \\ 0 \\ \omega_0 \end{pmatrix} = \begin{pmatrix} 0 \\ -myz\omega_0 \\ my^2\omega_0 \end{pmatrix}$$

$$\Rightarrow \mathbf{L} = -myz\omega_0\hat{j} + my^2\omega_0\hat{k}, \tag{12.3.12}$$

which is identical to (12.3.11) if we notice that at the instant shown in Figure 12.3, the cylindrical unit vector $\hat{\rho}$ is equivalent to the Cartesian unit vector $\hat{j}$ and $y = R$. According to (11.3.21), the torque, $\boldsymbol{\tau} = d\mathbf{L}/dt$, can be obtained by noticing that in (12.3.11) $d\hat{k}/dt = 0$ and $d\hat{\rho}/dt = \hat{\varphi}$, where the rest of the parameters are constant. Alternatively, we can use (7.6.7), where in the mass' own reference frame (the S' frame) the angular momentum is constant, so that according to the rest frame (the S frame) at the origin

$$\boldsymbol{\tau} = d\mathbf{L}/dt = \boldsymbol{\omega} \times \mathbf{L} = mR\omega_0^2\hat{k} \times (R\hat{k} - z\hat{\rho}) = -mR\omega_0^2 z\hat{\varphi}$$

$$= -I\omega_0^2(z/R)\hat{\varphi} = -I\omega_0^2\cot\theta\,\hat{\varphi}. \tag{12.3.13}$$

We find, therefore, that according to (12.3.11), the rotating mass's angular momentum is not co-linear with $\boldsymbol{\omega}$, and as a consequence there is a torque present on the system according to (12.3.13) due to the asymmetry in the mass distribution about the rotating axis. Thus, to keep the system rotating, as in Figure 12.4(a), such external torque is constantly needed. Figure 12.5 is a one-time frame from an animation using

Fixed-axis massless rod-point mass rotating system, $\theta = 25$

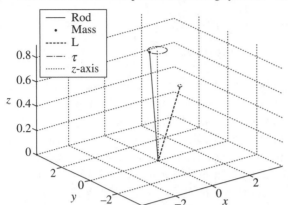

FIGURE 12.5 Snapshot of a fixed axis point mass at the end of a rotating massless rod.

the MATLAB script fixed_axis.m, which follows. When run, the script shows the mass' position, angular momentum, and torque as a function of time. The angle θ used for the animation is 25 degrees from the z-axis.

The simulation is performed using Cartesian coordinates, which are related to cylindrical coordinates by $\hat{\rho} = \hat{i}\cos\varphi + \hat{j}\sin\varphi$, $\hat{\varphi} = -\hat{i}\sin\varphi + \hat{j}\cos\varphi$, and $\hat{k} = \hat{k}$. When substituting these into (12.3.9, 12.3.11, and 12.3.13) we get, with $\varphi = \omega_0 t$, $\mathbf{r} = R\cos\omega_0 t\,\hat{i} + R\sin\omega_0 t\,\hat{j} + z\hat{k}$, $\mathbf{L} = I\omega_0(-\hat{i}\cot\theta\cos\omega_0 t - \hat{j}\cot\theta\sin\omega_0 t + \hat{k})$, $(\hat{j}\cot\theta\sin\omega_0 t + \hat{k})$, and $\boldsymbol{\tau} = I\omega_0^2\cot\theta(\hat{i}\sin\omega_0 t - \hat{j}\cos\omega_0 t)$, which are used in the script. The rest of the parameters have values, $\omega_0 = \pi$, $r = a = 1$, and $m = 1$, as shown in the script.

SCRIPT

```
%fixed_axis.m -animates the position of a rod-mass system, its angular momentum,
%and the related torque
clear; warning off;
w=pi;tau=2*pi/w;       %angular velocity and period
t=[0:tau/20:1.5*tau]; %time variable
N=length(t);           %number of points
th=25;                 %angle of rod tilt from z axis
a=1; thr=th*pi/180;    %rod length, angle inradians
s=sin(thr); c=cos(thr);R=a*s; z=a*c; co=cot(thr);%for orbit radius and height
m=1;I=m*R^2;A=I*w;     %mass, moment if inertia, angular momentum amplitude
v=max([R, z, A]);         %view window parameter
vxy=max([R, z, A*w*co]);%view window parameter
```

```
for i=1:N
clf
axis ([-vxy, vxy, -vxy, vxy, 0, v])
ct=cos(w*t(i)); st=sin(w*t(i)); %needed to get x, y coords vs time
x=R*ct; y=R*st;              %x, y particle coordinates versus time
line([0, x], [0, y], [0, z], 'color', 'black', 'linewidth', 1)%massless rod line
line([x, x], [y, y], [z, z], 'color', 'black', 'LineStyle', '.', 'linewidth', 1, ...
    'Marker', '.', 'MarkerSize', 20)        %the particle
Lx=-A*co*ct; Ly=-A*co*st; Lz=A;           %angular momentum components
Tx=A*w*co*st; Ty=-A*w*co*ct; Tz=0.0; %torque components
line([0, Lx], [0, Ly], [0, Lz], 'color', 'red', 'linewidth', 1.5, ...
    'LineStyle', '-') %L (ang. Mom.)
line([0, Tx], [0, Ty], [0, Tz], 'color', 'green', 'linewidth', 1.5, ...
    'LineStyle', '-.')%torque
line([0, 0], [0, 0], [0, z], 'color', 'm', 'LineStyle', ':', 'linewidth', 1.5)%z axis
line([Lx, Lx], [Ly, Ly], [Lz, Lz], 'color', 'red', 'Marker', 'd', 'MarkerSize', 5)
%Arrow for L
line([Tx, Tx], [Ty, Ty], [Tz, Tz], 'color', 'green', 'Marker', 'd', 'MarkerSize',
5)%Arrow for T
grid on
pause(.1)
end
h=legend('rod', 'mass', 'L', '\tau', 'z-axis', 2); set(h, 'FontSize', 14)
hold on
plot3(R*sin(w*t), R*cos(w*t), z*t./t, ':')% particle trace
str=cat(2, 'Fixed-axis massless rod-point mass rotating system', ...
        ', \theta = ', num2str(th, 3));
title(str, 'FontSize', 14)
xlabel('x', 'FontSize', 14), ylabel('y', 'FontSize', 14), zlabel('z', 'FontSize', 14)
%r=[x, y, z], L=[Lx, Ly, Lz], dot(r, L) %use to check if r, L are perpendicular
```

It is possible to consider a situation in which there are two masses located at positions with mirror symmetry, about the z-axis, as shown in Figure 12.4(c); that is, $\mathbf{r}_1 = R\hat{\rho}_1 + z\hat{k}$, $\mathbf{r}_2 = -R\hat{\rho}_1 + z\hat{k}$. By proceeding in the same way as for the one-mass case, for the case of two masses find that the total angular momentum is given by

$$\mathbf{L} = \mathbf{L}_1 + \mathbf{L}_2 = R\omega_0[R(m_1 + m_2)\hat{k} + z(m_2 - m_1)\hat{\rho}_1], \qquad (12.3.14)$$

with an associated torque

$$\boldsymbol{\tau} = R\omega_0^2 z(m_2 - m_1)\hat{\varphi}, \qquad (12.3.15)$$

which shows that if the masses are equal, $m_1 = m_2 = m$, then the total angular momentum, $\mathbf{L} \to 2mR^2\boldsymbol{\omega}$, is co-linear with $\boldsymbol{\omega}$ and in consequence the torque vanishes. Thus, any asymmetries in the rotation of a system about an axis may lead to undesired torques.

D. Rotation About a Symmetry Axis

Here we consider a fixed axis that is an axis of symmetry. For example, consider a uniform density rigid body as in Figure 12.6, where we start by imagining the system to be composed of many discrete particles each of mass m_i located at $\mathbf{r}_i$ from the origin. The axis of rotation is one of symmetry and the rotation's velocity is $\boldsymbol{\omega}$.

The kinetic energy of the rotating body can be written as a sum of the kinetic energies of each of the particles in the body

$$T = \frac{1}{2}\sum_i m_i v_i^2,
\tag{12.3.16}$$

where $\mathbf{v}_i = \boldsymbol{\omega} \times \mathbf{r}_i$ or $v_i = \omega r_i \sin\theta = \omega r_{\perp,i}$, where $r_{\perp,i}$ is the perpendicular distance between the ith particle and the rotation axis, so that (12.3.16) becomes

$$T = \frac{1}{2}I\omega^2 = \frac{L^2}{2I},
\tag{12.3.17a}$$

where the angular momentum is co-linear with $\boldsymbol{\omega}$

$$\mathbf{L} = I\boldsymbol{\omega},
\tag{12.3.17b}$$

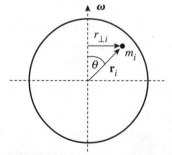

I FIGURE 12.6   Uniform density rigid body composed of many discrete particles.

and where for the discrete system of particles the moment of inertia is

$$I \equiv \sum_i m_i r_{\perp, i}^2, \tag{12.3.18a}$$

which is identical to (12.3.3c). Further, if in this equation we make the replacements $m_i \to dm$ and $r_{\perp, i} \to r_\perp$, then we can get the moment of inertia associated with a continuous mass distribution

$$I \equiv \int r_\perp^2 \, dm, \tag{12.3.18b}$$

where $r_\perp$ is the perpendicular distance from the axis of rotation to the element of mass dm.

EXAMPLE 12.2

(a) A uniform density solid disk, oriented in the xy-plane, of radius R and mass M rotates about an axis passing through its center of mass with angular velocity $\boldsymbol{\omega} = \omega_0 \hat{k}$, as shown in Figure 12.7(a). Obtain the angular momentum of the disk. What would the difference in the result be if the disk were replaced by a cylinder of the same mass and radius and rotating about an axis parallel to its length and passing through its center of mass?

(b) Assume the disk is held by a fixed frictionless spindle passing through its rotational z-axis. While rotating, a particle of mass m traveling with velocity $\mathbf{v} = v_0 \hat{i} - v_z \hat{k}$, where $v_0 \gg v_z$, makes a collision with the disk in such a

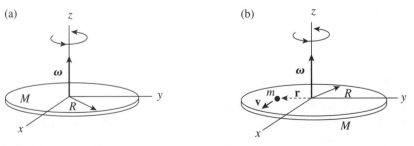

FIGURE 12.7 (a) A spinning disk. (b) A point mass collides with the spinning disk.

way as to remain lodged at position $\mathbf{r} = -(R/2)\hat{j}$ onto the disk, as shown in Figure 12.7(b). What is the resulting angular speed of the disk-particle system?

Solution

(a) The moment of inertia of the disk can be obtained from (12.3.18b) where for the disk we let $dm = \sigma dA$, with $dA = r\,dr\,d\theta$, with $\sigma = M/A$, $A = \pi R^2$, so that since the perpendicular distance of a mass element dm to the axis of

rotation is $r_\perp = r$, then $I = \sigma \int\limits_0^R r^2 dr \int\limits_0^{2\pi} d\theta = MR^2/2$. The disk's angular

momentum is co-linear with $\boldsymbol{\omega}$, so from (12.3.17b), $\mathbf{L} = I\boldsymbol{\omega} = MR^2\omega_0\hat{k}/2$. The moment of inertia, I, found here is equivalent to a single component, I_{zz}, of the moment of inertia tensor in (12.3.8), and no other moments were needed in this symmetric rotation example. The moment of inertia of a cylinder rotating about an axis parallel to its length and passing through its center of mass has the same value of inertia as a disk of the same mass and radius rotating as above; thus the angular momentum would be the same.

(b) This is an example of an inelastic collision that involves angular momentum conservation. The initial angular momentum of the system is the sum of the disk's angular momentum $I_{disk}\boldsymbol{\omega}_{disk}$ and the particle's angular momentum with respect to the origin, $\mathbf{r} \times (m\mathbf{v}_{particle})$. The final angular momentum is $(I_{disk} + mr^2)\boldsymbol{\omega}_f$. Neglecting the particle velocity component along the $\hat{k}$ direction, and setting these two equal, we have

$$M(R^2/2)\omega_0\hat{k} + mv_0(R/2)(-\hat{j} \times \hat{i}) = (MR^2/2 + m(R/2)^2)\boldsymbol{\omega}_f,$$

which gives, for the final angular velocity of the combination in the $\hat{k}$ direction, a magnitude of $\omega_f = (MR\omega_0 + mv_0)/[R(M + m/2)]$.

E. Gyroscope Precession (Simple Approach)

It is commonly known that a spinning gyroscope has complicated motions, which are due to the complex relationship between the applied torque and the angular momentum of the spinning wheel, whose axis is fixed at one end. Consider Figure 12.8, in cylindrical coordinates, as an example model of a gyroscope whose wheel, with mass m and moment of inertia I, is initially set spinning in the $\hat{\rho}$ direction with angular frequency magnitude ω_p.

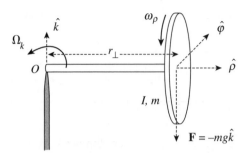

I FIGURE 12.8   Model of a gyroscope in cylindrical coordinates.

The force due to gravity acts to produce a torque on the spinning wheel, whose center of mass lies at $\mathbf{r} = r_\perp \hat{\rho}$ from the pivot point O, where $r_\perp$ is the moment arm, as seen in the figure. The gyroscope has a precession as well as a nutation associated with its motion. The precession is a motion that develops with an angular frequency Ω_k in the $\hat{k}$ direction. The nutation is a more subtle oscillation that develops with an angular frequency in the $\hat{\varphi}$ direction. Although a more complete description of the motion will be treated later in the text, we can get an understanding of the precessional motion by simply using Equation (7.6.7) for a rotating S' frame. We imagine that the S' frame rotates with the wheel and that the angular momentum of the spinning wheel (L_ρ) is very large so that the angular momentum associated with the precession is small. We have from (11.3.21) our relation between the applied torque and the change in angular momentum as

$$\frac{d\mathbf{L}}{dt} = \left(\frac{d\mathbf{L}}{dt}\right)_{rot} + \boldsymbol{\omega} \times \mathbf{L} \approx \boldsymbol{\omega} \times \mathbf{L} = \boldsymbol{\tau}, \tag{12.3.19}$$

where have taken $(d\mathbf{L}/dt)_{rot} \approx 0$ in accordance with our assumptions. Using cylindrical coordinates we write $\boldsymbol{\omega} \approx \omega_\rho \hat{\rho} + \Omega_k \hat{k}$, $\mathbf{L} \approx L_\rho \hat{\rho}$ where $L_\rho = I\omega_\rho$, and the applied torque is obtained from Figure 12.8 as $\boldsymbol{\tau} = \mathbf{r} \times \mathbf{F} = r_\perp \hat{\rho} \times (-mg\hat{k}) = mgr_\perp \hat{\varphi}$. Using these approximations into (12.3.19), get

$$\boldsymbol{\omega} \times \mathbf{L} = \Omega_k L_\rho \hat{\varphi} = mgr_\perp \hat{\varphi} \quad \Rightarrow \quad \Omega_k = \frac{mgr_\perp}{L_\rho}, \tag{12.3.20}$$

for the precession frequency. The same result can also be obtained if from (12.3.19) we write $d\mathbf{L} = L_\rho \, d\varphi \, \hat{\varphi} = \tau \, dt \, \hat{\varphi}$, since the angular momentum changes in the $\hat{\varphi}$ direction, and then solve for $\Omega_k \equiv d\varphi/dt$, with the magnitude of the torque given by

$\tau = mgr_\perp$. The precession frequency is inversely proportional to ω_ρ, so that large gyroscopic spins lead to slow precession.

■ 12.4   Further Inertia Properties

A. Moments of Inertia Theorems

Moments of inertia for various objects rotating about a particular symmetry axis are given in Table 12.1. The question arises as to what is the moment of inertia of similar objects about axes other than the ones given in the table. The answer to this question is provided by two theorems, i.e., the parallel and the perpendicular axes theorems, which we discuss next.

i. Parallel Axis Theorem

For the purposes of this subsection, we work with a continuous three-dimensional mass distribution of mass M, as shown in Figure 12.9, and the moment of inertia of the body rotating, say, about the z-axis, which from (12.3.18b) becomes

$$I_{zz} \equiv \int r_\perp^2 \rho dV, \tag{12.4.1}$$

Table 12.1   Moments of inertia of some homogeneous rigid bodies each of mass.

Hoop or cylindrical shell $I = MR^2$	Hollow cylinder $I = M(R_1^2 + R_2^2)/2$	Long thin rod $I = ML^2/12$	Solid sphere $I = 2MR^2/5$
Solid cylinder or disk $I = \frac{1}{2}MR^2$	Rectangular plate $I = M(a^2 + b^2)/12$	Long thin rod $I = ML^2/3$	Thin spherical shell $I = \frac{2}{3}MR^2$

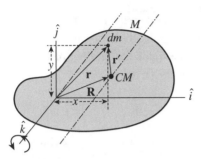

where $r_\perp$ is the perpendicular distance of a mass element dm, located at $\mathbf{r} = x\hat{i} + y\hat{j} + z\hat{k}$, to the axis of rotation, z; ρ is the rigid body density, and the center of mass is located at $\mathbf{R} = X\hat{i} + Y\hat{j} + Z\hat{k}$.

In Figure 12.9, the position of dm can be written as $\mathbf{r} = \mathbf{R} + \mathbf{r}'$, where $\mathbf{r}' = x'\hat{i} + y'\hat{j} + z'\hat{k}$,   or $\mathbf{r} = x\hat{i} + y\hat{j} + z\hat{k} = (x' + X)\hat{i} + (y' + Y)\hat{j} + (z' + Z)\hat{k}$. Therefore, the perpendicular distance of dm to the z-axis is $r_\perp^2 = (x' + X)^2 + (y' + Y)^2$ and (12.4.1) becomes

$$I_{zz} \equiv \int ((x' + X)^2 + (y' + Y)^2)\rho dx'dy'dz'$$

$$= \int (x'^2 + y'^2 + 2Xx' + 2Yy' + X^2 + Y^2)\rho dx'dy'dz', \qquad (12.4.2)$$

since the center of mass coordinates are constant. Furthermore, the integrals such as $\int x'\rho dx'dy'dz' = \int (x - X)\rho dV = \int x dm - X\int dm = MX - XM = 0,$ and

similarly $\int y'\rho dx'dy'dz' = 0$. Finally, since $r_{\perp CM}^2 \equiv X^2 + Y^2$ is the parallel distance of the axis of rotation from the center of mass rotation axis, and $r'^2_\perp = x'^2 + y'^2$ is the perpendicular distance of the mass element dm from the center of mass axis, (12.4.2) becomes

$$I_{zz} = \int r'^2_\perp \rho dm + Mr_{\perp CM}^2 = I_{CM, zz} + Mr_{\perp CM}^2, \qquad (12.4.3)$$

which is the mathematical expression of the parallel axis theorem. To state it in words, the moment of inertia of a rigid body about any axis is equal to the moment of inertia about a parallel axis passing through the center of mass plus the product of the rigid body mass and the square of the distance between the two axes.

EXAMPLE 12.3

Use the parallel axis theorem to obtain Table 12.1's result for the moment of inertia of a long thin rod rotating about an axis passing through one of its ends and perpendicular to the rod's length.

....

Solution

The rod's moment of inertia about the center of mass and parallel to the given axis is obtained from Table 12.1 as $I_{CM} = ML^2/12$. The distance to the end of the rod from the center of mass is $L/2$, to obtain $I_{end} = I_{CM} + M(L/2)^2 = ML^2/12 + M(L/2)^2 = ML^2/3$.

---

ii. Perpendicular Axis Theorem

Here we consider a thin flat plate (infinitesimally small dz) rotating about the z-axis as shown in Figure 12.10.

The moment of inertia is given by

$$I_{zz} \equiv \int (x^2 + y^2)\rho dV. \tag{12.4.4}$$

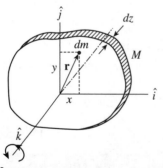

FIGURE 12.10   Thin, flat plate rotating about the z-axis.

If we consider the object's rotations about the x- and y-axes, the associated moments of inertia are

$$I_{xx} \equiv \int y^2 \rho dV \text{ and } I_{yy} \equiv \int x^2 \rho dV. \tag{12.4.5}$$

We see, therefore, that for a plate I_{zz} can be expressed in terms of the perpendicular axes moments as

$$I_{zz} = I_{xx} + I_{yy}, \tag{12.4.6}$$

which is the relation that expresses the perpendicular axis theorem. The theorem says that the moment of inertia of any plate (or plane lamina) about an axis normal to the plane of the lamina is equal to the sum of the moments of inertia about any two mutually perpendicular axes passing through the point of intersection (with the given axis) and lying in the plane of the lamina.

EXAMPLE 12.4

Find the moment of inertia of a hoop of mass M and radius a about an axis parallel to the plane of the hoop (xy-plane) passing (a) through its center, and (b) through its edge.

....

Solution

(a) Referring to Figure 12.11, the moment of inertia perpendicular to the plane of the hoop and passing through its center is $I_{CM,zz} = Ma^2$. The moments of inertia about the x- and y-axes, parallel to the hoop's plane and passing through its center, by symmetry are equal $I_{CM,xx} = I_{CM,yy}$; thus, $I_{CM,zz} = 2\,I_{CM,xx}$, so $I_{CM,xx} = Ma^2/2$.

(b) By the parallel axis theorem, $I_{edge,xx} = I_{CM,xx} + Ma^2 = 3\,Ma^2/2$.

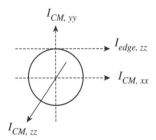

I FIGURE 12.11

EXAMPLE 12.5

A uniform semicircular disk, of mass M_s and radius a, lies in the xy-plane. Find the moment of inertia (a) about an axis passing through the disk's flat side whose orientation is along the y-axis, (b) about the z-axis of the semicircular disk that passes through its origin (or center of curvature), and (c) about the x-axis and passing through the origin.

Solution

(a) The moment of inertia can be obtained from $I_{yy} = \displaystyle\int_0^a x^2 dm = \sigma \int_0^a x^2 dA,$

where $\sigma = M_s/(\pi a^2/2)$, and $dA = 2ydx$, as shown in Figure 12.12. Since $y = a\sin\theta$, $x = a\cos\theta$, then $y = a\sqrt{1 - (x/a)^2}$, so that we have

$$I_{yy} = \frac{2M_s}{\pi a^2}2a\int_0^a x^2\sqrt{1 - (x/a)^2}dx. \tag{12.4.7}$$

This integral is involved, so that we will obtain it numerically. To do so, let's change units as follows. Let $x \equiv a\bar{x}$, and $dx = ad\bar{x}$, where since the range of x is $0 \le x \le a$, then the range of $\bar{x}$ is $0 \le \bar{x} \le 1$, and the preceding moment becomes

$$I_{yy} = M_s a^2 J, \text{ where } J \equiv \frac{4}{\pi}\int_0^1 \bar{x}^2\sqrt{1 - (\bar{x})^2}d\bar{x}. \tag{12.4.8}$$

The integral J is now purely numerical and can be carried out very simply using the MATLAB script moment_sdisk.m. The result of the run is saved in file

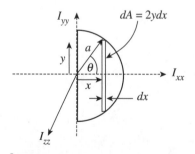

I_{yy} $dA = 2ydx$

y a

θ

x $\longrightarrow$ I_{xx}

dx

I_{zz}

| FIGURE 12.12

`momen_sdisk.txt`. Both files follow. To the accuracy of the MATLAB numerical calculation $J = 1/4$, and so

$$I_{yy} = M_s a^2 / 4. \tag{12.4.9}$$

It turns out that the preceding integral can be performed analytically using the integrals from Appendix B, and the result is exactly 1/4. Alternatively, if the MATLAB's symbolic toolbox is available, typing "syms x; int(4*x^2*sqrt(1-x^2)/pi, 0, 1)" in the command line gives the value "1/4."

SCRIPT

```
%moment_sdisk.m
clear;
f=inline('x.^2.*(1-x.^2).^(0.5)*4/pi');   %define the integrand
J=quad(f, 0.0, 1.0, 1.e-5);               %Simpson quadrature integration
fprintf('The integral is %4.3f', J);      %print the result to three decimals
```

OUTPUT

```
moment_sdisk.txt
>> moment_sdisk
The integral is 0.250
```

It is interesting to note that the preceding moment of inertia is exactly 1/2 the moment of inertia about an axis parallel to the plane of the disk of mass $M = 2M_s$ with the same radius and passing through its center.

(b) Because as in part (a) we expect that the moment of inertia I_{zz} of the semicircular disk is 1/2, that of an equivalent configuration for a disk with twice the mass, then

$$I_{zz} = (2M_s a^2 / 2)/2 = M_s a^2 / 2. \tag{12.4.10}$$

(c) Using the perpendicular axis theorem, get
$$I_{xx} = I_{zz} - I_{yy} = M_s a^2 / 2 - M_s a^2 / 4 = M_s a^2 / 2.$$

B. Principal Axes

The problem that exists with systems whose angular momentum does not lie in the direction of the rotational frequency, ω, is that undesired vibrations (due to $\tau = d\mathbf{L}/dt$) develop that cause the system to wobble and could, subsequently, become unstable and collapse. In order to guarantee that $\mathbf{L}$ is co-linear with ω, it is necessary that the rotation occur about a principal axis. For example, suppose the moment of inertia tensor were diagonal, with all the products of inertia vanishing, then from (12.3.7), using numerical indices

$$\{I\} \rightarrow I = \begin{pmatrix} I_1 & 0 & 0 \\ 0 & I_2 & 0 \\ 0 & 0 & I_3 \end{pmatrix}. \tag{12.4.11}$$

A coordinate system with an inertia tensor of this form is said to have coordinates along principal axes, and the moments of inertia, I_i, are known as principal moments of inertia. The directions of these axes are along principal unit vectors, say, $\hat{e}_1$, $\hat{e}_2$, and $\hat{e}_3$, so that (12.3.8) gives

$$\mathbf{L} = L_1\hat{e}_1 + L_1\hat{e}_2 + L_3\hat{e}_3 = I_1\omega_1\hat{e}_1 + I_2\omega_2\hat{e}_2 + I_3\omega_3\hat{e}_3, \tag{12.4.12}$$

where in general, $\omega = \omega_1\hat{e}_1 + \omega_2\hat{e}_2 + \omega_3\hat{e}_3$ has components of different magnitudes and is the case for $\mathbf{L}$. This means that even (12.4.11 and 12.4.12) are not a guarantee that $\mathbf{L}$ will be co-linear with ω; however, if the rotation is about one of the principal axes, say, $\omega \rightarrow \omega_0\hat{e}_1$ with $\omega_2 = \omega_3 = 0$, then (12.4.12) gives $L_1 = I_1\omega_0$, $L_2 = L_3 = 0$, and $\mathbf{L} \rightarrow I_1\omega$. Thus, we say that for rotations about a principal axis, the angular momentum is co-linear with the angular velocity.

The question arises: How are the principal axes found? One way is by inspection; that is, if the body has symmetry, it is possible to choose the axis of rotation so as to make the products of inertia vanish. For example, a right cylinder (cone or top) has a symmetry axis that is also a principal axis and lies along the length of the cylinder and passes through the center of mass. The other two axes, perpendicular to it, have equal moments of inertia. It turns out that principal axes are always mutually perpendicular. Another example is a rectangular solid. The principal axes lie perpendicular to the faces, if the origin is chosen to coincide with the center of mass.

If the principal axes can't be found by inspection, it is possible to find them by requiring that

$$\mathbf{L} = \{I\}\omega = I\omega \quad \Rightarrow \quad (\{I\} - I)\omega = 0, \tag{12.4.13}$$

where in this equation I is a 3×3 diagonal matrix whose matrix elements are to be found. This is equivalent to the matrix equation

$$\begin{pmatrix} I_{xx} - I & I_{xy} & I_{xz} \\ I_{yx} & I_{yy} - I & I_{yz} \\ I_{zx} & I_{zy} & I_{zz} - I \end{pmatrix} \begin{pmatrix} \omega_1 \\ \omega_2 \\ \omega_3 \end{pmatrix} = 0, \tag{12.4.14}$$

which is an equation for the eigenvalues or principal moments of inertia, I, and the eigenvectors associated with the ωs. A trivial but not useful solution exists when $\omega_1 = \omega_2 = \omega_3 = 0$. In a similar way to what was done in the case of the bimodal spring system of Section 4.13, a more useful but nontrivial solution to (12.4.14) exists if the determinant of the coefficients vanishes; i.e.,

$$\begin{vmatrix} I_{xx} - I & I_{xy} & I_{xz} \\ I_{yx} & I_{yy} - I & I_{yz} \\ I_{zx} & I_{zy} & I_{zz} - I \end{vmatrix} = 0. \tag{12.4.15}$$

Thus, there are three roots or eigenvalues corresponding to three possible values of I, each representing a principal moment of inertia. The three roots correspond to the solutions of a cubic equation. Mathematically, one of the cubic roots is always real, and the other two roots may involve complex numbers. However, if the diagonalization is to be meaningful, we must always obtain real values for the principal moments. By substituting a corresponding value of I back into (12.4.14), three equations result for the components of ω (ω_1, ω_2, ω_3), whose ratios give the direction cosines; that is, ω and its components obey the following rules:

$$\omega = \sum_{i=1}^{3} (\omega \cdot \hat{e}_i) \hat{e}_i = \omega \sum_{i=1}^{3} \cos\theta_i \hat{e}_i,$$

$$\omega_i = \omega \cdot \hat{e}_i = \omega \cos\theta_i, \tag{12.4.16}$$

$$\omega \cdot \omega = \omega^2 \Rightarrow \sum_{i=1}^{3} \cos\theta_i = 1,$$

where θ_i is the angle of the chosen axis relative to the original axis and, therefore, specifies the direction of the corresponding principal axis, as shown in Figure 12.13.

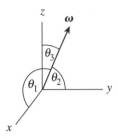

| FIGURE 12.13

EXAMPLE 12.6

Consider a uniform cube of mass M and sides a, and find the inertia tensor, the principal axes, and their associated moments of inertia about a corner of the cube.

Solution

Let the cube's origin be located at one of the corners of the cube, as shown in Figure 12.14. The moment of inertia about the x-axis is

$$I_{xx} = \int \rho(y^2 + z^2)dxdydz = Ma^{-3}\left(\int_0^a dx\right)\left(\int_0^a y^2 dy \int_0^a dz + \int_0^a z^2 dz \int_0^a dy\right)$$

$$= Ma^{-3}(a)\left(\frac{a^4}{3} + \frac{a^4}{3}\right) = 2Ma^2/3, \tag{12.4.17a}$$

and because of symmetry, we also have $I_{zz} = I_{yy} = I_{xx}$. For the products of inertia

$$I_{xy} = -\int \rho xy \, dxdydz = -Ma^{-3}\int_0^a xdx \int_0^a ydy \int_0^a dz = -Ma^2/4, \tag{12.4.17b}$$

| FIGURE 12.14

and also due to symmetry, $I_{yx} = I_{xy} = I_{zx} = I_{xz} = I_{yz} = I_{yz}$. Letting $c \equiv Ma^2$, the inertia tensor can be written as

$$\{I\} = \begin{pmatrix} 2c/3 & -c/4 & -c/4 \\ -c/4 & 2c/3 & -c/4 \\ -c/4 & -c/4 & 2c/3 \end{pmatrix},$$

(12.4.18)

and (12.4.15) becomes

$$\begin{vmatrix} 2c/3 - I & -c/4 & -c/4 \\ -c/4 & 2c/3 - & -c/4 \\ -c/4 & -c/4 & 2c/3 - I \end{vmatrix} = 0,$$

(12.4.19)

or

$$0 = \left(\frac{2c}{3} - I\right)\left[\left(\frac{2c}{3} - I\right)\left(\frac{2c}{3} - I\right) - \frac{c^2}{16}\right] + \frac{c}{4}\left[-\frac{c}{4}\left(\frac{2c}{3} - I\right) - \frac{c^2}{16}\right] - \frac{c}{4}\left[\frac{c^2}{16} + \frac{c}{4}\left(\frac{2c}{3} - I\right)\right]$$

$$= \left(\frac{2c}{3} - I\right)\left[\frac{4c^2}{9} - \frac{4cI}{3} + I^2 - \frac{3c^2}{16}\right] - \frac{c^3}{32} = I^3 - 2cI^2 + \frac{165}{144}Ic^2 - \frac{121}{864}c^3$$

(12.4.20)

$$= \left(I - \frac{c}{6}\right)\left(I - \frac{11c}{12}\right)\left(I - \frac{11c}{12}\right) = 0,$$

which indicates that the principal axes' moments of inertia are $I_1 = c/6 = Ma^2/6$, $I_2 = I_3 = 11\,Ma^2/12$, with I_1 being the principal moment of inertia about the principal axis of symmetry because two of the three moments are equal; that is, two eigenvalues are degenerate. Notice that if we find the direction of the principal axis of symmetry, we will also know the other two axes because they in general can be chosen to be mutually perpendicular to this one. Thus, we substitute $I = I_1$ back into (12.4.13 and 12.4.14) with $\{I\}$ as given by (12.4.18); that is,

$$(\{I\} - I_1)\boldsymbol{\omega} = 0 \quad \Rightarrow \quad \begin{pmatrix} 2c/3 - c/6 & -c/4 & -c/4 \\ -c/4 & 2c/3 - c/6 & -c/4 \\ -c/4 & -c/4 & 2c/3 - c/6 \end{pmatrix}\begin{pmatrix} \omega_1 \\ \omega_2 \\ \omega_3 \end{pmatrix} = \begin{pmatrix} 0 \\ 0 \\ 0 \end{pmatrix},$$

(12.4.21)

which gives the three equations

$$2\,\omega_1 - \omega_2 - \omega_3 = 0, \quad -\omega_1 + 2\,\omega_2 - \omega_3 = 0, \quad -\omega_1 - \omega_2 + 2\,\omega_3 = 0. \quad (12.4.22)$$

By inspection, one can see that a solution of these three equations is $\omega_1 = \omega_2 = \omega_3 \equiv \omega_0$, which means that our sought rotation is of the form

$$\boldsymbol{\omega} = \omega_1 \hat{i} + \omega_2 \hat{j} + \omega_3 \hat{k} = \omega_0 \hat{r}; \qquad (12.4.23)$$

and since the axis we seek, about which the body has moment of inertia I_1, is in the direction of this rotation, the principal axis of symmetry lies along cube diagonal. The other two axes can be perpendicular to this. To find them, we can write

$$\hat{r} \cdot \mathbf{B} = (\hat{i} + \hat{j} + \hat{k}) \cdot (B_x \hat{i} + B_y \hat{j} + B_z \hat{k}) = 0 = B_x + B_y + B_z, \qquad (12.4.24)$$

where $\mathbf{B}$ is any possible vector perpendicular to the symmetry axis whose components satisfy this equation. The possible choices of $\mathbf{B}$ are as follows:

$$\begin{matrix} B_x = 0 \\ B_y = -B_z \end{matrix} \Big\} \Rightarrow \mathbf{B}_1 = \begin{cases} (0, 1, -1) \\ (0, -1, 1) \end{cases}, \begin{matrix} B_y = 0 \\ B_x = -B_z \end{matrix} \Big\} \Rightarrow \mathbf{B}_2 = \begin{cases} (1, 0, -1) \\ (-1, 0, 1) \end{cases}, \begin{matrix} B_z = 0 \\ B_y = -B_x \end{matrix} \Big\} \Rightarrow \mathbf{B}_3 = \begin{cases} (1, -1, 0) \\ (-1, 1, 0) \end{cases}.$$
$$(12.4.25)$$

Here we notice that the components do include their mirror images (− sign) because (12.4.24) does include that possibility. Thus, here we have shown three possible solutions. Of these three, notice that because $\mathbf{B}_3$ does not involve the z coordinate, it must touch the two cube edges associated with the four cube faces. The other two vectors $\mathbf{B}_1$ and $\mathbf{B}_2$ must touch the cube edges associated with the bottom and top cube faces and, therefore, are related by rotation and are equivalent. Thus, the two principal axes that are perpendicular to the symmetry axis are $\mathbf{B}_1 (\mathbf{B}_2)$ and $\mathbf{B}_3$. The moments of inertia of the cube about these axes are degenerate; that is, $I_2 = I_3 = 11 Ma^2/12$, as stated before. It is possible to reproduce the above results with a MATLAB script det_soln.m that follows. Its corresponding output is stored in file det_soln.txt. The inertia tensor (A), roots (R), the eigenvectors (columns of P), and eigenvalues (Q) are identical to this understanding.

SCRIPT

```
%det_soln.m - symbolically finds roots of a matrix determinant as well as
%eigenvectors, eigenvalues of the cube's inertia tensor
%type "help symbolic" within MATLAB'S command line for help on symbolic functions
clear; format compact;
syms c i a x y z M r;    %symbolic variables
r=M/a^3;                 %cube density
%matrix elements in units of c use 3D symbolic integration
%diagonal elements - think of c=M*a^2 - our inertia unit
```

```
Ixx=r*int(int(int((y^2+z^2), x, 0, a), y, 0, a), z, 0, a)*c/(M*a^2);
Iyy=r*int(int(int((x^2+z^2), x, 0, a), y, 0, a), z, 0, a)*c/(M*a^2);
Izz=r*int(int(int((x^2+y^2), x, 0, a), y, 0, a), z, 0, a)*c/(M*a^2);
%products of inertia
Ixy=-r*int(int(int((x*y), x, 0, a), y, 0, a), z, 0, a)*c/(M*a^2);
Ixz=-r*int(int(int((x*z), x, 0, a), y, 0, a), z, 0, a)*c/(M*a^2);
Iyz=-r*int(int(int((y*z), x, 0, a), y, 0, a), z, 0, a)*c/(M*a^2);
Iyx=Ixy; Izx=Ixz; Izy=Iyz;   %use symmetry for the rest
A={[Ixx, Ixy, Ixz];[Iyx, Iyy, Iyz];[Izx, Izy, Izz]}%inertia tensor
I= i*triu(tril(ones(3)));     %create a unit matrix 3x3
M=A-I;                        %whose determinant we seek
R=simplify(factor(det(M)))    %to see the roots
[P, Q]=eig(A)                 %P=column eigen vectors, Q=eigenvalues
```

OUTPUT

```
det_soln.txt - output from det_soln.m
>> det_soln
A =
[  2/3*c, -1/4*c, -1/4*c]
[ -1/4*c,  2/3*c, -1/4*c]
[ -1/4*c, -1/4*c,  2/3*c]
R =
1/864*(c-6*i)*(11*c-12*i)^2
P =
[ -1, -1,  1]
[  0,  1,  1]
[  1,  0,  1]
Q =
[ 11/12*c,       0,      0]
[       0, 11/12*c,      0]
[       0,       0,  1/6*c]
>>
```

Finally, it is possible to visualize the cube and the principal axes discussed previously, as shown in Figure 12.15. The main body diagonal (red line) is the principal axis of symmetry. The two other principal axes are shown along with their mirror images (same color, solid lines), with the legend showing the corresponding directions as discussed before. The axes perpendicular to the principal axis of symmetry form a plane perpendicular to the body diagonal as shown. The two axes associated with the degenerate eigenvalues can be anywhere on that plane, and although they don't have to be chosen perpendicular, they can be picked so that all are mutually perpendicular. It turns out that for the case when

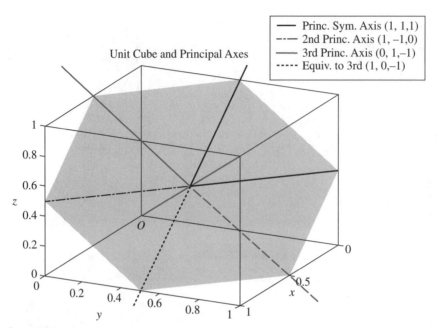

FIGURE 12.15 Uniform cube of mass M and sides a, showing the principal axes associated with rotations about a corner of the cube.

the inertia eigenvalues are nondegenerate, their associated principal axes are mutually perpendicular, due to the absence of symmetry. The ones we found here, because they intersect six cube edges, have vertices that form a hexagon shape. The MATLAB script cube_princ_ax.m is used to create the figure of the cube as well as the axes according to the recipe described in this example.

It should be noticed that the axes, except for the principal axes of symmetry, drawn in Figure 12.15 have been shifted by 1/2 the body diagonal for better viewing purposes, but otherwise they all normally intersect at the cube's corner of origin.

SCRIPT

```
%cube_princ_ax.m
%draws a cube with origin at a corner, the principle axes with the
%axis of symmetry
clear;
axis ([0, 1, 0, 1, 0, 1])
%grid on;
```

```
box on
view([1 1/2 1/2])%viewpoint in x, y, z cartesian coords
%view(azimuth, elevation)%if desired, maximum is 90 degrees each)
%principal axes
v=1/2; %view best if translate axes by v, other than the symmetry axis
a=line([0, 1], [0, 1], [0, 1], 'color', 'r', 'linewidth', 2);%symmetry axis
%first perpendicular axis and its mirror image
b(1)=line([0+v, 1+v], [0+v, -1+v], [0+v, 0+v], 'LineStyle', '-.', 'color'...
    , 'b', 'linewidth', 2);
line([0+v, -1+v], [0+v, 1+v], [0+v, 0+v], 'LineStyle', '-', 'color', 'b', ...
    'linewidth', 2);
%2nd perpendicular axis and its mirror image
b(2)=line([0+v, 0+v], [0+v, 1+v], [0+v, -1+v], 'LineStyle', '-', ...
    'color', 'g', 'linewidth', 2);
line([0+v, 0+v], [0+v, -1+v], [0+v, 1+v], 'LineStyle', '-', 'color'...
    , 'g', 'linewidth', 2);
%3rd perpendicular + mirror image, equivalent to 2nd by rotation symmetry
b(3)=line([0+v, 1+v], [0+v, 0+v], [0+v, -1+v], 'LineStyle', ':', 'color', 'k', ...
    'linewidth', 2);
line([0+v, -1+v], [0+v, 0+v], [0+v, 1+v], 'LineStyle', '-', 'color', ...
    'k', 'linewidth', 2);
title('Unit Cube and Principal Axes', 'FontSize', 13)
xlabel('x', 'FontSize', 14, 'Position', [7.6 4.55 3.5]);
ylabel('y', 'FontSize', 14, 'Position', [7.6 3.6 3.15]);
zlabel('z', 'FontSize', 14), text(.1, .04, 0, '0')
h=legend([a, b], 'Princ. Sym. Axis (1, 1, 1)', '2nd Princ. Axis (1, -1, 0)', ...
    '3rd Princ. Axis (0, 1, -1)', 'Equiv. to 3rd (1, 0, -1)', 0);
set(h, 'FontSize', 8, 'Position', [0.68 0.83 0.26 0.14])
hold on %let's draw a polygon at the axes intersection with cube edges
x = [1;1/2;0;0;1/2;1;]; y=[0;0;1/2;1;1;1/2;]; z=[1/2;1;1;1/2;0;0];
h=fill3(x, y, z, [0.75 0.75 0.75]);        %draws the hexagon
set(h, 'EdgeAlpha', [.3], 'FaceAlpha', [0.5]) %hexagon edges, transparent
```

EXAMPLE 12.7

Consider a 1 kg uniform square plate of 1 m sides. Find the inertia tensor, the principal axes, and their associated moments of inertia about a corner of the plate.

Solution

This example is somewhat similar to Example 12.6, and the main difference is that it is instead a two-dimensional example. Consequently, the inertia tensor does not involve integrations over the z coordinate, and $dm = \sigma dA = (M/a^2)dxdy$. The tensor takes the form

$$= \frac{M}{a^2} \begin{pmatrix} \int_0^a \int_0^a y^2 dx dy & -\int_0^a \int_0^a xy dx dy & 0 \\ -\int_0^a \int_0^a yx dx dy & \int_0^a \int_0^a x^2 dx dy & 0 \\ 0 & 0 & \int_0^a \int_0^a (x^2 + y^2) dx dy \end{pmatrix}.$$ (12.4.26)

From here the process proceeds as in Example 12.6; that is, we need to find the principle moments of inertia (eigenvalues) and the principle axes' directions (get direction cosines from the eigenvectors). To speed the process, we can perform the integrals in (12.4.26) numerically using a trapezoidal or Simpson's rule, given in Appendix B. The rest of the procedure can be carried out numerically as well. The MATLAB script det_soln2_2d.m, with the help of another script inert_el.m, which follows, does this process. The script is set to use Simpson's rule but, when uncommented, the trapezoidal rule (script trap.m) can be used as well. Generally, Simpson's rule (script simp.m) for integration is more accurate. The script obtains the eigenvectors and eigenvalues and prints the direction cosines as well as the principal moments of inertia. The output from det_soln2_2d.m is stored in the file det_soln2_2d.txt as given following the script listings.

SCRIPT

```
%det_soln2_2d.m - for a 2 dimensional solid
%numerically finds roots of a matrix determinant as well as
%eigenvectors, eigenvalues of the rectangle's inertia tensor
clear; format compact;
a=1;b=1;                        %solid plate sides
M=1; rho=M/(a*b);               %solid mass, density
ax=0; ay=0;                     %origin at corner as lower limits
bx=a+ax; Nx=11; dx=(bx-ax)/(Nx-1);%x upper limit, x spacing
by=b+ay; Ny=11; dy=(by-ay)/(Ny-1);%y upper limit, y spacing
ie=6;                           %matrix elements needed
x=[ax:dx:bx];y=[ay:dy:by];      %x, y, z grid
%below, trap is MATLAB's trapezoid rule
%simp is the more accurate simpson rule
    for j=1:Ny
        for i=1:Nx
            for m=1:ie
```

```
                fx(m, i)=rho*inert_el(m, x(i), y(j), 0);%z=0 in 2d
            end
        end
        for m=1:ie
            %fy(m, j)=trap(fx(m, :), dx);%trapezoidal rule
            fy(m, j)=simp(fx(m, :), dx); %Simpson rule
        end
    end
%finally integrate over the y coord to get the moments
for m=1:ie
    %ff=trap(fy(m, :), dy);%trapezoidal
    ff=simp(fy(m, :), dy); %Simpson
        if      m==1 Ixx=ff;
        elseif m==2 Ixy=ff;
        elseif m==3 Ixz=ff;
        elseif m==4 Iyy=ff;
        elseif m==5 Iyz=ff;
        elseif m==6 Izz=ff;
        end
%fprintf('m= %2i, The integral is %4.3f\n', m, ff)
end
Iyx=Ixy; Izx=Ixz; Izy=Iyz;   %use symmetry for the rest
A={[Ixx, Ixy, Ixz];[Iyx, Iyy, Iyz];[Izx, Izy, Izz]};%inertia tensor
disp 'Inertia Tensor'
disp(rats(A))                     %display A in string fraction form
I= i*triu(tril(ones(3)));         %create a unit matrix 3x3
M=A-I;                            %whose determinant we seek
[P, Q]=eig(A);                      %P=column eigenvectors, Q=eigenvalues
ac=round(100*acos(P)*180/pi)/100; %angles related to eigenvectors to 2 dec.
disp 'Direction cosines matrix'
fprintf('cos(%4.2d) cos(%4.2d) cos(%4.2d)\n', ac)%direction cosines
disp 'Principal moments'
S(1:3)=diag(Q);disp(rats(S)) %principal moments in string fraction form
%can check for orthogonaly of P1 and P2
%dot(P(1:3, 1), P(1:3, 2)), dot(P(1:3, 1), P(1:3, 3)), dot(P(1:3, 2), P(1:3, 3))
%can check that the eigenvectors add to 1
%sum(P(1:3, 1).^2), sum(P(1:3, 2).^2), sum(P(1:3, 3).^2)
```

FUNCTION

```
%inert_el.m
function inerel=inert_el(m, x, y, z)
%Inertia function integrands in cartesian coords
if m==1 inerel=y^2+z^2;
elseif m==2 inerel=-x*y;
elseif m==3 inerel=-x*z;
```

```
elseif m==4 inerel=x^2+z^2;
elseif m==5 inerel=-y*z;
elseif m==6 inerel=x^2+y^2;
else
    disp ' only 6 elements are needed '
    break
end
```

FUNCTION

```
%simp.m
function simpu=simp(f, inc)
%Simpson's rule for numerical integration
%f is an odd array of evaluated functions in steps inc
ip=length(f);        %must be an odd number
s1=sum(f(2:2:ip-1));%sums all even terms
s2=sum(f(3:2:ip-2));%sums all odd term does not include f(1) and f(ip)
simpu=(4.*s1+2.*s2+f(1)+f(ip))*inc/3.0;%finally add f(1) and f(ip)
```

FUNCTION

```
%trap.m - uses the native trapz
function trapez=trap(f, inc)
%trapezoid method for numerical integration
%f is an array of evaluated functions in steps inc
trapez=inc*trapz(f);
```

OUTPUT

```
det_soln2_2d.txt: Output from det_soln2_2d.m
>> det_soln2_2d
Inertia Tensor
     1/3        -1/4         0
    -1/4         1/3         0
      0           0         2/3
Direction cosines matrix
cos( 135) cos( 135) cos( 90)
cos( 135) cos( 45) cos( 90)
cos( 90) cos( 90) cos( 00)
Principal moments
     1/12        7/12        2/3
>>
```

From the preceding output, we see that the moment of inertia tensor has diagonal elements 1/3, 1/3, and 2/3 in units of kgm^2 and products of inertia of $-1/4$. The principal moments are in the fractions 1/12, 7/12, and 2/3, all in units of kgm^2. According to the direction cosines, we have the corresponding principal axes vectors

$$(\cos(135), \cos(135), \cos(90)) = (-1/\sqrt{2}, -1/\sqrt{2}, 0)$$
$$(\cos(135), \cos(45), \cos(90)) = (-1/\sqrt{2}, 1/\sqrt{2}, 0) \qquad (12.4.27)$$
$$(\cos(90), \cos(90), \cos(0)) = (0, 0, 1)$$

for the three principal axes, respectively.

■ 12.5   Kinetic Energy

In this section, the equations associated with a rigid body's kinetic energy are obtained. We consider a rigid body composed of N particles rotating with $\boldsymbol{\omega}$ about the origin, as in Figure 12.3. The rotational kinetic energy of the system can be written as

$$T_{rot} = \frac{1}{2}\sum_i m_i v_i^2 = \frac{1}{2}\sum_i m_i \mathbf{v}_i \cdot \mathbf{v}_i, \qquad (12.5.1)$$

where as before, the sum is over the N particles. Since $\mathbf{v}_i = \boldsymbol{\omega} \times \mathbf{r}_i$, then

$$T_{rot} = \frac{1}{2}\sum_i (\boldsymbol{\omega} \times \mathbf{r}_i) \cdot m_i \mathbf{v}_i = \frac{1}{2}\boldsymbol{\omega} \cdot \sum_i \mathbf{r}_i \times \mathbf{p}_i, \qquad (12.5.2)$$

where $\mathbf{p}_i = m_i \mathbf{v}_i$ is the momentum of the ith particle and where the back cover's triple product rule, $(\mathbf{A} \times \mathbf{B}) \cdot \mathbf{C} = \mathbf{A} \cdot (\mathbf{B} \times \mathbf{C})$ has been used. With the help of (12.3.1) the general form of the rotational energy is

$$T_{rot} = \frac{1}{2}\boldsymbol{\omega} \cdot \mathbf{L}. \qquad (12.5.3)$$

In terms of the moment of inertia (12.3.3b and 12.3.8) tensor, we can write this as

$$T_{rot} = \frac{1}{2}\boldsymbol{\omega} \cdot \sum_i m_i[r_i^2 \boldsymbol{\omega} - (\mathbf{r}_i \cdot \boldsymbol{\omega})\mathbf{r}_i], \qquad (12.5.4a)$$

or

$$T_{rot} = \frac{1}{2}(\omega_x \quad \omega_y \quad \omega_z)\begin{pmatrix} I_{xx} & I_{xy} & I_{xz} \\ I_{yx} & I_{yy} & I_{yz} \\ I_{zx} & I_{zy} & I_{zz} \end{pmatrix}\begin{pmatrix} \omega_x \\ \omega_y \\ \omega_z \end{pmatrix} = \frac{1}{2}\omega^T\{I\}\omega = \frac{1}{2}\sum_{\mu\nu} I_{\mu\nu}\omega_\mu\omega_\nu$$

(12.5.4b)

where ω^T is the transpose of the ω vector, and as before, the Greek letter indices μ, ν run over the coordinates. If the rigid body rotates in a coordinate system where all the products of inertia vanish, i.e., the coordinates system axes are principal axes, then $\mathbf{L} \rightarrow I_1\omega_1\hat{e}_1 + I_2\omega_2\hat{e}_2 + I_3\omega_3\hat{e}_3$, $\omega \rightarrow \omega_1\hat{e}_1 + \omega_2\hat{e}_2 + \omega_3\hat{e}_3$, and (12.5.3) gives

$$T_{rot} \rightarrow \frac{1}{2}(I_1\omega_1^2 + I_2\omega_2^2 + I_2\omega_2^2).$$

(12.5.5)

If in addition to rotating, the rigid body's center of mass has translational kinetic energy, the total kinetic energy is the sum of the translational and rotational energies, i.e., $T_{total} = T_{trans} + T_{rot}$. Letting M be the rigid body mass, $\mathbf{v}_{cm}$ its center of mass velocity, and $\mathbf{p}_{cm}$ the center of mass linear momentum, the translational energy can be written as $T_{trans} = Mv_{cm}^2/2 = \mathbf{v}_{cm} \cdot \mathbf{p}_{cm}/2$, and the total energy can be written compactly as

$$T = T_{rot} + T_{trans} = \frac{1}{2}\omega \cdot \mathbf{L} + \frac{1}{2}\mathbf{v}_{cm} \cdot \mathbf{p}_{cm}.$$

(12.5.6)

EXAMPLE 12.8

Consider the 1 kg uniform square plate of 1 m sides of Example 12.7 rotating with angular velocity magnitude of $\sqrt{2}$ rad/s. Find the angular momentum and rotational energy about the origin located at a plate corner (a) for a rotation about the diagonal and passing through the origin, and (b) for rotations about each of the principal axes.

Solution

(a) Since the inertia tensor is known from Example 12.7, we can write
$$\omega = \sqrt{2}(\hat{i}\cos 45 + \hat{j}\cos 45) \text{ in vector form and perform Equations (12.3.8)}$$

for the angular momentum and (12.5.4) for the rotational energy. The result for the angular momentum and the energy are

$$\mathbf{L} = (0.083\hat{i} + 0.083\hat{j})\,\text{kgm}^2/\text{s} \text{ and } T = 0.083 \text{ J, respectively.}$$

(b) For the rotations about the principal axes, we can use the principal moment of inertia tensor of Example 12.7, where the inertia tensor is diagonal with moments 1/12, 7/12, and 2/3. We have three possible rotation directions given by (12.4.27), so that each set of direction cosines is multiplied by the magnitude of the angular speed to get the desired $\boldsymbol{\omega}$. The angular momentum and energy are calculated as in part (a) for each rotation. The results are $\mathbf{L} = (-0.083, -0.583, 0)\,\text{kgm}^2/\text{s}$, $T = 0.333$ J, for the first principal axis rotation, $\mathbf{L} = (-0.083, 0.583, 0)\,\text{kgm}^2/\text{s}$, $T = 0.333$ J, for the second principal axis rotation, and $\mathbf{L} = (0, 0, 0.943)\,\text{kgm}^2/\text{s}$, $T = 0.667$ J for the last principal axis rotation.

These results have been obtained by the use of the following MATLAB script r_energy.m, with the output stored in file r_energy.txt which is also listed below.

SCRIPT

```
%r_energy.m  - finds the angular momentum and energy of a rigid body about an
%axis of rotation given the angular speed. Can also find the L and E
%for rotations about the principal axis, given their inertia and directions.
clear; format compact;
h=pi/180;          %factor to convert degrees to radians
mw=sqrt(2);w=[cos(45*h);cos(45*h);0]*mw;  %rot speed given
fprintf('w= %5.3f %5.3f %5.3f, magnitude=%5.3f\n', w, mw)
I={[1/3, -1/4, 0];[-1/4, 1/3, 0];[0, 0, 2/3]}   %inertia tensor about origin
L=I*w;T=w'*I*w/2; %angular momentum, energy about the given axis
fprintf('L= %5.3f %5.3f %5.3f, T=%5.3f\n', L, T)
Ip={[1/12, 0, 0];[0, 7/12, 0];[0, 0, 2/3]}      %principal axes moments
w1=[cos(135*h);cos(135*h);cos(90*h)]*mw ; %1st principal axis rotation
fprintf('1st p axis rotation: w1= %5.3f %5.3f %5.3f\n', w1)
%sqrt(w1'*w1)  %can check that the magnitude of rotation is unchanged
L1=Ip*w1;      %angular momentum about 1st princ. axis
T1=w1'*Ip*w1/2;%kinetic energy about 1st princ. axis
fprintf('L1= %5.3f %5.3f %5.3f, T1=%5.3f\n', L1, T1)
w2=[cos(135*h);cos(45*h);cos(90*h)]*mw;   %2nd principal axis rotation
fprintf('2nd p axis rotation: w2= %5.3f %5.3f %5.3f\n', w2)
%sqrt(w2'*w2)  %can check that the magnitude of rotation is unchanged
L2=Ip*w2;      %angular momentum about 2nd princ. axis
```

```
T2=w2'*Ip*w2/2;%kinetic energy about 1st princ. axis
fprintf('L2= %5.3f %5.3f %5.3f, T2=%5.3f\n', L2, T2)
w3=[cos(90*h);cos(90*h);cos(0*h)]*mw;    %3rd principal axis rotation
fprintf('3rd p axis rotation: w3= %5.3f %5.3f %5.3f\n', w3)
%sqrt(w3'*w3)  %can check that the magnitude of rotation is unchanged
L3=Ip*w3;       %angular momentum about 3rd princ. axis
T3=w3'*Ip*w3/2;%kinetic energy about 1st princ. axis
fprintf('L3= %5.3f %5.3f %5.3f, T3=%5.3f\n', L3, T3)
```

OUTPUT

```
r_energy.txt - output from r_energy.m
>> r_energy
w= 1.000 1.000 0.000, magnitude=1.414
I =
     0.3333    -0.2500         0
    -0.2500     0.3333         0
          0          0    0.6667
L= 0.083 0.083 0.000, T=0.083
Ip =
    0.0833          0         0
         0     0.5833         0
         0          0    0.6667
1st p axis rotation: w1= -1.000 -1.000 0.000
L1= -0.083 -0.583 0.000, T1=0.333
2nd p axis rotation: w2= -1.000 1.000 0.000
L2= -0.083 0.583 0.000, T2=0.333
3rd p axis rotation: w3= 0.000 0.000 1.414
L3= 0.000 0.000 0.943, T3=0.667
>>
```

■ 12.6 Euler's Equations

In this section we obtain the equations of motion for a rotating rigid body under the action of external forces.

A. Torque

The equation of motion of a rotating body under the action of a torque, as viewed from an inertial reference frame (S frame) or a fixed coordinate system, is given by (11.3.21)

$$\boldsymbol{\tau} = \left(\frac{d\mathbf{L}}{dt}\right)_{fixed}, \tag{12.6.1}$$

where $\boldsymbol{\tau}$ is the applied torque and $\mathbf{L}$ is the rigid body's angular momentum. For a rigid body, $\mathbf{L}$ is most conveniently expressed in terms of principal axes, which are fixed in the body and rotate with it. Therefore, the rotating body is not an inertial system and it is necessary to consider an S' frame rotating with the body, similar to what we did in Section 12.3C, but here we include the fact that, in general, the body's angular momentum can change as the body rotates. Accordingly, we use (7.6.7) in its full form and write (12.6.1) as

$$\boldsymbol{\tau} = \left(\frac{d\mathbf{L}}{dt}\right)_{fixed} = \left(\frac{d\mathbf{L}}{dt}\right)_{rot} + \boldsymbol{\omega} \times \mathbf{L}. \tag{12.6.2}$$

Writing $\mathbf{L} = \{I\}\boldsymbol{\omega}$, we see that

$$\left(\frac{d\mathbf{L}}{dt}\right)_{rot} = \left(\frac{d\{I\}}{dt}\right)_{rot} \boldsymbol{\omega} + \{I\}\left(\frac{d\boldsymbol{\omega}}{dt}\right)_{rot}. \tag{12.6.3}$$

However,

$$(d\boldsymbol{\omega}/dt)_{fixed} = (d\boldsymbol{\omega}/dt)_{rot} + \boldsymbol{\omega} \times \boldsymbol{\omega} = (d\boldsymbol{\omega}/dt)_{rot}$$

because $\boldsymbol{\omega} \times \boldsymbol{\omega} = 0$, and also since S' rotates with the body, its inertia moments don't change, so that $(d\{I\}/dt)_{rot} = 0$. Therefore, the equations of motion (12.6.2) can be written solely in terms of the fixed inertial reference frame as

$$\boldsymbol{\tau} = \left(\frac{d\mathbf{L}}{dt}\right) = \{I\}\left(\frac{d\boldsymbol{\omega}}{dt}\right) + \boldsymbol{\omega} \times \mathbf{L} = \{I\}\left(\frac{d\boldsymbol{\omega}}{dt}\right) + \boldsymbol{\omega} \times [\{I\}\boldsymbol{\omega}], \tag{12.6.4}$$

where the subscripted "fixed" is no longer needed and has been dropped. Once again, working with rotations about the principal axes

$$\mathbf{L} = \{I\}\boldsymbol{\omega} = L_1\hat{e}_1 + L_2\hat{e}_2 + L_3\hat{e}_3 = I_1\omega_1\hat{e}_1 + I_2\omega_2\hat{e}_2 + I_3\omega_3\hat{e}_3, \tag{12.6.5a}$$

and

$$\boldsymbol{\omega} = \omega_1\hat{e}_1 + \omega_2\hat{e}_2 + \omega_3\hat{e}_3, \tag{12.6.5b}$$

so that

$$\boldsymbol{\omega} \times \mathbf{L} = \begin{vmatrix} \hat{e}_1 & \hat{e}_2 & \hat{e}_3 \\ \omega_1 & \omega_2 & \omega_3 \\ I_1\omega_1 & I_2\omega_2 & I_3\omega_3 \end{vmatrix} = \hat{e}_1(I_3 - I_2)\omega_2\omega_3 - \hat{e}_2(I_3 - I_1)\omega_1\omega_3 + \hat{e}_3(I_2 - I_1)\omega_1\omega_2,$$

(12.6.5c)

in addition to

$$\{I\}\left(\frac{d\boldsymbol{\omega}}{dt}\right) = I_1\dot{\omega}_1\hat{e}_1 + I_2\dot{\omega}_2\hat{e}_2 + I_3\dot{\omega}_3\hat{e}_3,$$

(12.6.5d)

and $\boldsymbol{\tau} = \tau_1\hat{e}_1 + \tau_2\hat{e}_2 + \tau_3\hat{e}_3$, we finally obtain the Euler's torque equations for the rotational motion of a rigid body along the principal axes of the body

$$\begin{aligned} \tau_1 &= I_1\dot{\omega}_1 + (I_3 - I_2)\omega_2\omega_3 \\ \tau_2 &= I_2\dot{\omega}_2 + (I_1 - I_3)\omega_1\omega_3, \\ \tau_3 &= I_3\dot{\omega}_3 + (I_2 - I_1)\omega_1\omega_2 \end{aligned}$$

(12.6.6)

which were first obtained by Leonhard Euler (1707–1783) in the mid-1700s. Notice that, according to these equations, even in the absence of a torque, the angular velocity can change, and this is due to $\boldsymbol{\tau} = 0 \Rightarrow \mathbf{L} = $ constant; that is, the angular velocities must change in order for the angular momentum to remain constant. This is so unless the rotation takes place about a principal axis. For example, if initial rotation occurs about a principal axis, say, $\hat{e}_1$, then we must have $\omega_1 \neq 0$, and $\omega_2 = \omega_3 = 0$, in which case (12.6.6) leads to $I_1\dot{\omega}_1 = I_2\dot{\omega}_2 = I_3\dot{\omega}_3 = 0$, or that the angular velocity components along the principal axis remain constant, equal to their initial values in agreement with the previous statement. Similar arguments apply to rotations along any of the other two principal axes.

B. Connection to Kinetic Energy

From Equation (12.6.4) we can obtain the rate of change of energy as follows:

$$\boldsymbol{\omega} \cdot \boldsymbol{\tau} = \boldsymbol{\omega} \cdot \left(\{I\}\left(\frac{d\boldsymbol{\omega}}{dt}\right) + \boldsymbol{\omega} \times \mathbf{L}\right) = \boldsymbol{\omega} \cdot \{I\}\left(\frac{d\boldsymbol{\omega}}{dt}\right) + \boldsymbol{\omega} \cdot (\boldsymbol{\omega} \times \mathbf{L}); \quad (12.6.7)$$

however, using (12.6.5c) it is easy to see that the last term vanishes, and from (12.6.5d) we can see that $\boldsymbol{\omega} \cdot \{I\}(d\boldsymbol{\omega}/dt) = (d\boldsymbol{\omega}/dt) \cdot \{I\}\boldsymbol{\omega} = d(\boldsymbol{\omega} \cdot \mathbf{L})/dt/2$. Finally, using (12.5.3) we see that (12.6.7) becomes

$$\boldsymbol{\omega} \cdot \boldsymbol{\tau} = \frac{d}{dt}T_{rot}; \tag{12.6.8}$$

that is, dotting the rigid body's angular velocity with the torque is equivalent to the time rate of change of the rotational kinetic energy of the rigid body.

EXAMPLE 12.9

Solve Euler equations of motion for a symmetric top under the special circumstance that there are no external torques acting on it.

....

Solution

A symmetric top is shown in Figure 12.16. Because of symmetry, $\hat{e}_3$ is a principal axis of symmetry, so that the moment of inertia, I_3, about $\hat{e}_3$ is a symmetric principal moment of inertia. The other two principal moments of inertia are equal $I_1 = I_2$, since rotations about the principal axes $\hat{e}_1$ and $\hat{e}_2$ are equivalent. With this understanding, from (12.6.6), the equations of motion for the top become

$$0 = I_1\dot{\omega}_1 + (I_3 - I_1)\omega_2\omega_3, \quad 0 = I_1\dot{\omega}_2 + (I_1 - I_3)\omega_1\omega_3, \quad I_3\dot{\omega}_3 = 0. \tag{12.6.9}$$

The last of these tells us that

$$\omega_3 = \text{constant.} \tag{12.6.10}$$

I FIGURE 12.16

The first two of (12.6.9) can be rearranged to read

$$\dot\omega_1 + \gamma\omega_2\omega_3 = 0 \quad \dot\omega_2 - \gamma\omega_1\omega_3 = 0, \tag{12.6.11a}$$

where

$$\gamma \equiv (I_3 - I_1)/I_1. \tag{12.6.11b}$$

These two equations are a coupled system of equations for the unknowns ω_1 and ω_2. Taking their time derivative, we have

$$\ddot\omega_1 + \gamma\dot\omega_2\omega_3 = 0 \quad \ddot\omega_2 - \gamma\dot\omega_1\omega_3 = 0, \tag{12.6.12}$$

and substituting $\dot\omega_1$ and $\dot\omega_2$ from (12.6.11) into (12.6.12), we obtain two-second order differential equations for ω_1 and ω_2, along with the corresponding solutions

$$\begin{aligned}\ddot\omega_1 + (\gamma\omega_3)^2\omega_1 = 0 &\Rightarrow \omega_1 = c\cos(\gamma\omega_3 t + \varphi)\\ \ddot\omega_2 + (\gamma\omega_3)^2\omega_2 = 0 &\Rightarrow \omega_2 = c'\cos(\gamma\omega_3 t + \varphi')\end{aligned}. \tag{12.6.13}$$

These results can be more properly written if we notice that according to (12.6.12), $\dot\omega_1 = -\gamma\omega_2\omega_3$, and using this along with (12.6.13) we must have

$$\dot\omega_1 = -\gamma\omega_3 c\sin(\gamma\omega_3 t + \varphi) = -\gamma\omega_3 c'\cos(\gamma\omega_3 t + \varphi'), \tag{12.6.14}$$

so that we can let $c' = c$, and $\varphi' = \varphi - \pi/2$, to make it work. Substituting all this back into (12.6.13) we finally obtain

$$\omega_1 = c\cos(\Omega_b t + \varphi), \quad \omega_2 = c\cos(\Omega_b t + \varphi - \pi/2) = c\sin(\Omega_b t + \varphi), \tag{12.6.15a}$$

with the values of c and φ depending on the initial conditions and where we have defined the body precession frequency

$$\Omega_b = \gamma\omega_3 = \gamma\omega\cos\varphi_b, \tag{12.6.15b}$$

where we have used $\omega_3 = \omega\cos\varphi_b$ according to Figure 12.17 and explained below. This result means that ω_1 and ω_2 describe a circle of radius c. This is a circle made by the vector $\boldsymbol{\omega}$ about the symmetry axis $\hat{e}_3$ in which the constant com-

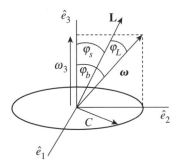

FIGURE 12.17 Precession of $\overline{\omega}$ as it traces out a cone, the *body cone,* whose half angle is φ_b with respect to $\hat{e}_3$. The angle between **L** and $\boldsymbol{\omega}$ is φ_L. The space cone half angle that **L** makes with respect to $\hat{e}_3$ is φ_s.

ponent ω_3 points. The precession of $\boldsymbol{\omega}$ at the rate Ω_b depends on the value of γ from (12.6.11b) according to the moment of inertia values. The higher the value of γ, the higher is Ω_b. In precessing, $\boldsymbol{\omega}$ traces out a cone, the *body cone,* whose half angle φ_b, shown in Figure 12.17 with respect to $\hat{e}_3$, can be found.

In Figure 12.17, $\omega = \sqrt{\omega_1^2 + \omega_2^2 + \omega_3^2} = \sqrt{c^2 + \omega_3^2}$, where the circle is of radius $c = \omega \sin\varphi_b$, and $\omega_3 = \omega \cos\varphi_b$, so that

$$\tan\varphi_b = \frac{c}{\omega_3}. \tag{12.6.16}$$

Also shown in Figure 12.17 is the angular momentum, **L**, which does not point in the same direction as $\boldsymbol{\omega}$. Using the preceding results into (12.6.5a), we can find the angle between **L** and $\boldsymbol{\omega}$ as

$$\cos\varphi_L = \frac{\boldsymbol{\omega} \cdot \mathbf{L}}{\omega L} = \frac{I_1 c^2 + I_3 \omega_3^2}{\sqrt{(c^2 + \omega_3^2)(I_1^2 c^2 + I_3^2 \omega_3^2)}} = \frac{2T_{rot}}{\omega L}, \tag{12.6.17}$$

where the magnitude of the angular momentum

$$L = \sqrt{I_1^2 c^2 + I_3^2 \omega_3^2} \tag{12.6.18}$$

has been used, which is constant, as is the angle φ_L. This is as it should be for T_{rot} to be conserved. In the present example, there are no external torques present, and from the fixed coordinate system viewpoint the angular momentum is constant in direction and magnitude, as dictated by the left-hand side of (12.6.2). We will come back to this point later. From the point of view of the rotating coordinate system, however, according to (12.6.3) the angular momentum can change. In fact, one can see that although the magnitude of $\mathbf{L}$ from (12.6.18) is constant and the angle of (12.6.17) between $\boldsymbol{\omega}$ and $\mathbf{L}$ remains also constant, the direction of $\mathbf{L}$ does change due to the precession frequency (Ω_b) of its L_1 and L_2 components (see Equation 12.6.5a) coming from the behavior of $\omega_1(t)$ and $\omega_2(t)$; thus, from the viewpoint of the rotating coordinate system, both $\mathbf{L}$ and $\boldsymbol{\omega}$ precess at the same rate while maintaining the same angle from each other. Also, because of their separation angle φ_L, $\mathbf{L}$ and $\boldsymbol{\omega}$ trace out different cones. Although the cone traced out by $\boldsymbol{\omega}$ is the body cone, the cone traced out by $\mathbf{L}$ is the *space cone*. The space cone half angle that $\mathbf{L}$ makes with respect to $\hat{e}_3$ is φ_s and can be found as follows:

$$\cos\varphi_s = \frac{\mathbf{L} \cdot \hat{e}_3}{L} = \frac{I_3\omega_3}{\sqrt{I_1^2 c^2 + I_3^2 \omega_3^2}} \quad \Rightarrow \quad \tan\varphi_s = \frac{I_1 c}{I_3\omega_3} = \frac{I_1}{I_3}\tan\varphi_b,$$

$$(12.6.19)$$

where (12.6.16) has been used. From (12.6.19) we see that if $I_3 > I_1$ then $\varphi_s < \varphi_b$ and the space cone lies within the body cone, else if $I_3 < I_1$ then $\varphi_s > \varphi_b$ and the space cone lies outside the body cone.

As mentioned before, from the viewpoint of the fixed coordinate system, the angular momentum is fixed due to the absence of external torques. Thus, according to the fixed coordinate system, the spin vector that points in the direction of $\boldsymbol{\omega}$, where $\boldsymbol{\omega}$ is itself a frequency, must precess around $\mathbf{L}$ with a frequency $\Omega_L \equiv |\Omega_b + \boldsymbol{\omega}|$. Also since $\Omega_b = \Omega_b \hat{e}_3$ and $\boldsymbol{\omega}$ is given by (12.6.5b) along with the components given in (12.6.10 and 12.6.13), we can then obtain the relation

$$\Omega_L = |\Omega_b + \boldsymbol{\omega}| = \frac{L}{I_1} = \Omega_b \frac{\sin\varphi_b}{\sin\varphi_L}, \qquad (12.6.20a)$$

where L is the magnitude of the angular momentum of (12.6.18), $I_1 = I_2$ is one of the principal moments of inertia, and φ_b, φ_L are given by (12.6.16 and 12.6.17). Because Ω_b is given by (12.6.15b) with γ given by (12.6.11b), and since from

Figure 12.17, $\varphi_L = \varphi_b - \varphi_s$, then with the help of (12.6.19), (12.6.20a) can also be written as

$$\Omega_L = \omega \frac{\sin \varphi_b}{\sin \varphi_s}. \tag{12.6.20b}$$

We can use MATLAB to simulate all of the above concepts. This is shown in Figure 12.18.

Figure 12.18(a) shows the positions of $\mathbf{L}$ and $\boldsymbol{\omega}$ as a function of time, as described by formula (12.6.5) along with (12.6.15), where the space and the body cones are respectively shown. The angular momentum is at an angle of $\varphi_s \approx 40.9$ degrees and $\boldsymbol{\omega}$ makes an angle of $\varphi_b \approx 45$ degrees, both measured from the ω_3 axis of symmetry. Since $I_3 > I_1$ the space cone lies within the body cone. The angle made between $\mathbf{L}$ and $\boldsymbol{\omega}$ is about $\varphi_L \approx 4.1$ degrees. The figure shows the precession of $\mathbf{L}$ and $\boldsymbol{\omega}$ about the ω_3 axis ($\hat{e}_3$) with frequency Ω_b. The script torque_free.m should be run to see the motion in real time. The moment of inertia values used in this simulation are $I_1 = I_2 = 1.3 \text{ kgm}^2$ and $I_3 = 1.5 \text{ kgm}^2$. Figure 12.18(b) shows a similar simulation with slightly different parameters, for viewing purposes, as shown on the figure. In addition to showing $\mathbf{L}$ and $\boldsymbol{\omega}$, the angular frequency is made to precess with a frequency of Ω_L rather than with Ω_b in order to simulate the motion seen by the S frame ($\hat{i}$, $\hat{j}$, $\hat{k}$) where $\mathbf{L}$ remains fixed. We have swapped the positions of the ω_3 axis with $\mathbf{L}$ and shifted $\boldsymbol{\omega}$ by the angle shift φ_s that $\mathbf{L}$ underwent. Thus, we performed operations according to the following. We changed $c \to c' = \omega \sin (\varphi_b - \varphi_s)$, and $\omega_3 \to \omega'_3 = \omega \cos (\varphi_b - \varphi_s)$, so that similar to (12.6.15), $\omega_1 \to c' \cos (\Omega_L t + \varphi) = \omega'_1$, and $\omega_2 \to c' \sin (\Omega_L t + \varphi) = \omega'_2$, which preserves the magnitude of $\boldsymbol{\omega} = (\omega'_1, \omega'_2, \omega'_3)$, but with a frequency Ω_L. The angular momentum magnitude is to remain the same, but pointing in the $\hat{k}$ direction, so $\mathbf{L} \to (0, 0, L)$; also since $\hat{e}_3$ must move with the body, $\hat{e}_3 \to e_{31}\hat{i} + e_{32}\hat{j} + e_{33}\hat{k}$, where $e_{31} = \sin \varphi_s \cos (\Omega_L + \varphi)$, $e_{32} = \sin \varphi_s \sin (\Omega_L + \varphi)$, and $e_{31} = \cos \varphi_s$. Here we suppose that the body coordinates are $\hat{e}_1$, $\hat{e}_2$, and $\hat{e}_3$ (S' frame), but only $\hat{e}_3$ is shown in Figure 12.18(b). Thus, Figure 12.18(b) shows the precession of $\boldsymbol{\omega}$ about $\mathbf{L}$ with a frequency Ω_L. There, also, $\hat{e}_3$ rotates with Ω_L to stay with the body. The angles between these vectors do not change after the transformation from the S' to the S

(a)   Torque Free Motion of a Top, $I_1 = 1.3$ kgm^2, $I_3 = 1.5$ kgm^2, $\omega = 1.41$ rad/s, $L = 1.98$ kgm^2/s

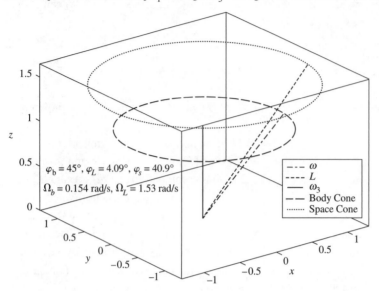

(b)   Torque Free Motion of a Top, $I_1 = 1.3$ kgm^2, $I_3 = 1.8$ kgm^2, $\omega = 1.41$ rad/s, $L = 2.22$ kgm^2/s

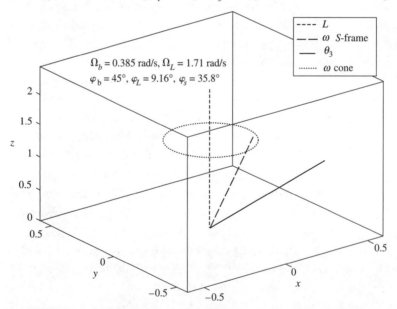

FIGURE 12.18   (a) Positions of **L**, and $\boldsymbol{\omega}$ as a function of time as described by formulas (12.6.5) along with (12.6.15) where the space and the body cones are respectively shown. (b) Similar simulation with slightly different parameters, for viewing purposes, and in addition to showing **L** and $\boldsymbol{\omega}$, the angular frequency is made to recess with a frequency of Ω_L rather than with Ω_b in order to simulate the motion seen by the S frame $(\hat{i}, \hat{j}, \hat{k})$ where **L** remains fixed.

view. The script `torque_free_s.m` performs this, which should be run to see the simulation in real time. Both MATLAB scripts follows. To summarize, $\boldsymbol{\omega}$ makes a body cone as it goes around $\hat{e}_3$ with frequency Ω_b (Figure 12.18a), as seen in the S' frame; it makes a space cone as it goes around $\mathbf{L}$ with frequency Ω_L (Figure 12.18b), as seen from the S frame.

SCRIPT

```
%torque_free.m - plots the angular frequency and momentum for torque free
%motion of a top versus time in the body (S') frame
clear; ap=180/pi;                           %angle conversion
I1=1.3; I2=I1; I3=1.5; gam=(I3-I1)/I1;      %principal moments of inertia
w3=1;fi=0; c=1;                             %initial values, c, fi, w3
omb=gam*w3;                                 %precessional freq.
tmax=2*pi/omb; N=40; ts=tmax/N;             %time limits
t=[0:ts:tmax];                              %time range
w1=c*cos(omb*t+fi);w2=c*sin(omb*t+fi);      %w1, w2 versus time
L1=I1*w1; L2=I2*w2; L3=I3*w3;               %Ang. Mom. components
v1=max(L1); v2=max(L2); v3=max(L3*(1+0.1)); %view window
fi_b=atan(c/w3);                            %angle between w and w3
fi_L=acos((I1*c^2+I3*w3^2)/...              %angle between w and L
        (sqrt((c^2+w3^2)*(I1^2*c^2+I3^2*w3^2))));
fi_s=atan(I1*tan(fi_b)/I3);                 %angle between L and w3
omL=omb*sin(fi_b)/sin(fi_L);                %prec. freq. of w about L
%omL=sqrt(I1^2*c^2+I3^2*w3^2)/I1            %another formula for omL
w=sqrt(c^2+w3^2);                           %magnitude of w
L=sqrt(I1^2*c^2+I3^2*w3^2);                 %magnitude of L
for i=1:N
    clf
    axis ([-v1, v1, -v2, v2, 0, v3])
    h(1)=line([0, w1(i)], [0, w2(i)], [0, w3], 'color', 'k', ...
            'LineStyle', '-.', 'linewidth', 1.5);           %w line
    h(2)=line([0, L1(i)], [0, L2(i)], [0, L3], 'color', 'r', ...
            'LineStyle', '-', 'linewidth', 1.5);            %L line
    h(3)=line([0, 0], [0, 0], [0, w3], 'color', 'b', 'linewidth', 1.5);%w3 line
    box on
    pause(.1)
end
hold on
h(4)=plot3 (w1, w2, w3*(t+0.01)./(t+0.01), 'k-.'); %plot the w cone (body)
h(5)=plot3 (L1, L2, L3*(t+0.01)./(t+0.01), 'r:'); %plot the L cone (space)
hh=legend(h, '\omega', 'L', '\omega_3', 'Body Cone', 'Space Cone', 4);
```

```
set(hh, 'FontSize', 8, 'Position', [0.7 0.3 0.2 0.17])
str1=cat(2, 'Torque Free Motion of a Top, I_1=', num2str(I1, 3), ...
         'kgm^2, I_3=', num2str(I3, 3), 'kgm^2', ', \omega=', num2str(w, 3)...
         , 'rad/s, L=', num2str(L, 3), 'kgm^2/s');
title(str1, 'FontSize', 10), xlabel('x', 'FontSize', 14)
ylabel('y', 'FontSize', 14), zlabel('z', 'FontSize', 14)
str2=cat(2, '\phi_b=', num2str(fi_b*ap, 3), '^o, \phi_L= ', ...
         num2str(fi_L*ap, 3), '^o, \phi_s=', num2str(fi_s*ap, 3), '^o');
str3=cat(2, '\Omega_b=', num2str(omb, 3), 'rad/s, \Omega_L=', ...
         num2str(omL, 3), 'rad/s');
text (-v1*(1-0.1), v2, v3*(1-0.75), str2, 'FontSize', 11)
text (-v1*(1-0.1), v2, v3*(1-0.9), str3, 'FontSize', 11)
```

SCRIPT

```
%torque_free_s.m - plots the angular frequency and momentum for
%torque free motion of a top versus time in the body (S') as well
%as in the space frame (S)
clear; ap=180/pi;                         %angle conversion
I1=1.3; I2=I1; I3=1.8; gam=(I3-I1)/I1;    %principal moments of inertia
w3=1; fi=0; c=1;                          %initial values, c, fi, w3
fi_b=atan(c/w3);                          %angle between w and w3
fi_L=acos((I1*c^2+I3*w3^2)/...            %angle between w and L
     (sqrt((c^2+w3^2)*(I1^2*c^2+I3^2*w3^2)))));
fi_s=atan(I1*tan(fi_b)/I3);               %angle between L and w3
omb=gam*w3;                               %precessional freq.
omL=omb*sin(fi_b)/sin(fi_L);              %prec. freq. of w about L
tmax=.9*2*pi/omb; N=100; ts=tmax/N;       %time limits
t=[0:ts:tmax];                            %time range
w1=c*cos(omb*t+fi);w2=c*sin(omb*t+fi);    %w1, w2 versus time
L1=I1*w1; L2=I2*w2; L3=I3*w3;             %Ang. Mom. components
w=sqrt(c^2+w3^2);                         %magnitude of w
L=sqrt(I1^2*c^2+I3^2*w3^2);               %magnitude of L
%=== shift old e3 by fi_s to swap places with old L, preserve magnitude
e31=1*sin(fi_s)*cos(omL*t+fi);            %must move with omL
e32=1*sin(fi_s)*sin(omL*t+fi);            %must move with omL
e33=1*cos(fi_s);
%new w and components, shift by fi_s, preserve magnitude, change rate
c=w*sin(fi_b-fi_s);
w3=w*cos(fi_b-fi_s);
w1=c*cos(omL*t+fi);w2=c*sin(omL*t+fi);    %use omL precession rate
%new L now in old w3 direction and fixed, preserve magnitude
L1=0; L2=0; L3=L;
v1=max(e31); v2=max(e32); v3=max(L3*(1+0.1));  %view window
for i=1:N
    clf
    axis([-v1, v1, -v2, v2, 0, v3])
```

```
   hold on
   h(1)=line([0, L1], [0, L2], [0, L3], 'color', 'r', ...        %L line
           'LineStyle', '-', 'linewidth', 1.5);
   h(2)=line([0, w1(i)], [0, w2(i)], [0, w3], 'color', 'm', ...   %S frame w line
           'LineStyle', '-', 'linewidth', 1);
   h(3)=line([0, e31(i)], [0, e32(i)], [0, e33], 'color', 'b', ...%e3 line
               'linewidth', 1.5);
   box on
   pause(.05)
end
h(4)=plot3 (w1, w2, w3*(t+0.01)./(t+0.01), 'm:');       %plot the S w cone
hh=legend(h, 'L', '\omega S-frame', 'e_3', '\omega cone', 1);
str1=cat(2, 'Torque Free Motion of a Top, I_1=', num2str(I1, 3), ...
           'kgm^2, I_3=', num2str(I3, 3), 'kgm^2');
str1=cat(2, 'Torque Free Motion of a Top, I_1=', num2str(I1, 3), ...
           'kgm^2, I_3=', num2str(I3, 3), 'kgm^2', ', \omega=', num2str(w, 3), ...
           'rad/s, L=', num2str(L, 3), 'kgm^2/s');
title(str1, 'FontSize', 10), xlabel('x', 'FontSize', 14)
ylabel('y', 'FontSize', 14), zlabel('z', 'FontSize', 14)
str2=cat(2, '\phi_b=', num2str(fi_b*ap, 3), '^o, \phi_L= ', ...
           num2str(fi_L*ap, 3), '^o, \phi_s=', num2str(fi_s*ap, 3), '^o');
str3=cat(2, '\Omega_b=', num2str(omb, 3), 'rad/s, \Omega_L=', ...
           num2str(omL, 3), 'rad/s');
text (-v1*(1-0.4), v2, v3*(1-0.15), str2, 'FontSize', 11)
text (-v1*(1-0.4), v2, v3*(1-0.05), str3, 'FontSize', 11)
```

EXAMPLE 12.10

Obtain and solve Euler equations of motion for an ellipsoidal body under the special circumstance that there are no external torques acting on it and whose rotational motion takes place about the center of mass. The ellipsoid's semimajor axes are $a = 3$ m, $b = 2$ m, $c = 1$ m, and its density is $\rho = (1/8)$ kg/m^3.

Solution

Euler's Equations (12.6.6) for this case become

$$\dot{\omega}_1 = -\gamma_1 \omega_2 \omega_3 \quad \dot{\omega}_2 = -\gamma_2 \omega_1 \omega_3 \quad \dot{\omega}_3 = -\gamma_3 \omega_1 \omega_2, \qquad (12.6.21)$$

where $\gamma_1 = (I_3 - I_2)/I_1$, $\gamma_2 = (I_1 - I_3)/I_2$, and $\gamma_3 = (I_2 - I_1)/I_3$. We need to obtain the principal moments of inertia. To obtain the inertia tensor, we can employ a numerical procedure to perform the moments. For example, since the ellipsoid equation is

$$((x - x_0)/a)^2 + ((y - y_0)/b)^2 + ((z - z_0)/c)^2 = 1, \qquad (12.6.22)$$

where we take the origin $(x_0, y_0, z_0) = (0, 0, 0)$, an inertia tensor component is

$$I_{xx} = \rho \int_{-a}^{a} \int_{-y_L}^{y_L} \int_{-z_L}^{z_L} (y^2 + z^2)\,dxdydz,$$

(12.6.23)

where in the limits $y_L = b\sqrt{1 - (x/a)^2}$ and $z_L = c\sqrt{1 - (x/a)^2 - (y/b)^2}$. The rest of the inertia components can similarly be set up. Because the standard xyz-coordinate system whose origin lies at the center of mass can be used as principal axes here, we should expect the inertia tensor will be diagonal, each element being a principal moment. The mass of the ellipsoid is $M = \rho 4\pi abc/3 = \pi$. Below, we first obtain the inertia tensor using the MATLAB script `ellipso.m`. and numerically confirm that indeed the tensor is diagonal as well as the expected Cartesian directions for the principal axes. The script `ellipso.m` makes use of the script `inert_el2.m` to carry out the integrations; both are listed below. The output of this script is in file `ellipso.txt`, also listed below. This script produces the ellipsoid Figure 12.19, which also shows the principal inertia moments numerically calculated using Simpson's rule as $I_1 = I_{xx} \approx 3.2$ kgm^2, $I_2 = I_{yy} = 6.3$ kgm^2, and $I_3 = 8.2$ kgm^2. The accuracy of the calculation is gauged by the ability to obtain the ellipsoid's mass. The script obtains a value close to π with a percent error of about 0.3%, as shown in the output.

Ellipsoid: $a = 3$, $b = 2$, $c = 1$, $\rho = 0.125$, $M = 3.14$, $I_{xx} = 3.2$, $I_{yy} = 6.3$, $I_{zz} = 8.2$

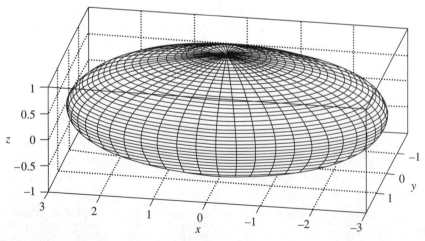

▌FIGURE 12.19   Ellipsoid used in Example 12.10.

SCRIPT

```
%ellipso.m calculates ellipsoid inertia tensor, & mass
%numerically, also plots it in 3d
%Ellipsoid: (x-x0)^2/a^2+(y-y0)^2/b^2+(z-z)^2/c^2=1
%uses Simpson's rile for 3d integration
clear;
a=3;b=2;c=1;                        %semimajor axes
rho=1/8;                            %density
ax=-a;                              %x lower limit
bx=a; Nx=35; dx=(bx-ax)/(Nx-1);    %x upper limit, points, spacing
x=[ax:dx:bx];                       %x grid
Ny=35;Nz=35;                        %y, z points
ie=7;                               %matrix elements calculated + volume
for k=2:Nx                                %x loop - begin at 2 so ay ~= 0
    ay=-b*sqrt(1-(x(k)/a)^2);             %y lower limit,
    by=b*sqrt(1-(x(k)/a)^2);              %y upper limit,
    if real(by) ~=0                       %check if array exists
      dy=(by-ay)/(Ny-1);                  %y spacing
      y=[ay:dy:by];                       %y grid
      for j=2:Ny                          %y loop - begin at 2 so az ~= 0
        az=-c*sqrt(1-(x(k)/a)^2-(y(j)/b)^2);  %z lower limit
        bz=c*sqrt(1-(x(k)/a)^2-(y(j)/b)^2);   %z upper limit
            if real(bz) ~= 0              %check z array exists
              dz=(bz-az)/(Nz-1);          %z spacing
              z=[az:dz:bz];               %z grid
              for i=1:Nz                  %z loop
                for m=1:ie
                  fz(m, i)=rho*inert_el2(m, x(k), y(j), z(i));
                end
              end
            end                           %end z loop
          end                             %end 2nd if
        for m=1:ie
           fy(m, j)=simp(fz(m, :), dz);       %Simpson rule
        end
      end                                 %end y loop
    end                                   %end 1st if
    for m=1:ie
       fx(m, k)=simp(fy(m, :), dy);           %Simpson rule
    end
end                                       %end x loop
%finally integrate over the x coord to get the moments
for m=1:ie
    ff=simp(fx(m, :), dx);                    %Simpson rule
       if    m==1 Ixx=ff;
       elseif m==2 Ixy=ff;
       elseif m==3 Ixz=ff;
```

```
        elseif m==4 Iyy=ff;
        elseif m==5 Iyz=ff;
        elseif m==6 Izz=ff;
        elseif m==7 Mass=ff;
        end
%fprintf('m= %2i, The integral is %4.3f\n', m, ff)
end
Iyx=Ixy; Izx=Ixz; Izy=Iyz;                    %use symmetry for the rest
A={[Ixx, Ixy, Ixz];[Iyx, Iyy, Iyz];[Izx, Izy, Izz]};%inertia tensor
disp(['Semimajor axes (m): a, b, c=', rat(a), ' ', rat(b), ' ', rat(c)])
disp 'Inertia Tensor in kgm^2'
disp(rats(A))                                 %display A in string fraction form
M=rho*4*pi*a*b*c/3;                           %actual ellipsoid mass
p_e=(Mass-M)*100/M;                           %%error on the mass
str1=cat(2, 'density=', num2str(rho, 3), 'kg/m^3, Mass=', num2str(Mass, 3), ...
           ' kg, % mass error=', num2str(p_e, 3));
disp(str1)
%draw the ellipsoid centered at x0, y0, z0, & semimajor axes a, b, c
x0=0; y0=0; z0=0;N=50;
[x, y, z]=ellipsoid(x0, y0, z0, a, b, c, N); %uses 50 mesh points
h=mesh(x, y, z, 'EdgeColor', [0.0 0.0 0.9]);
axis equal, grid on, box on, view ([-0.1 0.5 0.2])
xlabel('x', 'FontSize', 14), ylabel('y', 'FontSize', 14), zlabel('z', 'FontSize', 14)
Title(['Ellipsoid: a=', rat(a), ', b=', rat(b), ', c=', rat(c), ', \rho=', ...
      num2str(rho, 3), ', M=', num2str(M, 3), ', I_{xx}=', num2str(A(1, 1), 2), ...
      ', I_{yy}=', num2str(A(2, 2), 2), ', I_{zz}=', num2str(A(3, 3), 2)], ...
      'FontSize', 13)
```

FUNCTION

```
%inert_el2.m
function inerel2=inert_el2(m, x, y, z)
%Inertia function integrands in cartesian coords
%if m=7 it does the volume integrand
if     m==1 inerel2=y^2+z^2;
elseif m==2 inerel2=-x*y;
elseif m==3 inerel2=-x*z;
elseif m==4 inerel2=x^2+z^2;
elseif m==5 inerel2=-y*z;
elseif m==6 inerel2=x^2+y^2;
elseif m==7 inerel2=1.0;
else
    disp ' only 7 integrands are needed '
    break
end
```

```
                           OUTPUT

ellipso.txt: Output from ellipso.m
>> ellipso
Semimajor axes (m): a, b, c=3 2 1
Inertia Tensor in kgm^2
    145/46              0               0
       0            1589/251           0
       0               0           1242/151
density=0.125kg/m^3, Mass=3.15 kg, % mass error=0.296
>>
```

The results of the principal moments from the above script, ellipso.m, can now be used into another script, torquef2.m, which solves the equations of motion (12.6.21). The script solves the equations numerically in an array approach as we've done before. The script torquef2_der.m is the MATLAB function that calculates the derivatives, i.e., the right-hand side of the following equations

$$\frac{d\omega(1)}{dt} = -\gamma_1\omega(2)\omega(3) \quad \frac{d\omega(2)}{dt} = -\gamma_2\omega(1)\omega(3) \quad \frac{d\omega(3)}{dt} = -\gamma_3\omega(1)\omega(3), \quad (12.6.24)$$

with the $\gamma's$ as in (12.6.21) depending on the principal moments as shown in the script torquef2.m. The results are shown in Figure 12.20 along with the parameters used according to the above discussion.

Figure 12.20(a) contains plots of $\omega_1(t)$, $\omega_2(t)$, and $\omega_3(t)$, respectively, with their phase space plot in three dimensions on the lower right. Notice that the frequencies appear to be harmonic, although not exactly sinusoidal. The angular momentum and angular velocity vectors are simulated in Figure 12.20(b) in real time. The angular momentum has been divided by the average value of the three principal inertia moments for scaling reasons. The script torquef2.m performs the full simulation with the aid of the script torquef2_der.m.

The main difference between this system and that of Example 12.9 is that the principal inertia moments are all different. They are responsible for producing the interesting shape of the body and space phase shapes of Figure 12.20(b), which are no longer conic. The $\boldsymbol{\omega}$ phase space of Figure 12.20(b) is similar to the lower-right phase space of Figure 12.20(a). The main difference is that Figure 12.20(b) is a real-time simulation to visualize the wobbling behavior. The scripts used are listed on pages 441 and 442.

(a)

Ellipsoid System: $I_1 = 6.3, I_2 = 6.3, I_3 = 8.2$ $\gamma_1 = 0.601, \; \gamma_2 = 0.801, \; \gamma_3 = 0.386$

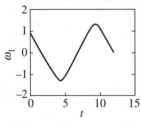

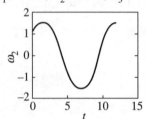

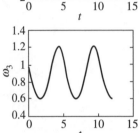

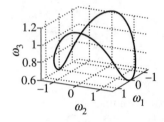

(b)   Ellipsoid System: $\gamma_1 = 0.601, \; \gamma_2 = 0.801, \; \gamma_3 = 0.386$

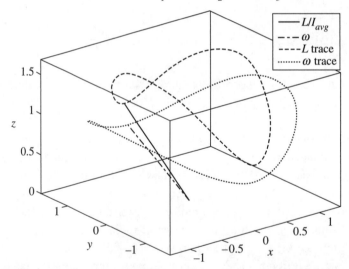

FIGURE 12.20   (a) Plots of $\omega_1(t)$, $\omega_2(t)$, and $\omega_3(t)$ from Equations (12.6.24) for the ellipsoid of Figure 12.19. (b) Snapshot of the real-time simulation of the angular momentum and angular velocity vectors.

SCRIPT

```
%torquef2.m
%program to solve Euler's equations for an ellipsoid without
%torques
clear;
I1=145/46; I2=1589/251; I3=1242/151;               %calculated by ellipso.m
gam1=(I3-I2)/I1; gam2=(I1-I3)/I2; gam3=(I2-I1)/I3;  %gammas
tmax=12;ts=0.05;                 %simulation run time and tome interval
tr=[0.0:ts:tmax];                %time range
w10=1; w20=1; w30=1;             %initial positions
ic1=[w10;w20;w30;];              %initial conditions:
%Use MATLAB's Runge-Kutta (4, 5) formula (uncomment/comment as needed)
%opt=odeset('AbsTol', 1.e-8, 'RelTol', 1.e-5);          %user set Tolerances
%[t, w]=ode45('torquef2_der', tr, ic1, opt, gam1, gam2, gam3);%with set tolerance
[t, w]=ode45('torquef2_der', tr, ic1, [], gam1, gam2, gam3);  %default tolerance
subplot(2, 2, 1), plot(t, w(:, 1), 'k')
xlabel('t', 'FontSize', 14), ylabel('\omega_1', 'FontSize', 14)
title(['Ellipsoid System: I_1=', num2str(I2, 2), ', I_2='...
       , num2str(I2, 2), ', I_3=', num2str(I3, 2)], 'FontSize', 12)
subplot(2, 2, 2), plot(t, w(:, 2), 'b')
xlabel('t', 'FontSize', 14), ylabel('\omega_2', 'FontSize', 14)
Title(['\gamma_1=', num2str(gam1, 3), ', \gamma_2='...
       , num2str(gam2, 3), ', \gamma_3=', num2str(gam3, 3)], 'FontSize', 12)
subplot(2, 2, 3), plot(t, w(:, 3), 'r')
xlabel('t', 'FontSize', 14), ylabel('\omega_3', 'FontSize', 14)
subplot(2, 2, 4), plot3(w(:, 1), w(:, 2), w(:, 3), 'color', [0.8 0.4 0.1])
view([1 1/2 1/2])          %viewpoint in x, y, z cartesian coords
v1x=min(w(:, 1));v1y=min(w(:, 2));v1z=min(w(:, 3)); %for window view
v2x=max(w(:, 1));v2y=max(w(:, 2));v2z=max(w(:, 3));
axis ([v1x, v2x, v1y, v2y, v1z, v2z])               %window size
grid on;
%box on
xlabel('\omega_1', 'FontSize', 14), ylabel('\omega_2', 'FontSize', 14)
zlabel('\omega_3', 'FontSize', 14)
%================== Simulation next ================================
figure
Iv=(I1+I2+I3)/3;                          %Average Moment
L1=I1*w(:, 1)/Iv; L2=I2*w(:, 2)/Iv; L3=I3*w(:, 3)/Iv;  %Ang. Mom./Iv
v1a=max(L1); v2a=max(L2); v3a=max(L3);
v1b=max(w(:, 1)); v2b=max(w(:, 2)); v3b=max(w(:, 3));
v1=max(v1a, v1b);v2=max(v2a, v2b);v3=max(v3a, v3b);    %view window
N=length(tr);
for i=1:5:N
   clf
   axis([-v1, v1, -v2, v2, 0, v3])
```

```
    hold on
    h(1)=line([0, L1(i)], [0, L2(i)], [0, L3(i)], 'color', 'k', ...  %L/Iv line
            'LineStyle', '-', 'linewidth', 2);
    h(2)=line([0, w(i, 1)], [0, w(i, 2)], [0, w(i, 3)], 'color', 'b', ...%w line
        'LineStyle', '-.', 'linewidth', 2);
    box on
    pause(.05)
end
h(3)=plot3(L1, L2, L3, 'k:');                  %plot the L/Iv trace
h(4)=plot3(w(:, 1), w(:, 2), w(:, 3), 'b:');     %plot the w trace
hh=legend(h, 'L/I_{avg}', '\omega', 'L trace', '\omega trace', 1);
xlabel('x', 'FontSize', 14), ylabel('y', 'FontSize', 14)
zlabel('z', 'FontSize', 14)
title(['Ellipsoid System: \gamma_1=', num2str(gam1, 3), ', \gamma_2='...
    , num2str(gam2, 3), ', \gamma_3=', num2str(gam3, 3)], 'FontSize', 13)
```

FUNCTION

```
%torquef2_der.m: returns the derivatives for the torque free ellipsoid
function derivs = torquef2_der( t, w, flag, gam1, gam2, gam3)
% w(1):w1, w(2):w2, w(3):w3
derivs =[-w(2).*w(3)*gam1;-w(3).*w(1)*gam2;-w(1).*w(2)*gam3;];
```

■ 12.7 Eulerian Angles

A. Fixed and Body Coordinate Systems

It is convenient to work with a standard way to describe the rotation of a body relative to a fixed coordinate system. There is no unique way of doing it, but Leonhard Euler developed a method that is in common use today. It is based on plane rotations about three axes by angles φ, θ, and ψ, which are called Eulerian angles, as shown in Figure 12.21.

The fixed coordinate system is labeled the $x-y-z$ system. The principal axes associated with the body form the body's coordinate system $1-2-3$ ($\hat{e}_1$, $\hat{e}_2$, $\hat{e}_3$, respectively) and rotate with it. The z'''-axis coincides with the 3-axis of the body, which is also its symmetry axis. The $x''(= x')$ axis is unique in that it is the intersection of the body's $1-2$ plane and the fixed $x-y$ plane; it is also known as the *line of nodes*. The Eulerian angles are understood according to the following rotations. If initially the body's principal axes are aligned with the fixed axes, a rotation about the z-axis by angle φ transforms $x \rightarrow x'$, $y \rightarrow y'$, and $z = z'$. A subsequent rotation by angle θ about the x'-axis transforms the axes as $x' = x''$, $y' \rightarrow y''$, and $z' \rightarrow z''$. Following this, a final rotation through angle ψ about the z''-axis transforms the axes as

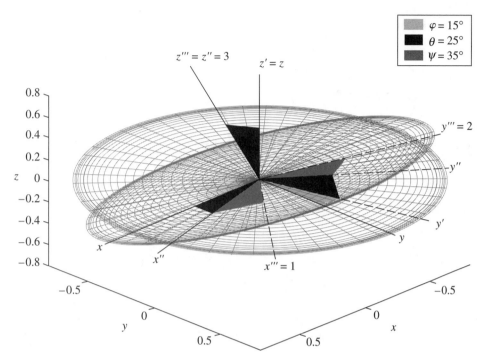

FIGURE 12.21 Consecutive plane rotations about the z-axis, the x'-axis, and the z''-axis by the respective Eulerian anges, φ, θ, and ψ.

$x'' \rightarrow x''' \equiv 1$, $y'' \rightarrow y''' \equiv 2$, and $z'' = z''' \equiv 3$. Angle ψ also represents the rotation of the body about its 3-axis. The Eulerian angles, the fixed $x-y$ and body $1-2$ planes, and the various primed coordinates described above are depicted in Figure 12.21, where rotations by angles $\varphi = 15$ degrees, $\theta = 25$ degrees, $\psi = 35$ degrees were made. In the figure, the counterclockwise rotation about the z-axis by angle $\varphi = 15$ degrees is colored blue; the counterclockwise rotation about the x'-axis by angle $\theta = 25$ degrees is colored black; and the counterclockwise rotation about the z''-axis by angle $\psi = 35$ degrees is colored red. The green axes are the body axes that lie in the body $1-2$ plane, where the 1-axis is at angle ψ from the line of nodes. The x'', y'', z'' axes system provides a connection between the fixed and the body axes. The figure is reproduced by running the MATLAB script euler_ang.m listed under the rotation matrix description below.

To understand the various transformations shown in Figure 12.21, let $\mathbf{r} = \hat{r}$ describe the initial alignment of the body's axes with the fixed axes. This is represented by the column vector

$$\mathbf{r} = \begin{pmatrix} \hat{i} \\ \hat{j} \\ \hat{k} \end{pmatrix}. \tag{12.7.1}$$

Similar to Section 5.4, the first rotation about the z-axis by angle φ is described by the rotation matrix

$$R_\varphi = \begin{pmatrix} \cos\varphi & \sin\varphi & 0 \\ -\sin\varphi & \cos\varphi & 0 \\ 0 & 0 & 1 \end{pmatrix}, \tag{12.7.2}$$

so that with the result of the rotation operation being $\mathbf{r}' = R_\varphi\mathbf{r}$, then

$$\mathbf{r}' = \begin{pmatrix} \hat{i}' \\ \hat{j}' \\ \hat{k}' \end{pmatrix} = \begin{pmatrix} \cos\varphi & \sin\varphi & 0 \\ -\sin\varphi & \cos\varphi & 0 \\ 0 & 0 & 1 \end{pmatrix} \begin{pmatrix} \hat{i} \\ \hat{j} \\ \hat{k} \end{pmatrix} = \begin{pmatrix} \hat{i}\cos\varphi + \hat{j}\sin\varphi \\ -\hat{i}\sin\varphi + \hat{j}\cos\varphi \\ \hat{k} \end{pmatrix}. \tag{12.7.3}$$

Similarly, we can obtain $\mathbf{r}'' = R_\theta\mathbf{r}'$ and $\mathbf{r}''' = R_\psi\mathbf{r}'' = \mathbf{r}_{123}$, where

$$R_\theta = \begin{pmatrix} 1 & 0 & 0 \\ 0 & \cos\theta & \sin\theta \\ 0 & -\sin\theta & \cos\theta \end{pmatrix}, \quad R_\psi = \begin{pmatrix} \cos\psi & \sin\psi & 0 \\ -\sin\psi & \cos\psi & 0 \\ 0 & 0 & 1 \end{pmatrix}, \quad \mathbf{r}_{123} \equiv \begin{pmatrix} \hat{e}_1 \\ \hat{e}_2 \\ \hat{e}_3 \end{pmatrix},$$

$$\tag{12.7.4}$$

for rotations by angles θ and ψ about the x'- and z''-axes, respectively, as discussed before, and $\mathbf{r}_{123}$ are the body axes. The final result from the three rotations is $R_{\psi\theta\varphi} = R_\psi R_\theta R_\varphi$, or

$$\begin{pmatrix} \hat{i}''' \\ \hat{j}''' \\ \hat{k}''' \end{pmatrix} = R_{\psi\theta\varphi}\begin{pmatrix} \hat{i} \\ \hat{j} \\ \hat{k} \end{pmatrix} = \begin{pmatrix} \cos\varphi\cos\psi - \sin\varphi\cos\theta\sin\psi & \sin\varphi\cos\psi + \cos\varphi\cos\theta\sin\psi & \sin\theta\sin\psi \\ -\cos\varphi\sin\psi - \sin\varphi\cos\theta\cos\psi & -\sin\varphi\sin\psi + \cos\varphi\cos\theta\cos\psi & \sin\theta\cos\psi \\ \sin\varphi\sin\theta & -\cos\varphi\sin\theta & \cos\theta \end{pmatrix}\begin{pmatrix} \hat{i} \\ \hat{j} \\ \hat{k} \end{pmatrix}.$$

$$\tag{12.7.5}$$

The matrices of (12.7.2 and 12.7.4) have been used by the script `euler_ang.m` to perform the operations described above and make possible the creation of Figure 12.21. The listing of the script follows.

SCRIPT

```
%euler_ang.m  - shows Euler angles: ph, th, ps; planes, line of nodes
clear, rc=pi/180; v=1+.1; ph=15; th=25; ps=35; %rc=deg-rad, v=multiplier, angles
phi=ph*rc; the=th*rc; psi=ps*rc;              %angles ph, th, ps in radians
axis ([-.8, .8, -.8, .8, -.8, .8]), view([1, 1, 1]), hold on; %window view
x=1;y=1;z=1; r={[x;0;0;], [0;y;0;], [0;0;z]};   %starting cartesian axes
line([0, r(1, 1)], [0, r(1, 2)], [0, r(1, 3)], 'color', 'b', 'LineStyle', '-')
```

```
line([0, r(2, 1)], [0, r(2, 2)], [0, r(2, 3)], 'color', 'b', 'LineStyle', '-')
line([0, r(3, 1)], [0, r(3, 2)], [0, r(3, 3)], 'color', 'b', 'LineStyle', '-')
text(r(1, 1)*v, r(1, 2)*v, r(1, 3)*v, 'x', 'FontSize', 11)
text(r(2, 1)*v, r(2, 2)*v, r(2, 3)*v, 'y', 'FontSize', 11)
text(r(3, 1)*v, r(3, 2)*v, r(3, 3)*v, 'z', 'FontSize', 11)
xlabel('x', 'FontSize', 14, 'Position', [0 1 -.8]);
ylabel('y', 'FontSize', 14, 'Position', [1 0 -.8]);
zlabel('z', 'FontSize', 14)
%--- phi-rotation about z by ph degrees
rphi={[cos(phi);-sin(phi);0;], [sin(phi);cos(phi);0;], [0;0;1;]};%z matrix rotation
rp=rphi*r;
line([0, rp(1, 1)], [0, rp(1, 2)], [0, rp(1, 3)], 'color', 'k', 'LineStyle', '--')
line([0, rp(2, 1)], [0, rp(2, 2)], [0, rp(2, 3)], 'color', 'k', 'LineStyle', '--')
line([0, rp(3, 1)], [0, rp(3, 2)], [0, rp(3, 3)], 'color', 'k', 'LineStyle', '--')
text(rp(1, 1)*v, rp(1, 2)*v, rp(1, 3)*v, 'x\prime', 'FontSize', 11)
text(rp(2, 1)*v, rp(2, 2)*v, rp(2, 3)*v, 'y\prime', 'FontSize', 11)
text(rp(3, 1)*v, rp(3, 2)*v, rp(3, 3)*v, 'z\prime=z', 'FontSize', 11)
x1=[0;.5;rp(1, 1)/2]; y1=[0;0;rp(1, 2)/2]; z1=[0;0;rp(1, 3)/2];
h(1)=fill3(x1, y1, z1, 'b'); %color ph angle from x-axis
x1=[0;0;rp(2, 1)/2]; y1=[0;.5;rp(2, 2)/2]; z1=[0;0;rp(2, 3)/2];
fill3(x1, y1, z1, 'b');      %color ph angle from y-axis
%--- theta-rotation about x' by th degrees from y-axis
rthe={[1;0;0;], [0;cos(the);-sin(the);], [0;sin(the);cos(the);]};%x' matrix rotation
rpp=rthe*rp;
line([0, rpp(1, 1)], [0, rpp(1, 2)], [0, rpp(1, 3)], 'color', 'r', 'LineStyle', '-.')
line([0, rpp(2, 1)], [0, rpp(2, 2)], [0, rpp(2, 3)], 'color', 'r', 'LineStyle', '-.')
line([0, rpp(3, 1)], [0, rpp(3, 2)], [0, rpp(3, 3)], 'color', 'r', 'LineStyle', '-.')
text(rpp(1, 1)*v, rpp(1, 2)*v, rpp(1, 3)*v, 'x\prime\prime', 'FontSize', 11)
text(rpp(2, 1)*v, rpp(2, 2)*v, rpp(2, 3)*v, 'y\prime\prime', 'FontSize', 11)
text(rpp(3, 1)*v, rpp(3, 2)*v, rpp(3, 3)*v, 'z\prime\prime', 'FontSize', 11)
x1=[0;rp(3, 1)/2;rpp(3, 1)/2]; y1=[0;rp(3, 2)/2;rpp(3, 2)/2];
z1=[0;rp(3, 3)/2;rpp(3, 3)/2];
h(2)=fill3(x1, y1, z1, 'k'); %color th angle from - z axis
x1=[0;rp(2, 1)/2;rpp(2, 1)/2]; y1=[0;rp(2, 2)/2;rpp(2, 2)/2];
z1=[0;rp(2, 3)/2;rpp(2, 3)/2];
fill3(x1, y1, z1, 'k');      %color th angle from - y' axis
%--- psi-rotation about z'' by ps degrees
rpsi={[cos(psi);-sin(psi);0;], [sin(psi);cos(psi);0;], [0;0;1;]};%z'' matrix-rotation
rppp=rpsi*rpp;
line([0, rppp(1, 1)], [0, rppp(1, 2)], [0, rppp(1, 3)], 'color', 'g', 'LineStyle', '--')
line([0, rppp(2, 1)], [0, rppp(2, 2)], [0, rppp(2, 3)], 'color', 'g', 'LineStyle', '--')
line([0, rppp(3, 1)], [0, rppp(3, 2)], [0, rppp(3, 3)], 'color', 'g', 'LineStyle', '--')
text(rppp(1, 1)*v, rppp(1, 2)*v, rppp(1, 3)*v, 'x\prime\prime\prime=1', 'FontSize', 11)
text(rppp(2, 1)*v, rppp(2, 2)*v, rppp(2, 3)*v, 'y\prime\prime\prime=2', 'FontSize', 11)
text(rppp(3, 1)*v, rppp(3, 2)*v, rppp(3, 3)*v, ...
    'z\prime\prime\prime=z\prime\prime=3', 'FontSize', 11)
x1=[0;rpp(1, 1)/2;rppp(1, 1)/2];y1=[0;rpp(1, 2)/2;rppp(1, 2)/2];
z1=[0;rpp(1, 3)/2;rppp(1, 3)/2];
```

```
h(3)=fill3(x1, y1, z1, 'r'); %color ps angle from x''-axis
x1=[0;rpp(2, 1)/2;rppp(2, 1)/2];y1=[0;rpp(2, 2)/2;rppp(2, 2)/2];
z1=[0;rpp(2, 3)/2;rppp(2, 3)/2];
fill3(x1, y1, z1, 'r');        %color ps angle from y'' axis
str1=cat(2, '\phi=', num2str(ph, 2), '^o'); str2=cat(2, '\theta=', num2str(th, 2), '^o');
str3=cat(2, '\psi=', num2str(ps, 2), '^o');
h=legend(h, str1, str2, str3, 1); set(h, 'FontSize', 11)
%Draw and rotate planes (rotation uses a minus sign due to internal construction)
[xx, yy, zz]=ellipsoid(0, 0, 0, x, y, 0, 50); %flat unit ellipsoids are the desired planes
mesh(xx, yy, zz, 'EdgeColor', [0 0 1], 'FaceAlpha', [0], 'EdgeAlpha', [0.1]); % draw plane
%draw, rotate planes: first about z axis by (-ph) - use rphi rot. matrix
xp=rthe(1, 1)*xx-rthe(1, 2)*yy-rthe(1, 3)*zz; %minus sign on off diagonals due to -ph
yp=-rthe(2, 1)*xx+rthe(2, 2)*yy-rthe(2, 3)*zz;
zp=-rthe(3, 1)*xx-rthe(3, 2)*yy+rthe(3, 3)*zz;
%next rotation: about x' axis by (-th) - use rthe rot. matrix, then draw plane
xpp=rphi(1, 1)*xp-rphi(1, 2)*yp-rphi(1, 3)*zp;%minus sign on off diagonals due to -th
ypp=-rphi(2, 1)*xp+rphi(2, 2)*yp-rphi(2, 3)*zp;
zpp=-rphi(3, 1)*xp-rphi(3, 2)*yp+rphi(3, 3)*zp;
mesh(xpp, ypp, zpp, 'EdgeColor', [1 0 0], 'FaceAlpha', [0], 'EdgeAlpha', [.1]); %draw plane
```

B. Angular Velocities

The angular velocity, $\boldsymbol{\omega}$, of the body is the vector sum of the three angular velocities associated with the rate of change of the Eulerian angles; that is,

$$\boldsymbol{\omega} = \dot{\boldsymbol{\varphi}} + \dot{\boldsymbol{\theta}} + \dot{\boldsymbol{\psi}}. \tag{12.7.6a}$$

We seek to express the components of $\boldsymbol{\omega}$ in terms of the body coordinates. We think of $\boldsymbol{\omega}$ as a superposition of the rotation along the z''' (3) axis and the x'', y'', z'' system with angular velocity $\boldsymbol{\omega}''$, where

$$\boldsymbol{\omega}'' = \dot{\boldsymbol{\varphi}} + \dot{\boldsymbol{\theta}}. \tag{12.7.6b}$$

Referring to Figure 12.22(a), we write the components of the $\dot{\boldsymbol{\varphi}}$ and $\dot{\boldsymbol{\theta}}$ velocities in the double primed system as follows:

$$\dot{\varphi}_{x''} = 0 \quad \dot{\varphi}_{y''} = \dot{\varphi}\sin\theta \quad \dot{\varphi}_{z''} = \dot{\varphi}\cos\theta, \quad \dot{\theta}_{x''} = \dot{\theta} \quad \dot{\theta}_{y''} = 0 \quad \dot{\theta}_{z''} = 0. \tag{12.7.7}$$

Using these into (12.7.6a and 12.7.6b), we can express $\boldsymbol{\omega}$ in component form

$$\boldsymbol{\omega} = \omega_{x''}\hat{i}'' + \omega_{y''}\hat{j}'' + \omega_{z''}\hat{k}'', \tag{12.7.8a}$$

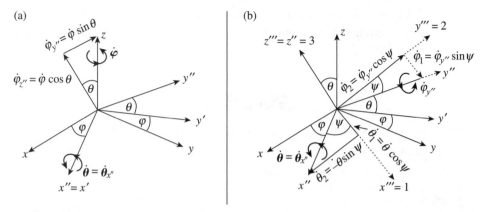

FIGURE 12.22 (a) $\dot{\varphi}$ and $\dot{\theta}$ velocities in the double primed system (b) components of ω_1, ω_2, ω_3 according to Equation (12.7.9).

where

$$\omega_{x''} = \dot{\theta}_{x''} \quad \omega_{y''} = \dot{\varphi}\sin\theta \quad \omega_{z''} = \dot{\varphi}\cos\theta + \dot{\psi}, \tag{12.7.8b}$$

because $\dot{\psi}$ is already along z'' (or 3). The final step is to express $\boldsymbol{\omega}$ in terms of body coordinates. To this end we write

$$\boldsymbol{\omega} = \omega_1\hat{e}_1 + \omega_2\hat{e}_2 + \omega_3\hat{e}_3, \tag{12.7.9a}$$

so that using (12.7.7) and referring to Figure 12.22(b), we see that

$$\begin{aligned}
\omega_1 &= \dot{\varphi}_1 + \dot{\theta}_1 + \dot{\psi}_1 = \dot{\varphi}_{y''}\sin\psi + \dot{\theta}\cos\psi + 0 = \dot{\varphi}\sin\theta\sin\psi + \dot{\theta}\cos\psi \\
\omega_2 &= \dot{\varphi}_2 + \dot{\theta}_2 + \dot{\psi}_2 = \dot{\varphi}_{y''}\cos\psi - \dot{\theta}\sin\psi + 0 = \dot{\varphi}\sin\theta\cos\psi - \dot{\theta}\sin\psi, \\
\omega_3 &= \dot{\varphi}_3 + \dot{\theta}_3 + \dot{\psi}_3 = \dot{\varphi}\cos\theta + 0 + \dot{\psi} = \dot{\varphi}\cos\theta + \dot{\psi}
\end{aligned}$$

$$\tag{12.7.9b}$$

where in Figure 12.22(b), $\dot{\theta}_2 = -\dot{\theta}\sin\psi$ because it points opposite to the $\hat{e}_2$ axis direction; also $\dot{\varphi}_3 = \dot{\varphi}_{z''}$ given in (12.7.7), and as implied before, $\dot{\psi}_1 = \dot{\psi}_2 = 0$, $\dot{\psi}_3 = \psi$.

Thus, (12.7.9) are the angular velocity and its components as expressed in the $1-2-3$ coordinate system relative to the angular velocities seen in the fixed space coordinate system, that is, the Eulerian angles.

EXAMPLE 12.11

Make use of the above Eulerian angles to obtain the rate of precession of $\boldsymbol{\omega}$ about the fixed angular momentum of the system for the case of the symmetric top of Example (12.9) in the absence of any torques.

Solution

Referring to Figure 12.23, because there are no torques present we take $\mathbf{L}$ along the fixed z-axis with components along y'' and z'', so that θ is fixed and the motion takes place along the $y''-z''$ plane shown.

Thus the component of $\boldsymbol{\omega}$ along the y''-direction is $\omega_{y''} = \omega\sin\varphi_b$, as shown in the figure. Setting this equal to the corresponding component in (12.7.8b), we find that the rate of precession

$$\dot{\varphi} = \omega\frac{\sin\varphi_b}{\sin\theta}. \tag{12.7.10}$$

This result is identical to that of (12.6.20b) where in the present notation $\varphi_s = \theta$, i.e., the angle between the body's $\hat{e}_3$-axis and the fixed space z-axis.

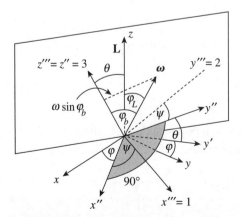

FIGURE 12.23 L along the fixed z-axis with components along y' and z', and the component of $\boldsymbol{\omega}$ along the y'' direction.

C. One Point Fixed Symmetric Top Revisited

In Section 12.3.E, we looked at the precessional motion of a gyroscope under the action of a torque using a simple approach. This spinning gyroscope is a version of a symmetric top with one point fixed and acted on by a torque due to gravity. In this section, we will use the Eulerian angles to study the motion of such an object in more detail. In Figure 12.24, a symmetric top is shown with its symmetry axis (3), which is its spin axis as well as its principal axis of symmetry, at angle θ from the fixed z-axis. The force due to gravity acts on its center of mass located at distance ℓ from the origin about which the torque acts. We let I_3 be the moment of inertia about the 3 axis. The moments of inertia about the 1, 2 axes are equal, so that $I_1 = I_2 = I$, where $I_3 \neq I$. Next, we consider the system's energy and its dynamic equations of motion.

i. Energy

The kinetic energy of the spinning top shown in Figure 12.24 can be written in terms of the angular velocities along the $1-2-3$ axes, Equations (12.7.9). We have

$$T = \frac{1}{2}[I_1(\omega_1^2 + \omega_2^2) + I_3\omega_3^2] = \frac{1}{2}I_1(\dot{\varphi}^2\sin^2\theta + \dot{\theta}^2) + \frac{1}{2}I_3(\dot{\varphi}\cos\theta + \dot{\psi})^2,$$

(12.7.11)

and taking the origin at the stationary tip, from which the top's center of mass is at a distance ℓ, the potential energy is

$$V(\theta) = mg\ell\cos\theta.$$

(12.7.12)

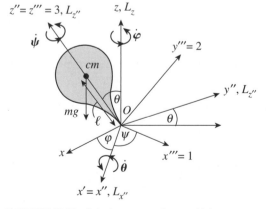

FIGURE 12.24 Spinning symmetric top with its symmetry axis (3), which is its spin axis as well as its principal axis of symmetry, at angle θ from the fixed z-axis.

ii. Dynamics

The torque experienced by the spinning top of Figure 12.24 is about the x''-axis, that is, in the direction of $\dot{\boldsymbol{\theta}}$, so that we don't directly use Euler's Equations (12.6.6) written for the $1-2-3$ axes. Instead we work with Euler's equations in the $x''-y''-z''$ system to write the torque and its components as

$$\boldsymbol{\tau} = \tau_{x''}\hat{i}'' + \tau_{y''}\hat{j}'' + \tau_{z''}\hat{k}'', \quad \tau_{x''} = mg\ell\sin\theta, \quad \tau_{y''} = 0, \quad \tau_{z''} = 0. \tag{12.7.13}$$

From (12.7.6 and 12.7.7) the corresponding double-prime angular velocity is

$$\boldsymbol{\omega}'' = \dot{\boldsymbol{\varphi}} + \dot{\boldsymbol{\theta}} = (\dot{\varphi}_{x''} + \dot{\theta}_{x''})\hat{i}'' + (\dot{\varphi}_{y''} + \dot{\theta}_{y''})\hat{j}'' + (\dot{\varphi}_{z''} + \dot{\theta}_{z''})\hat{k}''$$
$$= \dot{\theta}\hat{i}'' + \dot{\varphi}\sin\theta\hat{j}'' + \dot{\varphi}\cos\theta\hat{k}'' \equiv \omega''_{x''}\hat{i}'' + \omega''_{y''}\hat{j}'' + \omega''_{z''}\hat{k}''. \tag{12.7.14}$$

The principal moments of inertia are $I_{x''x''} = I_{y''y''} = I$ and $I_{z''z''} = I_s \equiv I_3$, so the angular momentum and its components can be obtained as

$$\mathbf{L} = L_{x''}\hat{i}'' + L_{y''}\hat{j}'' + L_{z''}\hat{k}'', \quad L_{x''} = I_{x''x''}\omega''_{x''} = I\dot{\theta}, \tag{12.7.15a}$$
$$L_{y''} = I_{y''y''}\omega''_{y''} = I\dot{\varphi}\sin\theta, \quad L_{z''} = I_{z''z''}\omega''_{z''} = I_3(\dot{\varphi}\cos\theta + \dot{\psi}) \equiv I_s\omega_s,$$

where we have further defined the angular speed purely along the spin axis (3) as

$$\omega_s = \dot{\varphi}\cos\theta + \dot{\psi}. \tag{12.7.15b}$$

Using (7.6.7) or following a process similar to that of Section 12.6, the torque is related to the angular momentum as

$$\boldsymbol{\tau} = \left(\frac{d\mathbf{L}}{dt}\right)_{fixed} = \left(\frac{d\mathbf{L}}{dt}\right)_{rot} + \boldsymbol{\omega}'' \times \mathbf{L}. \tag{12.7.16}$$

Thus, using (12.7.13 to 12.7.16), we can obtain the set of equations that determine the dynamics of the top's motion as

$$mg\ell\sin\theta = I\ddot{\theta} + I_s\omega_s\dot{\varphi}\sin\theta - I\dot{\varphi}^2\cos\theta\sin\theta, \tag{12.7.17a}$$

$$0 = I\frac{d}{dt}(\dot{\varphi}\sin\theta) - I_s\omega_s\dot{\theta} + I\dot{\theta}\dot{\varphi}\cos\theta, \tag{12.7.17b}$$

and

$$0 = I_s \dot{\omega}_s. \tag{12.7.17c}$$

Since the moments of inertia are assumed to be constant, (12.7.17c) means that the angular momentum along the 3-axis is constant, or

$$I_s \dot{\omega}_s = 0 = \frac{d}{dt}(I_s \omega_s) = \frac{d}{dt}L_{z''} \Rightarrow L_{z''} = \text{constant}. \tag{12.7.18}$$

Additionally, if we multiply (12.7.17b) by $\sin\theta$ followed by some rearrangement, the equation is equivalent to

$$\frac{d}{dt}(I\dot{\varphi}\sin^2\theta + I_s\omega_s\cos\theta) = 0, \tag{12.7.19}$$

where (12.7.17c) has also been used. At this point, with the use of Figure 12.24, we notice that the total contribution to the angular momentum in the z direction can be calculated by summing the contributions from the double-prime angular momentum components of (12.7.15) in the z-direction, as follows:

$$L_z = (L_{x''})_z + (L_{y''})_z + (L_{z''})_z = L_{y''}\sin\theta + L_{z''}\cos\theta = I\dot{\varphi}\sin^2\theta + I_s\omega_s\cos\theta, \tag{12.7.20}$$

where the $L_{x''}$ component, being perpendicular to the $y''-z''$ plane, does not have a contribution in the z-direction. Thus, (12.7.19 and 12.7.20) mean that, in addition to (12.7.18), a similar result for the z-component of the fixed axes angular momentum

$$\frac{d}{dt}L_z = 0 \Rightarrow L_z = \text{constant}, \tag{12.7.21}$$

is implied by (12.7.17b). For later use, it is also important to write expressions for the x and y components of the angular momentum. These are

$$L_x = L_{x''}\cos\varphi - L_{y''}\cos\theta\sin\varphi + L_{z''}\sin\theta\sin\varphi = I\dot{\theta}\cos\varphi - (I\dot{\varphi}\cos\theta - I_s\omega_s)\sin\theta\sin\varphi$$
$$L_y = L_{x''}\sin\varphi + L_{y''}\cos\theta\cos\varphi - L_{z''}\sin\theta\cos\varphi = I\dot{\theta}\sin\varphi + (I\dot{\varphi}\cos\theta - I_s\omega_s)\sin\theta\cos\varphi. \tag{12.7.22}$$

The above expressions for L_x, L_y, and L_z can be confirmed by performing the rotation operation $\mathbf{r}'' = R_\theta R_\varphi \mathbf{r}$ outlined in (12.7.1 to 12.7.4), and guided by this process one can identify the components of L'' that point in the direction of the fixed Cartesian axes. Furthermore, the components of the body z''-axis angular momentum, $L_{z''}$, can be identified from the above expressions (12.7.20 and 12.7.22) as

$$(L_{z''})_x = L_{z''} \sin\theta \sin\varphi, \quad (L_{z''})_y = -L_{z''} \sin\theta \cos\varphi, \quad (L_{z''})_z = L_{z''} \cos\theta$$

$$(12.7.23)$$

where $L_{z''}$ is given by (12.7.15a).

EXAMPLE 12.12

Use the dynamics equations of motion (12.7.17) to obtain the precession frequency of the simple gyroscope of Section 12.3E.

Solution

In the limit of $\theta = 90° = $ constant, then we also take $\dot\theta = \ddot\theta = 0$ so that (12.7.17a) implies that $mg\ell = I_s \omega_s \dot\varphi$, while (12.7.15b) gives $\omega_s \to \dot\psi$, and so $\dot\varphi = mg\ell / I_s\dot\psi$. This result is identical to the precessional frequency of the gyroscope (12.3.20) if we notice that in the present case $r_\perp = \ell\sin\theta \to \ell$, and $I_s\dot\psi$ is the spin angular momentum about the symmetric body axis.

EXAMPLE 12.13

Assuming θ remains fixed at an angle other than 90 degrees, obtain an expression for the precessional frequency of a top.

Solution

From (12.7.17a), taking $\dot\theta = \ddot\theta = 0$, we have $mg\ell = I_s \omega_s \dot\varphi - I\dot\varphi^2 \cos\theta$, which is a quadratic equation for $\dot\varphi$ with solutions

$$\dot\varphi = \frac{I_s\omega_s \pm \sqrt{(I_s\omega_s)^2 - 4Img\ell\cos\theta}}{2I\cos\theta}. \tag{12.7.24a}$$

This means that there are two rates of precession, a fast one ($+$ sign) and a slow one ($-$ sign), and it depends on the initial conditions as to which one takes place. A physical situation, however, demands that for this case $\dot\varphi$ be either zero or positive, which means that we must have

$$I_s^2\omega_s^2 \geq 4Img\ell\cos\theta \tag{12.7.24b}$$

for stability.

D. Further Aspects of the Spinning Point Fixed Symmetric Top

i. Nutation

In general, a spinning top does not only precess with a precession rate described by $\dot{\varphi}$ as discussed in the previous subsection, but θ is not always fixed; in fact, θ does change as a function of time, while the top precesses. This behavior is known as nutation. To understand nutation, let's look at the total energy of the spinning top under the applied torque as shown in Figure 12.24. The total energy of the system from (12.7.11 and 12.7.12) is $E = T + V$ or

$$E = \frac{1}{2}I(\dot{\varphi}^2 \sin^2\theta + \dot{\theta}^2) + \frac{1}{2}I_s\omega_s^2 + mg\ell\cos\theta, \tag{12.7.25}$$

where we have used $\omega_s = \dot{\varphi}\cos\theta + \dot{\psi}$ for the top spin, and $I = I_1$, $I_s = I_3$, as before. Since from (12.7.18) the spin energy $I_s\omega_s^2/2 = L_{z''}^2/2\,I_s$ is a constant of the motion, we can let this be a reference energy and work with the also-constant energy

$$E' \equiv E - \frac{1}{2}I_s\omega_s^2 = T' + V_{eff}(\theta), T' \equiv \frac{1}{2}I\dot{\theta}^2 \tag{12.7.26}$$

where

$$V_{eff}(\theta) \equiv V(\theta) + \frac{1}{2}I\dot{\varphi}^2\sin^2\theta = mg\ell\cos\theta + \frac{(L_z - L_{z''}\cos\theta)^2}{2\,I\sin^2\theta}, \tag{12.7.27}$$

is an effective potential, and where from (12.7.20) we have also used

$$\dot{\varphi} = \frac{L_z - L_{z''}\cos\theta}{I\sin^2\theta}, \tag{12.7.28}$$

Thus since from (12.7.18 and 12.7.21) L_z and $L_{z''}$ are both constant and (12.7.27) is purely a function of θ. This effective potential can be studied as a function of θ. The idea is that, based on (12.7.26), the spinning top has a harmonic behavior versus theta. Similar to a harmonic oscillator of Chapter 2 or planetary motion in Chapter 8, we can write a differential equation for $\dot{\theta}(t)$

$$\dot{\theta}(t) = \sqrt{\frac{2}{I}(E' - V_{eff}(\theta))} = \frac{d\theta(t)}{dt}. \tag{12.7.29}$$

This differential equation for $\dot{\theta}(t)$ is another way of expressing the earlier result of (12.7.17). This equation is useful in that it gives us a complementary understanding.

For example, since E' is constant, and since $V_{eff}(\theta)$ can't be greater than E', then there must be turning points associated with any oscillatory behavior. The turning points are the values of θ at which $\dot{\theta}$ is zero, i.e.,

$$E' - V_{eff}(\theta) = 0. \tag{12.7.30}$$

These turning points are an indication that the top nutates between, what is most realistically physical, two different values of θ (later referred to as θ_1 and θ_2), as we shall see in a later simulation.

EXAMPLE 12.14

Consider the effective potential given in (12.7.27). Use the small oscillations method of Chapter 3 in order to obtain an approximate relation for the nutation frequency of the above symmetric point fixed top about an angle of $\theta_0 = \pi/2$, assuming that its spin rate is large.

....

Solution

The idea is to use (3.3.4) where, because the top is spinning about the symmetry axis, the mass is replaced by the corresponding perpendicular inertia moment I so that the nutation frequency is

$$\Omega_n = \sqrt{\frac{\kappa}{I}}, \tag{12.7.31a}$$

where here

$$\kappa = \frac{d^2 V_{eff}(\theta)}{d\theta^2}\bigg|_{\theta = \theta_0}, \tag{12.7.31b}$$

is the effective harmonic oscillator constant associated with the spinning top. The effective potential $V_{eff}(\theta)$ is given by (12.7.27), and its second derivative evaluated at $\theta_0 = \pi/2$ is obtained as

$$V''_{eff}(\theta_0 = \pi/2) = \frac{(L_{z''}^2 + L_z^2)}{I}\bigg|_{\theta = \theta_0} = \frac{(I_s^2 \omega_s^2 + I^2 \dot{\varphi}^2)}{I}, \tag{12.7.32}$$

where we have used $L_{z''} = I_s \omega_s$, $L_z = I\dot{\varphi}\sin^2\theta + I_s\omega_s\cos\theta$ from (12.7.18 and 12.7.20). From Example 12.12, we know that at $\theta = \pi/2$, $\dot{\psi} = \omega_s$, and the precession rate is $\dot{\varphi} = mg\ell/I_s\omega_s$, so that if the spin rate ω_s is high, then we can neglect $\dot{\varphi}$ in (12.7.32) to obtain from (12.7.31)

$$\kappa \approx \frac{I_s^2\omega_s^2}{I}, \quad \text{and} \quad \Omega_n \approx \frac{I_s}{I}\omega_s, \tag{12.7.33}$$

which means that if the spin angular momentum is high, the nutation rate (Ω_n) is high but the precession rate ($\dot{\varphi}$) is low.

EXAMPLE 12.15

Numerically solve the equations of motion (12.7.17) of a spinning fixed-point symmetric top and obtain plots of the Eulerian angles as well as their rates of change versus time, and explain the results. Give a plot of the energy (12.7.26) and the effective potential (12.7.27) associated with it, and explain and identify the turning points about which the top nutates. Finally, simulate the motion of the top's total angular momentum (12.7.20 and 12.7.22), as well as the body z'' angular momentum (12.7.23) a function of time for the first 10 seconds and explain the observed behavior. Assume the following top parameters: $I = 1$ kgm^2, $I_s = 1.5$ kgm^2, $M = 1$ kg, $g = 9.8$ m/s^2, $\ell = .1$ m. Also use the following initial conditions: $\varphi_0 = 0$ rad, $\theta_0 = 0.6\pi/2$ rad, $\psi_0 = 0$ rad, $\dot{\varphi}_0 = 0$ rad/s, $\dot{\theta}_0 = 0$ rad/s, and keep in mind the stability condition (12.7.24b) for ω_s.

Solution

The equations of motion (12.7.17) can be solved numerically as a system of equations using MATLAB when set up as shown below. First, the equation for $\dot{\theta}$ from (12.7.17a) along with one for θ are written as

$$\frac{d\dot{\theta}}{dt} = (mg\ell - (I_s\omega_s - I\dot{\varphi}\cos\theta)\dot{\varphi})\sin\theta/I, \quad \frac{d\theta}{dt} = \dot{\theta}, \tag{12.7.34a}$$

which are coupled to the equations for φ and $\dot{\varphi}$, which follow. The equation for $\dot{\varphi}$ from (12.7.17b) is obtained by performing the only time derivative shown and rearranging the result so that we have the second system

$$\frac{d\dot{\varphi}}{dt} = \frac{(I_s\omega_s - 2I\dot{\varphi}\cos\theta)\dot{\theta}}{I\sin\theta}, \quad \frac{d\varphi}{dt} = \dot{\varphi}. \tag{12.7.34b}$$

Finally from (12.7.17c), because $I_s\omega_s$ is constant and because I_s is also constant, then ω_s is itself also constant, so we can use (12.7.15b) to obtain a single first-order differential equation for ψ

$$\frac{d\psi}{dt} = \omega_s - \dot{\varphi}\cos\theta, \qquad (12.7.34c)$$

and which can also be used to obtain $\dot{\psi}$ by virtue of its definition. Thus, there are five first-order differential equations for φ, θ, and ψ that include their derivatives. The equations for φ and θ are coupled to each other, and the equation for ψ depends on them. In MATLAB we proceed as in previous examples and let the variables take on the array form $w(1, 2, 3, 4, 5, 6) = (\varphi, \dot{\varphi}, \theta, \dot{\theta}, \psi, \dot{\psi})$. The five right-hand sides of (12.7.34) are placed in the script top_der.m, and the script top.m uses the previously employed Runge–Kutta method, calls top_der.m, and obtains the solution of $w(1..5)$ as a function of time. Once the first five equations are fully solved, the last array element of $w(6)$, which corresponds to $\dot{\psi}$, is done separately by performing (12.7.34c). This process has been very successful in obtaining the Eulerian angles and their rates as a function of time with the given initial conditions as shown in Figure 12.25(a), and more details are found in the MATLAB script top.m. The value of $\omega_s = 1.5 + \sqrt{4Imgl\cos\theta_0}/I_s$ in rad/s is used, big enough to satisfy the stability condition (12.7.24b). The first-column panels of the figure contain the solutions $\varphi(t)$, $\theta(t)$, and $\psi(t)$ in radians, and the second-column panels contain the corresponding rates versus time in radians per second. The rates $\dot{\varphi}(t)$ and $\dot{\psi}(t)$ are seen to oscillate in time, but they are always positive so that $\varphi(t)$ and $\psi(t)$ increase indefinitely, indicating perpetual precession, and spin respectively, because there is no friction present in the current model. In contrast, $\dot{\theta}(t)$ oscillates between positive and negative values, which is the reason why $\theta(t)$ varies in time between two different values (θ_1 and θ_2), one higher than the other. This is the nutation effect we discussed previously in connection with Equation (12.7.30) and that is indeed the same $\theta(t)$ involved in the solution to the differential Equation (12.7.29), although we have obtained it through the numerical solution of Equation (12.7.34) here.

This brings us to the next aspect of the example, that is, the effective potential. Figure 12.25(b) shows the effective potential versus θ for the same initial parameters as in Figure 12.25(a). Also shown in Figure 12.25(b) is the energy E' (flat blue line) and the turning point values θ_1 and θ_2 at which the potential (green

(a)

Top Euler Angles: $\varphi_0 = 0$, $\theta_0 = 0.95$, $\psi_0 = 0$ rad $d\varphi/dt_0 = 0$, $d\theta/dt_0 = 0$, $\omega_s = 2.5$ rad/s

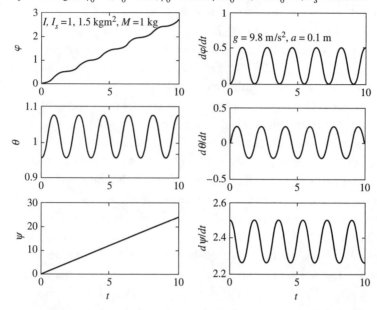

(b)   Top $V_{eff}(\theta)$: $\varphi_0 = 0$, $\theta_0 = 0.95$, $\psi_0 = 0$ rad $d\varphi/dt_0 = 0$, $d\theta/dt_0 = 0$, $\omega_s = 2.5$ rad/s

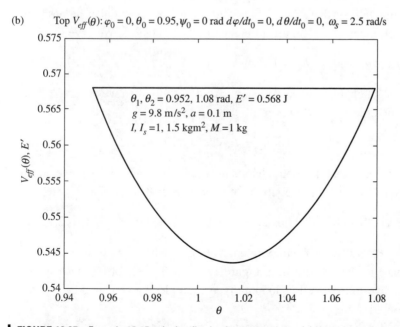

FIGURE 12.25 Example 12.15 spinning fixed point symmetric top (a) Numerical solution of the equations of motion (12.7.17), (b) plot of the energy (12.7.26) and the effective potential (12.7.27), and (c) a snapshot of the simulated motion of the top's total angular momentum (12.7.20, 12.7.22) as well as the body z'' angular momentum (12.7.23) as a function of time. (continues on page 459)

curve) equals the energy E'. This energy is confirmed to be constant, as expected. The turning points correspond to zero values in the $\dot{\theta}(t)$ curve of Figure 12.25(a).

The final part of this example is the angular momenta. Having already obtained $\varphi(t)$, $\theta(t)$, and $\psi(t)$ and their derivatives, it is just a matter of evaluating the Expressions (12.7.20 and 12.7.22) for the total angular momentum and (12.7.23) for the body z''-axis angular momentum. This is shown in Figure 12.25(c) on page 459, which is a snapshot of the simulation. The script top.m should be run to view the motion in real time. The motion in real time is a plot of the angular momenta components as a function of time and demonstrates the top's precession and nutation. The nutation of the body $z''(3)$-axis angular momentum shown in the figure has a cusp-like shape, as indicated by the trace drawing. The red vertical line is the z-component of the angular momentum, which is constant. The projection of the body $z''(3)$-axis angular momentum onto the z-axis is shown by the starred markers, whose separation changes in time to indicate that its magnitude changes as the body nutates. While the body nutation is seen in the oscillatory character of $\dot{\theta}(t)$, $\dot{\varphi}(t)$ also experiences an interesting behavior as a function of θ. This is explained by looking at the value of $\dot{\varphi}$ at either θ_1 or θ_2. Equation (12.7.28) is a convenient way of looking at what happens to $\dot{\varphi}(t)$ as a function of θ. This equation tells us that when $\theta = \theta_{1,2}$, the value of $\dot{\varphi}$ can either be positive, negative, or zero, depending on the difference $L_z - L_{z''}\cos\theta_{1,2}$. We have that if

$$L_z - L_{z''}\cos\theta_1 > 0, \text{ or } L_z - L_{z''}\cos\theta_2 > 0; \tag{12.7.35a}$$

that is, $\dot{\varphi}$ does not change sign as θ changes from θ_1 to θ_2, then we observe wave-like monotonic precession. However, if

$$L_z - L_{z''}\cos\theta_1 < 0(\,>0), \text{ or } L_z - L_{z''}\cos\theta_2 > 0(\,<0), \tag{12.7.35b}$$

which simply means that if $\dot{\varphi}$ does change in sign in the range as θ changes from θ_1 to θ_2, then the observed precession is of a looping-type shape. Finally, if

$$L_z - L_{z''}\cos\theta_1 = 0(\,\neq 0), \text{ or } L_z - L_{z''}\cos\theta_2 \neq 0(\,= 0), \tag{12.7.35c}$$

that is, if $\dot{\varphi} = 0$ at either of the turning points θ_1 or θ_2, then the precession is of a cusp-like shape as the one shown in Figure 12.25(c). The script top.m is capable of showing wave-like as well as looping-type examples by changing $\dot{\varphi}_0$, that is, the initial condition for $\dot{\varphi}$.

The MATLAB scripts top.m and top_der.m used in this example to create Figure 12.24 follow.

(c)   Top motion: $\varphi_0 = 0, \theta_0 = 0.95, \psi_0 = 0$ rad $d\varphi/dt_0 = 0$, $d\theta/dt_0 = 0$, $\omega_s = 2.5$ rad/s

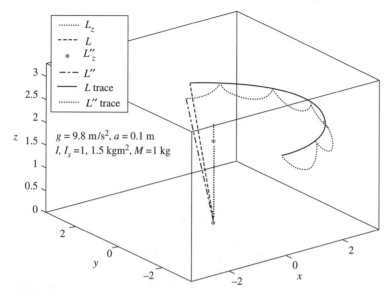

| FIGURE 12.25   (continued).

SCRIPT

```
%top.m, program to solve Euler's equations, plot them and their
%respective angular speeds, as well as angular momentum properties
clear;
I=1.0; Is=1.5; g=9.8; M=1; a=.1;     %inertia , gravity, mass, lever arm
tau0=M*g*a;                          %torque value at theta=0
tmax=10; ts=0.05;                    %simulation run time and tome interval
tr=(0.0:ts:tmax);N=length(tr);       %time range, array size
ph0=0; th0=.6*pi/2+0.01; ps0=0;      %init angle values in rad
ws=1.5+sqrt(4*I*tau0*cos(th0))/Is;   %init spins ang. speed
ph0d=0.0; th0d=0; %cusps example     %init angular speeds in rad/sec
%ph0d=0.2; th0d=0;%waves (monotonic) example
%ph0d=0.8; th0d=0;%Looping example
ic1=[ph0;ph0d;th0;th0d;ps0];         %initial conditions:
%Use MATLAB's Runge-Kutta (4, 5) formula (uncomment/comment as needed)
%opt=odeset('AbsTol', 1.e-8, 'RelTol', 1.e-5);     %user set Tolerances
%[t, w]=ode45('top_der', tr, ic1, opt, I, Is, ws, tau0);%with set tolerance
[t, w]=ode45('top_der', tr, ic1, [], I, Is, ws, tau0);  %default tolerance
w(:, 6)=ws-w(:, 2).*cos(w(:, 3));               %derivative of psi
%Next: plots of the Eulerian angles and their derivatives
subplot(3, 2, 1), plot(t, w(:, 1), 'k'), ylabel('\phi', 'FontSize', 14)
title(['Top Euler Angles: \phi_0=', num2str(ph0, 2), ', \theta_0=', ...
       num2str(th0, 2), ', \psi_0=', num2str(ps0, 2), ' rad '], 'FontSize', 11)
str1=cat(2, 'I, I_s=', num2str(I, 2), ', ', num2str(Is, 2), ' kgm^2', ...
```

```
                ', M=', num2str(M, 2), 'kg');
text(0.1, max(w(:, 1)*(1-0.1)), str1)
subplot(3, 2, 2), plot(t, w(:, 2), 'b'), ylabel('d\phi/dt', 'FontSize', 14)
title(['d\phi/dt_0=', num2str(ph0d, 2), ', d\theta/dt_0=', num2str(th0d, 2), ...
        ', \omega_s=', num2str(ws, 2), ' rad/s'], 'FontSize', 11)
str2=cat(2, 'g=', num2str(g, 2), ' m/s^2, a=', num2str(a, 2), ' m');
text(0.1, max(w(:, 2))*(1+0.1), str2)
%w(1):phi, w(2):phi_dot, w(3):theta, w(4):theta_dot, w(5):psi, w(6):psi_dot
subplot(3, 2, 3), plot(t, w(:, 3), 'r'), ylabel('\theta', 'FontSize', 14)
subplot(3, 2, 4), plot(t, w(:, 4), 'r'), ylabel('d\theta/dt', 'FontSize', 14)
subplot(3, 2, 5), plot(t, w(:, 5), 'r')
xlabel('t', 'FontSize', 14), ylabel('\psi', 'FontSize', 14)
subplot(3, 2, 6), plot(t, w(:, 6), 'r')
xlabel('t', 'FontSize', 14), ylabel('d\psi/dt', 'FontSize', 14)
%Angular momentum components Lx, Ly, Lz, L''z, L''y, L''z
Lx=I*w(:, 4).*cos(w(:, 1))-(I*w(:, 2).*cos(w(:, 3))-Is*ws).*sin(w(:, 3))...
    .*sin(w(:, 1));
Ly=I*w(:, 4).*sin(w(:, 1))+(I*w(:, 2).*cos(w(:, 3))-Is*ws).*sin(w(:, 3))...
    .*cos(w(:, 1));
Lz=I*w(:, 2).*sin(w(:, 3)).^2+Is*ws*cos(w(:, 3));
%Lz" along x, y, z
Lzppx=Is*ws*sin(w(:, 3)).*sin(w(:, 1));
Lzppy=-Is*ws*sin(w(:, 3)).*cos(w(:, 1));
Lzppz=Is*ws*cos(w(:, 3));
%Effective Potential and Nutation angle Turning points
figure
Vef=M*g*a*cos(w(:, 3))+(Lz-Lzppz).^2./(2*I*sin(w(:, 3)).^2);
Ekp=0.5*I*w(:, 4).^2; Ep=Ekp+Vef;              %kinetic, and prime energy
% [thm1, i1]=min(Ekp);[thm2, i2]=min(Ekp(i1+1:N));%aprox Ekp zeros, near ends
% th1=w(i1, 3); th2=w(i2+i1, 3);                %theta values at which Ekp=0
th1=w(1, 3); th2=w(N, 3);                        %Ekp zeros-2nd way better
plot(w(:, 3), [Ep, Vef]), xlabel('\theta', 'FontSize', 12)
ylabel('V_{eff}(\theta), E\prime', 'FontSize', 12)
str3=cat(2, '\theta_1, \theta_2 = ', num2str(th1, 3), ', ', ...
          num2str(th2, 3), ' rad', ', E\prime=', num2str(Ep(1), 3), ' J');
d1=abs(th2-th1); Vm=min(Vef); d2=Ep(1)-Vm;       %viewing scaling factors
text(min(w(:, 3))*(1+0.15*d1), Ep(1)*(1-0.4*d2), str1)
text(min(w(:, 3))*(1+0.15*d1), Ep(1)*(1-0.25*d2), str2)
text(min(w(:, 3))*(1+0.15*d1), Ep(1)*(1-0.1*d2), str3)%nutation angle points
title(['Top V_{eff}(\theta): \phi_0=', num2str(ph0, 2), ', \theta_0=', ...
          num2str(th0, 2), ', \psi_0=', num2str(ps0, 2), ' rad, d\phi/dt_0='...
        , num2str(ph0d, 2), ', d\theta/dt_0=', num2str(th0d, 2), ...
          ', \omega_s=', num2str(ws, 2), ' rad/s'], 'FontSize', 11)
% ================= Simulation next =================================
figure
va1x=min(Lx);va1y=min(Ly);va1z=min(Lz);          %get largest # for window view
va2x=max(Lx);va2y=max(Ly);va2z=max(Lz);
vb1x=min(Lzppx);vb1y=min(Lzppy);vb1z=min(Lzppz);
```

```
vb2x=max(Lzppx);vb2y=max(Lzppy);vb2z=max(Lzppz);
v1x=min(va1x, vb1x);v1y=min(va1y, vb1y);v1z=min(va1z, vb1z);
v2x=max(va2x, vb2x);v2y=max(va2y, vb2y);v2z=max(va2z, vb2z);
v=max([abs(v1x), v2x, abs(v1y), v2y, abs(v1z), v2z]);
axis([-v, v, -v, v, 0, v])%window size
 for i=1:N
    clf
    axis([-v, v, -v, v, 0, v])
    hold on
    h(1)=line([0, 0], [0, 0], [0, Lz(i)], 'color', 'r');%Lz line
    h(2)=line([0, Lx(i)], [0, Ly(i)], [0, Lz(i)], 'color', 'k', ...        %L
        'LineStyle', '-', 'linewidth', 2);
    h(3)=line([0, 0], [0, 0], [0, Lzppz(i)], 'color', 'm', 'Marker', '*');   %Lzppz
    h(4)=line([0, Lzppx(i)], [0, Lzppy(i)], [0, Lzppz(i)], 'color', 'b', ...%Lzpp
        'LineStyle', '-.', 'linewidth', 2);
    box on
    pause(.05)
 end
h(5)=plot3(Lx, Ly, Lz, 'k:');                          %plot the L trace
h(6)=plot3(Lzppx, Lzppy, Lzppz, 'b:');                 %plot the L trace
hh=legend(h, 'L_z', 'L', 'L\prime\prime_z', 'L\prime\prime', 'L trace', ...
    'L\prime\prime trace', 2); set(hh, 'FontSize', 11)
xlabel('x', 'FontSize', 14), ylabel('y', 'FontSize', 14)
zlabel('z', 'FontSize', 14)
title(['Top motion: \phi_0=', num2str(ph0, 2), ', \theta_0=', num2str(th0, 2), ...
    ', \psi_0=', num2str(ps0, 2), ' rad, d\phi/dt_0=', num2str(ph0d, 2), ...
    ', d\theta/dt_0=', num2str(th0d, 2), ', \omega_s=', num2str(ws, 2), ...
    ' rad/s'], 'FontSize', 12)
text(-v*(1-0.1), v, v*(1-.6), str1)
text(-v*(1-0.1), v, v*(1-0.5), str2)
```

```
                            FUNCTION
```

```
%top_der.m: returns the derivatives for the
function ders = top_der(t, w, flag, I, Is, ws, tau0)
%w(1):phi, w(2):phi_dot, w(3):theta, w(4):theta_dot, w(5):psi
%main program produces w(6):psi_dot
ders=[w(2);(Is*ws-2*I*w(2).*cos(w(3))).*w(4)./(I*sin(w(3)));...
    w(4);(tau0-(Is*ws-I*w(2).*cos(w(3))).*w(2)).*sin(w(3))/I;...
    ws-w(2).*cos(w(3));];
```

■ Chapter 12 Problems

12.1 Obtain the center of mass coordinates (x_{cm}, y_{cm}, z_{cm}) of an octant of a uniform density, solid sphere of radius R.

12.2 Find the center of mass of a semicircular ring of radius R and of uniform density and oriented in the upper half of the xy-plane.

12.3 For the two-mass rotating system of Section 12.3C, derive Equation (12.3.14) using (a) the $\mathbf{L} = \mathbf{r} \times \mathbf{p}$ method, (b) the $\mathbf{L} = \{I\}\boldsymbol{\omega}$ method, and (c) derive the resulting torque Equation (12.3.15).

12.4 One end of a cylinder is pivoted to the floor and precesses like a top. The cylinder is 1 m in length and 50 cm in diameter and is tilted at angle θ from the vertical while spinning clockwise (from an overhead view) at 25 rev/s about its own axis. Find the magnitude and direction of its precession frequency.

12.5 Obtain an expression for the moment of inertia of a rectangular plate, of sides a and b and mass M, rotating about an axis passing through one of its corners and perpendicular to the plane of the plate.

12.6 By using formula (12.3.18b), prove that the moment of inertia of a hoop about an axis passing through its center and perpendicular to its plane is given by $I = Ma^2$, where M and a are its mass and radius, respectively.

12.7 Find the moment of inertia of a rectangular plate, of width a, length b, and mass M rotating about an axis passing through its center and parallel to the plane of the plate as well as to its width.

12.8 Analytically show that the integral $J \equiv \dfrac{4}{\pi} \displaystyle\int_0^1 \overline{x}^2 \sqrt{1 - (\overline{x})^2}\, d\overline{x}$ of (13.3.28) works out to $1/4$.

12.9 Using the formula (12.3.18b), obtain the z-axis moment of inertia for the semicircular loop shown in Figure 12.12 of Example 12.5.

12.10 Obtain the moment of inertia about an axis passing through the edge of the semicircular loop of Example 12.5 and parallel to the x-axis.

12.11 A section (shaded) of a uniform semicircular disk of mass M_s and radius a lies in the xy-plane shown in Figure 12.26. Find the moment of inertia (I_{yy}) about an axis passing through the disk's flat side whose orientation is along the y axis. Give your result in terms of Ma^2 where M is the mass of a full disk. Also, what is the value of M_s in terms of M and the segment's area A_s in terms of a full disk's area?

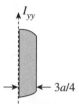

I_{yy}

← 3a/4

I FIGURE 12.26

12.12 Consider a uniform cube of mass M and sides a, and find the principal axes and their associated moments of inertia about the cube's center of mass.

12.13 Consider a 1 kg uniform rectangular solid of volume 1 m × 2 m × 3 m. Find the inertia tensor, the principal axes, and their associated moments of inertia about the solid's center of mass.

12.14 Consider a 1 kg uniform square plate of 1 m × 2 m sides. Find the inertia tensor, the principal axes, and their associated moments of inertia about the origin, located at the center of its long side's edge.

12.15 A 1 kg uniform cube of 1 m sides rotates with angular velocity magnitude of $\sqrt{2}$ rad/s. Find the angular momentum and rotational energy about the cube's center of mass (a) for a rotation about the diagonal and passing through the origin, and (b) for rotations about each of the principal axes.

12.16 Derive Equation (12.6.17).

12.17 Derive Equation (12.6.19).

12.18 Derive Equation (12.6.20a).

12.19 Show that Ω_L from Equation (12.6.20a) can be written as

$$\Omega_L = \omega\sqrt{1 + \left[(I_3/I_1)^2 - 1\right]\cos^2\varphi_b}.$$

12.20 Derive Equation (12.6.20b).

12.21 Earth is close to a symmetric top with moments of inertia $I_3/I_1 \approx 1.00327$, and the angle that its rotational frequency makes with the North Pole is $\varphi_b \approx 0.2$ arc sec. Find the associated precessional period (a) as seen by an observer on Earth, and (b) as seen by an observer from outer space. This effect is known as the Chandler wobble.

12.22 Obtain and solve Euler equations of motion for an ellipsoidal body under the special circumstance that there are no external torques acting on it and that its rotational motion takes place about the center of mass. The ellipsoid's semi-major axes are $a = 3\,\text{m}$, $b = 1\,\text{m}$, $c = 1\,\text{m}$, and its density is $\rho = (1/8)\,\text{kg/m}^3$. Be sure to give plots of $\omega_1(t)$, $\omega_2(t)$, and $\omega_2(t)$, as well as their phase space. Include a simulation of the comparison between $\mathbf{L}(t)$ and $\boldsymbol{\omega}(t)$.

12.23 Derive Equation (12.7.5).

12.24 Using the angular speeds along the $1-2-3$ axes of (12.7.9), show that the kinetic energy of the spinning top results in Equation (12.7.11).

12.25 Derive Equation (12.7.17).

12.26 Derive Equation (12.7.19).

12.27 A top is set spinning in a vertical position. In order for it not to topple over, it must be set spinning sufficiently fast. How fast is sufficiently fast?

12.28 Derive the expressions for L_x, L_y, and L_z shown in (12.7.20 and 12.7.22). (*Hint:* Use the rotation matrices R_θ, R_φ of [12.7.2 and 12.7.4].)

12.29 Derive Equation (12.7.32).

12.30 Repeat the calculations of Example 12.15, but modify the initial parameters in order to obtain a looping type of precession.

■ Additional Problem

12.31 Consider a uniform solid sphere of mass M and radius R. Obtain its associated moment of inertia shown in Table 12.1.

13 | Lagrangian Dynamics

■ 13.1 Introduction

Newton's laws of motion have proven to be very useful in solving problems in classical mechanics as well as in providing a general and deeper understanding of nature and all its laws. However, applying Newton's laws to complicated problems can become a difficult undertaking, especially if a description of the motion is needed for systems that either move in a complicated manner, or other coordinates than Cartesian coordinates are used, or if the systems are constrained to move on a given surface, or even for systems than involve several objects. Lagrangian dynamics refers to the application of a different process to obtain the equations of motion of a given system. The essential development of the Lagrangian approach is due to Joseph Louis de Lagrange (1736–1813), and it is a different way of solving the Newton's $\mathbf{F} = m\mathbf{a}$ problem; indeed, Lagrange's dynamics is a reformulation of Newtonian mechanics and finds itself generally and vastly employed in physics and engineering applications. The Lagrangian method has an added benefit; it not only encompasses Newton dynamics, but it also leads to the concept of the Hamiltonian or the energy of a system and a process by means of which to calculate it. The result makes it possible to obtain the Hamiltonian of a complicated system. The Hamiltonian is a cornerstone in the field of quantum mechanics.

■ 13.2 Generalized Coordinates

The first step in the Lagrangian formulation is to develop an understanding of coordinates. In the Lagrangian approach, the coordinates such as position, angular displacements, and so on are described by the *generalized coordinates*, $q_1, q_2, \ldots, q_N$. The time derivatives of these coordinates are the *generalized velocities*, $\dot{q}_1, \dot{q}_2, \ldots, \dot{q}_N$. In the Lagrangian method, the generalized coordinates and velocities are treated as independent variables. For a system of N particles, there are $3N$ coordinates, which specify the state or configuration of a system. If there are constraints, such as for example, a body is rigid, or the N particles are constrained to move on a surface, then the number of coordinates that specify the system is less than $3N$. In general, there are $n \leq 3N$ generalized coordinates, so $q_1, q_2, \ldots, q_n$ will entirely specify the configuration of the system. Suppose we have two particles without constraints. Then the generalized coordinates are $q_1, q_2, q_3, q_4, q_5, q_6$, which take the place of the equivalent $x_1, y_1, z_1, x_2, y_2, z_2$ Cartesian coordinates for the particles, respectively. If each q_i is independent of any other q_j, we say the system has $n = 3N$ degrees of freedom. When the number of degrees of freedom of an N particle system in three dimensions is less than $3N$, the system is said to be constrained. Here, a *holonomic* system is a system that has holonomic constraints. A holonomic constraint refers to expressing the constraint as a function of the coordinates and possibly time, such as $f(q_1, q_2, \ldots, q_n, t) = 0$, but not as a function of the velocities. Holonomic constraints have the effect of reducing the number of degrees of freedom of the system. The simple pendulum, for example, has a holonomic constraint in that the magnitude of the particle's position vector equals the length of the string ($\sqrt{x^2 + y^2} = L$.) We restrict ourselves to this kind of problem here. Constraints that cannot be expressed as functions of the coordinates and possibly time are said to be *nonholonomic*.

The generalized coordinates can be written as a function of the rectangular coordinates and time, that is,

$$q_1 = q_1(x_1, x_2, \ldots, x_{3N}, t)$$
$$q_2 = q_2(x_1, x_2, \ldots, x_{3N}, t)$$
$$\ldots \qquad\qquad\qquad\qquad\qquad\qquad (13.2.1a)$$
$$q_n = q_n(x_1, x_2, \ldots, x_{3N}, t),$$

for N particles, where we have used x_1, x_2, x_3 for the x, y, z coordinates of the first particle, x_4, x_5, x_6 for the second particle, and so on. In a similar way, the rectangular coordinates can be written as functions of the generalized coordinates

$$x_1 = x_1(q_1, q_2, \ldots, q_n, t)$$
$$x_2 = x_2(q_1, q_2, \ldots, q_n, t)$$
$$\vdots \qquad\qquad\qquad\qquad\qquad\qquad (13.2.1b)$$
$$x_{3N} = x_{3N}(q_1, q_2, \ldots, q_n, t).$$

For a holonomic system, the number of degrees of freedom is $n < 3N$. The preceding time dependence appears if one of the coordinate systems is moving relative to the other. When the time variable does not appear explicitly in (13.2.1), such a system is referred to as *scleronomic* and the system is independent of time. If the time does appear explicitly, the system is called *rheonomic* and is, therefore, time dependent.

In this formulation, a small coordinate change needs to be understood. For example, if a system undergoes a small change from one configuration to another, then the generalized coordinates change, that is,

$$q_1 \rightarrow q_1 + \delta q_1, \quad q_2 \rightarrow q_2 + \delta q_{2,...}, \quad q_n \rightarrow q_n + \delta q_n, \tag{13.2.2a}$$

and similarly

$$x_1 \rightarrow x_1 + \delta x_1, \quad x_2 \rightarrow x_2 + \delta x_{2,...}, \quad x_n \rightarrow x_n + \delta x_{3N}. \tag{13.2.2b}$$

In (13.2.2a), since $q = q(x_1, x_2, \ldots, x_{3N})$, then an infinitesimal change in a generalized coordinate can be expressed in terms of the change in rectangular coordinates, or

$$\delta q_1 = \frac{\partial q_1}{\partial x_1}\delta x_1 + \frac{\partial q_1}{\partial x_2}\delta x_2 + \cdots + \frac{\partial q_1}{\partial x_{3N}}\delta x_{3N}$$

$$\vdots$$

$$\delta q_n = \frac{\partial q_n}{\partial x_1}\delta x_1 + \frac{\partial q_n}{\partial x_2}\delta x_2 + \cdots + \frac{\partial q_n}{\partial x_{3N}}\delta x_{3N},$$

or

$$\delta q_k = \sum_{i=1}^{3N} \frac{\partial q_k}{\partial x_i}\delta x_i, \tag{13.2.2c}$$

and similarly an infinitesimal change in a rectangular coordinate can be expressed in terms of the change in generalized coordinates

$$\delta x_1 = \frac{\partial x_1}{\partial q_1}\delta q_1 + \frac{\partial x_1}{\partial q_2}\delta q_2 + \cdots + \frac{\partial x_1}{\partial q_n}\delta q_n$$

$$\vdots$$

$$\delta x_{3N} = \frac{\partial x_{3N}}{\partial q_1}\delta q_1 + \frac{\partial x_{3N}}{\partial q_2}\delta q_2 + \cdots + \frac{\partial x_{3N}}{\partial q_n}\delta q_n,$$

or,

$$\delta x_i = \sum_{k=1}^{n} \frac{\partial x_i}{\partial q_k} \delta q_k. \tag{13.2.2d}$$

EXAMPLE 13.1

Consider a central force problem in which the polar coordinates r, θ are associated with the generalized coordinates q_1, q_2, and obtain the change of the generalized coordinates in terms of the change in rectangular coordinates.

Solution

Let's express the generalized coordinates in terms of rectangular coordinates. We have $q_1 \equiv r = q_1(x_1, x_2) = \sqrt{x_1^2 + x_2^2}$, and $q_2 \equiv \theta = q_2(x_1, x_2) = \tan^{-1}(x_2/x_1)$, where x_1, x_2 stand for the x, y Cartesian coordinates, respectively. The change in q_1 is

$$\delta q_1 = \frac{\partial q_1}{\partial x_1} \delta x_1 + \frac{\partial q_1}{\partial x_2} \delta x_2 = \frac{1}{2} (x_1^2 + x_2^2)^{-1/2} 2 \, x_1 \delta x_1 + \frac{1}{2} (x_1^2 + x_2^2)^{-1/2} 2 \, x_2 \delta x_2$$

$$= \frac{x_1}{q_1} \delta x_1 + \frac{x_2}{q_1} \delta x_2, \tag{13.2.3}$$

and the change in q_2 is $\delta q_2 = \dfrac{\partial q_2}{\partial x_1} \delta x_1 + \dfrac{\partial q_2}{\partial x_2} \delta x_2$. With the partial derivative

$$\frac{\partial q_2}{\partial x_1} = \frac{1}{1 + (x_2/x_1)^2} \frac{\partial}{\partial x_1} \left(\frac{x_2}{x_1} \right) = -\frac{x_2}{x_1^2 + x_2^2} = -\frac{x_2}{q_1^2},$$

and similarly

$$\frac{\partial q_2}{\partial x_1} = \frac{1}{1 + (x_2/x_1)^2} \frac{\partial}{\partial x_2} \left(\frac{x_2}{x_1} \right) = \frac{x_1}{q_1^2},$$

we have for δq_2

$$\delta q_2 = -\frac{x_2}{q_1^2} \delta x_1 + \frac{x_1}{q_1^2} \delta x_2. \tag{13.2.4}$$

■ 13.3   Generalized Forces

The concept of a generalized force refers to a function that is expressed in terms of generalized coordinates and that would take the place of a standard force or a torque if we were working with Cartesian coordinates in linear motion or polar coordinates, say, in rotational motion. For example, in Cartesian coordinates as a particle moves from position $\mathbf{r}$ to $\mathbf{r} + \delta\mathbf{r}$, the amount of work δW done by a force $\mathbf{F}$ is written as

$$\delta W = \mathbf{F} \cdot \delta\mathbf{r} = F_x\delta_x + F_y\delta_y + F_z\delta_z = \sum_i F_i\delta x_i, \tag{13.3.1}$$

where for a single particle $i = 1, 2, 3$ to represent x, y, and z. For the case of N particles, the index would take on the values $i = 1, 2, \ldots, 3N$. Now, because it is more convenient to express δW in terms of generalized coordinates, then with the help of (13.2.2d), we can rewrite (13.3.1) as

$$\delta W = \sum_k \left(\sum_i F_i \frac{\partial x_i}{\partial q_k} \right) \delta q_k. \tag{13.3.2}$$

We next define the generalized force as

$$Q_k \equiv \sum_i F_i \frac{\partial x_i}{\partial q_k}, \tag{13.3.3}$$

and the work done in terms of generalized coordinates and the generalized forces becomes

$$\delta W = \sum_k Q_k\delta q_k. \tag{13.3.4}$$

Thus the amount of work done by a force $\mathbf{F}$ in moving a particle from position $\mathbf{r}$ to $\mathbf{r} + \delta\mathbf{r}$ is equivalent to the work done by the generalized forces Q_k acting through the generalized displacement q_k as it increases to $q_k + \delta q_k$. Note that if q_k is a linear distance, then Q_k is a force, but if q_k is an angular displacement Q_k is a torque. For a conservative force we have $\mathbf{F} = -\boldsymbol{\nabla}V$ so that, using rectangular coordinates, we write each component as $F_i = \partial V/\partial x_i$, and (13.3.3) becomes

$$Q_k \equiv - \sum_i \frac{\partial V}{\partial x_i} \frac{\partial x_i}{\partial q_k}; \tag{13.3.5a}$$

however, since

$$\frac{\partial V}{\partial q_k} = \frac{\partial V}{\partial x_1}\frac{\partial x_1}{\partial q_k} + \frac{\partial V}{\partial x_2}\frac{\partial x_2}{\partial q_k} + \cdots + \frac{\partial V}{\partial x_{3N}}\frac{\partial x_{3N}}{\partial q_k} = \sum_i \frac{\partial V}{\partial x_i}\frac{\partial x_i}{\partial q_k}, \tag{13.3.5b}$$

then we see that (13.3.5a) becomes

$$Q_k = -\frac{\partial V}{\partial q_k}, \tag{13.3.5c}$$

where in general, $V \equiv V(q_1, q_2, \ldots, q_n)$. For the case of a nonconservative force we can make the replacement

$$F_i \rightarrow F'_i + F_i = F'_i - \partial V/\partial x_i, \tag{13.3.6}$$

where F' is the nonconservative part of the force, in (13.3.3), to get with the help of (13.3.5b)

$$Q_k = \sum_i F_i \frac{\partial x_i}{\partial q_k} = \sum_i F'_i \frac{\partial x_i}{\partial q_k} - \sum_i \frac{\partial V}{\partial x_i}\frac{\partial x_i}{\partial q_k} = Q'_k - \frac{\partial V}{\partial q_k}, \tag{13.3.7}$$

where Q'_k is the nonconservative part of the generalized force.

■ 13.4 Lagrange's Equations

In this section, we are interested in obtaining the dynamical equations of motion of a system of particles in terms of the generalized coordinates. We can think of two ways of carrying out this task. The first is to substitute the generalized coordinates of Section 13.2 directly into Newton's second law, $\mathbf{F} = m\mathbf{a}$. The second way is to do it through the system's energy. Because it is often the case that the energy of the system can be identified, this second way tends to be the preferred and most convenient way to proceed. We begin by working with the kinetic energy, which we write in rectangular coordinates as

$$T = \frac{1}{2}\sum_i m_i \dot{x}_i^2, \tag{13.4.1}$$

which for N particles involves $3N$ terms. As in (13.2.1), $x_i = x_i(q_1, q_2, \ldots, q_n, t)$, so that

$$\dot{x}_i = \frac{dx_i}{dt} = \sum_k \frac{\partial x_i}{\partial q_k}\dot{q}_k + \frac{\partial x_i}{\partial t}, \tag{13.4.2}$$

where we have used $\dot{q}_k \equiv dq_k/dt$. Furthermore, we notice that

$$\frac{\partial \dot{x}_i}{\partial \dot{q}_k} = \frac{\partial}{\partial \dot{q}_k}\frac{dx_i}{dt} = \frac{\partial}{\partial \dot{q}_k}\left(\sum_{k'}\frac{\partial x_i}{\partial q_{k'}}\dot{q}_{k'} + \frac{\partial x_i}{\partial t}\right) = \sum_{k'}\frac{\partial x_i}{\partial q_{k'}}\frac{\partial \dot{q}_{k'}}{\partial \dot{q}_k} = \frac{\partial x_i}{\partial q_k}, \qquad (13.4.3)$$

where we have used $\dfrac{\partial}{\partial \dot{q}_k}\dfrac{\partial x_i}{\partial t} = 0$; also since the x_i's do not depend on $\dot{q}_i$'s,

$\dfrac{\partial}{\partial \dot{q}_k}\dfrac{\partial x_i}{\partial q_{k'}} = 0$, and finally we notice that $\dfrac{\partial}{\partial \dot{q}_k}\dot{q}_{k'} = \delta_{kk'}$ so that the only the kth term survives the sum over k'. Thus, multiplying (13.4.3) by $\dot{x}_i$ and differentiating the result with respect to t, we get

$$\frac{d}{dt}\left(\dot{x}_i\frac{\partial \dot{x}_i}{\partial \dot{q}_k}\right) = \frac{d}{dt}\left(\dot{x}_i\frac{\partial x_i}{\partial q_k}\right)$$

$$= \frac{d}{dt}\left(\frac{\partial}{\partial \dot{q}_k}\left[\frac{1}{2}\dot{x}_i^2\right]\right) = \ddot{x}_i\frac{\partial x_i}{\partial q_k} + \dot{x}_i\frac{d}{dt}\frac{\partial x_i}{\partial q_k} = \ddot{x}_i\frac{\partial x_i}{\partial q_k} + \dot{x}_i\frac{\partial \dot{x}_i}{\partial q_k}. \qquad (13.4.4)$$

Multiplying the last line by m_i we see that

$$\frac{d}{dt}\left(\frac{\partial}{\partial \dot{q}_k}\left[\frac{1}{2}m_i\dot{x}_i^2\right]\right) = m_i\ddot{x}_i\frac{\partial x_i}{\partial q_k} + m_i\dot{x}_i\frac{\partial \dot{x}_i}{\partial q_k} = F_i\frac{\partial x_i}{\partial q_k} + \frac{\partial}{\partial q_k}\left(\frac{1}{2}m_i\dot{x}_i^2\right), \qquad (13.4.5)$$

since $F_i = m_i\ddot{x}_i$. Using (13.4.1) we see that summing (13.4.5) over i gives

$$\frac{d}{dt}\frac{\partial T}{\partial \dot{q}_k} = \sum_i F_i\frac{\partial x_i}{\partial q_k} + \frac{\partial T}{\partial q_k}, \qquad (13.4.6)$$

which with the use of (13.3.7) it becomes

$$\frac{d}{dt}\frac{\partial T}{\partial \dot{q}_k} = Q_k + \frac{\partial T}{\partial q_k}. \qquad (13.4.7)$$

This is the set of equations we sought. They represent a set of differential equations in terms of generalized coordinates and that are known as *Lagrange's equations*. It is useful to look at both conservative and nonconservative systems to simplify the notation. For conservative systems, (13.3.5c) allows us to rewrite (13.4.7) as

$$\frac{d}{dt}\frac{\partial T}{\partial \dot{q}_k} = -\frac{\partial V}{\partial q_k} + \frac{\partial T}{\partial q_k} = \frac{\partial L}{\partial q_k}, \quad L \equiv T - V, \qquad (13.4.8)$$

where L is known as the *Lagrangian* of the system and is equal to the difference between the kinetic and potential energies of the system. Because for conservative systems, the potential energy $V = V(x_1, x_2, \ldots, x_{3N}) = -\int_{r_i}^{r} \mathbf{F} \cdot d\mathbf{r}$ is independent of the path, and it is also independent of the velocities, $\partial V/\partial \dot{r}_i = 0$, or equivalently, $V \equiv V(q_1, q_2, \ldots, q_n)$ is independent of the generalized velocities, or $\partial V/\partial \dot{q}_k = 0$, then

$$\frac{\partial T}{\partial \dot{q}_k} = \frac{\partial L}{\partial \dot{q}_k}, \tag{13.4.9}$$

and (13.4.8) can be written as

$$\frac{d}{dt}\frac{\partial L}{\partial \dot{q}_k} = \frac{\partial L}{\partial q_k}, \tag{13.4.10}$$

which is a more compact form of Lagrange's equations for a conservative system. If instead of (13.3.5c) we use (13.3.7) into (13.4.7), we obtain

$$\frac{d}{dt}\frac{\partial L}{\partial \dot{q}_k} = \frac{\partial L}{\partial q_k} + Q'_k, \tag{13.4.11}$$

which are Lagrange's equations for the case of a nonconservative system.

EXAMPLE 13.2

Consider a particle of mass m moving with velocity v on a frictionless surface. Write Lagrange's equations and show that the solution reduces to the familiar form of Newton's first law.

Solution
We let $\dot{q} = v$, and $T = m\dot{q}^2/2$, with $V = 0$, so that the Lagrangian is $L = T - V = T$. This is a conservative system, so take $Q'_k = 0$, and the left- and right-hand sides of Lagrange's equations in (13.4.10) or (13.4.11) become, respectively,

$$\frac{d}{dt}\left(\frac{\partial L}{\partial \dot{q}}\right) = \frac{d}{dt}\left[\frac{\partial}{\partial \dot{q}}\left(\frac{1}{2}m\dot{q}^2\right)\right] = \frac{d}{dt}m\dot{q} = m\ddot{q}, \quad \frac{\partial L}{\partial q} = \frac{\partial}{\partial q}\left(\frac{1}{2}m\dot{q}^2\right) = 0. \tag{13.4.12a}$$

So equating these two results, we see that

$$\frac{d}{dt}\left(\frac{\partial L}{\partial \dot{q}}\right) = 0 \quad \Rightarrow \quad \frac{\partial L}{\partial \dot{q}} = m\dot{q} = p = \text{constant},$$ (13.4.12b)

which is the expression for linear momentum conservation and thus equivalent to Newton's first law.

EXAMPLE 13.3

Consider a particle of mass m attached to the end of a spring of constant k. Write the Lagrangian of the system and use Lagrange's equations to obtain the equation of motion of the system.

....

Solution

We take $Q'_k = 0$, $T = \frac{1}{2}m\dot{q}^2$, $V = \frac{1}{2}kq^2$, and $L = \frac{1}{2}m\dot{q}^2 - \frac{1}{2}kq^2$. Performing the left- and right-hand sides of (13.4.11), we have $\frac{d}{dt}\left(\frac{\partial L}{\partial \dot{q}}\right) = m\ddot{q}$, $\frac{\partial L}{\partial q} = -kq$, and setting these two results equal we find the equation of motion, $m\ddot{q} + kq = 0$, as expected. Furthermore, because this is the differential equation of a harmonic oscillator, from Chapter 3 we already know its solution is $q = C\sin(\omega t + \delta)$, where $C = \sqrt{(q_0^2 + (\dot{q}_0/\omega)^2)}$, $\delta = \arctan(\omega q_0/\dot{q}_0)$ and $\omega = \sqrt{k/m}$.

EXAMPLE 13.4

(a) For a conservative system, obtain a general expression for the energy of a system.

(b) Apply the result to the harmonic oscillator problem of Example 13.3.

(c) What is the generalized force of the harmonic oscillator?

....

Solution

(a) Let the Lagrangian be $L = L(q_1, q_2, \ldots, q_n, \dot{q}_1, \dot{q}_2, \ldots, \dot{q}_n, t)$, and taking its time derivative, get

$$\frac{dL}{dt} = \sum_i \frac{\partial L}{\partial q_i}\dot{q}_i + \sum_i \frac{\partial L}{\partial \dot{q}_i}\ddot{q}_i + \frac{\partial L}{\partial t}.$$ (13.4.13)

The first right-hand side term of this expression involves the RHS of (13.4.10), so that we can write

$$\frac{dL}{dt} = \sum_i \dot{q}_i \frac{d}{dt}\frac{\partial L}{\partial \dot{q}_i} + \sum_i \frac{\partial L}{\partial \dot{q}_i}\ddot{q}_i + \frac{\partial L}{\partial t} = \sum_i \frac{d}{dt}\left(\dot{q}_i \frac{\partial L}{\partial \dot{q}_i}\right) + \frac{\partial L}{\partial t},$$ (13.4.14)

and rearranging we see that

$$\frac{d}{dt}\left(\sum_i \left(\dot{q}_i \frac{\partial L}{\partial \dot{q}_i}\right) - L\right) = -\frac{\partial L}{\partial t} = \frac{dE}{dt},$$ (13.4.15)

where, as described below (see Euler's theorem and the Hamiltonian) we have identified the energy

$$E \equiv \sum_i \left(\dot{q}_i \frac{\partial L}{\partial \dot{q}_i}\right) - L.$$ (13.4.16)

Thus, if L does not depend explicitly on time, $\partial L/\partial t = 0$ and $E =$ constant or conserved.

(b) For a harmonic oscillator, from the previous example $L = \frac{1}{2}m\dot{q}^2 - \frac{1}{2}kq^2$, and

$$\sum_i \left(\dot{q}_i \frac{\partial L}{\partial \dot{q}_i}\right) \to \dot{q}\frac{\partial L}{\partial \dot{q}} = m\dot{q}^2,$$ so that (13.4.16) gives $E = m\dot{q}^2 - \frac{1}{2}m\dot{q}^2 + \frac{1}{2}kq^2 = \frac{1}{2}m\dot{q}^2 + \frac{1}{2}kq^2$ as expected. Furthermore, since $\partial L/\partial t = 0$, the energy is conserved.

(c) From (13.3.5c) $Q_k = -\frac{\partial V}{\partial q_k} \to -\frac{\partial}{\partial q}\left(\frac{1}{2}kq^2\right) = -kq$, which is the expression of Hooke's law.

EXAMPLE 13.5

Figure 13.1 shows two masses m_1, m_2 connected through a massless rope over a pulley. The system setup is known as Atwood's machine. At the instant shown, each mass moves with speed v, while the pulley rotates with angular speed $\omega = v/r$, where r is the radius of the pulley whose moment of inertia is I. At time $t = 0$ m_2 is located at a height L from the ground. Use Lagrange's equations to obtain the equations of motion of the system. Assume the rope does not stretch and that there is no slippage between the pulley and the rope.

Solution

Let $q = x$, $\dot{q} = v$, and $\omega = \dot{q}/r$, so that the kinetic energy of the system is $T = \dfrac{1}{2}(m_1 + m_2)\dot{q}^2 + \dfrac{1}{2}I\left(\dfrac{\dot{q}}{r}\right)^2$, and the potential energy is $V = m_1gq + m_2g(L - q)$. The

Lagrangian is $L = T - V = \dfrac{1}{2}\left(m_1 + m_2 + \dfrac{I}{r^2}\right)\dot{q}^2 - (m_1 - m_2)gq - m_2gL$. Applying

Lagrange's equations for conservative systems, we have $\dfrac{\partial L}{\partial q} = -(m_1 - m_2)g$, and

$\dfrac{d}{dt}\dfrac{\partial L}{\partial \dot{q}} = \left(m_1 + m_2 + \dfrac{I}{r^2}\right)\ddot{q}$. Equating these two, get $\left(m_1 + m_2 + \dfrac{I}{r^2}\right)\ddot{q} = (m_2 - m_1)g$, or

solving for the acceleration, obtain $\ddot{q} = (m_2 - m_1)g / \left(m_1 + m_2 + \dfrac{I}{r^2}\right)$. If $m_2 > m_1$, then

$\ddot{q} > 0$ or a clockwise acceleration, else if $m_2 < m_1$, $\ddot{q} < 0$ or counterclockwise acceleration.

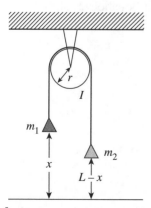

▌ FIGURE 13.1   The Atwood machine of Example 13.5.

A. Systems with Constraints

In Example 13.5 we see that because the rope connected both bodies as well as the pulley, we did not have to use more than one degree of freedom. The absence of slipping is one constraint, and the nonstretching of the rope is another. The apparent number of degrees of freedom of the system was decreased by the number of constraints. Thus

$$\begin{pmatrix} \text{apparent number} \\ \text{of degrees of freedom} \end{pmatrix} - \begin{pmatrix} \text{number of} \\ \text{constraints} \end{pmatrix} = \begin{pmatrix} \text{actual number of} \\ \text{degrees of freedom} \end{pmatrix}, \quad (13.4.17)$$

so that for the case of Atwood's machine in Example 13.5, there are $3 - 2 = 1$ actual degrees of freedom. These kinds of constraints are called *holonomic constraints*, and they tend to reduce the number of degrees of freedom. Had it not been for the holonomic constraints, three coordinates would have been needed to describe the system of Example 13.5. A *nonholonomic* type of constraint is one for which it is impossible to relate any of the coordinates with any other; one must, therefore, include the rate of change of such coordinates in the description of the system, and, by the same token, end up with more degrees of freedom. One such example is that of a mass that's constrained to move along a curved surface but, due to slippage, the mass can leave the surface, and be no longer constrained to move on the surface. We do not deal with nonholonomic constraints here.

B. Euler's Theorem and the Hamiltonian

We next consider a homogenous system in time; that is, we take the Lagragian for a closed system and show how the conservation of energy arises. Again, we begin by working with the kinetic energy in rectangular coordinates of Equation (13.4.1). Using (13.4.2) we notice that we can express

$$\begin{aligned} \dot{x}_i^2 &= \left(\sum_k \frac{\partial x_i}{\partial q_k} \dot{q}_k + \frac{\partial x_i}{\partial t} \right) \left(\sum_{k'} \frac{\partial x_i}{\partial q_{k'}} \dot{q}_{k'} + \frac{\partial x_i}{\partial t} \right) \\ &= \sum_{kk'} \frac{\partial x_i}{\partial q_k} \frac{\partial x_i}{\partial q_{k'}} \dot{q}_k \dot{q}_{k'} + 2 \frac{\partial x_i}{\partial t} \sum_k \frac{\partial x_i}{\partial q_k} \dot{q}_k + \left(\frac{\partial x_i}{\partial t} \right)^2, \end{aligned} \quad (13.4.18a)$$

which in the case of a scleronomic or time-independent system $\partial x_i / \partial t \to 0$, and (13.4.18a) becomes

$$\dot{x}_i^2 \to \sum_{kk'} \frac{\partial x_i}{\partial q_k} \frac{\partial x_i}{\partial q_{k'}} \dot{q}_k \dot{q}_{k'}. \quad (13.4.18b)$$

With this expression, the kinetic energy (13.4.1) becomes

$$T = \sum_{kk'} a_{kk'} \dot{q}_k \dot{q}_{k'}, \qquad (13.4.19)$$

where we have defined $a_{kk'} = a_{k'k} = (1/2) \sum_i m_i (\partial x_i / \partial q_k)(\partial x_i / \partial q_{k'})$. Differentiating (13.4.19) with respect to $\dot{q}_{k''}$, get

$$\frac{\partial T}{\partial \dot{q}_{k''}} = \sum_{kk'} a_{kk'} \frac{\partial \dot{q}_k}{\partial \dot{q}_{k''}} \dot{q}_{k'} + \sum_{kk'} a_{kk'} \frac{\partial \dot{q}_{k'}}{\partial \dot{q}_{k''}} \dot{q}_k$$

$$= \sum_{kk'} a_{kk'} \delta_{kk''} \dot{q}_{k'} + \sum_{kk'} a_{kk'} \delta_{k'k''} \dot{q}_k = 2 \sum_{k'} a_{k''k'} \dot{q}_{k'}, \qquad (13.4.20)$$

Multiplying this expression by $\dot{q}_{k''}$, summing over k'' and comparing the result with (13.4.19), get

$$\sum_{k''} \frac{\partial T}{\partial \dot{q}_{k''}} \dot{q}_{k''} = 2 \sum_{k''k'} a_{k''k'} \dot{q}_{k''} \dot{q}_{k'} = 2\,T. \qquad (13.4.21)$$

This is a special case of Euler's theorem, which states that if $f(y_k)$ is a homogeneous function of y_k (that is of degree n), then $\sum_k y_k \partial f / \partial y_k = nf$. The significance of (13.4.21) in relation to the energy function (13.4.16) can be seen if we recall that when the potential energy function is independent of the generalized velocities, $V = V(q_1, q_2, \ldots, q_n)$, then (13.4.9) applies and consequently (13.4.16) and (13.4.21) enable us to find that the energy function

$$E \equiv \sum_i \left(\dot{q}_i \frac{\partial L}{\partial \dot{q}_i} \right) - L = \sum_i \left(\dot{q}_i \frac{\partial T}{\partial \dot{q}_i} \right) - L = 2T - (T - V) = T + V, \quad (13.4.22)$$

is indeed the total energy which we later refer to as the Hamiltonian H. As mentioned in Example 13.4, if L does not depend explicitly on time, this quantity is conserved.

■ 13.5 Generalized Momentum and Ignorable Coordinates

If a coordinate q_k does not explicitly appear in the expression for kinetic and potential energy, then

$$\frac{\partial L}{\partial q_k} = 0, \tag{13.5.1}$$

so that by (13.4.10) we find that

$$\frac{d}{dt}\frac{\partial L}{\partial \dot{q}_k} = 0 \quad \Rightarrow \quad \frac{\partial L}{\partial \dot{q}_k} = \text{constant}, \tag{13.5.2}$$

so that q_k is an *ignorable coordinate* and $\dot{q}_k$ is an integral of the motion; that is, $\dot{q}_k$ is a conserved quantity.

EXAMPLE 13.6

A particle of mass m moves in the presence of a potential of the form $V(y, z)$; obtain the particle's equations of motion.

....

Solution

Since the potential is given, we write the Lagrangian as $L = \frac{1}{2}m(\dot{x}^2 + \dot{y}^2 + \dot{z}^2) - V(y, z)$, and proceed to obtain the sets of derivatives according to the RHS of (13.4.10)

$$\frac{\partial L}{\partial x} = 0 \quad \frac{\partial L}{\partial y} = -\frac{\partial V}{\partial y} = F_y \quad \frac{\partial L}{\partial z} = -\frac{\partial V}{\partial z} = F_z. \tag{13.5.3a}$$

Similarly, for the LHS of (13.4.10)

$$\frac{d}{dt}\frac{\partial L}{\partial \dot{x}} = m\ddot{x} \quad \frac{d}{dt}\frac{\partial L}{\partial \dot{y}} = m\ddot{y} \quad \frac{d}{dt}\frac{\partial L}{\partial \dot{z}} = m\ddot{z}. \tag{13.5.3b}$$

Equating these, we obtain

$$\frac{\partial L}{\partial x} = m\ddot{x} = 0 \Rightarrow m\dot{x} = p_x = const \quad m\ddot{y} = F_y \quad m\ddot{z} = F_z. \qquad (13.5.3c)$$

While the last two equations in (13.5.3c) give the dynamics of the motion in the y, z directions, the first implies that the momentum in the x direction is conserved.

In the preceding example, x is an ignorable coordinate, $\dot{x}$ is an integral of the motion and is constant; in addition, as we have seen, the momentum associated with the ignorable coordinate is conserved. Thus, we can define the generalized momentum p_k associated with the generalized coordinate q_k as

$$p_k \equiv \frac{\partial L}{\partial \dot{q}_k}, \qquad (13.5.4)$$

and if the generalized coordinate is ignorable, that is, (13.5.1) applies, then the generalized momentum is also conserved as indicated by (13.5.2). Finally, note that if $\dot{q}_k$ is a linear (angular) velocity, then p_k is a linear (angular) momentum. Because of its relationship to q_k, p_k is referred to as the momentum *conjugate* to q_k or *conjugate momentum*.

■ 13.6   More Examples of Lagrange's Equations

Next a few examples of the applications of Lagrange's equations are given. For some of the examples, the results are already known and demonstrate the formulation's power.

EXAMPLE 13.7

Obtain the equation of motion for a mass initially resting on the surface of a frictionless inclined plane at angle θ from the horizontal, as shown in Figure 13.2.

Solution

The kinetic energy is $T = \frac{1}{2}m\dot{x}^2$, and the potential energy is $V = mgh = mgx\sin\theta$. The Lagrangian becomes $L = m\dot{x}^2/2 - mgx\sin\theta$. The system is conservative, so we set $Q'_k = 0$ in (13.4.11), then $\frac{\partial L}{\partial x} = mg\sin\theta$ and $\frac{d}{dt}\frac{\partial L}{\partial \dot{x}} = m\ddot{x}$, and after setting these equal to each other, we obtain $\ddot{x} = -g\sin\theta$, as expected.

EXAMPLE 13.8

Use Lagrange's equations to obtain the equation of motion of a simple pendulum.

Solution

Referring to Figure 13.3, we write the kinetic energy, $T = I\omega^2/2 = mL^2\dot{\theta}^2/2$, the potential energy, $V = mgh = mgL(1 - \cos\theta)$, and so the Lagrangian is $L = mL^2\dot{\theta}^2/2 - mgL(1 - \cos\theta)$.

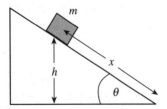

FIGURE 13.2   The inclined plane of Example 13.7.

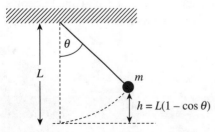

FIGURE 13.3   The simple pendulum of Example 13.8.

The system is assumed to be conservative, so that we have

$$Q'_k = 0, \quad \frac{\partial L}{\partial \theta} = mgL\sin\theta, \quad \frac{d}{dt}\frac{\partial L}{\partial \dot\theta} = mL^2\ddot\theta, \tag{13.6.1}$$

and setting the last two equal, we find

$$\ddot\theta + \frac{g}{L}\sin\theta = 0, \tag{13.6.2}$$

which is the pendulum equation of motion we worked with in Chapter 4.

EXAMPLE 13.9

Use Lagrange's formalism to obtain the equations of motion for a planet in orbit around the sun.

....

Solution

In this problem we use two generalized coordinates, $q_1 = r$ and $q_2 = \theta$. The kinetic energy of a planet in orbit around the sun has two translational motion terms. One is due to its radial movement and the other is due to its orbital motion; that is, $T = \frac{1}{2}m\dot r^2 + \frac{1}{2}mr^2\dot\theta^2$, where m is the planet's mass. The potential energy is $V(r) = -\kappa/r$, where $\kappa = Gmm_s$ with G, the universal gravitational constant, and m_s, the mass of the sun. With the above understanding, the Lagrangian becomes

$$L = \frac{1}{2}m\dot q_1^2 + \frac{1}{2}mq_1^2\dot q_2^2 + \frac{\kappa}{q_1}. \tag{13.6.3}$$

Once again, we assume the system is conservative and set $Q'_k = 0$. The RHS of (13.4.11) for each coordinate is

$$\frac{\partial L}{\partial q_1} = mq_1\dot q_2^2 - \frac{\kappa}{q_1^2}, \quad \frac{\partial L}{\partial q_2} = 0, \tag{13.6.4a}$$

while the LHS of (13.4.11) for each coordinate gives

$$\frac{d}{dt}\frac{\partial L}{\partial \dot{q}_1} = \frac{d}{dt}(m\dot{q}_1) = m\ddot{q}_1, \quad \frac{d}{dt}\frac{\partial L}{\partial \dot{q}_2} = \frac{d}{dt}(mq_1^2\dot{q}_2) = 2\,mq_1\dot{q}_1\dot{q}_2 + mq_1^2\ddot{q}_2.$$

(13.6.4b)

Setting the first (second) of (13.6.4a) with the first (second) of (13.6.4b) and replacing the coordinates, we obtain

$$m\ddot{r} = mr\dot{\theta}^2 - \frac{\kappa}{r^2}, \quad \frac{d}{dt}(mr^2\dot{\theta}) = 0 \Rightarrow p_\theta = mr^2\dot{\theta} = const,$$

(13.6.5)

where the last equation is consistent with our ignorable coordinate discussion of Section 13.5, i.e., $d[\partial L/\partial \dot{q}_2]/dt = 0 \Rightarrow \dot{q}_2 = const$. The first equation of (13.6.5) represents the motion of a planet under the central gravitational force from the sun, and the second represents the fact that the angular momentum, p_θ, is conserved in agreement with our results in Chapter 8 for this problem.

EXAMPLE 13.10

Obtain the equations of motion for a symmetric top under gravity and rotating about a fixed point located at its apex. Assume there are no nonconservative forces present and that the top's center of mass is located at a distance ℓ from its apex. It also has a symmetric principal moment of inertia I_3, and non-symmetric principal moments of inertia I_1. Refer to Chapter 12, Sections 6 and 7 for other details.

Solution
From Section 12.7, Equations (12.7.11 and 12.7.12), we already know the kinetic and potential energies, thus the Lagrangian is

$$L = \frac{1}{2}I_1(\dot{\varphi}^2\sin^2\theta + \dot{\theta}^2) + \frac{1}{2}I_3(\dot{\varphi}\cos\theta + \dot{\psi})^2 - mg\ell\cos\theta.$$

(13.6.6)

We naturally let the generalized coordinates be the Eulerian angles φ, θ, and ψ, so that the partial derivatives of the Lagrangian with respect to the generalized coordinates are

$$\frac{\partial L}{\partial \varphi} = 0, \quad \frac{\partial L}{\partial \theta} = I_1\dot{\varphi}^2\sin\theta\cos\theta - I_3(\dot{\varphi}\cos\theta + \dot{\psi})\dot{\varphi}\sin\theta + mg\ell\sin\theta, \quad \frac{\partial L}{\partial \psi} = 0, \quad (13.6.7a)$$

and the associated time derivatives for the partial derivatives of the Lagrangian with respect to the generalized velocities are

$$\frac{d}{dt}\frac{\partial L}{\partial \dot{\varphi}} = \frac{d}{dt}[I_1\dot{\varphi}\sin^2\theta + I_3(\dot{\varphi}\cos\theta + \dot{\psi})\cos\theta], \quad \frac{d}{dt}\frac{\partial L}{\partial \dot{\theta}} = I_1\ddot{\theta}, \quad \frac{d}{dt}\frac{\partial L}{\partial \dot{\psi}} = \frac{d}{dt}[I_3(\dot{\varphi}\cos\theta + \dot{\psi})].$$

(13.6.7b)

In the absence of nonconservative forces, given that the first and third derivatives in (13.6.7a) are zero, and which we equate to the first and third of (13.6.7b), respectively, we see that φ and ψ are ignorable coordinates, so that we have the associated generalized momenta

$$p_\varphi = I_1\dot{\varphi}\sin^2\theta + I_3(\dot{\varphi}\cos\theta + \dot{\psi})\cos\theta, \quad p_\psi = I_3(\dot{\varphi}\cos\theta + \dot{\psi}).$$

(13.6.8)

If we recall (12.7.15b), we see that p_φ corresponds to Equation (12.7.20), and p_ψ corresponds to Equation (12.7.18) for the momenta along the fixed z-axis and the body 3-symmetry axis, respectively, in the top problem, and where $I_1 = I_2 = I$, $I_3 = I_s$. These two momentum quantities are conserved, in agreement with what we found previously in Section 7 of Chapter 12. The results here, however, are produced in a natural way by the powerful Lagrange formalism. Finally, equating the second equation of (13.6.7a) with the second of (13.6.7b), we find the equation of motion for θ

$$I_1\ddot{\theta} = I_1\dot{\varphi}^2\sin\theta\cos\theta - I_3(\dot{\varphi}\cos\theta + \dot{\psi})\dot{\varphi}\sin\theta + mg\ell\sin\theta,$$

(13.6.9)

which is identical to the torque Equation (12.7.17a) for the top problem in Chapter 12, Section 7.

EXAMPLE 13.11

(a) Obtain the generalized coordinates for the two masses in the double-pendulum system shown in Figure 13.4.

(b) Write the Lagrangian of the system.

(c) Obtain the equations of motion for the system.

(d) For the case when $m_1 = m_2$, and $L_1 = L_2$, write the equations of motion for small oscillations.

(e) For the case of small oscillations, obtain the vibrational frequencies and the corresponding normal modes of vibration.

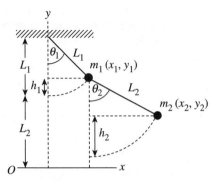

I FIGURE 13.4   The double pendulum of Example 13.11.

(f) Give a plot of the full range of motion of the double pendulum using the following parameters: $L_1, L_2 = 1, 2$ m, $m_1, m_2 = 1, 2$ kg, $g = 9.8$ m/s^2,   with   initial   conditions, $\theta_{10}, \theta_{20} = \pi/2, \pi/3$ rad, and $(d\theta_1/dt)_0 = 0 = (d\theta_2/dt)_0$.

Solution

(a) Referring to Figure 13.4, we let the positions of m_1 and m_2 be determined by

$$x_1 = L_1\sin\theta_1,\ y_1 = L_1\cos\theta_1, \tag{13.6.10a}$$

and

$$x_2 = x_1 + L_2\sin\theta_2,\ y_2 = y_1 + L_2\cos\theta_2, \tag{13.6.10b}$$

respectively. The corresponding velocities are

$$\dot{x}_1 = L_1\dot{\theta}_1\cos\theta_1,\ \dot{y}_1 = -L_1\dot{\theta}_1\sin\theta_1, \tag{13.6.11a}$$

and

$$\dot{x}_2 = L_1\dot{\theta}_1\cos\theta_1 + L_2\dot{\theta}_2\cos\theta_2,\ \dot{y}_2 = -L_1\dot{\theta}_1\sin\theta_1 - L_2\dot{\theta}_2\sin\theta_2. \tag{13.6.11b}$$

(b) The potential energy is written as

$$V = m_1gL_2 + m_1gh_1 + m_2g(h_1 + h_2), \tag{13.6.12a}$$

where from Figure 13.4 we see that

$$h_1 = L_1(1 - \cos\theta_1), \quad h_2 = L_2(1 - \cos\theta_2), \tag{13.6.12b}$$

and the zero of energy is chosen at a point $L_1 + L_2$ below the pendulum's fixed support. The kinetic energy is

$$T = \frac{1}{2}m_1(\dot{x}_1^2 + \dot{y}_1^2) + \frac{1}{2}m_2(\dot{x}_2^2 + \dot{y}_2^2), \tag{13.6.12c}$$

so that by substituting the generalized coordinates and velocities (13.6.10) into (13.6.12), respectively, we get the Lagrangian

$$L = T - V = \frac{1}{2}m_1 L_1^2\dot{\theta}_1^2 + \frac{1}{2}m_2[L_1^2\dot{\theta}_1^2 + L_2^2\dot{\theta}_2^2 + 2L_1 L_2\dot{\theta}_1\dot{\theta}_2\cos(\theta_1 - \theta_2)]$$
$$- m_1 g L_2 - (m_1 + m_2)g L_1(1 - \cos\theta_1) - m_2 g L_2(1 - \cos\theta_2). \tag{13.6.13}$$

(c) The equations of motion are obtained by equating the corresponding sets of Equations (13.4.10) for conservative systems, that is,

$$\frac{d}{dt}\left(\frac{\partial L}{\partial\dot{\theta}_1}\right) = \frac{\partial L}{\partial\theta_1}, \quad \frac{d}{dt}\left(\frac{\partial L}{\partial\dot{\theta}_2}\right) = \frac{\partial L}{\partial\theta_2}. \tag{13.6.14}$$

From the first of these, we get

$$(m_1 + m_2)L_1\ddot{\theta}_1 + m_2 L_2\ddot{\theta}_2\cos(\theta_1 - \theta_2) = -m_2 L_2\dot{\theta}_2^2\sin(\theta_1 - \theta_2) - (m_1 + m_2)g\sin\theta_1, \tag{13.6.15a}$$

and from the second we get

$$m_2 L_2\ddot{\theta}_2 + m_2 L_1\ddot{\theta}_1\cos(\theta_1 - \theta_2) = m_2 L_1\dot{\theta}_1^2\sin(\theta_1 - \theta_2) - m_2 g\sin\theta_2. \tag{13.6.15b}$$

(d) Letting $m_1 = m_2 = m$, and $L_1 = L_2 = L$, as well as taking $\sin\theta_i \approx \theta_i$, $\cos\theta_i \approx 1$, $\sin(\theta_1 - \theta_2) \approx 0$, and $\cos(\theta_1 - \theta_2) \approx 1$, for small oscillations, Equations (13.5.19) simplifies to

$$\ddot{\theta}_1 + \frac{\ddot{\theta}_2}{2} = -\frac{g}{L}\theta_1, \quad \ddot{\theta}_1 + \ddot{\theta}_2 = -\frac{g}{L}\theta_2. \tag{13.6.16}$$

(e) For small oscillations, the vibrational frequencies can be obtained if we assume solutions of the form

$$\theta_1 = A_1 \cos \omega t, \text{ and } \theta_2 = A_2 \cos \omega t, \tag{13.6.17}$$

which are then substituted into (13.6.16), to obtain the matrix equation

$$\begin{pmatrix} \left(\dfrac{g}{L} - \omega^2 \right) & -\dfrac{\omega^2}{2} \\[2ex] -\omega^2 & \left(\dfrac{g}{L} - \omega^2 \right) \end{pmatrix} \begin{pmatrix} A_1 \\ A_2 \end{pmatrix} = 0. \tag{13.6.18}$$

These equations have frequency solutions obtained by setting the determinant of the matrix to zero. If we do that, we get two mode frequencies

$$\omega_{1,2}^2 = \begin{cases} (2 - \sqrt{2})g/L \\ (2 + \sqrt{2})g/L \end{cases}, \tag{13.6.19}$$

one low (ω_1) and one high (ω_2). Substituting each of these into either of the equations in (13.6.18) we get the vibrational modes

$$(A_2/A_1)_{s,a} = \begin{cases} \sqrt{2} \\ -\sqrt{2} \end{cases}, \tag{13.6.20}$$

where the upper mode (s) corresponds to the lower vibrational frequency of the symmetric mode; that is, the bobs move in the same direction. The lower mode (a), corresponding to the higher frequency, is the antisymmetric mode, i.e., the bobs move in opposite directions. If we pick $A_1 \equiv 1$, and substitute (13.6.18 and 13.6.19) into (13.6.17), we get the time-dependent behavior of each bob corresponding to each mode of vibration

$$\text{symmetric: } \theta_{1s} = \cos \omega_1 t, \quad \theta_{2s} = \sqrt{2} \cos \omega_1 t, \tag{13.6.21}$$

$$\text{antisymmetric: } \theta_{1a} = \cos \omega_2 t, \quad \theta_{2a} = -\sqrt{2} \cos \omega_2 t.$$

(f) In order to produce a full plot of $\theta_1(t)$ and $\theta_2(t)$, we need to solve the full differential Equation (13.6.15) numerically. To this end, we rearrange (13.6.15a) to read

$$\ddot{\theta}_1 = -\alpha \ddot{\theta}_2 - \beta, \tag{13.6.22a}$$

where we have let

$$\alpha = \frac{m_2 L_2 \ddot{\theta}_2 \cos(\theta_1 - \theta_2)}{(m_1 + m_2)L_1}, \quad \beta = \frac{m_2 L_2 \dot{\theta}_2^2 \sin(\theta_1 - \theta_2) + (m_1 + m_2)g\sin\theta_1}{(m_1 + m_2)L_1}. \quad (13.6.22b)$$

We then substitute (13.6.22a) into (13.6.15b) and solve for $\ddot{\theta}_2$, to get

$$\ddot{\theta}_2 = \frac{\beta L_1 \cos(\theta_1 - \theta_2) + L_1 \dot{\theta}_1^2 \sin(\theta_1 - \theta_2) - g\sin\theta_2}{(L_2 - \alpha L_1 \cos(\theta_1 - \theta_2))}. \quad (13.6.23)$$

We next define the w array suitable for numerical computation using MATLAB, as we have done before

$$w(1, 2, 3, 4) \equiv (\theta_1, \dot{\theta}_1, \theta_2, \dot{\theta}_2), \quad (13.6.24a)$$

and set up the derivatives of this array as

$$\frac{dw(1, 2, 3, 4)}{dt} = (w(2), -\alpha\ddot{\theta}_2 - \beta, w(4), \ddot{\theta}_2), \quad (13.6.24b)$$

since $w(2) = d\theta_1/dt = dw(1)/dt$, $-\alpha\ddot{\theta}_2 - \beta = dw(2)/dt = \ddot{\theta}_1$ as given by (13.6.22a), $w(4) = d\theta_2/dt = dw(3)/dt$, and $\ddot{\theta}_2 = dw(4)/dt$. In (13.6.24b), wherever $\ddot{\theta}_2$ appears, we are to use the RHS of (13.6.23). The above differential equation set is to be used in conjunction with the initial conditions of the problem. The MATLAB scripts doublep.m and doublep_der.m incorporate these equations for the numerical computation using the Runge-Kutta scheme. The results for the initial conditions given are shown in Figure 13.5.

In the upper box of Figure 13.5(a) the angles $\theta_1(t)$ and $\theta_2(t)$ are plotted versus time. The deviation of these curves from harmonicity that occurs for not-so-small angles demonstrates the general complicated motion of this pendulum. The lower plot of Figure 13.5(a) is a phase space plot of θ_2 versus θ_1 and shows their behavior is somewhat chaotic. *Chaos* refers to the nonlinear and irregular, deterministic behavior of a system that is very sensitive to the initial conditions. More interesting behavior is expected by exploring the initial, $t = 0$, parameters. The actual motion of what the pendulum does for the initial conditions used is shown by the

(a)   Double Pendulum: $\theta_{10} = 0.785$, $\theta_{20} = 1.05$ rad, $d(\theta_1/dt)_0 = 0$, $d(\theta_2/dt)_0 = 0$

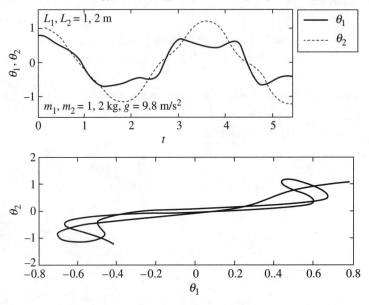

(b)   Double Pendulum: $\theta_{10} = 0.785$, $\theta_{20} = 1.05$ rad, $d(\theta_1/dt)_0 = 0$, $d(\theta_2/dt)_0 = 0$

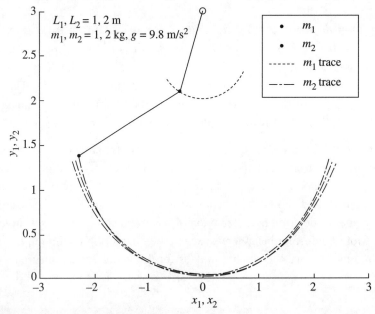

FIGURE 13.5   The double pendulum (a) angles $\theta_1(t)$ and $\theta_2(t)$ are plotted versus time (upper figure) and versus each other (lower figure); (b) simulation of the pendulum for the initial conditions shown. The coordinates of each mass are given by (13.6.10a, b).

simulation of Figure 13.5(b). The coordinates of each mass are given by (13.6.10a, b). The magnitude of each pendulum length should be constant as a function of time; that is, one expects $r_1 = \sqrt{x_1(t)^2 + y_1(t)^2}$ and $r_2 = \sqrt{x_2(t)^2 + y_2(t)^2}$ to be conserved and equal to L_1 and $L_1 + L_2$, respectively. This can be checked with the script by uncommenting the proper lines; it works well here, albeit with some small variations (less than 7%) due to the numerical process. One needs to run the script doublep.m in order to see the simulation in real time. The scripts doublep.m and doublep_der.m used are listed below. Finally, for simulation purposes, because the pendulum actually hangs, the script doublep.m shifts the y coordinates $y_1 \to L_1 + L_2 - y_1$, and $y_2 \to L_1 + L_2 - y_2$ with the y_i's as given in (13.6.10.)

SCRIPT

```
%doublep.m, program to solve the double pendulum equations
%of motion numerically and plot their solutions
clear;
L1=1; L2=2; m1=1; m2=2; g=9.8;      %lengths, masses, gravity
tau=sqrt(g/(L1+L2));                %a time unit
tmax=3*tau; ts=0.05;                %simulation run time and time interval
tr=(0.0:ts:tmax);N=length(tr);      %time range, array size
th10=pi/4; th20=pi/3;               %init angle values in rad
th10d=0.0; th20d=0.0;               %init angular speeds in rad/sec
ic1=[th10;th10d;th20;th20d];        %initial conditions:
%Use MATLAB's Runge-Kutta (4,5) formula (uncomment/comment as needed)
%opt=odeset('AbsTol',1.e-8,'RelTol',1.e-5);       %user set Tolerances
%[t,w]=ode45('doublep_der',tr,ic1,opt,L1,L2,m1,m2,g);%with set tolerance
[t,w]=ode45('doublep_der',tr,ic1,[],L1,L2,m1,m2,g); %default tolerance
%Next: plots of the angles versus time
%w(1):theta1, w(2):theta1_dot, w(3):theta2, w(4):theta2_dot
str1=cat(2,'Double Pendulum: \theta_{10}=',num2str(th10,3),...
        ', \theta_{20}=',num2str(th20,3),' rad, d(\theta_{1}/dt)_0=',...
      num2str(th10d,3),', d(\theta_{2}/dt)_0=',num2str(th20d,3));
str2=cat(2,'L_1, L_2=',num2str(L1,3),', ',num2str(L2,3),' m');
str3=cat(2,'m_1, m_2=',num2str(m1,3),', ',num2str(m2,3),' kg',...
        ', g=',num2str(g,3),' m/s^2');
subplot(2,1,1),
plot(t,w(:,1),'k-',t,w(:,3),'b:'),
xlabel ('t','FontSize',13), ylabel('\theta_1, \theta_2','FontSize',13)
title(str1,'FontSize',12)
ym=min([w(:,1);w(:,3)]); yp=max([w(:,1);w(:,3)]);    %window size
axis([0,tmax,ym*(1+0.3),yp*(1+0.3)])
text(.1,yp*(1+0.05),str2)
text(.1,ym*(1+0.05),str3)
```

```
h=legend('\theta_1','\theta_2',-1); set(h,'FontSize',13)
subplot(2,1,2), plot(w(:,1),w(:,3),'r')
xlabel ('\theta_1','FontSize',13),ylabel('\theta_2','FontSize',13)
% ================== Simulation next ==================================
%Coordinates versus time
x1=L1*sin(w(:,1)); y1=L1*cos(w(:,1));
x2=x1+L2*sin(w(:,3)); y2=y1+L2*cos(w(:,3));
%r1=sqrt(x1.^2+y1.^2);r2=sqrt(x2.^2+y2.^2); %length conservation check
%figure, plot(t,[r1,r2])                    %uncomment as desired
%support is at L1+L2, where y's are measured from
v=L1+L2; y1=v-y1; y2=v-y2;   %shift the y's for simulation purpose
figure, vx=max([v;x1;x2]);vy=max([v;y1;y2]);
axis([-vx,vx,0,vy])                          %window size
 for i=1:N
    clf
    axis([-vx,vx,0,vy])
    hold on
    plot(0,v,'ko');                          %pivot point
    line([0,x1(i)],[v,y1(i)],'color', 'k');  %arm1
    h(1)=plot(x1(i),y1(i),'k.');             %m1 at x1,y1
    line([x1(i),x2(i)],[y1(i),y2(i)],'color','b'); %arm2
    h(2)=plot(x2(i),y2(i),'b.');             %m2 at x2,y2
    pause(.05)
 end
h(3)=plot(x1,y1,'k:');                       %m1 trace
h(4)=plot(x2,y2,'b-.');                      %m2 trace
h=legend(h,'m_1','m_2','m_1 trace', 'm_2 trace');
set(h,'FontSize',13)
xlabel('x_1, x_2','FontSize',13), ylabel('y_1, y_2','FontSize',13)
title(str1,'FontSize',12),
text(-v*(1-0.1),v*(1-.05),str2)
text(-v*(1-0.1),v*(1-.1),str3)
```

FUNCTION

```
%doublep_der.m: returns the derivatives for the double pendulum problem
function ders = doublep_der(t,w,flag,L1,L2,m1,m2,g)
%w(1):theta1, w(2):theta1_dot, w(3):theta2, w(4):theta2_dot
tp=w(1)-w(3); cs=sin(tp); cc=cos(tp);
ta=m2*L2*cc/(m1+m2)/L1;
tb=(m2*L2*w(4).^2.*cs+(m1+m2)*g*sin(w(1)))/(m1+m2)/L1;
tc=(L1*tb.*cc+L1*w(2).^2.*cs-g*sin(w(3)))./(L2-ta.*L1.*cc);
ders=[w(2);-ta.*tc-tb;w(4);tc];
```

■ 13.7   Hamilton's Equations

From the preceding sections we can say that, in general, the Lagrangian is a function of the generalized coordinates, the generalized velocities, and time or $L = L(\dot{q}_k, q_k, t)$, where $k = 1, \ldots, n$, then a change in L is

$$dL = \sum_k \frac{\partial L}{\partial \dot{q}_k} d\dot{q}_k + \sum_k \frac{\partial L}{\partial q_k} dq_k + \frac{\partial L}{\partial t} dt. \qquad (13.7.1)$$

Using the definition of the generalized momentum (13.5.4) $p_k \equiv \partial L / \partial \dot{q}_k$, as well as Lagrange's Equation (13.4.10) for conservative systems, we see that

$$\frac{dp_k}{dt} = \frac{d}{dt} \frac{\partial L}{\partial \dot{q}_k} = \frac{\partial L}{\partial q_k} = \dot{p}_k, \qquad (13.7.2)$$

so that (13.7.1) can be rewritten as

$$dL = \sum_k p_k d\dot{q}_k + \sum_k \dot{p}_k dq_k + \frac{\partial L}{\partial t} dt. \qquad (13.7.3)$$

Similar to (13.4.22), we next define the energy function or Hamiltonian as

$$H \equiv \sum_k p_k \dot{q}_k - L, \qquad (13.7.4)$$

and its change is

$$dH = \sum_k (p_k d\dot{q}_k + \dot{q}_k dp_k) - dL = \sum_k (\dot{q}_k dp_k - \dot{p}_k dq_k) - \frac{\partial L}{\partial t} dt, \qquad (13.7.5)$$

where (13.7.3) has been used. If we write $H = H(p_k, q_k, t)$, then a change in H can also be rewritten as

$$dH = \sum_k \left(\frac{\partial H}{\partial p_k} dp_k + \frac{\partial H}{\partial q_k} dq_k \right) + \frac{\partial H}{\partial t} dt. \qquad (13.7.6)$$

Equating expressions (13.7.5 and 13.7.6), we obtain Hamilton's equations,

$$\dot{q}_k = \frac{\partial H}{\partial p_k}, \quad \dot{p}_k = -\frac{\partial H}{\partial q_k}, \quad \frac{\partial H}{\partial t} = -\frac{\partial L}{\partial t}, \tag{13.7.7}$$

which represent a different form of the previous information regarding generalized coordinates. These equations are also known as *Hamilton's canonical equations of motion*.

EXAMPLE 13.12

Prove that the Hamiltonian, H, indeed represents the total energy of a system and explain under what conditions it is conserved.

Solution
From (13.7.4) and $p_k \equiv \partial L/\partial \dot{q}_k$ of (13.5.4), we have

$$H = \sum_k \frac{\partial L}{\partial \dot{q}_k}\dot{q}_k - L. \tag{13.7.8}$$

Writing the kinetic energy as

$$T = \frac{1}{2}\sum_i m_i \dot{q}_i^2, \tag{13.7.9a}$$

and the potential energy as

$$V(q_1, \dots, q_n), \tag{13.7.9b}$$

so that

$$L = \frac{1}{2}\sum_i m_i \dot{q}_i^2 - V(q_1, \dots, q_n), \tag{13.7.10a}$$

we have

$$\frac{\partial L}{\partial \dot{q}_k} = m_k \dot{q}_k. \tag{13.7.10b}$$

Substituting the above expressions into (13.7.8) and noting that from (13.7.9a) $2T = \sum_i m_i \dot{q}_i^2$, then H becomes

$$H = \sum_k (m_k \dot{q}_k)\dot{q}_k - T + V = T + V, \tag{13.7.11}$$

which is the total energy of the system. In particular, we note that if in (13.7.7), $\partial H/\partial t = 0$, then the total energy H is conserved.

EXAMPLE 13.13

Write the Hamiltonian for a simple spring-mass harmonic oscillator, and from it obtain the equations of motion.

....

Solution

We let the generalized coordinate be $q = x$, and the generalized velocity be $\dot{q} = \dot{x}$, so that the Hamiltonian is

$$H = T + V = \frac{1}{2}m\dot{q}^2 + \frac{1}{2}kq^2. \tag{13.7.12a}$$

The Lagrangian for this system is

$$L = T - V = \frac{1}{2}m\dot{q}^2 - \frac{1}{2}kq^2. \tag{13.7.12b}$$

From (13.5.4), $p \equiv \partial L/\partial \dot{q} = m\dot{q}$, so that (13.7.12a) becomes

$$H = T + V = \frac{p^2}{2m} + \frac{1}{2}kq^2. \tag{13.7.12c}$$

Using Hamilton's equations (13.7.7), we have for the single particle's generalized quantities

$$\dot{p} = -\frac{\partial H}{\partial q} = -kq, \quad \dot{q} = \frac{\partial H}{\partial p} = \frac{p}{m}. \tag{13.7.13}$$

While the first represents Newton's first law, the second represents the relation between momentum and velocity, as expected.

EXAMPLE 13.14

Use the Lagrangian of Example 13.9 to obtain the Hamiltonian corresponding to a planet in orbit around the sun.

....

Solution

Let's rewrite the Lagrangian as

$$L = \frac{1}{2}m\dot{r}^2 + \frac{1}{2}mr^2\dot{\theta}^2 + \frac{\kappa}{r}, \tag{13.7.14}$$

and using (13.7.4), we have

$$H = p_r\dot{r} + p_\theta\dot{\theta} - \left(\frac{1}{2}m\dot{r}^2 + \frac{1}{2}mr^2\dot{\theta}^2 + \frac{\kappa}{r}\right). \tag{13.7.15a}$$

With the help of (13.5.4), we have

$$p_r = \frac{\partial L}{\partial \dot{r}} = m\dot{r}, \quad p_\theta = \frac{\partial L}{\partial \dot{\theta}} = mr^2\dot{\theta}, \tag{13.7.15b}$$

and when these are substituted into the above H, it becomes

$$H = m\dot{r}^2 + mr^2\dot{\theta}^2 - \left(\frac{1}{2}m\dot{r}^2 + \frac{1}{2}mr^2\dot{\theta}^2 + \frac{\kappa}{r}\right) = \frac{1}{2}m\dot{r}^2 + \frac{1}{2}mr^2\dot{\theta}^2 - \frac{\kappa}{r}. \tag{13.7.16}$$

The first two terms are the total kinetic energy, T, and the last term is the potential energy, V, of the planet in its orbit around the sun, so that once again, $H = T + V$, as expected. This energy is, of course, the same energy function employed in Section 8.4, although a different notation was employed there.

Finally, we note that Example 13.12 allows us to obtain Newton's second law from Lagrange's equations in a simple way as follows. Continuing to use the Lagrangian form (13.7.10a), then using (13.3.5c), we have that

$$\frac{\partial L}{\partial q_k} = -\frac{\partial V}{\partial q_k} = Q_k, \tag{13.7.17}$$

and using Lagrange's Equation (13.4.10) for conservative systems, we can equate the time derivative of (13.7.10b) with (13.7.17) to get

$$Q_k = \frac{d}{dt}\frac{\partial L}{\partial \dot{q}_k} = m\ddot{q}_k, \qquad (13.7.18)$$

which, in the present notation, is equivalent to Newton's second Law of motion.

■ 13.8 Hamilton's Variational Principle of Least Action and Simulation

Sir William Rowan Hamilton (1805–1865) was an Irish mathematician who studied dynamical systems and, in contrast to Newton's laws before him, he proposed a variational principle, which in turn, essentially, encompasses Newton's laws of motion. According to Hamilton, the actual motion followed by a mechanical system as it moves from a starting point to a final point within a given time will be the motion that provides an extremum for the time integral of the Lagrangian; that is,

$$I = \delta \int_{t_1}^{t_2} L \, dt = 0, \qquad (13.8.1a)$$

where δ is an infinitesimal variation of any system parameter. This is known as Hamilton's variational principle of least action, the action being the integral

$$S \equiv \int_{t_1}^{t_2} L \, dt. \qquad (13.8.1b)$$

It means that since a system can be described in terms of the Lagrangian, $L = L(\dot{q}_k, q_k)$, then the evolution of a system in time is determined by (13.8.1a). For example, suppose that at $t = t_1$, a system is found in a state that can be described by the generalized variables, $q_1(t_1), \ldots, q_n(t_1)$ and $\dot{q}_1(t_1), \ldots, \dot{q}_n(t_1)$; as the system evolves from this state to another at $t = t_2$ for which the generalized variables have the values $q_1(t_2), \ldots, q_n(t_2)$, and $\dot{q}_1(t_2), \ldots, \dot{q}_n(t_2)$, there are many possible paths that can be taken to reach the final state. Figure 13.6 shows three possible paths in the evolution process.

The central idea is that according to Hamilton's principle the actual path taken will be that for which (13.8.1) is satisfied. In this section, we show how it is possible

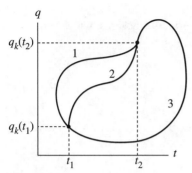

| FIGURE 13.6   Three possible paths in the evolution process of the action integral $S \equiv \int_{t_1}^{t_2} L \, dt$.

to derive Lagrange's equations of motion starting from Hamilton's variational principle. Let $L = L(\dot{q}_k, q_k)$, so that (13.8.1) becomes

$$
I = \delta \int_{t_1}^{t_2} L(\dot{q}_k, q_k) dt = \int_{t_1}^{t_2} \delta L(\dot{q}_k, q_k) dt
$$

$$
= \int_{t_1}^{t_2} \sum_k \left(\frac{\partial L}{\partial q_k} \delta q_k + \frac{\partial L}{\partial \dot{q}_k} \delta \dot{q}_k \right) dt = \int_{t_1}^{t_2} \sum_k \left(\frac{\partial L}{\partial q_k} \delta q_k + \frac{\partial L}{\partial \dot{q}_k} \frac{d}{dt} \delta q_k \right) dt = 0, \quad (13.8.2)
$$

where in the last expression $\delta \dot{q}_k = d \delta q_k / dt$ has been used. The last integral can be carried out by parts, $\int u \, dv = uv - \int v \, du$, using $u = \partial L / \partial \dot{q}_k$, $dv = d[\delta q_k dt]/dt$, or, for each k

$$
\int_{t_1}^{t_2} \frac{\partial L}{\partial \dot{q}_k} \frac{d}{dt} \delta q_k dt = \frac{\partial L}{\partial \dot{q}_k} \delta q_k \Big|_{t_1}^{t_2} - \int_{t_1}^{t_2} \delta q_k \left(\frac{d}{dt} \frac{\partial L}{\partial \dot{q}_k} \right) dt. \quad (13.8.3)
$$

Furthermore, because at the boundary points, t_1, t_2, there is no longer any variation in the motion, the quantity $\dfrac{\partial L}{\partial \dot{q}_k} \delta q_k \Big|_{t_1}^{t_2} = 0$. Substituting these results into (13.8.2), get

$$
I = \int_{t_1}^{t_2} \sum_k \left(\frac{\partial L}{\partial q_k} - \frac{d}{dt} \frac{\partial L}{\partial \dot{q}_k} \right) \delta q_k dt = 0. \quad (13.8.4)
$$

In order for the integral over time to vanish, we must have

$$\sum_k \left(\frac{\partial L}{\partial q_k} - \frac{d}{dt} \frac{\partial L}{\partial \dot{q}_k} \right) \delta q_k = 0,$$ (13.8.5)

and because the q_k's are independent their variations are also independent, so that for each k in the sum, we must have

$$\frac{\partial L}{\partial q_k} - \frac{d}{dt} \frac{\partial L}{\partial \dot{q}_k} = 0,$$ (13.8.6)

which are Lagrange's equations of (13.4.10) for conservative systems. Note that since Newton's second law of motion can be obtained from the Lagrangian formulation, as shown in the previous section, then it follows that Hamilton's principle, while being a different approach, is equivalent to solving Newton's second law of motion. It is interesting to note that the variational principle idea also plays a role in optics associated with Fermat's principle of least time. Other versions of the least-action principle have found themselves of use in general relativity, quantum mechanics, and particle physics (see Edwin F. Taylor, "A call to action, " *American Journal of Physics* Vol. 71, 423, 2003).

At this point it is useful and illuminating to simulate Hamilton's least action principle. We will work with the motion of a single particle free falling near Earth's surface, in one dimension. We know all there is to know about such motion. The path taken by the particle is

$$y = y_0 + v_0 t - \frac{1}{2} g t^2,$$ (13.8.7a)

where y_0 is the initial position at time $t = t_0 \equiv 0$, with v_0 the initial speed, and g the acceleration due to gravity. If in addition to knowing the initial position of the particle, we also know its final position, y_f, the above analytic path tells us that the time at which y_f is reached is

$$t_f = \frac{v_0}{g} + \sqrt{\left(\frac{v_0}{g}\right)^2 - 2\frac{(y_f - y_0)}{g}}.$$ (13.8.7b)

The potential and kinetic energies as well as the Lagrangian of the particle are fully known, i.e.,

$$T = \frac{1}{2} m v^2, \quad V = mgy, \quad L = T - V,$$ (13.8.8)

so that the action for the particle is, from (13.8.1b)

$$S = \int_{t_0}^{t_f} L dt = \int_{t_0}^{t_f} \left(\frac{1}{2} m v^2 - m g y \right) dt. \tag{13.8.9}$$

According to Hamilton's least action principle (13.8.1a), S is stationary with respect to an infinitesimal change in the independent coordinates. Thus if we calculate (13.8.9) many times under small changes in the independent coordinate, y, then the true path of the particle will have been found when the change does not affect S and, therefore, by (13.8.1a), the action is an extremum (minimum, or saddle point). One way of performing the simulation is through a Monte-Carlo (MC) procedure. That is, given the initial and final positions of a particle, a guess is made as to the path of the particle; subsequently, small changes in the path are made, one at a time, through a random procedure. For example, after a change in a given step is made, the action (13.8.9) is calculated, and if the change in the step leads to a lower value of S, then the step is accepted; otherwise the step is discarded. The process is repeated until any change in the path produces no change in S, then by Equation (18.8.1a) Hamilton's principle of least action is satisfied and the final path is the correct path that agrees with that predicted by Newton's equations. In our simulation, the algorithm employed, given the initial conditions, follows. (i) Guess a path, $y_{n=1, k=1}(t)$, of N steps, each labeled in the range, $1 < k < N$, where n is a trial index. (ii) Calculate the action $S_{n=1, k=1}$, of the starting trial. (iii) Modify the kth step in the path in the range, $2 < k < N - 1$, say $k = 2$, in a random way, such as $y_k = y_k + 2\, dy(R - 1/2)$, where dy is the maximum allowed change in y, and R is a MATLAB-generated random number in the range $0 \le R \le 1$. (iv) Calculate $S_{n=1, k}$. (v) Compare the change in S. If $S_{n=1, k}$ is less than $S_{n=1, k=1}$, accept the step. (vi) If the step is accepted, modify the original path based on the accepted step, so that the new path is $y_{n=1, k}(t)$. If the step is not accepted, leave the path alone. (vii) Increase the index k to $k + 1$ and repeat steps (iii)–(iv) until $k = N - 1$. (viii) The final resulting path from the first finished trial is $y_{n=1, k=N}(t)$. The process (i)–(viii) is repeated as $n \to n + 1$ for as many trials as it takes to get $dS = S_{n, N-1} - S_{n-1, N-1} \approx 0$, or smaller than a certain tolerance, and the final path is $y_n \equiv y_{n, k=N}$. Below the tolerance used is 1×10^{-7}. In the limit as $n \to \infty$, we expect $dS \to 0$ and $y_n \to y$, where y is the analytic solution. The action integral (13.8.9) is carried out according to the rectangular rule of Appendix B, as follows:

$$S = \int_{t_0}^{t_f} L dt \approx \Delta t \sum_{k=1}^{N-1} L_k, \tag{13.8.10}$$

where $\Delta t = (t_f - t_0)/(N - 1)$, $L_k \equiv L(t_k)$,   and $t_k = t_0 + (k - 1)\Delta t$.   The Lagrangian (13.8.8) at each kth step is approximated by

$$L(t_k) \approx \frac{1}{2}m\left(\frac{y_{k+1} - y_k}{\Delta t}\right)^2 - mgy_k, \tag{13.8.11a}$$

where we have replaced the particle's velocity v at a given time by the approximate discrete difference of the coordinate over time. The initial trajectory guess is an interpolated straight line between initial and final points

$$y_k = y_0 + \frac{(y_f - y_0)}{(t_f - t_0)}(t_k - t_0). \tag{13.8.11b}$$

The results of the simulation are shown in Figure 13.7. The upper figure dots correspond to the simulation, which required 471 trials to obtain $dS \approx 10^{-9}$, making a total of 1639 random accepted steps or successful exchanges. The straight line is the original guess and the dashed line is the analytic result (13.8.7a). The lower figure shows how S gets lowered from trial to trial until its variation becomes small as the obtained dotted curve in the upper figure becomes closer to the analytic result. The

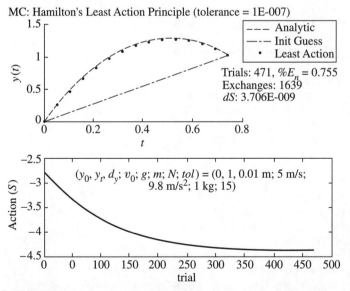

FIGURE 13.7   Simulation of Hamilton's least action principle for the case of the motion of a single particle free falling near Earth's surface, in one dimension.

simulation demonstrates and reinforces the validity of Hamilton's principle of least action.

The upper figure also shows the progress made during each trial through an estimate of the error between the simulated and the analytical curves according to

$$E_n = \frac{1}{N} \sqrt{\sum_{k=1}^{N} \frac{(y_n(t_k) - y(t_k))^2}{y(t_k)^2 + \varepsilon}}, \tag{13.8.12}$$

where y_n is the simulation's resulting nth trial path, y is the path of the analytic solution (13.8.7a), and ε is a very small number to avoid dividing by zero. In the limit as $n \to \infty$, because $y_n \to y$, one expects $E_n \to 0$. For our case of 471 trials, we have an error of 0.755%. The calculation's input parameters, $(y_0, y_f, dy; v_0; g; m; N) = (0, 1, 5\text{ m}; 5\text{ m/s}; 9.8\text{ m/s}^2; 1\text{ kg}; 15)$, are shown in the action, or lower plot of Figure 13.7. The number of time intervals chosen is $N = 15$, which can be increased, at a calculational time cost. Following is the MATLAB script least_action.m used to create Figure 13.7, and which incorporates the above-mentioned algorithm. When run, the script shows the real-time progress made toward a final solution at the end of each simulated trial.

SCRIPT

```
%least_action.m - simulates Hamilton's Least Action principle for a
%particle under the action of gravity. The trajectory is compared
%with what is expected analytically
clear; tol=1.e-7;   %clear, and tolerance reasonably small
v0=5; g=9.8; m=1;   %initial speed, gravity and mass, given
y0=0; yf=1;         %initial, final height, given
t0=0; N=15; dy=.01; %initial time, # of points, max change in y allowed
tf=v0/g+sqrt((v0/g)^2-2*(yf-y0)/g); %final time estimate from analytic soln
if (v0/g)^2-2*(yf-y0)/g < 0,        %work with proper v0, y0, yf in this case
    disp 'v0 is too small, or yf is too large, stopped'
    return
end
dt=(tf-t0)/(N-1);           %time step
t=t0:dt:tf;                 %time array
yan=y0+v0*t-g*t.^2/2;       %exact analytic trajectory
yg=y0+(yf-y0)*(t-t0)/(tf-t0); %trajectory guess: interpolate end points
%Kinetic and Potential energies follow
T(1:N-1)=m*((yg(2:N)-yg(1:N-1))/dt).^2/2; V(1:N-1)=m*g*yg(1:N-1);
L=T-V;              %initial Lagrangian (Kinetic minus Potential energies)
S1=dt*sum(L);       %initial action, rectangular rule - use N-1 terms
yn=yg;              %the initial trajectory is the guess
rand('state',0)     %reset random numbers for reproducible runs , else comment
```

```
dS=abs(S1); n=0;    %change in the action, initially it is equal to S1
je=0;               %exchange counter for each trial
subplot(2,1,1)
while dS > tol,     %simulation stops when the change in the action is tiny
 cla;
 n=n+1;             %trial number
 yp=yn;             %trajectory memory, use the saved trajectory
 for k=2:N-1        %y(1), y(N) don't change so start at k=2, end at k=N-1
yp(k)=yp(k)+2*dy*(rand(size(yp(k)))-0.5);%modify kth step in trajectory
                               %memory, use Monte-Carlo step
 T(1:N-1)=m*((yp(2:N)-yp(1:N-1))/dt).^2/2; V(1:N-1)=m*g*yp(1:N-1);%energies
 L=T-V;        %Lagrangian (Kinetic minus Potential energies)
 S2=dt*sum(L);%new action integral by rectangular rule - use N-1 terms
  if S2 < S1       %accept step if the action S decreases
     dS=abs(S2-S1);%variation of the action (must be small before stopping)
     S1=S2;        %save the lower value of the action
     yn=yp;        %modify trajectory based on accepted step, and keep it
     je=je+1;      %count exchanges made
  else
     yp=yn;        %if the step is not accepted, discard memory changes made,
                   %and refresh the trajectory memory
  end
 end
S(n)=S1;            %keep track of the action for each trial
pen=100*sqrt(sum((yn-yan).^2./(yan.^2+.001^2)))/N;%percent error estimate
hold on
plot(t,yan,'k-',t,yg,'b-.',t,yn(:),'r.')%plot analytic,guess, and simulated
str1=cat(2,'Trials: ',num2str(n,4),', %E_n=',num2str(pen,3));
str2=cat(2,'Exchanges:',num2str(je,4));
str3=cat(2,'dS:',num2str(dS,4));
text(t0,max(yan)*(1-0.01),str1);
text(t0,max(yan)*(1-0.13),str2);
text(t0,max(yan)*(1-0.25),str3);
pause (.01)
end
h=legend('Analytic','Init Guess','Least Action',-1); set(h,'FontSize',11)
xlabel('t','FontSize',14),ylabel('y(t)','FontSize',14)
title(['MC: Hamilton''s Least Action Principle',' (tolerance=',... %MC=Monte-Carlo
      num2str(tol,2),')'],'FontSize',12)
subplot(2,1,2), plot(1:n,S(1:n))          %plot the action for each trial
xlabel('trial','FontSize',14),ylabel('Action (S)','FontSize',14)
str4=cat(2,'(y_0, y_f, dy; v_0; g; m; N; tol) = (',num2str(y0,2),', ',...
    num2str(yf,2),', ',num2str(dy,2),'m; ',...
    num2str(v0,2),'m/s; ',num2str(g,2),'m/s^2; ',num2str(m,2),'kg; ',...
    num2str(N,2),')');
text(round(n/40),S(round(n/40)),str4);
```

■ Chapter 13 Problems

13.1 Consider the central force Example 13.1 in which the polar coordinates r, θ are associated with the generalized coordinates q_1, q_2. Letting x_1, x_2 stand for the Cartesian coordinates, respectively, express the changes of x_1 and x_2 in terms of the changes in generalized coordinates.

13.2 Two particles of masses m_1 and m_2 move with velocities $\mathbf{v}_1$ and $\mathbf{v}_2$. They are acted upon by ordinary forces derivable from a potential energy function $V(x_1, y_1, z_1, x_2, y_2, z_2)$. Use Lagrange's equations to obtain the equations of motion of the system of particles.

13.3 Use the Lagrangian formalism to generalize the result of Example 13.9, and obtain the equations of motion for a body under a general potential of the form $V(r, \theta)$.

13.4 Give the details involved in arriving at the Lagrangian expression (13.6.13) of the double pendulum.

13.5 Show the steps leading to the equations of motion (13.6.15) of the double pendulum.

13.6 Derive the double pendulum small oscillations result shown in Equation (13.6.16). (*Hint:* Use the information provided in the double pendulum example.)

13.7 Show the steps leading to Expressions (a) (13.6.19) and (b) (13.6.20) of the double pendulum in the small oscillations approximation.

13.8 Simulate the double-pendulum system of Example 13.11 for two different initial conditions; the first with $\theta_{10}, \theta_{20} = \pi/2, \pi/2$ rad, and the second with $\theta_{10}, \theta_{20} = \pi/2, -\pi/2$ rad. Explain your observations regarding the behavior of the two systems and their initial energy. Use the following values for the rest of the parameters: $L_1 = L_2 = 1$ m, $m_1 = m_2 = 1$ kg, $g = 9.8$ m/s^2, $(d\theta_1/dt)_0 = 0 = (d\theta_2/dt)_0$, and use a suitable simulation time interval. (Reference: T. Shinbrot, C. Grebogi, J. Wisdom, and J.A. Yorke, "Chaos in a double pendulum," *American Journal of Physics* 60, 491, June 1992.)

13.9 Use Equation (13.7.4) to obtain the Hamiltonian for the double pendulum of Example 13.11. Comment on your result as regards the relationship between the potential and kinetic energies with the Hamiltonian.

13.10 Obtain Hamilton's equations of motion for a planet in orbit around the sun and explain each of the terms that appear in the resulting equations.

13.11 In terms of the Eulerian angles, write the Hamiltonian for a symmetric top under gravity and rotating about a fixed point located at its apex.

13.12 Perform a simulation of Hamilton's least action principle for the motion of a single 1 kg particle moving under the action of gravity, on the surface of the Earth, in one dimension in a frictionless environment, given that its takeoff and landing points are on the ground with a takeoff initial speed of 7 m/s. Compare the simulation with the corresponding analytic result, and stop the simulation until the error (13.8.12) is less than or equal to about 0.27%. (*Hint:* Modify the tolerance and be careful not to use all zeros as initial guesses.)

13.13 Use Hamilton's least action principle to simulate one period of the trajectory of a 1 kg particle attached to the end of a spring with a spring constant of $k = 9$ N/m. At $t = 0$, the particle is located at its equilibrium position but has a speed of 3 m/s. Compare the simulation to the analytical solution. The simulation should produce an error less than 2.3% with a $dS \approx 1 \times 10^{-8}$. (*Hint:* Use a large number of points, say $N = 50$, and use a guess made of three carefully chosen straight lines. Hint: Note that in this problem, in addition to the initial and final positions, the initial speed does produce a third well-known point near the initial point through the approximate time derivative.)

■ Additional Problems

13.14 For a particle moving in an external general potential, use Lagrange's equations to derive Newton's second law of motion.

13.15 A particle is attached to the end of a spring with constant k, as shown in Figure 13.8. Assuming the spring is massless, obtain the particle's equations of motion.

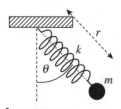

❘ FIGURE 13.8

13.16 A simple pendulum of length l is free to swing, but it is also held by a support of mass m_1 that is free to slide in the x direction, as shown in Figure

13.9. Letting the pendulum's mass be m_2 (a) write the kinetic energy, (b) the potential energy, and (c) the Lagrangian of the system. (d) Obtain the Lagrange's equations of motion of the pendulum with the sliding support. (e) If the angle the pendulum makes with the vertical is assumed to be small, solve the resulting equations of motion.

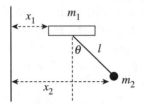

| FIGURE 13.9

13.17 A block of m slides down a frictionless inclined plane at angle θ from the horizontal, as shown in Figure 13.10. If the inclined plane is of mass M and length l and it is free to slide frictionlessly on the floor, use a Lagrangian development and obtain the acceleration of each block. Assume that m is instantaneously located at a distance s from the inclined plane's top corner and along the plane.

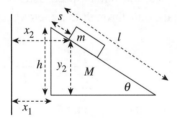

| FIGURE 13.10

13.18 A mass m is attached to a cord that's wrapped around a pulley which is supported by the ceiling. If the mass is let go, obtain (a) Lagrange's equations, (b) the acceleration of the mass, (c) the pulley's angular acceleration, and (d) the tension on the cord. Assume the pulley has moment of inertia I and radius R and that the cord does not stretch nor does it slip.

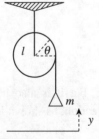

| FIGURE 13.11

13.19 Write the Lagrangian for the spring mass system of Figure 4.13(a) of Chapter 4 and obtain Lagrange's equations of motion. Does the result agree with the equations of motion obtained in Chapter 4, Section 13?

13.20 A spring of constant k contains mass m_1 at one end and mass m_2 at the opposite end, as shown in Figure 13.12. The two-mass-spring system is able to free fall. (a) Obtain the Lagrangian of the system. (b) Obtain Lagrange's equations of motion for the variables x_1, y_1, x_2, and y_2 versus time. (c) Given the initial conditions $x_{10} = -3L/4$, $y_{10} = 0 = \dot{x}_{10}$, $\dot{y}_{10} = 4\,\text{m/s}$ and $x_{20} = -x_{10}$, $y_{20} = 0 = \dot{x}_{20} = \dot{y}_{20}$, where the spring's unstretched length is $l = 1\,\text{m}$, use $m_1 = 1\,\text{kg}$, $m_2 = 5\,\text{kg}$, $g = 9.8\,\text{m/s}^2$, and $k = 25\,\text{N/m}$ to develop a simulation of the system as it free falls. (*Hint:* the correct Lagrangian of the system, when reduced to one-dimensional motion in the x direction, will have the form $L = \dfrac{1}{2}(m_1\dot{x}_1^2 + m_2\dot{x}_2^2) - \dfrac{1}{2}k(x_2 - x_1 - l)^2$.)

I FIGURE 13.12

MATLAB Tutorial

■ A.1 Introduction

The acronym MATLAB stands for matrix laboratory. It is a commercial software tool for the purpose of performing numerical computation and data visualization. It can be obtained from *http://www.mathworks.com/*. There are several versions available. The student version is the least expensive. This textbook, for the most part, makes use of the numerical capabilities of the bare bone MATLAB. There are toolboxes available for needs beyond the textbook, but at an additional cost. As used in the text, MATLAB can be used by means of the "command window" or through the use of "scripts" or "m" files. The command window accepts simple calculator-type commands, which MATLAB's engine performs after the user presses "Enter." The scripts are simple text documents written with MATLAB commands before invoking the engine on them. In this tutorial, we will give simple examples of how to run MATLAB and how to use the scripts that accompany the text. For further help, refer to the text's commented scripts, as well as other tutorials available through Web search engines.

■ A.2 Tutorial Notation

Within the command window, MATLAB's prompt ">>" indicates the place where the commands are entered. Anything typed at the prompt is user input that the engine will try to act on once the user presses Enter. MATLAB is case sensitive and allows for many different ways to define variables. For example, below we show a simple example of addition of three different variables.

```
>> a=2
a =
     2
>> B=3
B =
     3
>> b=4
b =
     4
```

```
>> a+b+B
ans =
     9
```

In the above example, we defined the variable and then added them. We could have also added the three numbers together in one stroke, as follows:

```
>> 2+3+4
ans =
     9
```

Both answers are equivalent, but the first case enables the user to use and reuse variables without needing to retype them.

■ A.3 General Features

MATLAB allows the use of comments. These are preceded by the percent sign "%" and can be placed anywhere on a line. For example:

```
>> a=3 %defines a
a =
     3
```

We could have made the same definition and suppressed the output if a semicolon ";" has been placed immediately after the command. Also, more than one command can be placed on the same line when separated by a comma or a semicolon. For example, the command

```
>> a=3; b=4; %define a and b
```

produces no output, but does define the variables. This can be checked by looking at the defined variables:

```
>> who
```

Your variables are:

```
a    b
```

To check on the value of a variable, just type its name and press Enter. For example,

```
>> a
a =
     3
```

To clear the value of a variable, such as a, just type "clear a" after the prompt. To clear all the variables, type "clear all" after the prompt. Typing "clear" also works, and typing "clc" and pressing Enter clears the command window. Typing "ctrl-c" stops the execution of a script. (Press the Ctrl key while simultaneously pressing the c key.)

MATLAB has "workspace window" that shows all the defined variables it recognizes during the current session. To view the workspace window, choose that option under the "View" drop-down menu.

Define a row vector c with components 1, 3, and 5 like this:

```
>> c=[1 2 3]
c =
     1     2     3
```

To define a column vector d with components 2, 4, and 7, separate the elements with semicolons; for example:

```
>> d=[2;4;7]
d =
     2
     4
     7
```

We can easily perform the matrix product of these two vectors according to the rules of matrices. For example, c*d should produce a single element 1×1, but the product d*c should produce a 3×3 matrix. Let's see,

```
>> c*d
ans =
    31
```

```
>> d*c
ans =
     2     4     6
     4     8    12
     7    14    21
```

We can take the transpose of this matrix as follows:

```
>> (d*c)'
ans =
     2     4     7
     4     8    14
     6    12    21
```

or we could just do "transpose(d*c)" at the prompt. We can learn about the details of the stored variable by typing "whos" at the prompt.

```
>> whos
  Name      Size        Bytes  Class
  ans       3x3            72  double array
  c         1x3            24  double array
  d         3x1            24  double array
  e         1x3            24  double array

Grand total is 18 elements using 144 bytes
```

We can also ask for help at any time. For example, typing "help" at the MATLAB prompt gives a list of all help topics available. Typing "help whos" at the prompt outputs the description for the command "whos" and so on. MATLAB's power lies in the way it uses arrays to perform numerical tasks.

MATLAB has a unique multiplication feature. To see this, first let's create a square matrix.

```
>> A=[1 3 5;2 4 6; 7 9 1]
A =
     1     3     5
     2     4     6
     7     9     1
```

If we desired to square the matrix, we would simply do as follows:

```
>> A^2
ans =
     42    60    28
     52    76    40
     32    66    90
```

which is the simple matrix product of A*A. However, the multiplication operation preceded by a dot has a different result.

```
>> A.*A
ans =

      1     9    25
      4    16    36
     49    81     1
```

This is *not* a matrix product; instead, it is a dot-multiplication. The result is a matrix that contains each of the elements of A squared. The usefulness of this feature in MATLAB is realized when we, for example, wish to plot a function. Suppose we want a plot of the function y=1/x on the interval [2, 6]. Because y is a function of x, then we could first create an array containing the variable x from 2 to 7 in steps of, for example, 0.25. After that, we could invert an element of the array at a time and store it in y, then plot y versus x. MATLAB does this very simply as follows. First, use the "colon" loop idea for the x array as follows:

```
>> x=[2:0.25:6];
```

where we have suppressed the output, but basically we have created an array of x from 2 to 6 in steps of 0.25 for a total of 136 different values of x. This is checked by looking at the workspace window or by typing "whos" at the prompt. We next use MATLAB's dot-division, which is similar to the dot-multiplication explained above. All we do is place a "." before the "/" symbol as follows:

```
>> y=1./x;
```

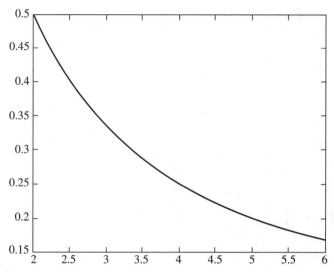

| FIGURE A.1

which is ready for MATLAB's plot command,

```
>> plot (x,y)
```

The resulting graph is shown in Figure A.1.

It is possible to place labels on a graph as well, which is shown in many of the text's scripts. The help menu can be accessed through MATLAB's command window for further details as well.

While we are on the subject of vectors, it is useful to point out that we can also do vector dot and cross products. For example, let's define the following vectors,

```
>> A=[1,2,3]
A =
     1     2     3

>> B=[4,5,6]
B =
     4     5     6
```

Next, let's find the dot and cross products of A and B,

```
>> dot(A,B)
ans =
    32
>> cross(A,B)
ans =
    -3     6    -3
```

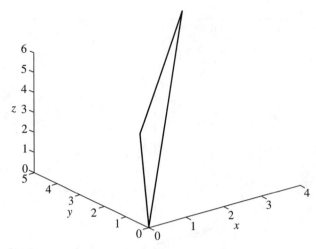

| FIGURE A.2

Let's find the angle between A and B and convert it to degrees

```
>> dot(A,B)/(sqrt(dot(A,A))*sqrt(dot(B,B)))*180/pi
ans =
    55.8423
```

Here, the operation dot(A,A) is identical to sum(A.*A). We can also draw these vectors in three dimensions, in addition to connecting them. This is done below.

```
>> axis([0,4,0,5,0,6])                      %determines the axes ranges to use
>> line([0,1],[0,2],[0,3],'color','r')      % the A vector line in red
>> line([0,4],[0,5],[0,6],'color','b')      % the B vector line in blue
>> line([1,4],[2,5],[3,6],'color','k')      % the connecting vector between A and B in black
>> xlabel('x'),ylabel('y'),zlabel('z')      % put labels for the axes
```

The graph obtained by the above operations is shown in Figure A.2.

■ A.4 Operands

MATLAB's full range of operands in listed below. These are important when typing performing commands in a script or in a command window during a working session.

Operand	Description	Operand	Description	Operand	Description
+	addition	~=	not equal	,	separator
-	subtraction	>	greater than	;	end row
*	scalar multiplication	>=	greater than or equal	()	subscript enclosure, expression precedence
/	scalar right division	|	or	[]	matrix
\	matrix left division	&	and	Ctrl-c	abort
^	scalar power	~	not		
.*	array multiplication	>>	MATLAB prompt	;	suppress output
./	array division	'	transpose		
.^	array exponentiation	%	comment, format specification		
=	assignment	.	decimal point		
<	less than	...	line continuation		
<=	less than or equal	\n	new line format specification		
==	logical equal	:	vector generation		

■ A.5 M-Files and Functions

An m-file is a text file that contains MATLAB commands in a suitable sequence for it to interpret and execute. Lines that do not refer to standard MATLAB language are commented. A simple m-file can be created from the command window by typing the command "diary test.m" followed by several commands wished to be included or tested. The file can then be closed by typing "diary off" and be accessed through the MATLAB editor or any other text editor. With the editor, MATLAB output lines, which are not standard commands, can be deleted to keep just the basic typed commands, and the file can be resaved in the working directory. By invoking the name of the file within the MATLAB command window, that is, by typing "test", MATLAB's engine will proceed to interpret it and produce the needed output. Any errors in the script will be seen in the command window and can be easily fixed with the text editor. This textbook contains scripts that are ready to run in the manner explained.

MATLAB has built-in functions as described in the next section; however, a user can also build functions. A user function in MATLAB is also an m-file that a user builds in order to perform operations that are repetitive in nature. For example, suppose that we needed the numerical derivative of the log(x) at x; we could create a script file named "logder.m", which is a function, as follows:

```
function [F] = logder(x,del)
% logder.m calculates the derivative of the log(x) at x within an interval
[x,x+del]
% where the value of del is provided by the user
F = (log(x+del)-log(x))/del;
```

Once the text file with the preceding lines is saved with the name lodger.m, it can be invoked through the MATLAB command window or through a line in a script file. For example, suppose that we needed the derivative of the log(x) at x=5. Assuming the above file is saved in the working directory, we could type the line "logder(5,1.e-3)" at the command window. The session would proceed as follows:

```
>> logder(5,1.e-3)
ans =
    0.2000
```

Furthermore, typing "help logder" at the command window produces the output:

```
"logder.m calculates the derivative of the log(x) at x within an interval [x,x+del]
where the value of del is provided by the user",
```

which are basically the lines that are commented in the user-built script itself. Thus it is important to write comments within the functions to recall their usage, as well as to make sure that such functions exist or are present in the working directory. The above is not the only way to create functions. Another way is to create functions within a script with the help of the "inline" command. For example, f=inline('exp(x)') defines the function f(x)=exp(x) within a script.

■ A.6 Built-in Functions

A listing of some of MATLAB's built-in functions follows. These are important when performing commands for mathematical operations in a script or in a command window's working session.

Function	Description	Function	Description	Function	Description
sin	trig sine	ceil	round toward positive infinity	diag	extract diagonal of a matrix or make a diagonal matrix
cos	trig cosine	max	largest component	triu	upper triangular part of a matrix
tan	trig tangent	min	smallest component	tril	lower triangular part of a matrix
asin	inverse sine	sign	signum	size	size of a matrix
acos	inverse cosine	length	length of a vector	det	determinant of a square matrix
atan	inverse tangent	sort	sort in ascending order	inv	inverse of a matrix
exp	exponential	sum	sum of elements	rank	rank of a matrix
log	natural logarithm	prod	product of elements	rref	reduced row echelon form
abs	absolute value	median	median value	eig	eigenvalues and eigenvectors
sqrt	square root	mean	mean value	poly	polynomial
rem	remainder	std	standard deviation	lu	LU factorization
round	round to nearest integer	eye	identity matrix	qr	QR factorization
floor	round toward negative infinity	zeros	matrix of zeros	quad	numerical integration
mod	modulus	ones	matrix of ones	fzero, roots	zeros of functions and roots of polynomials
inline	create a function on the fly	rand	randomly generated matrix	ode23, ode45	differential equation solvers

To use these functions, one can invoke help. For example, if we needed help on the function "mean," we could type "help mean" in the command window to get help on that function. In this particular case, the help command tells us that for vectors, MEAN(X) is the mean value of the elements in X. An example of this is as follows:

```
>> mean([1,2,3,4,5,6,7])
ans =
     4
```

The help command is very useful in MATLAB and should be used as often as possible, at least until one is familiar with it.

■ A.7 Plotting

Some of MATLAB's plotting command are listed below. These are useful in visualizing a function in two or three dimensions. The textbook contains scripts that make use of most of these commands.

Command	Description	Command	Description	Command	Description
plot	two-dimensional plot	semilogx	use log scale on the x-axis	surfl	three-dimensional shaded surface with lighting
subplot	table of plots	semilogy	use log scale on the y-axis	mesh	three-dimensional mesh plot
loglog	use log-log scales	surf	three-dimensional shaded surface	grid	adds grid lines

Please refer to the scripts in the text for the various uses of these commands.

■ A.8 Programming

Programming refers to a set of commands in a script for the express purpose of performing tedious calculations. Programming is most useful when loops and if-else statements are used in conjunction with built-in or user-built functions, as well as with command lines within m-files. Below are two examples that involve a loop and an if-else statement.

Loops
A simple loop in MATLAB is as follows:
```
for i=1:1:101
x(i)=(i-1)*2*pi/(101-1);
y(i)=sin(x(i));
end
```

The above lines basically create a 101-element array for x between 0 and 2 p for which the sine is evaluated and stored in the array for y. Of course, in MATLAB there are other ways of doing the same thing, but the above is an example of a loop use. We could add an if-else statement, say, for the purpose of converting half the above wave into a square box. This can be done as follows:

```
for i=1:1:101
x(i)=(i-1)*2*pi/(101-1);
if x(i)<=pi
```

```
y(i)=sin(x(i));
else
y(i)=sign(sin(x(i)));
end
end
```

One could easily follow up the above lines with a "plot(x,y)" command to visualize the results.

■ A.9   Zeros of Functions

MATLAB's built-in function "fzero" can be used to obtain the variable values at which a function takes a zero value. One example of this is to find the value of x at which x=exp(-x) is satisfied. This can be done as follows:

```
>> fzero('x-exp(-x)',.3)
ans =
    0.5671
```

where we have used 0.3 as an initial guess. See the help command for a full function description. One can also use built-in functions as well as user-built functions in place of the expression between the single quotes, as explained before. Another example is to obtain roots of polynomials. The built-in function "roots" can be employed for this purpose. Suppose we wanted to know the roots of the cubic polynomial x^3+3x^2-2x+5. We would do this as follows:

```
>> roots([1 3 -2 5])
ans =
  -3.8552
   0.4276 + 1.0555i
   0.4276 - 1.0555i
```

which gives one real and two complex conjugate roots.

■ A.10   Numerical Integration

Numerical integration is possible with MATLAB. The built quad function can be used for that purpose. One example for the integration of the sin function on [0, p] is as follows:

```
>> quad(@sin,0,pi)
ans =
    2.0000
```

One could easily replace the "sin" in the above line with the name of a user built-in function. Another example is to create a function on the fly with the "inline" command

```
>> g=inline('2*x.^2+x.^3.*sin(x)');
>> quad(g,0,pi)
ans =
   32.8276
```

Here the function 2*x.^2+x.^3.*sin(x) was integrated on [0, p].

■ A.11 Differential Equations

MATLAB has the capability of solving differential equations with the built-in function ode23 or ode45. For example, suppose we desire to solve the problem y'=(y-t^2)*exp(-t) for y(t) on the interval [0, 20] with the initial condition that at t=0, y=1. In MATLAB, this can be accomplished as follows:

```
f=inline('(y-t.^2).*exp(-t)');
[t,y]=ode23(f,[0,20],1);
plot(t,y)
```

Other more complicated cases make use of user-built functions. See the textbook's scripts for examples or type "help ode23" within the command window for further details. The "ode45" command is also available, as are other, more sophisticated methods, as explained in MATLAB's help facility.

■ A.12 Movies

MATLAB can make movies. Take a traveling wave as an example. First make a space and time array, plot the wave every time step, save the plot frame into an array M, pause for a short time to see the plot versus time on the screen, play the movie at 15 frames per second, and finally save it as an avi file to be used later as follows:

```
x=0:0.01:1;              %space array
t=0:0.01:1;              %time array
nt=length(t);            %time steps
for j=1:nt
    y=sin(2*pi*(x-t(j)));  %a traveling wave
    plot(x,y)              %plot for every time step
    M(j) = getframe;       %make movie frames if desired
    pause(0.05);           %pause for a short time
end
movie(M,1,15)            %play once, at 15 frames per minute
movie2avi(M,'wave');     %creates the file wave.avi, which can be played later
```

According to the help command for code that is compatible with all versions of MATLAB, including versions before MATLAB Release 11 (5.3), use:

```
M = moviein(n);
for j=1:n
  plot_command
  M(:,j) = getframe;
  end
movie(M)
```

■ A.13 Publish Code to HTML

It is possible to publish written code on a website. To do so, one needs to convert the script and its output into html format. To do so, suppose we have a script `file.m`. Then typing in the command line:

```
>>publish('file.m')
```

runs the script, waits for the output, and automatically converts the script and the output to HTML code in the subdirectory "html" under the main directory where the `file.m` is run from. Typing "help publish" gives further information on this command.

■ A.14 Symbolic Operations

MATLAB has symbolic functions capability. Typing "help symbolic" within the command line, a list of available functions is given if the *Symbolic Math Toolbox* is available with the version of the software. A simple example of a symbolic operation is to plot the function $y = 2t$. This is accomplished by typing the following lines:

```
>> syms t
>> y=2*t
y =
2*t
>> ezplot(t,y)
```

Another example is to plot the function $y = 2\exp(-at^2)$ and its derivative. This is accomplished as follows:

```
>> syms t y f a
>> y=2*exp(-a*t^2)
y =
2*exp(-a*t^2)
>> f=diff(y,t)
f =
-4*a*t*exp(-a*t^2)
>> a=0.5
```

```
a =
   0.5000
>> ezplot(t,eval(y),[-5 5])
>> hold on
>> ezplot(t,eval(f),[-5 5])
```

Other examples for symbolic operations are available within the textbook or within the help menu within MATLAB.

■ A.15 Toolboxes

MATLAB's basic capabilities can be extended through the use of toolboxes, which can be ordered separately. Toolboxes are separate sets of scripts created by www.mathworks.com in order to solve specific sets of problems. Currently, the toolboxes available include:

Math and Optimization, Statistics and Data Analysis, Control System Design and Analysis, Signal Processing and Communications, Image Processing, Test & Measurement, Computational Biology, Financial Modeling and Analysis, Application Deployment, Application Deployment Targets, Database Connectivity and Reporting, Fixed-Point Modeling, Event-Based Modeling, Physical Modeling, Simulation Graphics, Control System Design and Analysis, Signal Processing and Communications, Code Generation, Rapid Control Prototyping and HIL SW/HW Embedded Targets, Verification, Validation, and Testing. More detailed information on these toolboxes can be obtained through *http://www.mathworks.com.*

B | Useful Mathematical Formulas Used in the Text

■ Linear Algebra

Let $A \equiv \begin{pmatrix} a & b \\ c & d \end{pmatrix}$, the determinant of a 2×2 matrix given

$$\det(A) \equiv \begin{vmatrix} a & b \\ c & d \end{vmatrix} = ad - bc. \tag{B.1}$$

The inverse of A is $A^{-1} = Ad(A)/\det(A)$, where

$$Adj(A) \equiv \begin{pmatrix} d & -c \\ -b & a \end{pmatrix}^{T} = \begin{pmatrix} d & -b \\ -c & a \end{pmatrix};$$

thus,

$$A^{-1} = \begin{pmatrix} d & -b \\ -c & a \end{pmatrix} / (ad - bc). \tag{B.2}$$

■ Integrals

$$\int x^2 \sqrt{c^2 - x^2} = -\frac{x(c^2 - x^2)^{3/2}}{4} + \frac{c^2 x \sqrt{c^2 - x^2}}{8} + \frac{c^4}{8} \sin^{-1}(x/c) \tag{B.3}$$

$$\int \frac{du}{\sqrt{b^2 - u^2}} = \sin^{-1}(u/b) \tag{B.4}$$

$$\int \frac{dx}{(x^2 + a^2)^m} = \frac{x}{2(m-1)a^2(x^2 + a^2)^{m-1}} + \frac{2m - 3}{(2m - 2)a^2} \int \frac{dx}{(x^2 + a^2)^{m-1}} \tag{B.5}$$

$$\int (ax + b)^{m/2}dx = \frac{2(ax + b)^{(m+2)/2}}{(m + 2)a} \tag{B.6}$$

$$\int x(ax + b)^{m/2}dx = \frac{2(ax + b)^{(m+4)/2}}{(m + 4)a^2} - \frac{2b(ax + b)^{(m+2)/2}}{(m + 2)a^2} \tag{B.7}$$

■ Numerical Integration

Rectangular Rule

$$\int_a^b f(x)dx \approx h\{f(x_1) + f(x_2) + f(x_3) + \cdots + f(x_{N-1})\} \tag{B.8}$$

where $x_i = a + (i - 1)h$, for $1 \le i \le N$, $h = (b - a)/(N - 1)$.

Trapezoidal Rule

$$\int_a^b f(x)dx \approx \frac{h}{2}\{f(x_1) + 2[f(x_2) + f(x_3) + \cdots + f(x_{N-1})] + f(x_N)\}, \tag{B.9}$$

where the integer $N \ge 2$, $x_i = a + (i - 1)h$ for $1 \le i \le N$, and $h \equiv (b - a)/(N - 1)$.

Simpson's Rule

$$\int_a^b f(x)dx \approx \frac{h}{3}\left\{ \begin{matrix} f(x_1) + 2[f(x_2) + f(x_4) + f(x_6) + \cdots + f(x_{N-1})] \\ + 4[f(x_3) + f(x_5) + f(x_7) + \cdots + f(x_{N-2})] + f(x_N)) \end{matrix} \right\}, \tag{B.10}$$

where $N \ge 3$ is an odd integer, $x_i = a + (i - 1)h$ for $1 \le i \le N$, and $h \equiv (b - a)/(N - 1)$.

Plane Geometry

Ellipse

Figure C.1 shows some of the properties of the ellipse. The foci locations F, F' are the ellipses' foci. The distance $2b$ is the minor axis, and b is the semiminor axis. Similarly, the distance $2a$ is the major axis, and a is the semimajor axis.

The position $r(\theta)$ obeys the conic section formula,

$$r(\theta) = \frac{r_{min}(1 + e)}{(1 + e\cos\theta)}. \tag{C.1}$$

The value of the eccentricity determines the shape of the resulting curve. In particular,

$$e \begin{cases} < 1 \Rightarrow ellipse{:}(x/a)^2 + (y/b)^2 = 1, r + r' = 2\,a, e = \overline{FF'}/2\,a \\ = 0 \Rightarrow circle{:}(x)^2 + (y)^2 = a^2, r = r' = a \\ = 1 \Rightarrow parabola{:}y^2 = 4\,ax, r = r', e = r/r' \\ > 1 \Rightarrow hyperbola{:}(x/a)^2 + y^2/(F^2 - a^2) = 1, r' - r = \pm 2\,a, e = F/a \end{cases} \tag{C.2}$$

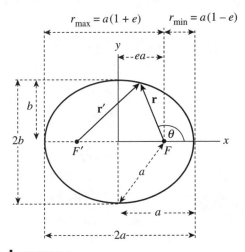

I FIGURE C.1

523

■ Trigonometry

The Pythagorean Formula (see Figure C.2)

$$a^2 + b^2 = c^2 \tag{C.3}$$

Trigonometric Relations (see Figure C.2)

$$\sin\theta = \cos\varphi = b/c = 1/\csc\theta, \quad \cos\theta = \sin\varphi = a/c = 1/\sec\theta, \quad \tan\theta = \cot\varphi = b/a = 1/\cot\theta \tag{C.4}$$

$$\sin^2\theta + \cos^2\theta = 1, \quad \sec^2\theta - \tan^2\theta = 1, \quad \csc^2\theta - \cot^2\theta = 1 \tag{C.5}$$

General Angle Relations

$$\sin(A \pm B) = \sin A \cos B \pm \cos A \sin B \tag{C.6}$$

$$\cos(A \pm B) = \cos A \cos B \mp \sin A \sin B \tag{C.7}$$

$$\cos A + \cos B = 2 \cos\left(\frac{A+B}{2}\right)\cos\left(\frac{A-B}{2}\right) \tag{C.8}$$

$$\cos A - \cos B = 2 \sin\left(\frac{A+B}{2}\right)\sin\left(\frac{B-A}{2}\right) \tag{C.9}$$

$$\cos\theta = 1 - 2 \sin^2\theta/2 \tag{C.10}$$

$$\sin 2\theta = 2 \cos\theta\sin\theta \tag{C.11}$$

$$\cos^2\theta = (1 + \cos 2\theta)/2 \tag{C.12}$$

$$\cos^3\theta = (3 \cos\theta + \cos 3\theta)/4 \tag{C.13}$$

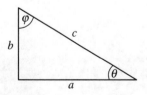

I FIGURE C.2 Right angle.

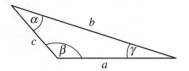

| FIGURE C.3 General triangle.

Law of Sines (see Figure C.3)

$$\frac{\sin\alpha}{a} = \frac{\sin\beta}{b} = \frac{\sin\gamma}{c} \qquad\qquad\qquad\qquad (\text{C.14})$$

Law of Cosines (see Figure C.3)

$$a^2 = b^2 + c^2 - 2\,bc\cos\alpha \qquad\qquad\qquad\qquad (\text{C.15})$$

Euler Identity

$$e^{ix} = \cos x + i\sin x \qquad\qquad\qquad\qquad (\text{C.16})$$

D

One-, Two-, and Three-Dimensional Elements

Common themes in formulas used in the text are integrals of the form

$$I = \int dm \text{ or } I = \int dq \tag{D.1}$$

where m is a mass and q is charge. Depending on the geometry, dm or dq can take on the expressions $\rho(r)\,dV$ or $\sigma(r)\,dA$ or $\lambda(r)\,dl$, where $\rho(r)$, $\sigma(r)$, and $\lambda(r)$ are the three-, two-, and one-dimensional densities, respectively. The quantities dV, dA, and dl are the volume, areal, and linear elements of interest, respectively, and are discussed next.

■ Simple Volume Density Elements

Referring to Figure D.1(a), for a sphere of radius R, we use spherical coordinates and $dV = r^2 \sin\theta\, d\theta\, d\varphi\, dr$ where $0 < \theta < \pi$, $0 < \varphi < 2\pi$, and $0 < r < R$.

Referring to Figure D.2, for a rectangular solid of sides a, b, and c, it is best to use Cartesian coordinates so that $dV = dx\,dy\,dz$ with the ranges of $0 < x < a$, $0 < y < b$, and $0 < z < c$.

Referring to Figure D.3, for a cylinder of radius R and length L, polar coordinates are appropriate and we can write $dV = \rho\,d\rho\,d\theta\,dl$, with $0 < \rho < R$, $0 < \theta < 2\pi$, and $0 < l < L$.

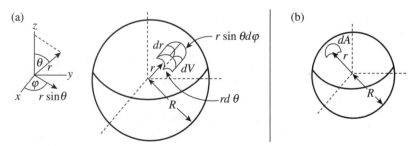

(a)

(b)

❙ FIGURE D.1

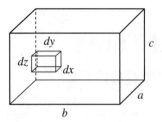

FIGURE D.2

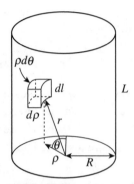

FIGURE D.3

■ Simple Areal Density Elements

For the case of spherical coordinates, we refer to Figure D.1(b), where the area element is easily identified using similar elements as in Figure D.1(a), so that since r is not varying, $r \to R$ and $dA = (Rd\theta)(R\sin\theta d\varphi) = R^2\sin\theta d\theta d\varphi$. In Cartesian coordinates, we can have various choices of area elements depending on which coordinates we are interested in; thus, from Figure D.2, $dA = dxdy, dxdz$, or $dydz$. For the cylinder, using Figure D.3, there are two possible area elements. One is for the caps, $dA = \rho d\rho d\theta$, and the other is the curved area $dA = Rd\theta dl$.

■ Simple Linear Density Elements

From Figure D.1, again taking $r \to R$, here the part of a sphere that would interest us could either be for the case of two possible paths, one involving a variation in θ, that is, $dl = Rd\theta$, which also plays a role in a two-dimensional disk or circle. The second path is the one for which φ can vary; that is, $dl = R\sin\theta d\varphi$. For the Cartesian case, from Figure D.2, there are three choices, depending on the coordinate that varies, so that $dl = dx, dy$, or dz. For cylindrical coordinates, using Figure D.3, we could either have the cylinder length as variable, in which case the element is simply dl, or we could have the curved path as the element if θ is the variable, in which case $\rho \to R$ and the path element of interest is just $Rd\theta$.

Answers to Selected Problems

Chapter 1

1. a. 2π **b.** 1

3. For small c, highest amplitude occurs at around $\omega \approx \omega_0 = \sqrt{k/m} = 1$

5. $c_1 = 0.313$ kg/s

7. b. $v_\infty = -2.71$ m/s

7. c. $a_n = F_n/m$, $v_{n+1} = v_n + a_n\Delta t$, $x_{n+1} = x_n + v_{n+1}\Delta t$,
$F_n = -mg - kv_n^3$, $t_n = n\Delta t$

9. a. $F_{net} = (m_1 + m_2)a$ **b.** $F - f_{k1} - F_{12} = m_1 a$, $F_{21} - f_{k2} = m_2 a$
c. $F_{12} = F_{21} = 6.44$ N

11. $m_0 = 0.0371$ kg, $k = 64.13$ N/m

Chapter 2

1. 16.75 hp

3. a. 16.53 m **b.** 3.67 s

5. $y = y_0 - gt^2/2$

8. $v = v_t \tan[\tan^{-1}(v_0/v_f) - t/\tau]$,

$$x = x_0 + \tau v_t \ln\left\{\frac{\cos[\tan^{-1}(v_0/v_f) - t/\tau]}{\cos[\tan^{-1}(v_0/v_f)]}\right\}$$

11. a. $t_{top} = -\dfrac{m}{c}\ln\left[\dfrac{mg}{mg + cv_0}\right]$

b. $t_{top} \approx -\dfrac{m}{c}\left[-\dfrac{cv_0}{mg} + \dfrac{c^2 v_0^2}{2\,m^2 g^2}\right] = \dfrac{v_0}{g} - \dfrac{c^2 v_0^2}{2\,mg^2}$

13. $v = \dfrac{F_0 t_2}{2\,m}$

15. $y(t = t_{top}) = y_0 - \left(\dfrac{m}{c}\right)^2 g \ln\left[1 + \dfrac{c v_0}{mg}\right] + \dfrac{m}{c} v_0$

17. a. $v(t) = v_0 + a_0 t + \dfrac{F_0}{m\omega}\sin\omega t$

■ Chapter 3

3. a. $\sqrt[6]{2}\,\sigma$ **b.** $\omega = \sqrt{72\,\varepsilon/(\sqrt[3]{2}\,\sigma m)}$

5. 0.12 Ns/m

7. $<p> = \dfrac{F_0^2 \gamma \omega^2}{m\left[(2\,\gamma\omega)^2 + (\omega_0^2 - \omega^2)^2\right]}; \ \omega = \omega_0$

9. $x(t) = u_0 e^{-\gamma t}\displaystyle\int e^{\gamma t} e^{\gamma t} + C e^{-\gamma t} = (u_0 t + C)e^{-\gamma t}$

12. a. $A = 0.1$ m   **b.** $\delta = -2.863° = -0.05$ rad $= -\varphi$ **c.** $p_{ave} = 6.3$ mW

■ Chapter 4

3. 0.9204 s

5. $\omega = \sqrt{4\,cg/m}$

7. Approximately expect $x_i/x_b \approx 0.1$

9. $\Delta y = 3.4*10^{-3}$ m, $\omega = 53.34$ Hz

11. b. $\theta = 15° e^{-\gamma t}\cos\omega t$ **c.** $c = 0.381\,\dfrac{\text{kg}}{\text{s}}$

13. $\omega = \sqrt{\dfrac{k_3(k_1 + k_2)}{m(k_3 + k_1 + k_2)}}$

15. $A_1^3 + b_1 A_1^2 + b_2 A_1 + b_3 = 0$, where

$$b_1 = -\dfrac{27}{28}\theta_0, \quad b_2 = -\dfrac{32\,\omega_0^2}{28\,a_3}, \quad b_3 = -b_2\theta_0$$

◼ Chapter 5

1. $0.242\,\hat{i} + 0.097\,\hat{j} + 0.966\,\hat{k}$

3. $-2\,\hat{i} + 5\,\hat{j} + 2\,\hat{k}$

5. $307,\,182\,\hat{k}\,AU\cdot\text{kg}\cdot\text{km/s}$

7. $(13.21\hat{i} - 15.02\hat{j})$ cm; $(45.06\hat{i} - 39.62\hat{j})$ cm/s; $(-118.86\hat{i} - 135.18\hat{j})$ cm/s

9. $8\,a^4$

11. $4/3$

◼ Chapter 6

2. $t_\theta = \tan\theta\, t_{\pi/2 - \theta}$

3. $x \approx \dfrac{2\,v_{0x}v_{oz}}{g} - \dfrac{8}{3}\dfrac{c v_{0z}^2 v_{0x}}{mg^2}$

7. $-\ell^2(1 + \ell)/2$

9. $2\,A\text{rexp}\left(-r^2\right)$

15. $y = \dfrac{eEL^2}{2\,m_e v_0^2}, B = E/v_0$

◼ Chapter 7

3. $\mathbf{r} = v_0\hat{t} + 2\,b\hat{k};\ \mathbf{v} = 2\,v_0\hat{j};\ \mathbf{a} = -v_0^2\hat{k}/b$

5. $1.88 \times 0^{-4}\,\text{N}$

7. 24 hr

11. $\varepsilon(\theta = 45°) = 0.0984°$

13. $x' = 6.34$ mm

15. $\tau_p = 41.84$ hr, $\tau_f = 10.99$ s, $\theta = 1.314°$

■ Chapter 8

4. Let $L = 0$, find $t = \sqrt{m/k} \sin^{-1}(x/\sqrt{2\,E/k})$, and rearrange.

5. *Hint:* Step gives $\theta(t) = \tan^{-1}(mC^2/Lt)$, and $r = \dfrac{\sqrt{(mC^2)^2 + (Lt)^2}}{mC}$.

8. $411, 563$ km

9. $\tau^2/a^3 = 4\,\pi^2/Gm_E$

11. $\sqrt{2\,GM/R}$

13. $r_{\min} = 4.447*10^9$ km, $v_{r=r\max} = 3.679$ km/s

15. $r_{\max} = a(1 + e)$

■ Chapter 9

1. $-Gm_0(m_1 + m_2)/x_1$

3. a. $-\dfrac{GM}{L}\ln\left\{\dfrac{L/2 + \sqrt{y^2 + (L/2)^2}}{-L/2 + \sqrt{y^2 + (L/2)^2}}\right\}$ **b.** $-GM/y;\ 0;\ \infty$

5. a. $-\dfrac{2\,GM}{R^2}[\sqrt{z^2 + R^2} - z]$ **b.** $-2\,GM/R,\ -GM/z$

7. a. $-GM/r$ **b.** $-GM[3 - (r/R)^2]/2\,R$

9. a. $-GM\hat{r}/R^2$ **b.** $-GMr\hat{r}/R^3$

11. $m_1 = 4\,m_s;\ m_2 = 1\,m_s$

13. $r_1 = 6.25\ AU;\ r_2 = 13.74\ AU;\ m_1 = 1\,m_s,\ m_2 = 2.2\,m_s$

15. $m_2 = 1\,m_s;\ a_1 = 6.25\ AU;\ a_2 = 13.75\ AU$

■ Chapter 10

1. $3.649*10^4$ N

3. b. $\Theta = 68.2°$ **c.** $e = -1.78;\ r_{\min} = 18.8\,a_b$

6. a. 11.74 fm **b.** 20.33 fm

Chapter 11

3. $(4\,R/3\,\pi)\hat{j}$

5. a. $\mathbf{v}_1(t) = 2\,t\hat{i};\ \mathbf{v}_2(t) = -\hat{i} + \hat{j} - \dfrac{1}{t^2}\hat{k};\ \mathbf{v}_3(t) = -2\,t\hat{j}$

6. c. $e_{k1} = 2\,m_1 t^2;\ e_{k2} = \dfrac{1}{m_2}\left(m_2^2 + \dfrac{m_2^2}{2\,t^4}\right);\ e_{k3} = 2\,m_3 t^2$

7. a. $\mathbf{L} = (-2\,m_2/t + m_3 t)\hat{i} + (-m_2/t + (-2 - t)m_2/t^2)\hat{j}$
$\qquad + (-2\,m_1 t + m_2(-2 - t) + m_2 t - 2*m_3 t)\hat{k}$

  c. $\boldsymbol{\tau} = (2\,m_2/t^2 + m_3)\hat{i} + (-2(-2 - t)m_2/t^3)\hat{j} + (-2\,m_1 - 2\,m_3)\hat{k}$

9. $\mathbf{v}_1 = -\sqrt{\dfrac{m_1}{m_2}v_0^2 - v_{0x}^2}\,\hat{j};\ \mathbf{v}_2 = \dfrac{m}{m_2}v_{0x}\hat{i} + \dfrac{m_1}{m_2}\sqrt{\dfrac{m_1}{m_2}v_0^2 - v_{0x}^2}\,\hat{j}$

13. $1.99*10^2$ s

15. a. $2*10^{13}$ **b.** $6\,\mu m$

19. $v_{1f} = b + \sqrt{b^2 - \dfrac{m_2}{m_1 + m_2}\left(\dfrac{c_x^2 + c_y^2}{m_1 + m_2} - \dfrac{2\,E}{m_1}\right)}$, where

$b = \dfrac{(c_x \cos\theta_1 + c_y \sin\theta_1)}{m_1 + m_2},\ c_x = m_1 v_{1ix} + m_2 v_{2ix},\ c_y = m_1 v_{1iy} + m_2 v_{2iy},$

$E = \dfrac{1}{2}(m_1 v_{1i}^2 + m_2 v_{2i}^2)$

$v_{2f} = \sqrt{\dfrac{2\,E}{m_2} - \dfrac{m_1 v_{1i}^2}{m_2}}$, and $\theta_2 = \tan^{-1}\left\{\dfrac{m_1 v_{1f}\sin\theta_1 - c_y}{c_x - m_1 v_{1f}\cos\theta_1}\right\}$

21. $\theta_2 = 43.16$ degrees; $v_i = 0.87*10^6$ m/s; $KE_i = 0.378*10^{12}u_0$ J; $KE_i = 0.75*10^{12}u_0$ J

Chapter 12

1. $x_{cm} = y_{cm} = z_{cm} = 3\,R/8$

2. $x_{cm} = 0;\ y_{cm} = R/\pi$

5. $I_{corner} = \dfrac{3}{4}M(a^2 + b^2)$

7. $I_{xx} = \dfrac{1}{12}Mb^2$

9. $I_{zz} = \dfrac{1}{2}M_s a^2$

11. $I_{yy} = 0.072\, Ma^2$; $M_s = 0.428\, M$; $A_s = 0.428\, A$

13. $I = \begin{pmatrix} 13/12 & 0 & 0 \\ 0 & 5/6 & 0 \\ 0 & 0 & 5/12 \end{pmatrix}$ kgm²; directions: (001), (010), (100) with

corresponding moments: 5/12; 5/6; 13/12, because the tensor is diagonal.

15. a. $\mathbf{L} = (0.167\,\hat{i} + 0.167\,\hat{j} + 0.167\,\hat{k})$ kgm²/s; $T = 0.250$ J

b. Rotation about x, $L_x = 0.236$ kgm², $T = 0.167$ J, and similarly about the y-z axes.

21. a. 305.8 d **b.** 0.9967 d

27. $\omega_s > \sqrt{4\,\mathrm{Img}\ell/I_s}$

■ Chapter 13

1. $\delta x_1 = \dfrac{x_1}{q_1}\delta q_1 - x_2\delta q_2$; $\delta x_2 = \dfrac{x_2}{q_1}\delta q_1 + x_1\delta q_2$

3. $m\ddot{r} = mr\dot{\theta}^2 + f_r$; $\dfrac{d}{dt}(mr^2\dot{\theta}) = f_\theta$, where $f_r = -\dfrac{\partial V}{\partial r}$, $f_\theta = -\dfrac{\partial V}{\partial \theta}$

9.
$H = \dfrac{1}{2}m_1 L_1^2\dot{\theta}_1^2 + \dfrac{1}{2}m_2 L_1^2\dot{\theta}_1^2 + \dfrac{1}{2}m_2 L_2^2\dot{\theta}_2^2 + m_2 L_1 L_2\dot{\theta}_1\dot{\theta}_2\cos(\theta_1 - \theta_2)$
$+ m_1 g L_2 + (m_1 + m_2)g L_1(1 - \cos\theta_1) + m_2 g L_2(1 - \cos\theta_2)$

11. $H = \dfrac{1}{2}I_1(\dot{\varphi}^2\sin^2\theta + \dot{\theta}^2) + \dfrac{1}{2}I_3(\dot{\varphi}^2\cos\theta + \dot{\psi})^2 + mg\ell\cos\theta$

13. The three straight lines are:

$$y_{guess} = \begin{cases} y_0 + \dfrac{(y_{h1} - y_0)}{(\tau/4 - t_0)}(t - t_0), & t_0 < t \le \tau/4 \\[2mm] y_{h1} + \dfrac{(y_{h2} - y_{h1})}{(3\tau/4 - \tau/4)}(t - \tau/4), & \tau/4 < t \le 3\tau/4 \\[2mm] y_{h2} + \dfrac{(y_f - y_{h2})}{(\tau - 3\tau/4)}(t - 3\tau/4), & 3\tau/4 < t \le \tau \end{cases}$$

where $\tau = t_f$, $y_{h1} = A$, and $y_{h2} = -A$. Also,

$$\frac{dy}{dt} \approx \frac{y_1 - y_0}{\Delta t} = v_0 \Rightarrow y_1 = y_0 + v_0 \Delta t.$$

15. $\ddot{r} - r\dot{\theta}^2 - g\cos\theta + \dfrac{kr}{m} = 0,\ r\ddot{\theta} + 2\,\dot{r}\dot{\theta} + g\sin\theta = 0$

17. $L = \dfrac{1}{2}M\dot{x}_1^2 + \dfrac{m}{2}(\dot{x}_1^2 + 2\,\dot{x}_1\dot{s}\cos\theta + \dot{s}^2) - mg(\ell - s)\sin\theta$, where x_1 is the

position of M. $\ddot{s} = \dfrac{(M + m)g\sin\theta}{(M + m\sin^2\theta)},\ \ddot{x}_1 = -\dfrac{mg\sin\theta\cos\theta}{(M + m\sin^2\theta)}.$

References

Arya, Atam P. 1998. *Introduction to Classical Mechanics*, 2nd ed. Upper Saddle River, NJ: Pearson Education, Prentice Hall.

Besancon, R. M., ed. 1990. *The Encyclopedia of Physics*, 3rd ed. Reinhold, NY: Van Nostrand.

Boas, May L. 2006. *Mathematical Methods in the Physical Sciences*, 3rd ed. New York: John Wiley.

Chow, Tai L. 1995. *Classical Mechanics*. New York: John Wiley.

Davis, A. Douglas. 1986. *Classical Mechanics*. New York: Harcourt Brace Jovanovich.

Dugas, Rene. 1988. *A History of Mechanics*. New York: Dover.

Fowles, Grant R. and George L. Cassaday. 2005. *Analytical Mechanics*, 7th ed. Belmont, CA: Thomson, Brooks/Cole.

French, A. P. 1971. *Newtonian Mechanics*. CITY: The MIT Introductory Series, Norton.

Giordano, Nicholas J. and Hisao Nakanishi. 2006. *Computational Physics*, 2nd ed. Upper Saddle River, NJ: Pearson/Prentice Hall.

Gould, H., J. Tobochnik, and W. Christian, W. 2007. *An introduction to Computer Simulation Methods: Applications to Physical Systems*, 3rd ed. New York: Pearson Addison Wesley.

Griffiths, D. J. 1999. *Introduction to Electrodynamics*, 3rd ed. Upper Saddle River, NJ: Prentice Hall.

Kaufmann III , William J. 1994. *Universe*, 4th ed. New York: W. H. Freeman.

Kleppner, Daniel and Robert J. Kolenkow. 1976. *An Introduction to Mechanics*. Boston: McGraw Hill.

Marion, J. B. and S. T. Thornton. 1988. *Classical Dynamics of Particles and Systems*, 3rd ed. New York: Harcourt Brace Jovanovich.

McLean, W. G. and E. W. Nelson. 1978. *Engineering Mechanics*. Outline Series in Engineering. New York: McGraw-Hill.

Rohlf, J. W. 1994. *Modern Physics from α to Z⁰*. New York: John Wiley & Sons Inc.

Spiegel, Murray R. 1967. *Theoretical Mechanics*, Schaum's Outline Series in Science. New York: McGraw-Hill.

Spiegel, Murray R. 1969. *Mathematical Handbook of Formulas and Tables*, Schaum's Outline Series. New York: McGraw-Hill.

Symon, Keith R. 1971. *Mechanics*, 3rd. ed. Reading, MA: Addison-Wesley.

Taylor , John R. 2005. *Classical Mechanics*. Sausalito, CA: University Science Books.

Television Networks, A&E, 235 East 45th Street, New York, NY 10017, (*http://www.biography.com/*).

Wells, Dare A. 1967. *Lagrangian Dynamics*, Schaum's Outline Series in Engineering. New York: McGraw-Hill.

Western, Arthur B. "Dramatic Demonstration of Energy Conservation Using Projectile Motion" submitted to The Physics Teacher [received date 8/4/95], Department of Physics and Applied Optics, *http://www.rose-hulman.edu/~western/slingsho.html.*

Index